Current Topics in Behavioral Neurosciences

Volume 30

About this Series

Current Topics in Behavioral Neurosciences provides critical and comprehensive discussions of the most significant areas of behavioral neuroscience research, written by leading international authorities. Each volume offers an informative and contemporary account of its subject, making it an unrivalled reference source. Titles in this series are available in both print and electronic formats.

With the development of new methodologies for brain imaging, genetic and genomic analyses, molecular engineering of mutant animals, novel routes for drug delivery, and sophisticated cross-species behavioral assessments, it is now possible to study behavior relevant to psychiatric and neurological diseases and disorders on the physiological level. The *Behavioral Neurosciences* series focuses on "translational medicine" and cutting-edge technologies. Preclinical and clinical trials for the development of new diagnostics and therapeutics as well as prevention efforts are covered whenever possible.

More information about this series at http://www.springer.com/series/7854

Markus Wöhr · Sören Krach
Editors

Social Behavior from Rodents to Humans

Neural Foundations and Clinical Implications

Springer

Editors
Markus Wöhr
Behavioral Neuroscience
Philipps-University of Marburg
Marburg
Germany

Sören Krach
Social Neuroscience Lab, Department of Psychiatry and Psychotherapy, Center of Brain, Behavior and Metabolism (CBBM)
Lübeck University
Lübeck
Germany

ISSN 1866-3370 ISSN 1866-3389 (electronic)
Current Topics in Behavioral Neurosciences
ISBN 978-3-319-47427-4 ISBN 978-3-319-47429-8 (eBook)
DOI 10.1007/978-3-319-47429-8

Library of Congress Control Number: 2016952898

Printed on acid-free paper

This Springer imprint is published by Springer Nature
The registered company is Springer International Publishing AG
The registered company address is: Gewerbestrasse 11, 6330 Cham, Switzerland

Preface

Sociability is the greatest advantage in the struggle for life
Pyotr Kropotkin, *Mutual Aid: A Factor of Evolution*

"Social Behavior from Rodents to Humans: Neural Foundations and Clinical Implications" provides a broad and accessible overview on the rapidly developing field of social neuroscience. Social neuroscience is an expanding research field devoted to understand how social behavior is implemented by biological factors, and how social experiences in turn impact the brain. A major goal of this volume is to integrate research findings on the neural basis of social behavior across different levels of analysis from rodent studies on molecular neurobiology to behavioral neuroscience to functional neuroimaging data on human social behavior.

Showcasing an array of cutting-edge research programs, leading investigators present new approaches in a field covering a wide range of behaviors, ranging from maternal care, sexual behavior and attachment to social observational learning, social decision making and altruism. In the first two parts, an introduction to the various different kinds of social behaviors in rodents, Part I, and humans, Part II, is provided, with special emphasis being laid onto newly developed paradigms to study communicative functions in rodents, but also proxemics and higher-order social cognition in humans, including second-person neurosciences and hyper-scanning. Importantly, chapters cover a broad range of research based on studies in healthy participants, their application and translation into a clinical context, as well as developmental aspects, as exemplified in Part III, devoted to clinical implications, with a special emphasis on autism spectrum disorder.

Most rodents are social animals, displaying a rich repertoire of social behaviors and living at least substantial parts of their lives in societies in which they use complex ways to communicate with each other, for instance during mating and while forming social bonds. Part I therefore starts off with a comprehensive and insightful overview by Lukas and de Jong (2017) on behavioral methods used to study components of this rich social behavior repertoire in a standardized manner, ranging from affiliative to aggressive encounters: friends or foes? Their overview describes traditional but also novel approaches, such as the recently developed female intruder test (de Jong et al. 2014). At the neurobiological level, Lukas and de Jong (2017) focus on the neuropeptides oxytocin and vasopressin, key players in regulating social behavior.

Because of the richness of the social behavior repertoire, the ability to recognize and distinguish between individuals—to identify friends and foes, is essential for survival. In the second chapter, Camats et al. (2017) provide an excellent overview on the present knowledge about social memory in rodents, comparing social recognition memory performance across different rodent taxa and behavioral test paradigms. As olfactory signals are of particular relevance for the formation of social recognition memory, they also summarize key components of the brain circuit processing olfactory information linked to social recognition memory. Finally, they give examples of factors that they and their colleagues have identified to interfere with the formation of social recognition memory, such as social isolation, but also elements of the experimental procedures, including anaesthesia or even transportation (Perna et al. 2015). Their careful consideration appears essential to avoid spurious results.

In the third chapter, Kiyokawa (2017) shows that olfactory signals are not only highly relevant for the formation of social recognition memory but also serve important communicative functions. Kiyokawa (2017) describes two types of olfactory communication in rodents. Summarizing own previous findings of an extensive serious of studies (Inagaki et al. 2014), he first describes pheromones, which are released in response to stress and serve as alarm signals to warn conspecifics about danger through activating the fear circuit in the brain. Secondly, he describes appeasing olfactory communication and how this is involved in the fascinating phenomenon of social buffering.

Besides olfactory signals, rodents emit ultrasonic vocalizations. Wöhr et al. (2017) provide an overview on the communicative functions of the two main types of ultrasonic vocalizations emitted by adult rats, 22-kHz calls serving an alarming function and 50-kHz calls serving an affiliative function. They further show that social experiences strongly affect the emission of ultrasonic vocalizations in the sender but also the behavioral responses displayed by the receiver. Behavioral responses elicited in the receiver are typically studied by means of a playback paradigm. Using this approach, they showed that not only juvenile social isolation (Seffer et al. 2015) but also environmental enrichment (Brenes et al. 2015) might have negative effects on ultrasonic communication.

Another approach to study the communicative function of ultrasonic vocalizations is described by Kisko et al. (2017), namely surgical devocalization. In their studies, Kisko et al. (2015; 2017) devocalized rats to test how the lack of ultrasonic vocalizations affects their social play behavior. They found that social play behavior was clearly reduced in pairs of devocalized juvenile rats. Furthermore, in adult rats, devocalization led to a marked increase in aggressive behaviors. Therefore, the two studies highlight the important communicative functions fulfilled by ultrasonic vocalizations and their role in modulating social play and aggression.

In his seminal work, Panksepp helped identify and illuminate the functional neuroanatomies and neurochemistries of seven primary processes, which he named SEEKING/Enthusiasm, RAGE/Anger, FEAR/Anxiety, sexual LUST/Passion, maternal CARE/Nurturance, separation-distress PANIC/Grief and PLAY/Social Joy. In his chapter, Panksepp (2017) provides a thought-provoking overview on how several of these systems figure heavily in social bonding. Considering depression as an example, he argues that sustained overactivity of the separation-distress PANIC/Grief system, reflecting the "excessive psychological pain of loneliness", can lead to a downward cascade known as psychological despair. Psychological despair, in turn, is characterized by abnormally low activity of the SEEKING/Enthusiasm system, also known as brain reward networks, resulting in amotivational states typical for depression. As 50-kHz calls are believed to reflect activity of the SEEKING/Enthusiasm system in rats ("rat laughter"; Panksepp 2005), Panksepp suggests measuring 50-kHz calls in rat models as a marker to assess treatment efficacy of novel pharmacological compounds for treating depression.

The apparent paradox that altruistic behaviors are expressed by the same individuals that also compete for resources is discussed by Lahvis (2017). He argues that natural selection does not only favor individuals to act in their own interest, implying a "competitive psychology", but also animals displaying helping behavior, presumably at their own expense, suggesting a more "compassionate psychology". He further argues that altruism can only be partially explained by ultimate mechanisms, such as kin selection and reciprocity, and suggests that an additional "stake in others" is necessary for the evolution of altruism: the so called "camaraderie effect". He sees the "camaraderie effect" as a by-product of two highly adaptive psychological experiences, namely social motivation and empathy, which both can be studied in rodents. In fact, there is evidence that rodents do not only derive pleasure from social interactions but also experience empathy (Chen et al. 2009) - the generation of an affective state more appropriate to the situation of another compared to one's own.

In the last chapter of Part I, Hernandez-Lallement et al. (2017) report on a fascinating new behavioral paradigm to assess pro-social decision-making in rats using a social reinforcement framework. In the pro-social choice task, an actor and a partner are trained together in a double T-maze. First, the actor makes a choice and enters one of two different compartments. Then, gets a reward, which is identical irrespective of which compartment it has chosen. However, entering one compartment triggers the delivery of an additional reward for the partner rat, whereas entering the alternative compartment does not yield any additional reward to the partner. By means of this new behavioral paradigm, Hernandez-Lallement et al. (2015) demonstrated that rats prefer mutual rewards, and now discuss factors that could drive social decision making in rodents, with a particular emphasis on inter-individual differences.

"Whether supportive, strategic, combative or romantic, social interactions are at the core of everyday experience" (McCall 2017). It is an essential fact of humankind that we either engage into social interactions or at least constantly simulate and represent ourselves in the context of our surrounding social world. The rise and significance of social media for humans in modern societies is another proof of this principle and a powerful cultural artifact that demonstrates how humans value and nurture the connections between them. Part II is devoted to the question of how the various facets of sociality and their underlying neural principles, that help us to engage with others, can be understood in the neurosciences of human social interactions. The opening chapter by Keysers and Gazzola (2017) "A Plea for Cross-species Social Neuroscience" thereby bridges Part I with Part II in arguing why it will be essential to integrate animal neuroscience and human neuroscience in order to deepen our understanding of the neural basics of social phenomena. Basing their plea on the spectacular discovery of motor mirror neurons in monkey research more than two decades ago (di Pellegrino et al. 1992), Keysers and Gazzola (2017) provide a convincing argument of how cohesive cross-species neurosciences and models for neural functioning that span over taxonomies of biology could greatly expand our understanding of embodied cognition.

In this line, the following Chapter, "Models, Mechanisms and Moderators Dissociating Empathy and Theory of Mind" by Kanske et al. (2017), tackles the fundamental question of how humans are able to access another person's mind. This a fundamental and necessary capability to socially interact in a meaningful way (Kanske et al. 2015). The authors hereby describe the most influential models for grasping others' perspectives based on two neural routes: an affective route which helps us to directly share others' emotions in an empathic manner and a cognitive route which helps us to represent and reason about others' mental states, called Theory of Mind (ToM). The authors provide a compelling example of how to dissociate these two routes and their underlying neural processes within one paradigm (Kanske et al. 2015) and elaborate how situational and personality factors impact their use and neural functioning during social interaction.

However, what is so appealing for us to engage into social interactions and why do humans and other social beings value and nurture social relations, at first hand? In their Chapter "Reward: From Basic Reinforcers to Anticipation of Social Cues" Rademacher et al. (2017) outline how the human reward system processes all kind of social stimuli that motivate behavior. Importantly, they provide strong empirical arguments showing the rewarding aspects of social connectedness and affiliation do already shape the reward signal in the anticipation of social interaction in structures of the striatum (Rademacher et al. 2010; Spreckelmeyer et al. 2009). The chapter thereby introduces into state-of-the-art neuroimaging paradigms to effectively study the effects that social incentives exert on humans. The authors close the chapter with the role of the neuropeptide oxytocin in mediating the rewarding quality of social interactions by activating dopaminergic reward pathways in response to social cues. This is arguably one of the most famous examples for how human and rodent neuroscience can expand on each other for understanding social cognition.

The question of whether these positive and rewarding aspects of social interaction ultimately lead to human cooperation is asked in the following chapter on "Human Cooperation and Its Underlying Mechanisms" by Strang and Park (2017). While people do share a lot of goods on a voluntary basis even with unrelated others, also defective or egocentric behavior is evident in society. This chapter provides and elaborate overview of current economic games that are useful to examine how cooperative, defective or altruistic behaviors are represented in the human brain. The authors discuss how social punishment in form of deliberate non-cooperation is applied to maintain social cooperation, and how emotions impact decision making.

Most emotions are triggered by or experienced within social encounters. Our bodies, and hence our brains, are no "completed beings" but always involve the opening towards another person or a group (Deleuze 1993). We feel sad because of a romantic relationship breakdown, we are proud about being honored at school or feel embarrassed because we are ridiculed in front of an audience. Chapter "The Social Neuroscience of Interpersonal Emotions" demonstrates that emotions are interpersonal in nature as we constantly represent ourselves in the context of our surrounding social world. In their review, Müller-Pinzler et al. (2017) describe innovative interpersonal set-ups to evoke authentic emotions in neuroscientific environments. Based on the most recent evidence on guilt, pride, anger, and embarrassment (Müller-Pinzler et al. 2015), they summarize the neural networks that underlie these most prototypical examples of interpersonal emotions. They complete this chapter by explaining the vicarious experience of these interpersonal emotions (Krach et al. 2011) based on the above introduced neural pathways of empathizing and mentalizing (Kanske et al. 2015; Paulus et al. 2014).

Offending social encounters usually trigger anger. In their Chapter "Deconstructing Anger in the Human Brain" Gilam and Hendler (2017) further elaborate the recent experimental advances of bringing realistic social interactions into a neuroimaging environment for studying this so-called basic emotion underlying many interpersonal conflicts. In order to capture the full picture of the human flexibility to control and regulate anger and even adapt anger to socially accepted norms, the authors ultimately stress that spontaneous and dynamic interactive paradigms need to be embedded into neuroscientific research in order to truly understand the interpersonal aspects of emotions.

The underlying neuroendocrine mechanisms of approach and avoidance behavior are targeted in greater detail in Chapter "On the Control of Social Approach–Avoidance Behavior: Neural and Endocrine Mechanisms" by Kaldewaij et al. (2017). While humans possess the capacity to control their behavioral action tendencies, the authors elaborate how evaluations of emotional situations happen in an automatic and implicit way. The chapter reviews current research on the way endogenous hormone levels of testosterone, oxytocin or cortisol modulate approach-avoidance behaviors.

To study social emotions or social phenomena in general, high degrees of experimental control and ecological validity are essential and are therefore addressed in detail in several chapters of the book. The "Science of Proxemics" by McCall, specifically, is a guide to new avenues for the study of social interactions in a multidimensional but highly controlled environment while at the same time one has to account for the "inherently nonlinear" phenomenology of social interactions (McCall 2017). The science of proxemics thereby introduces the principles for studying fundamental aspects of human social behavior in virtual environments. By tracking motion data, location information, and head orientation this innovative paradigm allows to identify characteristic patterns in the way people orient toward or away from each other and engage in mutual, joint or averted gaze (McCall & Singer 2015). The utility and validity of this approach is finally demonstrated by showing convincing modulations of these markers for social behavior through various experimental manipulations.

Part III is devoted to clinical implications. A particular emphasis is put on the case of autism spectrum disorder, a group of neurodevelopmental disorders characterized by social and communication deficits, paralleled by repetitive and stereotyped patterns of behavior. The first chapter by Schroeder et al. (2017) provides an overview on human core symptoms, candidate genes, and current experimental approaches to study pathomechanisms in rodents. The chapter focuses on genetic rodent models, with an exemplary description of top-ranked models, including Fmr1 mutant mice as a model for fragile X syndrome, the most common known single genetic cause of autism spectrum disorder, but also other models for studying promising candidate genes, such as Neuroligin3 and Neuroligin4 and Shank gene family members, including Shank2 (Schmeisser et al. 2012).

Pietropaolo et al. (2017) take it a step further in the second chapter and critically discuss current strategies for testing promising pharmacological and environmental treatment approaches in rodent models for autism spectrum disorder. In their very insightful discussion, they identify a number of problems inherent to the research field, which delay the study of such treatment approaches. The problems include the presence of a large variety of rodent models, the difficulty in choosing the most appropriate behavioral markers for assessing treatment efficacy, the limited knowledge on the neurobiological bases of autism spectrum disorder and of its etiology, and, finally, the complexity of the disorder itself. As an example of a promising approach, they describe beneficial effects of environmental enrichment on behavioral abnormalities, such as direct reciprocal social interaction deficits, in Fmr1 mutant mice (Oddi et al. 2015)—a model relevant for autism spectrum disorder, yet characterized by lower levels of complexity and clearer etiology.

The third chapter "Neuroimaging-based Phenotyping of the Autism Spectrum" by Bernhardt et al. (2017) questions whether a consistent neuroimaging phenotype would indeed adequately describe the very heterogeneous symptomatology of autism spectrum disorders. The authors review recent advances in our understanding of the structural and functional organization of regional and large-scale brain networks and their disturbances in autism spectrum disorders. A broad spectrum of analyses tools, ranging from diffusion tractography studies to structural neuroimaging covariance and graph-theoretical approaches are introduced and evaluated in their potential to identify biologically and clinically relevant endophenotypes.

The chapters on autism spectrum disorder are closed by a review on "Current Practice and Future Avenues in Autism Therapy" by Poustka and Kamp-Becker (2017). Poustka and Kamp-Becker (2017) provide an overview on established and evidence-based interventions and treatments in autism spectrum disorder. Taken the functional independence and quality of life of persons diagnosed with autism spectrum disorder into consideration, they evaluate the advantages and disadvantages of up-to-date social skill-based or behavior-based interventions, computer-supported trainings, as well as new avenues such as neurofeedback and real-time functional neuroimaging or adjuvant pharmacotherapy including the promising neuropeptide oxytocin.

Baez et al. (2017) introduce a framework for how basically all social cognition depends on contextual factors in their chapter on the "The Social Context Network Model in Psychiatric and Neurological Diseases". The effects of social context thereby rely on the functioning of central hubs in a fronto-temporo-insular brain network, which, if altered, result in impairments to adjust cognition and behavior according to contextual demands. To validate their approach Baez et al. (2017) apply the social context network model to various domains of social cognition and make a compelling case for how it helps to understand the peculiarities in frontotemporal dementia and autism spectrum disorder.

In "Social-Cognitive Deficits in Schizophrenia", Mier and Kirsch (2017) argue why and how the social neurosciences can inform about the emergence of schizophrenia, a severe, highly heritable psychiatric disorder. The authors introduce a neural system-based model of how psycho-social processes such as emotion and intention recognition and executive functioning contribute to the observed hypo- and hyperactivation and aberrant connectivity in schizophrenia.

Finally, future directions are discussed in a translational manner, including the effects of stress. In an elegantly written chapter, Tzanoulinou and Sandi (2017) describe how early life stress programs the social brain. They discuss the physiological and neurobiological consequences of stress during peri-adolescence in the context of rodent paradigms that model human adversity, such as social neglect and isolation, social abuse, and exposure to fearful experiences (Marquez et al. 2013). Furthermore, they discuss peri-adolescent stress as a potent component strongly influencing the social behavior repertoire even of individuals in close contact with stressed individuals and across generations. By revising the existing literature, defining open questions, and debating the adaptive function of observed changes in social behavior such as pathological aggression, they expand the framework in which interactions among peri-adolescent stress, the social brain, and behavior can be better conceptualized.

In closing, we would like to express our profound gratitude to Dr. Bart Ellenbroek, Editor-in-Chief of Current Topics in Behavioral Neuroscience, for allowing us the opportunity to compile this volume entitled "Social Behavior from Rodents to Humans: Neural Foundations and Clinical Implications". Our appreciation is extended to all reviewers for their time, effort and persistence. Finally, we would like to voice a special gratitude to the staff at Springer for their tireless help getting this project off the ground and to completion. In particular, we are indebted to Susanne Dathe and Amudha Vijayarangan for helping us to pursue and secure the appropriate venue for the project.

Marburg, Germany
Lübeck, Germany

Markus Wöhr
A. Özge Sungur
Frieder M. Paulus
Sören Krach

References

Baez S, García AM, Ibanez A (2017) The social context network model in psychiatric and neurological diseases. Curr Top Behav Neurosci 30:379–396
Bernhardt BC, Di Martino A, Valk SL, Wallace GL (2017) Neuroimaging-based phenotyping of the autism spectrum. Curr Top Behav Neurosci 30:341–355
Brenes JC, Lackinger M, Höglinger GU, Schratt G, Schwarting RKW, Wöhr M (2015) Differential effects of social and physical environmental enrichment on brain plasticity, cognition, and ultrasonic communication in rats. J Comp Neurol (in press)
Camats Perna J, Engelmann M (2017) Recognizing others: rodent's social memories. Curr Top Behav Neurosci 30:25–45
Chen Q, Panksepp JB, Lahvis GP (2009) Empathy is moderated by genetic background in mice. PLoS ONE 4: e4387
de Jong TR, Beiderbeck DI, Neumann ID (2014) Measuring virgin female aggression in the female intruder test (FIT): effects of oxytocin, estrous cycle, and anxiety. PLoS ONE 9: e91701
Deleuze G (1993) Spinoza und das Problem des Ausdrucks in der Philosophie. München: Wilhelm Fink
di Pellegrino G, Fadiga L, Fogassi L, Gallese V, Rizzolatti G (1992) Understanding motor events: a neurophysiological study. Exp Brain Res 91:176–180
Gilam G, Hendler T (2016) With love, from me to you: embedding social interactions in affective neuroscience. Neurosci Biobehav Rev 68:590–601
Gilam G, Hendler T (2017) Deconstructing anger in the human brain. Curr Top Behav Neurosci 30:257–273
Hernandez-Lallement J, van Wingerden M, Marx C, Srejic M, Kalenscher T (2015) Rats prefer mutual rewards in a prosocial choice task. Front Neurosci 8:443.
Hernandez-Lallement J, van Wingerden M, Schäble S, Kalenscher T (2017) A social reinforcement learning hypothesis of mutual reward preferences in rats. Curr Top Behav Neurosci 30:159–176
Inagaki H, Kiyokawa Y, Tamogami S, Watanabe H, Takeuchi Y, Mori Y (2014) Identification of a pheromone that increases anxiety in rats. Proc Natl Acad Sci USA 111:18751–18756
Kaldewaij R, Koch SBJ, Volman I, Toni I, Roelofs K (2017) On the control of social approach-avoidance behavior: neural and endocrine mechanisms. Curr Top Behav Neurosci 30:275–293
Kanske P, Böckler A, Trautwein FM, Singer T (2015) Dissecting the social brain: introducing the EmpaToM to reveal distinct neural networks and brain-behavior relations for empathy and theory of mind. Neuroimage 122:6–19
Kanske P, Böckler A, Singer T (2017) Models, mechanisms and moderators dissociating empathy and theory of mind. Curr Top Behav Neurosci 30:193–206
Keysers C, Gazzola V (2017) A Plea for Cross-species Social Neuroscience. Curr Top Behav Neurosci 30:179–191
Kisko TM, Euston DR, Pellis SM (2015) Are 50-kHz calls used as play signals in the playful interactions of rats? III. The effects of devocalization on play with unfamiliar partners as juveniles and as adults. Behav Processes 113:113–121
Kisko TM, Wöhr M, Pellis VC, Pellis SM (2017) From play to aggression: high-frequency 50-kHz ultrasonic vocalizations as play and appeasement signals in rats. Curr Top Behav Neurosci 30:91–108
Kiyokawa Y (2017) Social odors: alarm pheromones and social buffering. Curr Top Behav Neurosci 30:47–65
Krach S, Cohrs JC, de Echeverria Loebell NC, Kircher T, Sommer J, Jansen A, Paulus FM (2011) Your Flaws Are My Pain: Linking Empathy To Vicarious Embarrassment. PLoS ONE 6: e18675
Kropotkin P (1972) Mutual Aid: A Factor of Evolution. New York: New York University Press.

Lahvis GP (2017) Social reward and empathy as proximal contributions to altruism: the camaraderie effect. Curr Top Behav Neurosci 30:127–157

Lukas M, de Jong TR (2017) Conspecific interactions in adult laboratory rodents: friends or foes? Curr Top Behav Neurosci 30:3–24

Marquez C, Poirier GL, Cordero MI, Larsen MH, Groner A, Marquis J, Magistretti PJ, Trono D, Sandi C (2013) Peripuberty stress leads to abnormal aggression, altered amygdala and orbitofrontal reactivity and increased prefrontal MAOA gene expression. Transl Psychiatry 3: e216

McCall C (2017) Mapping social interactions: the science of proxemics. Curr Top Behav Neurosci 30:295–308

McCall C, Singer T (2015) Facing off with unfair others: introducing proxemic imaging as an implicit measure of approach and avoidance during social interaction. PLoS ONE 10: e0117532.

Mier D, Kirsch P (2017) Social-cognitive deficits in schizophrenia. Curr Top Behav Neurosci 30:397–409

Müller-Pinzler L, Gazzola V, Keysers C, Sommer J, Jansen A, Frässle S, Elnhäuser W, Paulus FM, Krach S (2015) Neural pathways of embarrassment and their modulation by social anxiety. Neuroimage 119:252–261

Müller-Pinzler L, Krach S, Krämer UM, Paulus FM (2017) The social neuroscience of interpersonal emotions. Curr Top Behav Neurosci 30:241–256

Oddi D, Subashi E, Middei S, Bellocchio L, Lemaire-Mayo V, Guzman M, Crusio WE, D'Amato FR, Pietropaolo S (2015) Early social enrichment rescues adult behavioral and brain abnormalities in a mouse model of fragile X syndrome. Neuropsychopharmacology 40:1113–1122

Panksepp J (2005) Psychology. Beyond a joke: from animal laughter to human joy? Science 308:62–3

Panksepp J (2017) The psycho-neurology of cross-species affective/social neuroscience: understanding animal affective states as a guide to development of novel psychiatric treatments. Curr Top Behav Neurosci 30:109–125

Paulus FM, Müller-Pinzler L, Jansen A, Gazzola V, Krach S (2014) Mentalizing and the Role of the Posterior Superior Temporal Sulcus in Sharing Others' Embarrassment. Cereb Cortex 25:2065–2075

Perna JC, Wotjak CT, Stork O, Engelmann M (2015) Timing of presentation and nature of stimuli determine retroactive interference with social recognition memory in mice. Physiol Behav 143:10–14

Pietropaolo S, Crusio WE, D'Amato FR (2017) Treatment approaches in rodent models for autism spectrum disorder. Curr Top Behav Neurosci 30:325–340

Poustka L, Kamp-Becker I (2017) Current practice and future avenues in autism therapy. Curr Top Behav Neurosci 30:357–378

Rademacher L, Krach S, Kohls G, Irmak A, Gründer G, Spreckelmeyer KN (2010) Dissociation of neural networks for anticipation and consumption of monetary and social rewards. Neuroimage 49:3276–3285

Rademacher L, Schulte-Rüther M, Hanewald B, Lammertz S (2017) Reward: from basic reinforcers to anticipation of social cues. Curr Top Behav Neurosci 30:207–221

Schmeisser MJ, Ey E, Wegener S, Bockmann J, Stempel AV, Kuebler A, Janssen AL, Udvardi PT, Shiban E, Spilker C, Balschun D, Skryabin BV, Dieck S, Smalla KH, Montag D, Leblond CS, Faure P, Torquet N, Le Sourd AM, Toro R, Grabrucker AM, Shoichet SA, Schmitz D, Kreutz MR, Bourgeron T, Gundelfinger ED, Boeckers TM (2012) Autistic-like behaviours and hyperactivity in mice lacking ProSAP1/Shank2. Nature 486:256–260

Schroeder JC, Reim D, Boeckers TM, Schmeisser MJ (2017) Genetic animal models for autism spectrum disorder. Curr Top Behav Neurosci 30:311–324

Seffer D, Rippberger H, Schwarting RKW, Wöhr M (2015) Pro-social 50-kHz ultrasonic communication in rats: post-weaning but not post-adolescent social isolation leads to social impairments - phenotypic rescue by re-socialization. Front Behav Neurosci 9:e102

Spreckelmeyer KN, Krach S, Kohls G, Rademacher L, Irmak A, Konrad K, Kircher T Gründer G (2009) Anticipation of monetary and social reward differently activates mesolimbic brain structures in men and women. Soc Cogn Affect Neurosci 4:158–65

Strang S, Park SQ (2017) Human cooperation and its underlying mechanisms. Curr Top Behav Neurosci 30:223–239

Tzanoulinou S, Sandi C (2017) The programming of the social brain by stress during childhood and adolescence: from rodents to humans. Curr Top Behav Neurosci 30:411–429

Wöhr M, Engelhardt KA, Seffer D, Sungur AÖ, Schwarting RKW (2017) Acoustic communication in rats: effects of social experiences on ultrasonic vocalizations as socio-affective signals. Curr Top Behav Neurosci 30:67–89

Contents

Part I Social Behavior in Rodents

Conspecific Interactions in Adult Laboratory Rodents: Friends or Foes? 3
Michael Lukas and Trynke R. de Jong

Recognizing Others: Rodent's Social Memories 25
Judith Camats Perna and Mario Engelmann

Social Odors: Alarm Pheromones and Social Buffering 47
Yasushi Kiyokawa

Acoustic Communication in Rats: Effects of Social Experiences on Ultrasonic Vocalizations as Socio-affective Signals 67
Markus Wöhr, K. Alexander Engelhardt, Dominik Seffer, A. Özge Sungur and Rainer K.W. Schwarting

From Play to Aggression: High-Frequency 50-kHz Ultrasonic Vocalizations as Play and Appeasement Signals in Rats 91
Theresa M. Kisko, Markus Wöhr, Vivien C. Pellis and Sergio M. Pellis

The Psycho-Neurology of Cross-Species Affective/Social Neuroscience: Understanding Animal Affective States as a Guide to Development of Novel Psychiatric Treatments 109
Jaak Panksepp

Social Reward and Empathy as Proximal Contributions to Altruism: The Camaraderie Effect 127
Garet P. Lahvis

A Social Reinforcement Learning Hypothesis of Mutual Reward Preferences in Rats 159
Julen Hernandez-Lallement, Marijn van Wingerden, Sandra Schäble and Tobias Kalenscher

Part II Social Behavior in Humans

A Plea for Cross-species Social Neuroscience 179
Christian Keysers and Valeria Gazzola

Models, Mechanisms and Moderators Dissociating Empathy and Theory of Mind 193
Philipp Kanske, Anne Böckler and Tania Singer

Reward: From Basic Reinforcers to Anticipation of Social Cues 207
Lena Rademacher, Martin Schulte-Rüther, Bernd Hanewald and Sarah Lammertz

Human Cooperation and Its Underlying Mechanisms 223
Sabrina Strang and Soyoung Q. Park

The Social Neuroscience of Interpersonal Emotions 241
Laura Müller-Pinzler, Sören Krach, Ulrike M. Krämer and Frieder M. Paulus

Deconstructing Anger in the Human Brain 257
Gadi Gilam and Talma Hendler

On the Control of Social Approach–Avoidance Behavior: Neural and Endocrine Mechanisms 275
Reinoud Kaldewaij, Saskia B.J. Koch, Inge Volman, Ivan Toni and Karin Roelofs

Mapping Social Interactions: The Science of Proxemics 295
Cade McCall

Part III Clinical Implications

Genetic Animal Models for Autism Spectrum Disorder 311
Jan C. Schroeder, Dominik Reim, Tobias M. Boeckers and Michael J. Schmeisser

Treatment Approaches in Rodent Models for Autism Spectrum Disorder 325
Susanna Pietropaolo, Wim E. Crusio and Francesca R. D'amato

Neuroimaging-Based Phenotyping of the Autism Spectrum 341
Boris C. Bernhardt, Adriana Di Martino, Sofie L. Valk and Gregory L. Wallace

Current Practice and Future Avenues in Autism Therapy 357
L. Poustka and I. Kamp-Becker

The Social Context Network Model in Psychiatric and Neurological Diseases 379
Sandra Baez, Adolfo M. García and Agustín Ibanez

Social-Cognitive Deficits in Schizophrenia 397
Daniela Mier and Peter Kirsch

The Programming of the Social Brain by Stress During Childhood and Adolescence: From Rodents to Humans 411
Stamatina Tzanoulinou and Carmen Sandi

Part I
Social Behavior in Rodents

Conspecific Interactions in Adult Laboratory Rodents: Friends or Foes?

Michael Lukas and Trynke R. de Jong

Abstract Interactions between adult conspecifics, including sexual behaviors, affiliation, and aggression are crucial for the well-being, survival, and reproduction of mammals. This holds true for any mammalian species, but certainly for humans: An inability to optimally navigate the social system can have a strong negative impact on physical and mental health. Translational rodent models have been used for decades to unravel the neural pathways and substrates involved in normal and abnormal conspecific interactions. Researchers in the field of translational social neuroscience face a double challenge: Not only do they need to pay considerable attention to the behavioral ecology of their model species or their ancestors, they also have to expect a relatively large variability in behavior and adjust their experimental design accordingly. In this chapter, we will lay out traditional and novel rodent models and paradigms to study sexual, affiliative, and aggressive interactions among adult conspecifics. We will discuss the merits and main findings and briefly consider the most promising novel directions. Finally, we review the modulatory involvement of two major players in mammal social interaction: the central oxytocin and vasopressin system.

Keywords Aggression · Sociality · Sexual behavior · Social preference · Social defeat · Oxytocin · Vasopressin

Contents

1 Introduction 4
2 Which Rodents Are We Talking About? 5
3 Friendly Encounters 6

M. Lukas (✉) · T.R. de Jong
Molecular and Behavioral Neurobiology, University of Regensburg, Regensburg, Germany
e-mail: michael.lukas@ur.de

T.R. de Jong
e-mail: trynke.de-jong@ur.de

Curr Topics Behav Neurosci (2016) 30: 3–24
DOI 10.1007/7854_2015_428
Published Online: 18 May 2016

3.1 Mating and Pair-Bonding ... 6
3.2 Sociality ... 8
4 Unfriendly Encounters ... 10
4.1 Social Aggression ... 11
4.2 Social Defeat ... 14
5 Oxytocin and Vasopressin as Key Regulators of Adult Social Encounters ... 16
5.1 Opposite-Sex Interactions ... 16
5.2 Same-Sex Interactions ... 18
6 Concluding Remarks ... 19
References ... 20

1 Introduction

Intraspecific social interactions in adults, such as sexual behavior, affiliation, and aggression, are crucial for the fitness of mammals. An individual that adequately decides who to mate with and who to fight against and who to approach and who to avoid strongly increases its chances to survive and reproduce. Conversely, wrong decisions can lead to considerable stress, social isolation, and even death. This holds true for any mammalian species, but certainly for humans (Strang and Park 2017): Our species heavily relies on complex social and sexual ties, negotiation skills, hierarchical structures, shared resources and territories that need to be defended against enemies. An inability to optimally navigate the social system, for example through inborn or developmental behavioral or cognitive impairments, may have a strong negative impact on physical and mental health (Mier and Kirsch 2017; Schroeder et al. 2015; Tzanoulinou and Sandi 2017).

Translational rodent models have been used for decades to unravel the neural pathways and substrates involved in normal and abnormal conspecific interactions. Various paradigms have been consolidated over time to study the neuroscience of social behaviors in rodents in a standardized manner. Traditionally, the emphasis has been placed on basic neuronal networks underlying a specific type of behavior, as well as hormonal or pharmacological manipulations of this behavior. More recently, reciprocal links with other behavioral systems have received increasing attention. Examples of these links are the effects of stress (acute, chronic, or early in life), trait or state anxiety, cognitive skills, and impulse control on positive and negative social interactions, and vice versa the effects of positive and negative social experiences on stress system (re-)activity, mood, and cognition (Tzanoulinou and Sandi 2015). Furthermore, the modeling and treatment of human psychopathologies characterized by social deficits, such as autism spectrum disorders or schizophrenia, using translational laboratory rodents has recently intensified (Pietropaolo et al. 2017; Schroeder et al. 2015).

The field of translational social neuroscience comes with certain challenges. Foremost, neuroscientists will benefit from knowing the behavioral ecology of their model animal or its ancestors far more intimately than is typically necessary for the study of depression and anxiety or learning and memory. Furthermore, experimental trials in social neuroscience often contain at least two individuals (e.g., mating partners, residents and intruders), each with their own and behavioral state contributing to considerable variability in the results. Avoiding any disturbances or inconsistencies that add to this variability is thus of the utmost importance.

In this chapter, we introduce and discuss both classic and novel rodent models and paradigms to study friendly and unfriendly interactions among adult conspecifics. We focus on inter-species variability in "social lifestyle," and how to relate social behavior in the field to experiments in the laboratory and, ultimately, translate the results to human interactions. Finally, we review the significant modulation of adult interactions by central oxytocin (OXT) and arginine-vasopressin (AVP), as these two neuropeptides are currently under intense investigation as targets to treat human social disorders.

2 Which Rodents Are We Talking About?

The execution of mounts and intromissions during mating, threats and attacks during fighting, or sniffing and adjacent lying during social approach/affiliation is fairly stereotypical across rodent species. On the other hand, the selection of the optimal behavioral response toward another animal (e.g. to approach or avoid, to mate or to fight) is strongly influenced by the "social lifestyle" of a species. While many rodent species have been used in descriptive studies of intraspecific social interactions, only a handful have been used to perform the bulk of translational neurobiological or neuroendocrinological experiments. Since none of these preferred model species display the full array of human sociality, they have been used in laboratories as representatives of separate components of our social life. Thus, pair-bonded and biparental prairie voles (*Microtus ochrogaster*) or California mice (*Peromyscus californicus*) are frequently used to study the neurobiological aspects of the formation and disruption of human family bonds (De Jong et al. 2013; Young et al. 2011), but are less useful to study sexual behavior. The neurobiology of aggression and social defeat is often studied in highly territorial, solitary living Syrian hamsters (*Mesocricetus auratus*, Ferris and Delville 1994) and territorial pair-bonded California mice (Trainor et al. 2011). The naked mole-rat (*Heterocephalus glaber*) is the preferred rodent species to study social organization and dominance hierarchies, based on its unique eusocial lifestyle (Mooney et al. 2015).

However, by far the most popular model species remain the promiscuous, uniparental, group-living, moderately hierarchical rats (*Rattus norvegicus*) and house mice (*Mus musculus*) (Berdoy and Drickamer 2007). Rats and mice can form large coherent social groups and are therefore the preferred species to study sociality, social memory/recognition (Camats Perna and Engelmann 2017), and even play behavior including accompanying ultrasonic vocalizations (Kisko et al. 2017; Wöhr et al. 2017). In addition, they adapt very well to their environment (the basis of their widespread distribution and tendency to live in urban areas (Feng et al. 2014)) and are thus optimally suited to investigate the effects of epigenetic variability and/or early-life stress on social behavioral profiles (Tzanoulinou and Sandi 2015). Moreover, the well-defined neurobiological pathways underlying stress, anxiety, addiction, and cognition in rats and mice are an excellent basis to investigate the reciprocal connections between these systems and the social behavior network.

Needless to say, the social and sexual behavior typically displayed by a species in the wild may differ profoundly from their behavior in the laboratory. The absence of predatory threats, adverse weather conditions, and food and water shortage combined with controlled housing in small same-sex/same-age groups or in isolation are likely to affect many neurobiological, neuroendocrinological, and behavioral parameters (Calisi and Bentley 2009; Keane et al. 2014). These effects are probably even stronger in animals that are the end product of generations of laboratory breeding, as indicated, for example, by the lower levels of aggression displayed by outbred Wistar rats compared to feral wild-type Groningen rats (Boer et al. 2003). Nevertheless, major hardwired species-specific behavioral strategies such as monogamy versus polygamy or gregarious versus solitary living usually remain intact in a laboratory environment.

3 Friendly Encounters

When two conspecific rodents have a shared goal, and are able to determine that they do so, their interactions will most likely be peaceful. An encounter between two adult conspecifics of opposite sexes will often lead to consensual mating and sometimes even the formation of an affiliate pair-bond, as this is likely to increase the reproductive success of both individuals. In group-living species, adults of the same sex may benefit from collaboration in order to find resources, defend a shared territory, escape predators, and keep warm at night. Displaying affiliative behavior when interacting will facilitate such collaborations. These friendly—sexual and affiliative—interactions among rodents have been studied in detail in various paradigms (see Fig. 1) to understand the equivalent behaviors in humans—especially to find treatments when they are impaired.

3.1 Mating and Pair-Bonding

Place a healthy adult male rodent in a cage with an estrous female conspecific, and the results can be easily predicted: The male will approach the female, sniff her anogenital region to confirm her reproductive state, and become—at some point—sexually aroused. This is defined as the *anticipatory* phase of copulation. The estrous female will respond with proceptive behaviors signaling her willingness to mate, including the stereotypical lordosis posture. This allows the pursuing male to mount her, insert his erect penis ("intromission"), and copulate until sperm is

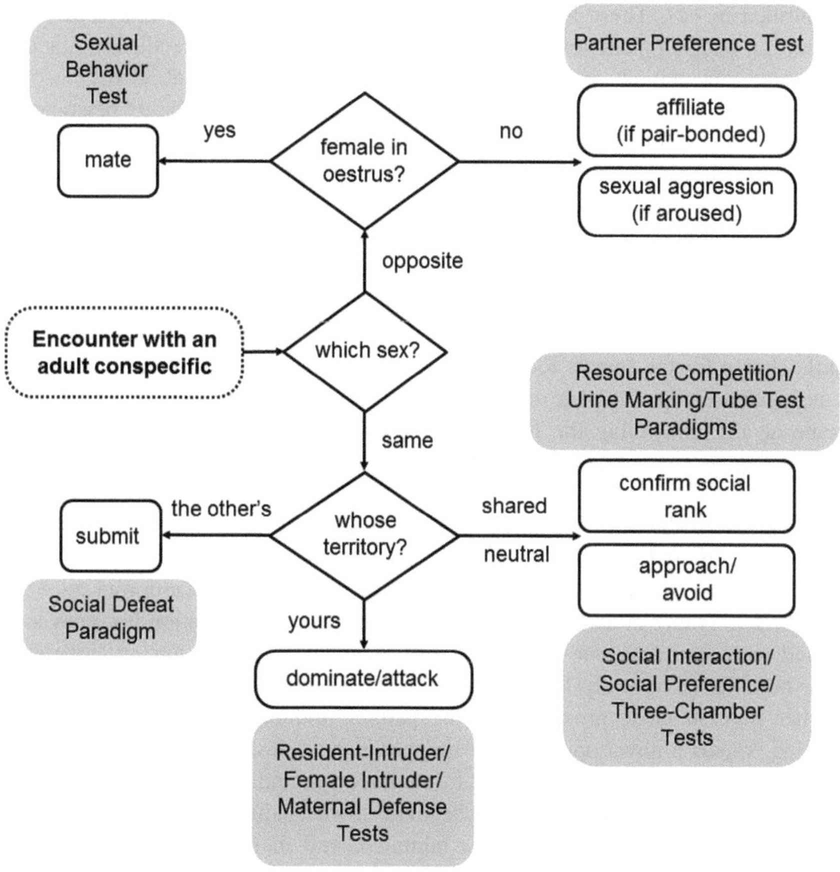

Fig. 1 Flowchart illustrating the various behavioral responses of adult rodents toward conspecifics (*white quadrangles*), and the associated experimental paradigms designed to measure variability in these responses under controlled conditions (*gray quadrangles*)

ejaculated. These behaviors comprise the *consummatory* phase of copulation. Ejaculation is always followed by a post-ejaculatory interval after which copulation is resumed—provided the female is still available.

The neurobiological pathways underlying these sexual behaviors (including the medial preoptic area, posteromedial bed nucleus of the stria terminalis, medial amygdala, and paraventricular and ventromedial hypothalamic nuclei), as well as pharmacological compounds affecting them, have been studied predominantly in rats (Veening and Coolen 2014). Rats normally display a consistent high level of sexual motivation and a rapid transition from the anticipatory to the consummatory phase of mating, typically resulting in 2–3 rounds of copulation in a standard 30-min test, in contrast to the much "slower" house mice. In addition, physical markers of copulatory behavior such as erection, ejaculation, and lordosis are easily quantified in rats. These features have proven very useful in the search for treatments of physiological sexual disorders in humans such as erectile dysfunction, premature ejaculation, or female arousal disorders (Giuliano et al. 2010).

While the relatively simple central pathways controlling copulatory behavior are by now well understood, the more complex neurobiology of sexual motivation is still under investigation. Even though male and estrous female conspecifics are likely to mate when they meet, they may chose not to. Even promiscuous rats can display preferences for specific individuals, and a minority of rats show very little interest in mating at all: They either avoid or show affiliative behavior toward an opposite-sex stimulus animal (Pfaus et al. 2003). It is of note here that in standard laboratory rat mating tests, the stronger male controls the interaction and leaves little choice for the female to cooperate. Only in special gated or bi-leveled cages can females escape the male and pace the rhythm of the copulation or decide not to mate at all. Unraveling the innate and experiential causes for this variability in motivation is an interesting avenue of investigation to model human sexual motivation.

Aside from the ability and the motivation to mate, there is a third important dimension in male–female conspecific interactions: the development of stable pair-bonds in monogamous species such as prairie voles and California mice (for review see Young et al. 2011). In these species, sexual motivation becomes associated with one particular individual, sometimes after a single mating experience. Pair-bonding is typically tested using a partner preference paradigm, in which a higher time in close proximity to the partner, induced by previous successful mating, versus a novel opposite-sex stranger is the defining parameter. In addition to the profound effects of pair-bond formation on an individual's interactions with its preferred partner, the changes in behavior toward non-preferred individuals are equally dramatic: Sexual interest is inhibited, and in some cases aggression is displayed.

3.2 *Sociality*

Placing two rodent conspecifics with the same sex together will usually not lead to mating (although exceptions do occur), but the resulting interactions can still be positive and rewarding. In particular, when the animals meet in a neutral territorial context, e.g., in a shared territory like a group cage or in an arena that is novel to both, most rodents will to some extent display affiliate interactions classified in the laboratory as social interaction and/or sociality.

Preference toward a conspecific over a nonsocial stimulus (selective social preference), as opposed to avoidance, is integral to the expression of social interaction, to facilitate the motivation to approach rather than avoid or behave quiescent in varying social contexts. The motivation to initially approach another "neutral" conspecific raised a lot of attention in the last years, as the usage of rodent animal models got more in the focus of research concerning neurodevelopmental disorders such as schizophrenia or autism spectrum disorders (Pietropaolo et al. 2015; Schroeder et al. 2017; Tzanoulinou and Sandi 2015). Important diagnostic criteria for autism spectrum disorders are deficits in social communication and social interaction across multiple contexts, including deficits in social–emotional reciprocity; deficits in nonverbal communicative behaviors; and deficits in developing, maintaining, and understanding relationships (Diagnostic and Statistical Manual of Mental Disorders, Fifth Edition 2013). Therefore, it is not surprising that there are many paradigms to model sociality and social communication available for basic and psychopharmacological animal research.

Preclinical and clinical research indicates that the prosocial neuropeptides OXT and AVP play a major role in promoting sociality (see Sect. 5.2.1 and Lukas and Neumann 2013), so far mainly localized in the lateral septum of male mice (Guzman et al. 2013; Zoicas et al. 2014). As social contact is known to be rewarding under normal circumstances, reduced seeking of social encounters hints toward a disruption of reward-seeking tendencies, if not in general then at least in a social context. Central reward mechanisms classically involve the mesolimbic dopamine pathway (Bromberg-Martin et al. 2010). Indeed, it is suggested that social "wanting" dysfunctions in autism spectrum disorders is caused by a disruption of the dopaminergic–oxytocinergic mesolimbic circuitry including the nucleus accumbens and the ventral tegmental area (Kohls et al. 2012).

As indicated above, the majority of "sociality" research is done in laboratory rats and mice (Lukas and Neumann 2013) as well as California mice (Trainor et al. 2011). Since these species are neither strictly solitary nor overly gregarious, they naturally express a relatively variable spectrum of affiliation and social avoidance dependent on the particular social context. In addition, the general sociality of the animals can be easily deflected to an extreme in case the experimental question demands it. One example would be the induction of strong social avoidance using social defeat (see Sect. 4.2) followed by the assessment of potential prosocial effects of a drug.

Beginning with rodent models of sociality, there are two main groups of test paradigms. The first group measures sociality via exposing a test animal to a freely moving or caged/separated stimulus animal alone. Social interaction with this stimulus animal is then determined in terms of time spent in direct social contact (sniffing), or in general proximity (e.g., adjacent lying or staying in contact zones). The second group of paradigms measures sociality of an experimental animal by comparing its interaction with or proximity to a social stimulus (caged animal) versus a nonsocial stimulus (empty cage). Within this second group, it is possible to differentiate between consecutive and simultaneous presentation of the social and the nonsocial stimulus. For a very detailed and precise methodological description of all the single paradigms available, we kindly refer to the review from Toth and Neumann (2013). It is a matter of discussion, which of the multitude of paradigms is the most suitable to measure sociality. To answer this question, one has to attend to the behavioral, environmental, and developmental context of the animals to be tested. As these tests all depend on a certain behavioral activity (movement/sniffing) of the animal, it is important to adjust the test circumstances. For instance, animals with a known general lower activity are more likely to show high levels of social interaction when confronted with a social stimulus alone, or a social and a nonsocial stimulus one after the other. An example for this would be the testing of aged rats: Due to their general low locomotor activity, it is possible that during the often standardized limited time the tested animals are not able to efficiently explore several chambers with social and nonsocial stimuli simultaneously. Therefore, interpretations of social versus nonsocial investigation/proximity may be difficult or misleading. In addition, testing highly aggressive animals such as male Syrian hamsters or sexually experienced male or lactating female rats with a freely moving conspecific most certainly results in aggressive interactions (see Sect. 4.1). Thus, using a paradigm with caged or separated stimulus animals would be appropriate here. In particular, challenging are test subjects that have a very high level of innate anxiety, as this may lead to reduced locomotor activity or freezing and consequently low levels of social interaction in an unfamiliar environment. In this case, direct comparisons to a nonsocial stimulus are advised. It is also possible that high trait or state anxiety induces proximity to a social stimulus as this provides relief from stress and anxiety in socially living animals, usually referred to as social buffering (Kikusui et al. 2006), resulting in unexpected high levels of sociality. Here, investigation of social proximity alone without analysis of actual interaction, such as sniffing, can be misleading.

If sociality is analyzed with one of the above-mentioned paradigms, it is possible to differentiate the outcome in roughly three categories: social preference (higher interest in social over nonsocial stimulus), loss of social preference (reduction of social interest to the level of nonsocial stimulus), and social avoidance/fear (less interest/avoidance of social compared to nonsocial stimuli). A differentiation that becomes especially important, if one is interested in changes of sociality following negative, "unfriendly" encounters.

4 Unfriendly Encounters

When two conspecifics have conflicting goals, encounters can become distinctly agonistic, ranging from mutual avoidance to fighting to the death. In territorial species or lactating females, the cost of losing a territory or a litter is so high that virtually all intruders will be attacked, no matter their size or the circumstances. In group-living species, more variability will occur depending on the individual's social status, social recognition (Camats Perna and Engelmann 2015), social personality, reproductive state, and circumstances in early life (Tzanoulinou and Sandi 2015) as well as recent (social) experiences.

4.1 *Social Aggression*

Social aggression, as opposed to predatory aggression, is part of the standard behavioral repertoire of rodents in conflict situations. In order to survive and to reproduce, animals need to obtain access to food, water, shelter, and mates, and to raise their offspring. Since crucial resources are often scarce in nature, especially in times of higher population densities or during breeding seasons, fierce competition can occur between two or more conspecifics.

4.1.1 Response to Intrusion

Reactive aggression as a result of one individual threatening to overtake a valuable resource from another is easily induced in the laboratory by first giving experimental animals "ownership" of a territory and a mate (in the resident-intruder test, see Haller 2014) or a litter (in the maternal defense test, see Bosch 2013) and then introducing a smaller/weaker same-sex conspecific intruder. This elicits a fairly stereotypical set of behaviors in the experimental animals, starting with extensive sniffing of the intruder followed by agonistic behaviors that limit the intruder in its movements without causing bodily harm. Examples of these are towering over the intruder, pushing the intruder down to the ground or against a wall, and threatening to attack. The encounter can culminate in one or more attacks, which in rodents typically consist of a rapid clinch fight and biting (Haller 2014). Recent observations have shown that virgin female rats display a similar behavioral repertoire in a resident-intruder setting (de Jong et al. 2014), despite earlier reports of very low levels of aggression among non-lactating females (see for example Blanchard et al. 1988). The moderate levels of pushing, threatening, and attacking female intruders by virgin female residents are not instigated by prolonged cohabitation with a male, but already appear after a short period (48 h) of single housing. Furthermore, this aggressive behavior rapidly increases upon repeated exposure to unknown intruders (once per day for 3–5 consecutive days), especially when the female residents remain isolated in this training period (unpublished observations). This makes the paradigm different from the classic resident-intruder test, in which males are pair-housed with a female for at least a week prior to intrusion and their female mates are only temporarily removed from the cage (i.e., during an intrusion). These findings suggest that in virgin female rats, the defense of resources such as food and water rather than an established territory and/or a mate is sufficient to instigate moderate aggression. The rapid escalation of defensive aggression seen in late pregnancy and early lactation is putatively the result of the profound neuromodulations in the peripartum period (Bosch 2013).

The neural circuitry controlling aggressive behaviors, with an emphasis on attacks, has been studied quite extensively in male rodents (especially rats, mice, Syrian hamsters, and California mice) using the resident-intruder test and includes the hypothalamus and the extended amygdala (Veening et al. 2005). The neural circuitry controlling aggressive decision making (i.e., when to be aggressive and what level of violence to use) is much less clear. In particular, in the relatively gregarious rat, behavioral responses toward an intruder can range from violent aggression with attacks to mainly non-violent agonistic threats or even tolerance and affiliation, for reasons that are still under investigation. The neuropeptides OXT and AVP play a modulating role here, as discussed in Sect. 5. In addition, there is an increasing interest in the role of cortical pathways exerting behavioral control or the lack thereof (i.e., impulsivity). In particular, the serotonin system is thought to be important for the inhibition of behavioral impulses, such as the aggressive attack of intruders (Takahashi et al. 2011).

The question has arisen whether the defense of resources, which in itself is a healthy and natural behavior among rodents, is an appropriate model for the pathological aggression in humans associated with, for example, intermittent explosive disorder, schizophrenia, borderline personality disorder, antisocial personality disorder, and conduct disorder (Haller 2014; Miczek et al. 2013). One approach to improve the translatability of rodent models is to selectively investigate individuals that display abnormal or extreme behaviors in the resident-intruder test, including (a) attacking without giving the intruder the chance to escape or admit defeat (i.e., very quickly and/or without giving proper warning in the form of threats) and (b) displaying unnecessarily violent attacks (e.g., targeting vulnerable body parts for maximal damage, or attacking anesthetized, already defeated or female intruders) (de Boer et al. 2009; Haller 2014). Although laboratory rats and mice rarely show such extreme/abnormal behaviors spontaneously, some experimental manipulations have been found to induce it. These manipulations include the knockout of genes such as the gene for monoamine oxidase A in mice (Cases et al. 1995), selective breeding examples include short attack latency mice (Natarajan et al. 2009) and low anxiety behavior rats (Beiderbeck et al. 2012), treatment with alcohol (Miczek et al. 2013), depletion or suppression of the central serotonin system (Audero et al. 2013), depletion of corticosterone by adrenalectomy (Haller et al. 2004), or stress during infancy or puberty, such as maternal separation, peripubertal isolation, or peripubertal stress (Sandi and Haller 2015).

A remaining issue is that these animal experiments still rely on reactive aggression in the resident-intruder test, whereas the most violent and treatment-resistant type of human aggression is proactive and calculating, and associated with low levels of empathy ("callous–unemotional traits") (Blair 2010). It is by now well understood that reactive and proactive aggression are two very different behaviors, generally distinguished by increased (reactive aggression) versus decreased (proactive aggression) arousal in response to threatening stimuli, as measured in terms of stress responses and activation of distinct amygdalar and cortical areas. Although animal models are starting to close this translational gap by focusing on individuals that show impaired intruder-induced arousal (Haller 2014), behavioral paradigms that truly model proactive, goal-oriented aggression and/or variability in callous–unemotional traits in rodents are still lacking.

4.1.2 Social Hierarchies

Resident-intruder style paradigms artificially promote aggressive behavior by giving the resident a clear upper hand: The resident has something valuable to defend, the encounter takes place on the resident's territory, and the intruder is unfamiliar and relatively weak. A different picture emerges when group-living rodents, such as laboratory rats, interact with familiar conspecifics with whom they share a territory and its resources (i.e., cage mates in the laboratory). Using violent aggression to continuously defend resources against group mates is no longer feasible and would drain all individuals of much-needed energy.

One solution in this situation is the formation of a stable social hierarchy, in which the more dominant males or females automatically receive first choice of resources (Broom 2002). This solution has evolved in various rodent species, but most strikingly in the naked mole-rats that form large colonies of up to 300 individuals headed by a single dominant and reproductively active female (the queen) and one to three reproductive males. All other males and females are reproductively suppressed and perform subordinate working tasks such as colony maintenance and defense. When the queen is removed, the colony becomes unstable and aggression peaks until a novel queen has established herself and the social order is restored (Mooney et al. 2015).

Rats and mice form social hierarchies as well, but these hierarchies are more fluid and less associated with division of labor and reproduction—at least in laboratory settings (Blanchard et al. 1988; Ziporyn and McClintock 1991). Determining which individuals are (relatively) dominant or subordinate in small groups of rats and mice is therefore more difficult than in naked mole-rats and requires fairly elaborate behavioral paradigms. These include food or water competition paradigms, in which the dominant animal most often obtains the largest amount of a limited resource (Malatynska and Knapp 2005; de Jong et al. 2012); urine marking paradigms, in which small urine marks spread throughout a cage signify dominance whereas large urine pools at the edges signify submissiveness (Desjardins et al. 1973); or "passing behavior" paradigms in which dominant individuals are more likely to take the right of way in cramped burrow systems (Ziporyn and McClintock 1991). Social hierarchies in various rodent species are currently used to measure the behavioral, neurobiological, and physiological correlates of having a dominant or subordinate status. Examples include the multiple markers of chronic stress and depression observed upon subordination in mice and rats (Langgartner et al. 2015; Malatynska and Knapp 2005), the similarities between dominant status and pathological mania (Johnson et al. 2012; Malatynska and Knapp 2005), and the variability in OXT receptor binding dependent on status in naked mole-rats (Mooney et al. 2015).

4.2 *Social Defeat*

While the preceding paragraphs focused on individuals displaying aggression during conspecific encounters, the focus of this paragraph lies on the "victims" or recipients of aggressive actions. In this context, social neuroscientists and behavioral pharmacologists are mainly interested in the behavioral and physiological consequences of social defeat in rodents, in order to model the implications for human victims of violence or abuse, e.g., patients with post-traumatic stress disorders following negative social experiences.

As already indicated in Sect. 2, the preferred model species in these contexts are laboratory rats, mice, and California mice due to their plasticity in social behavior following negative social experiences, for example, social defeat (Toth and Neumann 2013). To model defeat situations in laboratory rodents, the social defeat procedure that was initially established by Miczek (1979), and further developed by Heinrichs et al. (1992), is predominantly used. The original procedure has been adapted in various ways, but the overarching concept is to introduce a test subject into the home cage territory of dominant male or female conspecifics (see also resident-intruder paradigm in Sect. 4.1.1). This naturally elicits an aggressive response followed by the defeat of the subject. Social defeat results in various acute physiological alterations, e.g., increase in heart rate, plasma corticosterone, and plasma testosterone, as well as changes in social and emotional behaviors dependent on the type of defeat. Social defeat paradigms are performed using either single or repeated defeats by different aggressor animals. Single social defeat, which was mostly done in male rats, induces anxiety-related behavior and social avoidance. Chronic social defeat additionally induces depressive-like and anhedonic behaviors, reduces general locomotion, and disturbs sleep patterns and normal circadian rhythmicity (for review see Hollis and Kabbaj 2014). However, in line with the main topic of our chapter we want to further focus on behavioral effects on sociality/social encounters in the following paragraphs.

4.2.1 Male Defeat

Acute social defeat in male rats results in a loss of social preference toward this defeater rat (Lukas et al. 2011). However, repeated negative social encounters in rats and mice lead to a general loss of preference or even avoidance of conspecifics (Hollis and Kabbaj 2014). Intriguingly, one particular study was able to show generally reduced social interactions even after a single social defeat (Haller and Bakos 2002). As indicated above, in contrast to this single defeat study, Lukas et al. (2011) were only able to demonstrate a loss of social preference when the defeater rat was used as social stimulus, indicating social recognition in the defeated experimental rats. This phenomenon has also been observed in a study reporting social avoidance in golden hamsters (Lai et al. 2005). Eventually this specific social avoidance reaction could be referred to as a form of rodent "bullying" that could be further used for translational studies regarding victims of this form of aggression. Additionally, this implies that for eliciting social avoidance toward an unknown rat, i.e., general social avoidance, a stronger form of social defeat or repeated defeat by different defeaters may be needed. Since social defeat elicits nonsocial anxiety in rats and mice (Hollis and Kabbaj 2014), it is of course possible that a reduction of social interaction after social defeat is due to increased general anxiety after the defeat. This again emphasizes the importance of simultaneous or consecutive testing of these animals' interactions with nonsocial stimuli to check for the specificity of the social deficits.

Interestingly, in California mice even repeated social defeat by a same-sex conspecific did not result in social avoidance in male, but it did in female individuals (Trainor et al. 2011).

4.2.2 Female Defeat

Concerning social defeat-induced social avoidance, one expects of course similar effects in female individuals. Compared to studies in males, studies in females are rather underrepresented. The main reason for the lack of studies in female rodents may be the general assumption that female rodents have a very low level of intraspecies aggression. However, there are particular studies addressing social avoidance in female California mice (*P. californicus*, Trainor et al. 2011) and female Wistar rats (Lukas and Neumann 2014), utilizing the high levels of territorial aggression of female California mice and the increased levels of aggression of lactating dams. These studies demonstrate that a single 10-min exposure to social defeat by a resident female California mouse to another female California mouse as well as by a lactating Wistar dam to another virgin female Wistar rat causes a specific lack of social preference for, and consequently, social avoidance of, the individual defeater for at least 2 h (Lukas and Neumann 2014; Trainor et al. 2011). The loss of social investigation of the lactating defeater rat is not due to a general loss of social motivation as a result of the previous encounter, since exposure to cage mates still motivates social preference behavior.

This seems to be comparable to observations in defeated male rats that only avoid the former defeater male in the social preference test (above and Lukas et al. 2011). In addition to the induction of specific social avoidance, acute or repeated exposure to social defeat by a lactating dam has also been shown to induce anxiety (Neumann et al. 2001) and depressive-like behavior (Shimamoto et al. 2011) in virgin female rats. Thus, adverse social stimuli can induce both social and emotional disturbances in females.

It may be of interest to also look into behavioral alterations in terms of potential social avoidance following being defeated in the newly established female intruder test (de Jong et al. 2014), as this procedure may be more comparable to male social defeat in rats than maternal defeat by a lactating dam. In addition, it may be particularly interesting to study a common type of defeat in females: the experience of social and sexual aggression by males. Recent findings have shown that cohabitation with abnormally aggressive males, due to peripubertal stress, caused female rats to display signs of chronic stress and increased anxiety and depression-like behavior (Cordero et al. 2012). Furthermore, aggressive and highly sexually aroused male rats can use considerable force in their attempts to mate with an unwilling female (unpublished observations), and this is likely to induce changes in stress reactivity, anxiety, and mood as well as reproductive behavior and physiology in the female victims. Future research is needed to investigate this particular form of (sexual) defeat.

5 Oxytocin and Vasopressin as Key Regulators of Adult Social Encounters

The neuropeptides OXT and AVP become increasingly attractive as potential therapeutic targets concerning dysregulated social interactions. Neuropeptides of the arginine vasotocin family, including OXT and AVP, are ubiquitous within vertebrates and evolutionary highly conserved, both in structure and function. At the same time, the distribution and sensitivity of OXT and AVP receptors are remarkably plastic and vary depending on "social lifestyle" (e.g., of a species), sex, reproductive state, hierarchical rank, etc. (Goodson 2013). It is now generally accepted that OXT and AVP are key regulators of social behaviors, as revealed and confirmed with agonists and antagonists in both animals and humans using complementary behavioral paradigms. Despite this well-established importance of central OXT and AVP, the detailed mechanisms by which they modulate conspecific encounters in various contexts are surprisingly complex and our knowledge is far from complete. An illustration of this complexity is provided in Table 1, showing what little we know about the endogenous release of OXT and AVP in various social encounters. However, in the following paragraphs we try to carefully summarize the behavioral effects of endogenous and exogenous OXT and AVP during sociality, sexual behavior, pair-bonding and aggression (for a simplified overview, see Table 2).

5.1 Opposite-Sex Interactions

For opposite-sex interactions, the roles of OXT and AVP are relatively well described. Thus, when an adult male and female rat meet, OXT is released from the paraventricular hypothalamic nucleus in both sexes and promotes the appetitive as well as the consummatory phases of copulation (Veening and Coolen 2014; Pfaus et al. 2012; Nyuyki et al. 2011; Waldherr and Neumann 2007). In contrast, AVP is much less involved in copulation in the promiscuous rat, with no known or consistent role in males and a mild, probably indirect inhibitory role in females (Pedersen and Boccia 2006). In male and female pair-bonded prairie voles, the

Table 1 Local oxytocin and/or vasopressin release following friendly and unfriendly encounters in adult laboratory animals. Adapted and modified from Lukas and Neumann (2013)

Stimulus	Brain region	Species	Availability	References
Oxytocin release				
Social investigation				
Male	LS	Mouse	Increase	Zoicas et al. 2014
Aggression				
Male resident	LS	Rat	No change	Beiderbeck et al. 2007
Lactating resident	LS, CeA	Rat	No change	Bosch 2013
	BST, PVN	Rat	Increase	Bosch 2013
Social defeat				
Male intruder	PVN	Rat	No change	Wotjak et al. 1996
	LS, SON	Rat	Increase	Ebner et al. 2000 and Engelmann et al. 1999
Female intruder	LS, CeA	Rat	No change	Bosch 2013
	PVN	Rat	Increase	Bosch 2013
Mating				
Male	NAc	Vole	Increase	Ross et al. 2009
	PVN	Rat	Increase	Waldherr and Neumann 2007
Female	PVN	Rat	Increase	Nyuyki et al. 2011
Vasopressin release				
Aggression				
Male resident	LS[a], BST[b]	Rat	Increase	Beiderbeck et al. 2007 and Veenema et al. 2010
Lactating resident	LS, BST, PVN	Rat	Increase	Bosch 2013
Social defeat				
Male	LS, SON	Rat	No change	Ebner et al. 2000 and Wotjak et al. 1996
	PVN	Rat	Increase	Wotjak et al. 1996
Mating				
Male	VP	Vole	Increase	Morales et al. 2004

BST bed nucleus of stria terminalis; *CeA* central amygdala; *LS* lateral septum; *NAc* nucleus accumbens; *PVN* paraventricular nucleus; *SON* supraoptic nucleus; *VP* ventral pallidum; *W* Wistar; *SD* Sprague-Dawley
[a]High-aggressive rats
[b]Low-aggressive rats

release of OXT and AVP is no longer associated with mating per se, but it becomes associated with a particular preferred opposite-sex individual (Cho et al. 1999). This effect has been demonstrated using local gain-and-loss of function studies of OXT and AVP receptors which appear to be associated with partner preference in the nucleus accumbens and prefrontal cortex of female and in the ventral pallidum of male prairie voles (Young et al. 2011).

Table 2 Effects of oxytocin and vasopressin on adult social behaviors

	Males	Virgin females	Lactating females
Oxytocin			
Sociality	+	No effect	Not investigated
Sexual behavior	+	+	——
Pair-bonding	+	+	——
Aggression	−	−	+ (Site-specific)
Vasopressin			
Sociality	+ (Systemic)	No effect	Not investigated
Sexual behavior	No effect	——	——
Pair-bonding	+	+	——
Aggression	+ (Site-specific)	Not investigated	+ (Site-specific)

Systemic central application was so far ineffective; *site-specific* effect not in all brain regions tested

5.2 *Same-Sex Interactions*

5.2.1 Sociality

A somewhat more intricate picture emerges when adult same-sex interactions are examined. To start with OXT, this neuropeptide can be generally considered to facilitate sociality and social preference in male rats and mice. Thus, administration of synthetic OXT enhances male affiliate interactions, whereas OXT receptor antagonists inhibit them, as shown using social preference, social interaction, social fear, and social defeat paradigms (Lukas et al. 2011; Ramos et al. 2013; Guzman et al. 2013; Zoicas et al. 2014). The prosocial potential of central OXT is further supported by the impaired social motivation of male OXT receptor knockout mice (Sala et al. 2011). Interestingly, no significant prosocial role of OXT could be demonstrated in the social preference and social defeat paradigms in virgin female rats (Lukas and Neumann 2014).

Just recently it was demonstrated that similar to OXT, AVP also promotes male sniffing and adjacent lying when administered either intraperitoneally or via inhalation (Ramos et al. 2013, 2014). This is in line with the impaired social motivation displayed by AVP1b-receptor knockout mice (DeVito et al. 2009). Intriguingly, neither central blockade of AVP1a-receptors in male and female rats nor central application of synthetic AVP in female rats affected their social preference (Lukas and Neumann 2014; Lukas et al. 2011). Nevertheless, due to the difference in application in the aforementioned studies, resulting in way higher doses possible than with central infusion, it is still most likely that AVP exerts prosocial effects via central or even systemic activity.

5.2.2 Aggression

The administration of OXT (centrally or systemically) inhibits the display of aggression by a resident against an intruder in male rats (Calcagnoli et al. 2013) as well as against a female intruder in virgin female residents (de Jong et al. 2014). In contrast, in lactating female rats OXT appears to promote aggression (but note the dependence on trait anxiety and brain areas, Bosch 2013)—a further demonstration that the OXT system is dramatically altered in the peripartum period.

AVP is generally associated with increased aggression among males, as demonstrated in the resident-intruder test in rats, mice, prairie voles, and Syrian hamsters (Caldwell et al. 2008). While this positive correlation between aggression and AVP was confirmed within the lateral septum of male rats, a negative association (i.e., inhibited release of AVP in high aggressive individuals and anti-aggressive effects of AVP infusion) was found in the dorsolateral bed nucleus of the stria terminalis (Veenema et al. 2010). Together with the prosocial effects of AVP described in Sect. 5.2.1, it could be argued that AVP promotes same-sex conspecific interactions per se, whereas the "tone" of the interaction (friendly or unfriendly) depends on the species, the individual, the context, and the central location in which it is released. It is clear that the complex social role of AVP is most prominent in males: Manipulation of the AVP system does not affect social preference or social avoidance in virgin female rats (Lukas and Neumann 2014). In lactating female rats, AVP modulates maternal aggression in a complex manner, which similar to OXT depends on trait anxiety and the brain area in which neurotransmission is measured or manipulated (Bosch 2013). The role of AVP in virgin female aggression remains to be investigated.

6 Concluding Remarks

This chapter lays out that social neuroscientists have to be aware of the ecology and the "social lifestyle" of the different laboratory rodent species and choose the experimental model best fitting to the scientific question. Further, the choice of the experimental paradigm has to be adjusted to (1) the species that is used, (2) the physiological and developmental state (age, sex, hormonal status), (3) the emotional state that influences social behavior (anxiety, depression), and (4) the desired readout parameter as well as the expected effect (increase or decrease) of pharmacological or behavioral manipulation on this parameter. Thus, one has to admit and accept that it is not possible to test all the aspects of intraspecies encounters with just one species in one test paradigm. However, as this review demonstrates, the literature concerning the testing of rodent social behavior, including the development of new paradigms for various species, is growing continuously. Searching for new directions and at the same time recognizing the benefits and stability of traditional methodology hopefully allows us to find a proper solution for the various scientific demands in this field.

References

Audero E, Mlinar B, Baccini G, Skachokova ZK, Corradetti R, Gross C (2013) Suppression of serotonin neuron firing increases aggression in mice. J Neurosci Official J Soc Neurosci 33:8678–8688. doi:10.1523/JNEUROSCI.2067-12.2013

Beiderbeck DI, Neumann ID, Veenema AH (2007) Differences in intermale aggression are accompanied by opposite vasopressin release patterns within the septum in rats bred for low and high anxiety. Eur J Neurosci 26(12):3597–3605

Beiderbeck DI, Reber SO, Havasi A, Bredewold R, Veenema AH, Neumann ID (2012) High and abnormal forms of aggression in rats with extremes in trait anxiety–involvement of the dopamine system in the nucleus accumbens. Psychoneuroendocrinology 37(12):1969–1980

Berdoy M, Drickamer L (2007) Comparative social organization and life history of rattus and mus. In: Wolff JO, Sherman PW (eds) Rodent societies: an ecological and evolutionary perspective. University of Chicago Press, Chicago, pp 380–392

Blair RJR (2010) Neuroimaging of psychopathy and antisocial behavior: a targeted review. Curr Psychiatry Rep 12:76–82. doi:10.1007/s11920-009-0086-x

Blanchard RJ, Flannelly KJ, Blanchard DC (1988) Life-span studies of dominance and aggression in established colonies of laboratory rats. Physiol Behav 43:1–7. doi:10.1016/0031-9384(88)90089-3

Bosch OJ (2013) Maternal aggression in rodents: brain oxytocin and vasopressin mediate pup defence. Philos Trans R Soc Lond B Biol Sci 368(1631):20130085. doi:10.1098/rstb.2013.0085

Bromberg-Martin ES, Matsumoto M, Hikosaka O (2010) Dopamine in motivational control: rewarding, aversive, and alerting. Neuron 68(5):815–834

Broom M (2002) A unified model of dominance hierarchy formation and maintenance. J Theor Biol 219(1):63–72. doi:10.1006/jtbi.2002.3109

Calcagnoli F, Boer S, Althaus M, Boer J, Koolhaas J (2013) Antiaggressive activity of central oxytocin in male rats. Psychopharmacology 229:1–13

Caldwell HK, Lee HJ, Macbeth AH, Young WS 3rd (2008) Vasopressin: behavioral roles of an "original" neuropeptide. Prog Neurobiol 84(1):1–24. doi:10.1016/j.pneurobio.2007.10.007

Calisi RM, Bentley GE (2009) Lab and field experiments: are they the same animal? Horm Behav 56(1):1–10. doi:10.1016/j.yhbeh.2009.02.010

Camats Perna J, Engelmann M (2017) Recognizing others: rodent's social memories. Curr Top Behav Neurosci 30:25–45

Cases O, Seif I, Grimsby J, Gaspar P, Chen K, Pournin S, Muller U, Aguet M, Babinet C, Shih JC et al (1995) Aggressive behavior and altered amounts of brain serotonin and norepinephrine in mice lacking MAOA. Science 268(5218):1763–1766

Cho MM, DeVries AC, Williams JR, Carter CS (1999) The effects of oxytocin and vasopressin on partner preferences in male and female prairie voles (*Microtus ochrogaster*). Behav Neurosci 113(5):1071–1079

Cordero MI, Poirier GL, Marquez C, Veenit V, Fontana X, Salehi B, Ansermet F, Sandi C (2012) Evidence for biological roots in the transgenerational transmission of intimate partner violence. Transl Psychiatry 2:e106. doi:10.1038/tp.2012.32

de Boer SF, Caramaschi D, Natarajan D, Koolhaas JM (2009) The vicious cycle towards violence: focus on the negative feedback mechanisms of brain serotonin neurotransmission. Front Behav Neurosci 3:52. doi:10.3389/neuro.08.052.2009

de Boer SF, van der Vegt BJ, Koolhaas JM (2003) Individual variation in aggression of feral rodent strains: a standard for the genetics of aggression and violence? Behav Genet 33:485–501, doi:10.1023/A:1025766415159

de Jong TR, Korosi A, Harris BN, Perea-Rodriguez JP, Saltzman W (2012) Individual variation in paternal responses of virgin male California mice (Peromyscus californicus): behavioral and physiological correlates. Physiol Biochem Zool PBZ 85:740–751. doi:10.1086/665831

De Jong TR, Harris BN, Perea-Rodriguez JP, Saltzman W (2013) Physiological and neuroendocrine responses to chronic variable stress in male California mice (*Peromyscus californicus*): influence of social environment and paternal state. Psychoneuroendocrinology 38(10):2023–2033. doi:10.1016/j.psyneuen.2013.03.006

de Jong TR, Beiderbeck DI, Neumann ID (2014) Measuring virgin female aggression in the female intruder test (FIT): effects of oxytocin, estrous cycle, and anxiety. PLoS ONE 9(3):e91701. doi:10.1371/journal.pone.0091701

Desjardins C, Maruniak JA, Bronson FH (1973) Social rank in house mice: differentiation revealed by ultraviolet visualization of urinary marking patterns. Science 182:939–941

DeVito LM, Konigsberg R, Lykken C, Sauvage M, Young WS, Eichenbaum H (2009) Vasopressin 1b receptor knock-out impairs memory for temporal order. J Neurosci 29 (9):2676–2683. doi:10.1523/jneurosci.5488-08.2009

Diagnostic and Statistical Manual of Mental Disorders, Fifth Edition (2013) American Psychiatric Association, Washington, DC

Ebner K, Wotjak CT, Landgraf R, Engelmann M (2000) A single social defeat experience selectively stimulates the release of oxytocin, but not vasopressin, within the septal brain area of male rats. Brain Res 872(1–2):87–92

Engelmann M, Ebner K, Landgraf R, Holsboer F, Wotjak CT (1999) Emotional stress triggers intrahypothalamic but not peripheral release of oxytocin in male rats. J Neuroendocrinol 11:867–872

Feng C, Hackett P, DeMarco A, Chen X, Stair S, Haroon E, Ditzen B, Pagnoni G, Rilling J (2014) Oxytocin and vasopressin effects on the neural response to social cooperation are modulated by sex in humans. Brain Imaging Behav 1–11. doi:10.1007/s11682-014-9333-9

Ferris CF, Delville Y (1994) Vasopressin and serotonin interactions in the control of agonistic behavior. Psychoneuroendocrinology 19(5–7):593–601

Giuliano F, Pfaus J, Srilatha B, Balasubramanian S, Hedlund P, S-i Hisasue, Marson L, Wallen K (2010) Experimental models for the study of female and male sexual function. J Sex Med 7:2970–2995. doi:10.1111/j.1743-6109.2010.01960.x

Goodson JL (2013) Deconstructing sociality, social evolution and relevant nonapeptide functions. Psychoneuroendocrinology 38(4):465–478

Guzman YF, Tronson NC, Jovasevic V, Sato K, Guedea AL, Mizukami H, Nishimori K, Radulovic J (2013) Fear-enhancing effects of septal oxytocin receptors. Nat Neurosci 16:1185–1187

Haller J (2014) Neurobiological bases of abnormal aggression and violent behaviour. Springer, Vienna, doi:10.1007/978-3-7091-1268-7

Haller J, Bakos N (2002) Stress-induced social avoidance: a new model of stress-induced anxiety? Physiol Behav 77(2–3):327–332

Haller J, Halasz J, Mikics E, Kruk MR (2004) Chronic glucocorticoid deficiency-induced abnormal aggression, autonomic hypoarousal, and social deficit in rats. J Neuroendocrinol 16 (6):550–557

Heinrichs SC, Pich EM, Miczek KA, Britton KT, Koob GF (1992) Corticotropin-releasing factor antagonist reduces emotionality in socially defeated rats via direct neurotropic action. Brain Res 581(2):190–197

Hollis F, Kabbaj M (2014) Social defeat as an animal model for depression. ILAR J 55(2):221–232. doi:10.1093/ilar/ilu002

Johnson SL, Leedom LJ, Muhtadie L (2012) The dominance behavioral system and psychopathology: evidence from self-report, observational, and biological studies. Psychol Bull 138(4):692–743. doi:10.1037/a0027503

Keane B, Parsons S, Smucker BJ, Solomon NG (2014) Length polymorphism at the avpr1a locus is correlated with male reproductive behavior in a natural population of prairie voles (*Microtus ochrogaster*). Behav Ecol Sociobiol 68(12):1951–1964. doi:10.1007/s00265-014-1801-0

Kikusui T, Winslow JT, Mori Y (2006) Social buffering: relief from stress and anxiety. Philos Trans R Soc Lond B Biol Sci 361(1476):2215–2228

Kisko TM, Wöhr M, Pellis VC, Pellis SM (2017) From play to aggression: high-frequency 50-kHz ultrasonic vocalizations as play and appeasement signals in rats. Curr Top Behav Neurosci 30:91–108
Kohls G, Chevallier C, Troiani V, Schultz RT (2012) Social 'wanting' dysfunction in autism: neurobiological underpinnings and treatment implications. J Neurodev Disord 4(1):10
Lai W-S, Ramiro L-LR, Yu HA, Johnston RE (2005) Recognition of familiar individuals in golden hamsters: a new method and functional neuroanatomy. J Neurosci 25(49):11239–11247. doi:10.1523/jneurosci.2124-05.2005
Langgartner D, Füchsl AM, Uschold-Schmidt N, Slattery DA, Reber SO (2015) Chronic subordinate colony housing paradigm: a mouse model to characterize the consequences of insufficient glucocorticoid signalling. Front Psychiatry 6, doi:10.3389/fpsyt.2015.00018
Lukas M, Neumann ID (2013) Oxytocin and vasopressin in rodent behaviors related to social dysfunctions in autism spectrum disorders. Behav Brain Res 251:85–94
Lukas M, Neumann ID (2014) Social preference and maternal defeat-induced social avoidance in virgin female rats: sex differences in involvement of brain oxytocin and vasopressin. J Neurosci Methods 234:101–107. doi:10.1016/j.jneumeth.2014.03.013
Lukas M, Toth I, Reber SO, Slattery DA, Veenema AH, Neumann ID (2011) The neuropeptide oxytocin facilitates pro-social behavior and prevents social avoidance in rats and mice. Neuropsychopharmacology 36:2159–2168
Malatynska E, Knapp RJ (2005) Dominant-submissive behavior as models of mania and depression. Neurosci Biobehav Rev 29:715–737. doi:10.1016/j.neubiorev.2005.03.014
Miczek KA (1979) A new test for aggression in rats without aversive stimulation: differential effects of d-amphetamine and cocaine. Psychopharmacology 60(3):253–259
Miczek KA, de Boer SF, Haller J (2013) Excessive aggression as model of violence: a critical evaluation of current preclinical methods. Psychopharmacology 226(3):445–458. doi:10.1007/s00213-013-3008-x
Mier D, Kirsch P (2017) Social-cognitive deficits in schizophrenia. Curr Top Behav Neurosci 30:397–409
Mooney SJ, Coen CW, Holmes MM, Beery AK (2015) Region-specific associations between sex, social status, and oxytocin receptor density in the brains of eusocial rodents. Neuroscience 303:261–269. doi:10.1016/j.neuroscience.2015.06.043
Morales JC, Cole C, Neumann ID, Langraf R, Young LJ (2004) Vasopressin release in the ventral pallidum during mating in the monogamous male prairie vole. Soc Neurosci Abs 2141
Natarajan D, Vries Hd, Saaltink D-J, Boer SFd, Koolhaas JM (2009) Delineation of violence from functional aggression in mice: an ethological approach. Behav Genet 39:73–90. doi:10.1007/s10519-008-9230-3
Neumann ID, Toschi N, Ohl F, Torner L, Kromer SA (2001) Maternal defence as an emotional stressor in female rats: correlation of neuroendocrine and behavioural parameters and involvement of brain oxytocin. Eur J Neurosci 13(5):1016–1024
Nyuyki KD, Waldherr M, Baeuml S, Neumann ID (2011) Yes, i am ready now: differential effects of paced versus unpaced mating on anxiety and central oxytocin release in female rats. PLoS ONE 6(8):e23599
Pedersen CA, Boccia ML (2006) Vasopressin interactions with oxytocin in the control of female sexual behavior. Neuroscience 139(3):843–851. doi:10.1016/j.neuroscience.2006.01.002
Pfaus JG, Kippin TE, Coria-Avila G (2003) What can animal models tell us about human sexual response? Annu Rev Sex Res 14:1–63. doi:10.1080/10532528.2003.10559810
Pfaus JG, Kippin TE, Coria-Avila GA, Gelez H, Afonso VM, Ismail N, Parada M (2012) Who, what, where, when (and maybe even why)? How the experience of sexual reward connects sexual desire, preference, and performance. Arch Sex Behav 41:31–62. doi:10.1007/s10508-012-9935-5
Pietropaolo S, Crusio WE, D'Amato FR (2017) Treatment approaches in rodent models for autism spectrum disorder. Curr Top Behav Neurosci 30:325–340

Ramos L, Hicks C, Kevin R, Caminer A, Narlawar R, Kassiou M, McGregor IS (2013) Acute prosocial effects of oxytocin and vasopressin when given alone or in combination with 3,4-methylenedioxymethamphetamine (MDMA, /'Ecstasy/') in rats: involvement of the V1A receptor. Neuropsychopharmacology 38:2249–2259

Ramos L, Hicks C, Caminer A, McGregor IS (2014) Inhaled vasopressin increases sociability and reduces body temperature and heart rate in rats. Psychoneuroendocrinology 46:46–51. doi:10.1016/j.psyneuen.2014.04.013

Ross HE, Cole CD, Smith Y, Neumann ID, Landgraf R, Murphy AZ, Young LJ (2009) Characterization of the oxytocin system regulating affiliative behavior in female prairie voles. Neuroscience 162(4):892–903

Sala M, Braida D, Lentini D, Busnelli M, Bulgheroni E, Capurro V, Finardi A, Donzelli A, Pattini L, Rubino T, Parolaro D, Nishimori K, Parenti M, Chini B (2011) Pharmacologic rescue of impaired cognitive flexibility, social deficits, increased aggression, and seizure susceptibility in oxytocin receptor null mice: a neurobehavioral model of autism. Biol Psychiatry 69(9):875–882

Sandi C, Haller J (2015) Stress and the social brain: behavioural effects and neurobiological mechanisms. Nat Rev Neurosci 16:290–304. doi:10.1038/nrn3918

Schroeder JC, Reim D, Boeckers TM, Schmeisser MJ (2017) Genetic animal models for autism spectrum disorder. Curr Top Behav Neurosci 30:311–324

Shimamoto A, DeBold J, Holly E, Miczek K (2011) Blunted accumbal dopamine response to cocaine following chronic social stress in female rats: exploring a link between depression and drug abuse. Psychopharmacology 218(1):271–279. doi:10.1007/s00213-011-2364-7

Strang S, Park SQ (2017) Human cooperation and its underlying mechanisms. Curr Top Behav Neurosci 30:223–239

Takahashi A, Quadros IM, Almeida RMMd, Miczek KA (2011) Brain serotonin receptors and transporters: initiation vs. termination of escalated aggression. Psychopharmacology 213:183–212. doi:10.1007/s00213-010-2000-y

Toth I, Neumann I (2013) Animal models of social avoidance and social fear. Cell Tissue Res 354 (1):107–118. doi:10.1007/s00441-013-1636-4

Trainor BC, Pride MC, Villalon Landeros R, Knoblauch NW, Takahashi EY, Silva AL, Crean KK (2011) Sex differences in social interaction behavior following social defeat stress in the monogamous California mouse (*Peromyscus californicus*). PLoS ONE 6(2):e17405

Tzanoulinou S, Sandi C (2017) The programming of the social brain by stress during childhood and adolescence: from rodents to humans. Curr Top Behav Neurosci 30:411–429

Veenema AH, Beiderbeck DI, Lukas M, Neumann ID (2010) Distinct correlations of vasopressin release within the lateral septum and the bed nucleus of the stria terminalis with the display of intermale aggression. Horm Behav 58(2):273–281

Veening JG, Coolen LM (2014) Neural mechanisms of sexual behavior in the male rat: emphasis on ejaculation-related circuits. Pharmacol Biochem Behav 121:170–183. doi:10.1016/j.pbb.2013.12.017

Veening JG, Coolen LM, de Jong TR, Joosten HW, de Boer SF, Koolhaas JM, Olivier B (2005) Do similar neural systems subserve aggressive and sexual behaviour in male rats? Insights from c-Fos and pharmacological studies. Eur J Pharmacol 526:226–239. doi:10.1016/j.ejphar.2005.09.041

Waldherr M, Neumann ID (2007) Centrally released oxytocin mediates mating-induced anxiolysis in male rats. Proc Natl Acad Sci USA 104(42):16681–16684

Wöhr M, Engelhardt KA, Seffer D, Sungur AÖ, Schwarting RKW (2017) Acoustic communication in rats: effects of social experiences on ultrasonic vocalizations as socio-affective signals. Curr Top Behav Neurosci 30:67–89

Wotjak CT, Kubota M, Liebsch G, Montkowski A, Holsboer F, Neumann I, Landgraf R (1996) Release of vasopressin within the rat paraventricular nucleus in response to emotional stress: a novel mechanism of regulating adrenocorticotropic hormone secretion? J Neurosci 16 (23):7725–7732

Young KA, Gobrogge KL, Liu Y, Wang Z (2011) The neurobiology of pair bonding: insights from a socially monogamous rodent. Front Neuroendocrinol 32(1):53–69. doi:10.1016/j.yfrne.2010.07.006

Ziporyn T, McClintock MK (1991) Passing as an indicator of social dominance among female wild and domestic norway rats. Behaviour 118:26–41. doi:10.1163/156853991X00184

Zoicas I, Slattery DA, Neumann ID (2014) Brain oxytocin in social fear conditioning and its extinction: involvement of the lateral septum. Neuropsychopharmacology 39(13):3027–3035. doi:10.1038/npp.2014.156

Recognizing Others: Rodent's Social Memories

Judith Camats Perna and Mario Engelmann

Abstract We provide in this chapter a brief overview of the present knowledge about social memory in laboratory rodents with a focus on mice and rats. We discuss in the first part the relevance of the processing of olfactory cues for social recognition in these animals and present information about the brain areas involved in the generation of a long-term social memory including cellular mechanisms thought to underlie memory consolidation. In the second part, we suggest that sensory modalities beyond olfaction may also be important in contributing to the long-term social memory trace including audition and taction (and vision). The exposure to stimuli activating the auditory system and taction is able to produce interference phenomena at defined time points during the consolidation of social memory. This ability of such—nonsocial—stimuli may provide a new approach to dissect the brain processes underlying the generation of the social memory trace in further studies.

Keywords Social memory · Social discrimination · Olfaction · Auditory stimuli · Behaviour · Retroactive interference · Memory consolidation · Mice · Rats

Contents

1 Social Recognition Behaviours and Experimental Paradigms 26
1.1 Social Recognition Assessment 26
1.2 Social Recognition Paradigms 27

J. Camats Perna (✉)
AG Neuroendokrinologie & Verhalten, Otto-von-Guericke-Universität Magdeburg,
Institut für Biochemie und Zellbiologie, Leipziger Str. 44/Haus 1,
39120 Magdeburg, Germany
e-mail: judith.camats@med.ovgu.de

M. Engelmann
AG Neuroendokrinologie & Verhalten, Otto-von-Guericke-Universität Magdeburg,
Institut für Biochemie und Zellbiologie, Center for Behavioral Brain Sciences,
Leipziger Str. 44/Haus 1, 39120 Magdeburg, Germany

Curr Topics Behav Neurosci (2016) 30: 25–46
DOI 10.1007/7854_2015_413
Published Online: 07 December 2015

2 Morphological Substrate and Mechanisms Underlying Social Memory 31
2.1 Selected Brain Areas Involved in Social Recognition 32
2.2 Selected Cellular Mechanisms Activated During the Consolidation of Social Memories 35
3 Conditions that Influence Social Memory Recognition 36
3.1 Interference by Husbandry and Experimental Procedures 36
3.2 Interference Depends Upon the Nature of Stimuli and the Timing of Their Presentation 39
4 Conclusions 42
References 43

Most rodents are social animals, living substantial parts of their lives in societies in which they use complex ways to communicate with each other to form social bonds. The ability to recognize and distinguish between individuals is therefore vital for their lifestyle. Individual recognition is not only important for the formation of parent–offspring bonds but also forms the basis of territorial behaviour, identifying the individual or group, defending resources such as mates, food or nest sites and allowing the detection of intruders and the rejection of strange animals from a social group. Information from a range of senses can be used for discrimination familiar from unfamiliar conspecifics. For rodents, olfaction is their dominant sense and their social behaviour is heavily influenced by the chemosignals secreted by conspecifics. This chapter will focus on selected neurobiological aspects that provide the substrate to success in this complex behaviour, known as social recognition memory. In this context, the second part will be focused on the data obtained in interference studies that shed additional light on the importance of the different sensory modalities involved in social recognition in laboratory rodents.

1 Social Recognition Behaviours and Experimental Paradigms

1.1 Social Recognition Assessment

Social recognition can be assessed in the laboratory in tests that represent different variants of a basic design allowing the measure of familiarity recognition. Most of the studies published so far investigated social memory related to the storage of information about distinct qualities attributed to a specific individual, which allows the identification of this animal upon a subsequent encounter within a relevant time window. The sources for the chemosensory signals used by rodents primarily to identify conspecifics are body fluids as urine or secretions from skin, reproductive tract or specialized scent glands producing pheromones and other semiochemical compounds (Natynczuk and Macdonald 1994; Heiss et al. 2009). There is evidence that each individual has unique composition of its “smell”, often called “olfactory signature”. This olfactory signature is composed of volatile and nonvolatile compounds (Popik et al. 1991; Sawyer et al. 1984) which—after being detected in social encounters—are processed by two segregated neuronal pathways, the main and the accessory olfactory system (Noack et al. 2010), respectively. Several investigations in rats and mice proved the olfactory nature of the recognition cue, since lesions of the olfactory bulb and chemically induced anosmia impair their individual recognition (Popik et al. 1991; Noack et al. 2010).

1.2 *Social Recognition Paradigms*

Exploiting the spontaneous investigatory behaviour of the animals towards conspecifics including the innate drive to investigate unfamiliar over familiar items, several nonconditioned paradigms have been widely used to measure social recognition in rodents. Basically, all tests rely on the exposure of the subject under study (experimental subject) towards conspecific (stimulus animal) and the monitoring of the behaviour of the experimental subject. In using this principle of experimentation, animals can be tested repeatedly in social recognition memory, i.e. under different treatment conditions, which provides a high statistical power in data analysis and allows the detection of side effects that may affect the animal's behaviour.

It has to be mentioned that for all paradigms, different versions exist in different laboratories using different exposure times, different inter-trial intervals and/or different stimulus animals. As shown in Table 1, social experiments with gerbils and hamsters have also been performed although not as extensively as with other taxa. In most of them, social scents including urine or secrets from the ventral gland were used instead of stimulus animals. Therefore, these studies should be better called "chemico-sensory" rather than "social". As outlined below, presenting the odour alone omits the stimulation of other sensory modalities of the experimental subject that may play also an important role not only for the consolidation and durability of the memory trace, but also for the brain areas involved. Nevertheless, studies in gerbils and hamsters suggested that urine by itself is also used as odorant communication in rodents, as the **scent marking behaviour** corroborates. Scent marks deposited in the environment may communicate information on territory ownership, social, reproductive, health and nutritional status and enable recognition of individuals (Borelli et al. 2009). A good example for the behavioural relevance is **the health assessment of conspecifics** to minimize the potential exposure to parasites and to avoid contagion: rodents have the innate ability to discriminate between healthy and parasitized conspecifics. Such behaviour is referred as "sick conspecific avoidance" and can be evaluated using a social preference test (Boillat et al. 2015). A similar behaviour is observed in **inbreeding avoidance**, particularly evident across the mammalian taxa as inbreeding can also cause a reduction in fitness. It has been shown that the attraction of mice to the urinary odours of other mice is subject to a "parent-of-origin" effect which causes both males and females to prefer the odour of urine from mice of an unrelated strain to that of urine from mice of the same strain as their mothers (Isles et al. 2001).

Table 1 Social recognition memory performance shown by rodents in different tests

Test / Taxon	Habituation /Dishabituation test "social recognition test"			Social discrimination test			Volatile fraction cage	
	STM	ITM	LTM	STM	ITM	LTM	STM	LTM
Mice	✓	✓	✓	✓	✓	✓	✓	✓
Rats	✓	✓+✗	✓+✗	✓	✗	✗	✗	✗
Voles	✓	n.d.	n.d	✓	n.d	n.d	n.d	n.d
Gerbils	✓	n.d	✗	n.d	n.d	n.d	n.d	n.d
Hamsters	✓	✓	✓	n.d	n.d	n.d	n.d	n.d

Social recognition abilities of different rodent taxa are listed. Information has been classified dependent on the paradigm used to test the memory abilities and on the type of memory tested (short-term memory (STM), intermediate-term memory (ITM), long-term memory (LTM)). *Green check* tested and intact recognition memory. *Red cross* tested but no recognition memory. *n.d.* = no data available

An intensively studied model of social recognition refers to mate recognition in pair-bonding prairie voles in which long-term pair bonds between males and females are formed. Following mating, these animals display a well-characterized suite of behaviours including selective affiliation with the familiar partner and aggression towards unfamiliar conspecifics (Carter et al. 1995). The act of mating in conjunction with the exposure to the odour stimuli of the partner leads to a recognition memory for the partner as well as promoting the formation of a pair bond. Interestingly, these features of pair bonding resemble the imprinting. In the **partner preference test,** the experimental subject is paired with a sexually experienced male or female for 24 h and allowed to mate (learning). After a defined exposure interval, the second (memory) session offers experimental subjects a choice between two stimulus animals (the previously mated and an unfamiliar conspecific), and durations spent with each, measured as social proximity and immobile social contact, are used to calculate a preference score. Although—compared to rats and mice—little is known about general vole social recognition, the principle mechanisms described so far fit into the theoretical framework generated by rat and mouse studies. Indeed, the main and accessory olfactory bulbs are critical relay stations also to generate partner preference formation in prairie voles (Curtis et al. 2001) (Table 1).

The **habituation–dishabituation paradigm** was, and still is, one of the most widely used test to study social recognition. The experimental subject is exposed repeatedly to a given unfamiliar stimulus animal, these exposures are knows as habituation sessions and are separated by exposure intervals during which the experimental subject remains undisturbed in its cage. Throughout the time of each exposure, a trained investigator records the duration of the direct sniffing by the experimental subject towards the stimulus animal, mostly of its anogenital and perioral areas. The social investigative response declines upon the number of habituation sessions, since the familiarity towards the stimulus animal increases. In the last session, called dishabituation, the presentation of an unfamiliar stimulus animal is expected to reinstate the initial level of social investigation. Despite the popularity of this test, it presents some difficulties in data interpretation as repeated testing of the same animal can lead to nonspecific behavioural changes, such as sensitization to the testing procedure (Engelmann et al. 1995). Further, this test has only limited suitability to analyse the duration of the recognition memory performance.

Another and more direct way to assess social recognition is the use of the **social discrimination paradigm** (Engelmann et al. 1995, 2011). This paradigm evolved from the social recognition test (Thor and Holloway 1982) and consists of two sessions. During the first session, a given stimulus animal is introduced in the cage of the experimental subject, allowing the acquisition of its olfactory signature. In the subsequent session, separated by the desired exposure interval, two stimulus animals are introduced at the same time in the experimental subject's cage, the familiar stimulus animal (presented in the first session) together with an unfamiliar one. Depending on the exposure interval chosen, different memories can be tested (from immediate-term memory lasting minutes to long-term memory lasting several days). The main difference between social recognition test (assessed in the habituation/dishabituation paradigm and in the original social recognition procedures) and social discrimination test is that the later measures the presence or absence of recognition categorically: during choice not only the previously encountered conspecific is presented (=social recognition test) but also—simultaneously—a novel, previously not encountered conspecific. Thus, the experimental subject is allowed to discriminate between both stimulus animals simultaneously in one session. This provides an internal control under identical experimental conditions and allows separating specific from nonspecific effects in pharmacological studies, thereby reducing the number of sessions for a given experimental series. By using different exposure intervals, the social discrimination test enables the investigation of the impact of manipulations on the different "stages" of memory. Moreover, social discrimination allows the emergence of social memory in animals that appeared to possess no social recognition when tested in the habituation/dishabituation test, thus showing a higher sensitivity in assessing this type of memory performance (Engelmann et al. 1995).

Using the social discrimination test, the performance of mice and rats has been investigated and revealed interesting findings: mice show a memory performance that lasts at least 24 h, whereas rats form short-term social recognition memory only (Table 1). A more detailed analysis in male rats revealed recognition memory to be extinct after $\sim$45 min, whereas female rats, exposed to juveniles from both sexes, show a slightly, but significantly longer recognition lasting $\sim$2 h (Dantzer et al. 1987; Engelmann et al. 1998). Although a great amount of studies confirm the absence of long-term memory in rats, it must be noted that some recently published studies suggested that male rats retain social long-term recognition memory for at least one week, attributing the discrepancies with the rest of studies to the different housing conditions (Shahar-Gold et al. 2013) (see 3.1.2 for a more detailed discussion).

Different **modifications from the social discrimination test were described.** Originally, the stimulus animals were allowed to move freely in the experimental subject's cage, although there are some variants where they are confined in wired cups, frequently when a three-chambered apparatus is used, referred in the literature as social choice test, test for sociability or social novelty preference. Although this makes "preference" measurements easier, it limits the access of the experimental subjects to the nonvolatile fraction of the olfactory signature of the stimulus animal. However, in particular, rats need direct access to the conspecific's body surface to show a proper social memory performance (Engelmann et al. 2011). Therefore, confining the stimulus animals in wired cups is of limited suitability for testing this taxon. Another modification is provided by the **volatile fraction cage** (Engelmann et al. 2011). Here, the juveniles are confined in two tubes separated by two fences from the experimental subject's cage, preventing direct tactile contact. The tubes are connected to two fan units which provide an air stream towards the experimental subject's cage, facilitating the access only to the volatile fraction of the olfactory signature of the stimulus animals. The volatile fraction cage is functional for studies aimed at discriminating the relevance of each fraction of the olfactory signature in order to establish the social recognition ability (Noack et al. 2010). Using this test, differences between mice and rats in the processing of the different fractions of the olfactory signatures of respective conspecifics have been confirmed. Mice recognize juvenile conspecifics on the basis of both, the volatile and nonvolatile components of their olfactory signatures. However, mice are also able to form long-term memory by just having access to the volatile fraction. Rats, in contrast, require access to the nonvolatile fraction of the olfactory signature, which is predominantly processed by the accessory olfactory bulb and results in short-term recognition memory only (Noack et al. 2010) (Table 1). Thus, the ability to form a long-term social recognition memory might be linked to the processing of the volatile fraction of the olfactory signature of the conspecific which does not play a significant role for social recognition in rats.

Interestingly, the wealth of data suggests that rats and mice differ concerning the persistence of social recognition memory under similar test conditions. In a seminatural environment, rats (i.e. *Rattus norvegicus*) show a quite similar social behavioural profile to mice (i.e. *Mus musculus*) (Eibl-Eibesfeldt 1950, 1952). Therefore, it remains remarkable that under the reported experimental conditions, long-term social memory can be measured in mice only. Recent studies suggest that the lack of being able to monitor long-term memory in rats might be linked not only to the fraction of the olfactory signature used to recognize a conspecific (Noack et al. 2010) but also to the isolation of the experimental subjects during the exposure interval between learning and retrieval (see 3.1.2). Although—considering the impact of interference by encountering conspecifics (see 3.2)—upon the first view contra-intuitive, testing this hypothesis might provide new insight in the social memory formation in rats. A detrimental impact of isolation on adult rats has been described in the context with brain plasticity (Stranahan et al. 2006). Further studies have to reveal the detailed effect of isolation on the performance in the social memory tests discussed here.

Although it is well described that olfaction is the most important sense for rodents to enable social recognition, it is not the only one. **Ultrasonic vocalizations** have been reported in several rodent species, and the capability to hear and emit these calls has been intensively studied in laboratory mice and rats. Ultrasonic vocalization in pups is thought to modulate mother–offspring interaction during early postnatal days as they decreased as pups grow up. Adult mice and rats instead emit ultrasounds in different social contexts, with species differences being evident. 50–70 kHz vocalizations appeared to be closely linked to facilitate mating and coordinate sexual arousal and are highly present during social investigation/interaction. A recent study showed that vocalizations may contain signatures of individuality and kinship helping to avoid inbreeding, and introduced the possibility to use ultrasonic vocalizations as an index of social memory in female mice. It is of note that this behaviour, monitored as a decreased number of calls emitted by the female during the second encounter with the familiar female stimulus animal, vanished with an exposure interval of 60 min (Moles et al. 2007), allowing to test short-term memory only. The recording and possibly spectrographic analysis of the ultrasonic calls in mice is progressively gaining relevance in the context with social memory testing and could have great impact to bring new information on motivational aspects underlying social behaviour and subjective states related to social interaction.

2 Morphological Substrate and Mechanisms Underlying Social Memory

The ability to recognize, use and behave according to socially relevant information requires a neuronal system that not only processes the information of the perceived social cues but also links it to emotion, motivation and adaptive behaviour. The ability to generate these associations is essential for triggering what we call "memory". Present research aimed at analysing the brain areas involved in social memory focused on those areas which can easily linked to processing of olfactory cues. As outlined above, in rodents, the olfactory system is the most important sensory system to form social memories. The initial processing of conspecific social cues takes place in the olfactory bulb, a well-described structure, ideal to study the involvement of the different neural substrates from the initial sensory detection through to limbic and higher cortical processing areas, which modulate complex social behavioural responses such as recognition memory. This type of memory is strongly modulated by neurotransmitter systems which act on the transduction and encoding of social information, which at the same time can be modulated by stress and social experiences (van der Kooij and Sandi 2012). There are numerous reviews describing in particular the relevance of vasopressin and oxytocin signalling in social recognition memory, which we highly recommend for further reading about this issue (Neumann and Landgraf 2012; Ferguson et al. 2002; Hammock 2015; Wacker and Ludwig 2012).

2.1 Selected Brain Areas Involved in Social Recognition

In this section, social memory formation will be presented and discussed on the basis of the present knowledge about the brain areas involved. Further, more details will be given on how this initial encounter has led to the formation of a long-term social memory enabling subsequent recognition of the previous encountered social stimulus.

2.1.1 Olfactory Bulb

The origin of segregating volatile odour and pheromone detection in the context with social encounters in rodents, is based on different sensory neurons localized either in the main olfactory epithelium or the vomeronasal organ (the latter is predominantly sensitive to nonvolatile molecules such as pheromones) that provide input to the olfactory bulb. The rodent olfactory bulb in turn is considered as the origin of the two distinct olfactory pathways: the main olfactory pathway and the accessory olfactory pathway, which are thought to transmit differential information about volatile and nonvolatile olfactory stimuli, respectively (Martinez-Marcos 2009). From the olfactory bulb, projections reach secondary and tertiary areas such as the cortex or limbic brain areas including the hypothalamus (Fig. 1). The olfactory bulb is essential for social recognition memory as it provides the first level of processing the olfactory information used to build social memories.

2.1.2 Medial Amygdala

Different inputs mainly originating in the vomeronasal organ converge in the medial amygdala (MeA), which seems to act as a major site for the integration of accessory and main olfactory pathways. Efferences from the MeA signal back to the accessory olfactory bulb, thereby likely controlling the impact of the nonvolatile fraction of the conspecific's "olfactory signature" on approach-avoidance behaviour (Fig. 1). Using the social discrimination test, the MeA had been proven to be essential in processing the nonvolatile fraction of the olfactory signature since its blockage immediately before the memory session, but not the learning session,

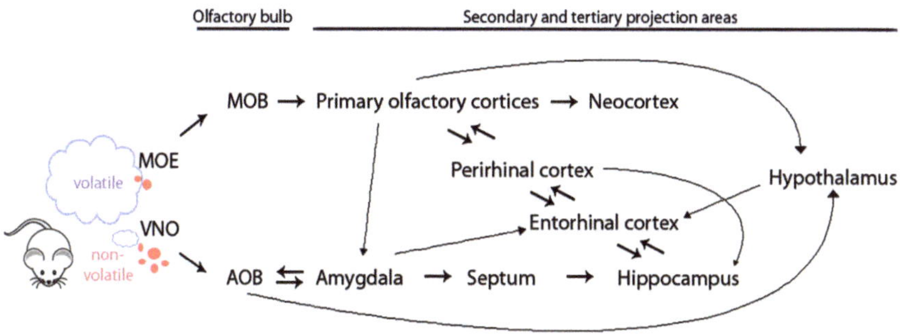

Fig. 1 Main brain circuit processing olfactory information linked to social recognition memory in the rodent brain. Nonvolatile stimuli are processed mainly by the vomeronasal organ (VNO) which projects to the accessory olfactory bulb (AOB) transmitting the information to higher limbic and cortical areas essential to form social recognition memory. Volatile stimuli, instead, are mainly processed by the main olfactory epithelium (MOE) that projects to the main olfactory bulb (MOB) and sends information to the primary olfactory cortices from where they will be transferred to tertiary projection areas, including the amygdala and the hippocampus

impaired social recognition memory in mice (Noack et al. 2015). Studies in hamsters showed an activation of the anterior MeA in response to both conspecific chemosensory stimuli (Meredith and Westberry 2004). In addition, this area is the site of action of different steroids and neuropeptides, therefore being sensitive to hormonal states and able to strongly modulate social recognition memory through neuropeptides such as oxytocin and vasopressin.

2.1.3 Entorhinal and Perirhinal Cortex

The **entorhinal cortex** functions as the gateway to the hippocampal formation, because its output, through the perforant pathway, is the major cortical source of input to the hippocampus. Furthermore, together with the subiculum, it also receives the major output from the hippocampus (Witter et al. 1989). The lateral entorhinal cortex is a component of the olfactory cortex, receiving inputs from both the main olfactory system and piriform cortex, and it also provides feedback to these areas, thereby possibly modulating their functions and the olfactory acuity for familiar odours. In addition to the wiring to the hippocampus (Fig. 1), the entorhinal cortex also receives inputs from the perirhinal cortex, amygdala, thalamus, hypothalamus and other modulatory areas. This suggests the entorhinal cortex as a brain area with an integrative function linked to the generation of olfactory cued social memory. Indeed in rodents, lesions of the entorhinal cortex resulted in deficits of short-term odour memory (Kaut and Bunsey 2001). An area closely linked to the entorhinal cortex is the **perirhinal cortex**, which surrounds the hippocampal formation and receives incoming sensory information from the olfactory cortices. The perirhinal cortex contributes to recognition memories that require long-term storage of conjunctive feature representations, such as the olfactory signature of a conspecific of mice and/or rats (Feinberg et al. 2012). Perirhinal cortex-lesioned animals demonstrate greater levels of impairment as the degree of feature ambiguity increases, together with impairments in distinguishing simultaneously presented stimuli. This suggests that this area might mediate the perceptual disambiguation of overlapping stimulus representations, in addition to support the generation of recognition memory. Individual recognition by male hamsters in the context of the Coolidge effect (i.e. ability to distinguish a novel from a familiar female) was found to be disrupted by lesions of the perirhinal and entorhinal cortices (Petrulis and Eichenbaum 2003). Additionally, neurons in the entorhinal cortex of hamsters were reported to be responsive to individual social odours (Petrulis et al. 2005), supporting its role in social recognition.

2.1.4 Hippocampus

The activation of immediate early genes has been used to study the involvement of specific brain regions in social recognition memory formation after an initial social encounter mimicking the learning session in a social memory test. Male mice (Ferguson et al. 2001; Richter et al. 2005; Engelmann 2009; Samuelsen and Meredith 2011) showed increased c-Fos synthesis in a number of brain regions including the MeA, the medial preoptic area and the piriform cortex, whereas the number of c-Fos-positive cells in the dorsal **hippocampal areas** was not significantly affected. Although lesions studies in rats tend to confirm a lack of hippocampal involvement in short-term social recognition memory (Bannerman et al. 2001; Squires et al. 2006), permanent hippocampal lesion in mice impaired social recognition memory for a juvenile 30 min after the first exposure without affecting immediate social recognition (Kogan et al. 2000). Recently, the hippocampal area CA2 has been suggested to be critical for this impairment (Hitti and Siegelbaum 2014), since inactivation of CA2 pyramidal cells or lesion in this region impairs social recognition memory without impacting other forms of hippocampus-dependent memory.

The discrepancy between the findings on c-Fos activation in the dorsal hippocampus of mice after the social recognition test and the effect of hippocampal lesions, challenges the interpretation of data from immediate early-gene activation in the context of memory formation and highlights the need to do more accurate quantifications. It has been demonstrated that distinct parts of the hippocampus are involved in different behaviours. This functional dissociation is supported by its anatomical connectivity and gene expression; therefore, a more detailed look at c-Fos synthesis by analysing each subarea might help to clarify its involvement. In addition, it is likely that specific brain regions are only temporarily involved in acquisition, consolidation and/or retrieval encoding, as this time-dependent contribution has been demonstrated for the hippocampus (Kogan et al. 2000).

Different results obtained from hippocampal lesions in rats and mice suggest that the involvement of this brain area in social recognition memory seems to differ between the taxa. Similarly as observed in mice, studies performed on Degus also reported deficits in social recognition caused by hippocampal lesions (Uekita and Okanoya 2011). However, no impairment was observed in hippocampal-lesioned male hamsters tested for the Coolidge effect (Petrulis and Eichenbaum 2003).

To sum up, multiple brain areas downstream of the olfactory bulb and piriform cortex are involved in the processing of the information about the perceived olfactory cues, including the corticomedial amygdala, entorhinal cortex, perirhinal cortex and the hippocampus. Most of these areas are critical for declarative memory such as recognition memory. However, the specific function including the nature of their contribution to recognition memory is not completely understood. Hence, further studies are necessary to reveal which role plays each of these areas in the complex process of social memory formation and, thus, to clarify the inconsistencies in the data available.

2.2 Selected Cellular Mechanisms Activated During the Consolidation of Social Memories

One of the hallmarks of recognition memory is that newly learned information is sensitive to disruption after acquisition. This labile state after learning suggests that a period of consolidation occurs, which may last for hours or even days, before the memory may be called "stable". It is well known that long-term, but not short-term, social memory requires consolidation and critically depends upon hippocampal functioning in mice (Squires et al. 2006; Kogan et al. 2000). Different studies investigated the molecular mechanisms underlying memory consolidation by producing irreversible lesions in the hippocampus or using drugs which interfere with protein synthesis. These studies revealed two distinct stages of protein synthesis being important for consolidation of olfactory recognition memory in mice: a short-term lasting stage, starting immediately after training and lasting for ~3 h, and a longer lasting stage, starting ~6 h after acquisition and lasting for ~12 h. The consolidation of the memory trace may have reached a reliable stability ~18 h after learning (Richter et al. 2005; Kogan et al. 2000; Wanisch et al. 2008). Analysis of immediate early-genes expression was used to investigate the brain areas involved in each of these stages in mice. The first stage coincided with an increase in the number of c-Fos immunoreactive cells in brain areas associated predominantly with the accessory olfactory bulb, such as the medial preoptic area and the medial nucleus of the amygdala, but also in the main olfactory bulb and piriform cortex (Richter et al. 2005). The relevance of the olfactory bulb was further supported, since the application of anisomycin (considered to act primarily as protein synthesis inhibitor blocking translation of the mRNA to the amino acid sequence) in this area, immediately and 6 h after the learning session, impaired social long-term memory formation in mice (Pena et al. 2014). During the second stage, the function of which depends upon the integrity of the first stage, proteins other than c-Fos are likely to be synthesized, and this process seems to be essential for olfactory engram formation, probably by enhancing intercellular communication (Richter et al. 2005). Although an increased c-fos transcription was not observed in the hippocampus, injection of anisomycin into mice dorsal hippocampus 3 h after the learning session also impaired long-term social recognition memory (Pena et al. 2014). This implies a distinct participation and provides additional information about the molecular basis of the social memory consolidation as tested in social recognition paradigm.

3 Conditions that Influence Social Memory Recognition

In the natural environment, learning episodes do not occur singly but are confronted with other, similar processes. Thus, memory formation takes place while other, potentially competitive, episodes induced by interfering conditions, and also requiring processing in the same areas of the central nervous system, are likely to happen at the same time. These interfering conditions can prevent the acquisition of information and/or impair or interrupt its consolidation. Usually, the laboratory conditions under which the memory tests take place try to avoid interference phenomena by isolating the animals to be tested in separated and undisturbed rooms. Such controlled and established conditions allow studying the effects of additional sensory modalities relevant for social long-term memory formation. Moreover, studies employing interfering conditions may help to better understand the mechanisms underlying the consolidation of the social memory trace. We will subsequently list some conditions that have been described to interfere with social recognition memory as tested in the laboratory.

3.1 Interference by Husbandry and Experimental Procedures

The disruption of cognitive function by stress-inducing elements from both housing and husbandry systems, as well as by experimental procedures, can additionally have potentially serious implications in the subject's welfare and consequently altering their performance in memory tests (van der Kooij and Sandi 2012; Mendl 1999). Thus, the careful consideration of these conditions and procedures is essential in order to avoid spurious results.

3.1.1 Transportation and Context

It is well established that husbandry procedures can disrupt social memory and induce behavioural changes (Burman and Mendl 2000). As social experiments are often run in rooms specially installed for behavioural testing, but separated from the animal facility, the study of the effects of animal **transportation** is highly relevant. Transportation immediately before the learning session did not affect the recognition performance measured 24 h later in mice (Engelmann et al. 2011). Moreover, studies in rats showed that transportation of the experimental subject rat 6 h after a 2-h learning session did not impair long-term social recognition memory; however, it did when the transportation was made 0.5 h after the learning session (Moura et al. 2011). Therefore, the timing of transportation seems to be a critical factor that researchers must consider when designing their social recognition memory experiments.

Linked to the transportation, a change of context is often also present. Most behavioural tests take place in different contexts from the one the subjects are familiarized with; thus, a possible interference effect induced by an unfamiliar context was also studied (Zheng et al. 2013; Burman and Mendl 2002). In rats and voles, the exposure to different **contextual cues** failed to impair the recognition of conspecifics odours, although it interfered with the ability to distinguish between the stimulus animals for some individuals. Thus, the process of learning social identity was robust on familiar territory and comparably variable when social scents were absent. There are no studies available that systematically investigated the possible impact of the testing context on social memory in other taxa.

3.1.2 Isolation

Solitary housing/isolation is a potent stressor for social species, whose effect has been widely studied also in mice and rats. In both taxa, extended isolation leads to the modification of different physiological parameters coinciding with alterations in behaviour including aggression, mating and anxiety-like behaviour. Even more severe consequences on the brain morphology and local gene expression can be found in animals which have been reared in isolation, including a reduction on medial prefrontal cortex volume and changes in the regulation of gene expression. Although the mechanisms by which isolation affects these parameters may be of high interest, we will focus subsequently on the effects of isolation on social recognition in the course of the acute experiment only.

Different studies demonstrated that chronic and acute social isolation disrupts long-term, but not short-term, social memory in mice. Short-term social memory was intact in rats after one week of isolation, suggesting a robust performance unaffected by housing conditions. However, recent studies using the habituation/dishabituation test observed that long-term social isolation during adolescence strongly influenced subsequent social behaviour both in mice and rats (Zhao et al. 2009). This implies that not only acute isolation may have an impact on the behavioural performance but also long-lasting isolation episodes that may have occurred previously in particular during rearing.

Interestingly, several attempts were made to increase the duration of social memory in rats tested in the social recognition test. Most of them involved changing the length of the learning session(s) and altering the housing conditions. Two successive five-minutes learning sessions lead to the presence of social recognition memory in rats

2 h later (Dantzer et al. 1987), although long-term memory was not observed even when the learning session was prolonged up to 0.5 h (Sekiguchi et al. 1991). Group-housed juvenile female rats were tested to discriminate between an unfamiliar social odour and an odour from a cage-mate, under different isolation conditions. Only the rats with relatively short isolation periods prior testing (1 h and 48 h) recognized the odour from the cage-mate, but not the rats isolated for 96 h (Burman and Mendl 2006). These results suggest rats to be able to show longer lasting memories for conspecifics and their odours with significantly longer "learning session" and short isolation periods. Another study reported that social recognition memory in rats may last at least 24 h after 2 h or longer exposure to the conspecific during the learning session, showing for the first time that male rats exhibit long-term social recognition memory (Moura et al. 2010). A subsequent study also showing long-term memory formation in rats was recently published (Shahar-Gold et al. 2013), reporting rapid and profound, but reversible, effects of housing conditions on social recognition memory in adult rats when the learning and the memory sessions took place in a neutral arena. Interestingly, their methodology differed from the other studies published by several parameters; (i) between learning and memory sessions, the experimentator returned the experimental subjects to the home cages and housed them together with their home cage-mates. (ii) The experimentator used a learning session in which the stimulus animals had unrestricted access to the stimulus animal. (iii) However, during the memory session, the stimulus animals were confined to transparent and slotted plastic corrals. As mentioned above, the use of these corrals makes it more difficult for the experimental subject to gain access to the nonvolatile fraction of the olfactory signature which is—according to other studies (Noack et al. 2010)—necessary for rats to recognize the previously encountered conspecific. Interestingly, mice cannot recognize a previously encountered conspecific when during learning both the volatile and non-volatile fraction of the olfactory signature were available (unrestricted, freely moving access) and during the memory session the experimental subjects have access to the volatile fraction of the olfactory signature of the (familiar) stimulus animal only (Noack et al. 2010). Thus, the manipulations used in the above-mentioned rat study (Shahar-Gold et al. 2013) do not easily explain the detection of long-term social memory in rats which differed from most of the previously published work, in which—moreover—the experimental subjects remained isolated between the learning and memory sessions, in order to avoid interference phenomena (see below). Nonetheless, it is important to note that all the referred studies, using indirect exposures and testing interference phenomena, were performed with animals isolated between the sessions. Besides, in all cases unfamiliar conspecifics were used as interference stimuli. Ongoing studies in our laboratory will further test the impact of social isolation on social recognition memory.

Taken together, the rapid and specific impairment of social recognition memory consolidation largely described in mice and rats suggests that molecular processes in the neuronal network underlying the consolidation of social memory are sensitive to manipulations by ongoing social activity. And probably this sensitivity is the cause for (at least some of) the discrepancies in this field due to the lack of standardized methods and analysis.

3.1.3 Anaesthesia

Some experimental manipulations involve for instance direct administration of substances in defined brain areas during the behavioural testing or briefly before or after a learning session. Therefore, the use of briefly acting anaesthetics is required to avoid stress that may interfere with the memory performance (see above). Moreover, a brief anaesthesia of the animal allows to monitor the successful treatment, rather than infusions in freely moving animals. The potency to interfere with memory varies among the different anaesthetics (Alkire and Gorski 2004), the dosage and the type of learning/memory task under study (Dutton et al. 2001). Among the different anaesthetics available, inhalation anaesthetics are particularly useful, and among them, isoflurane might be the substance of choice. Behavioural studies analysing hippocampus-dependent memory in animals exposed 24 h before the test for 15 min to 2.1 % isoflurane anaesthesia showed no impairment (Fidalgo et al. 2012). This is in line with observations that in adult mice a brief ($\sim$5 min), 1 % isoflurane anaesthesia immediately before the learning session failed to affect social recognition memory tested 24 h later (Engelmann et al. 2011). Nevertheless, due to the fact that different anaesthetics are used in the different laboratories, the different dose responses, the time point of administration (with respect to the experimental design) and the different learning tasks are employed; it is difficult to provide a general conclusion about the action of anaesthetics for social memory tests. However, it is important to note that also isoflurane anaesthesia might affect the outcome of social memory testing by directly affecting brain activity in an unintended manner: studies performed in mice hippocampal slices showed that high-dosage isoflurane anaesthesia (0.55 and 0.74 mM) blocked synaptic plasticity in the mouse hippocampus and impaired hippocampal long-term potentiation in a dose-dependent manner (Haseneder et al. 2009), anticipating severe effects on hippocampus-dependent memories such as social recognition memory under these extremely high doses. Therefore, a brief screening is suggested to monitor such effects in the defined experimental setup.

3.2 Interference Depends Upon the Nature of Stimuli and the Timing of Their Presentation

The nature of the **interference** stimuli and the timing of their presentation during consolidation determine retroactive interference for social recognition memory. Rats showed a short-term memory impairment by retroactive interference due to the exposure to another stimulus animal during the interval between the original learning and memory sessions in the social recognition test (Thor and Holloway 1982; Dantzer et al. 1987). Studies with mice showed that retroactive interference of social memory occurs only during the first 18 h after the original learning session (Engelmann 2009), if memory was tested 24 h later. The first 18 h after learning is a period in which the synthesis of proteins that are obviously required for the consolidation of long-term social recognition memory takes place (Wanisch et al. 2008; Richter et al. 2005). Retroactive interference experiments were done with the social discrimination task using different stimuli activating distinct sensory modalities including taction, audition, olfaction and vision. These stimuli were presented (for 1 min) at different time points after learning during the consolidation process. Interestingly, social recognition memory was sensitive to all stimuli presented within the first hours (up to 6 h in most cases), but not 22 h after the learning session (Perna et al. 2015) (Fig. 2). The insensitivity to all stimuli 22 h after learning might be linked to underlying consolidation processes of the information coding for the originally encountered stimulus animal that seems to be completed at this time point (Richter et al. 2005; Engelmann 2009; Wanisch et al. 2008). These findings highlight the wide diversity of stimuli able to impair social memory formation in mice, which therefore are worth of being considered when designing social memory tests.

Recently, experiments testing the persistence of the described interference effects during the ongoing consolidation of social memory were performed in our laboratory (Fig. 3). The aim of these experiments was to reveal whether a potential transient retrograde amnesia induced by isoflurane was able to suppress the interference phenomena induced by defined stimuli presented during the consolidation of long-term social memory. Whether this effect was dependent upon the activated sensory modalities was also tested by presenting different stimuli that potentially

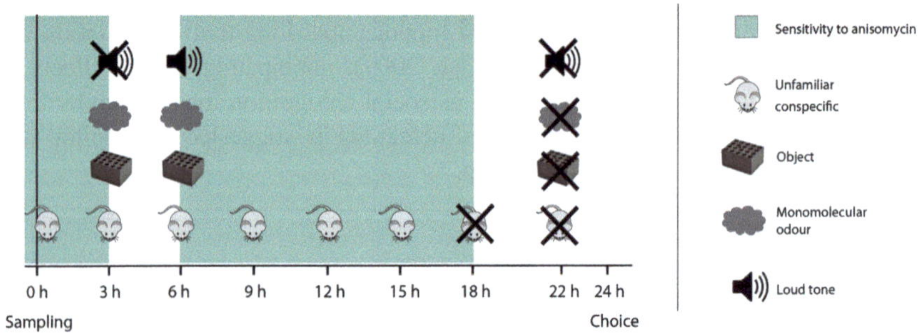

Fig. 2 Scheme illustrating the time-dependent ability of selected stimuli to interfere with the consolidation of long-term social memory in mice. The *green rectangles* represent the periods during which social memory consolidation is sensitive to the protein synthesis blocker anisomycin. In addition, interference effects are shown induced by exposure to (i) an unrestricted encounter with an unfamiliar conspecific (*pictogram: small mouse*), (ii) an object (*grey toy brick*), a monomolecular odour (*grey cloud*) and a loud tone (*black speaker*) at selected time points. Please note that the stimuli used provide different combinations of stimulating sensory modalities including (olfactory, tactile, visual and auditory) and that at least 22 h after the learning session memory performance is insensitive to them. Pure pictograms: induction of interference with social memory when presented at the given time point after learning; pictogram with *black cross*: failed to induce interference with social memory. All experiments were performed using the social discrimination test with an exposure interval of 24 h

induced interference with social memory. Among other, an object and an unfamiliar juvenile (interference stimuli) were presented 3 h after the learning session, followed by a brief (~3 min) 1 % isoflurane anaesthesia of the experimental subjects. Social recognition memory towards the familiar stimulus animal (Fig. 3a, juvenile 1) was tested 24 h after the learning session. Interestingly, the brief anaesthesia was able to

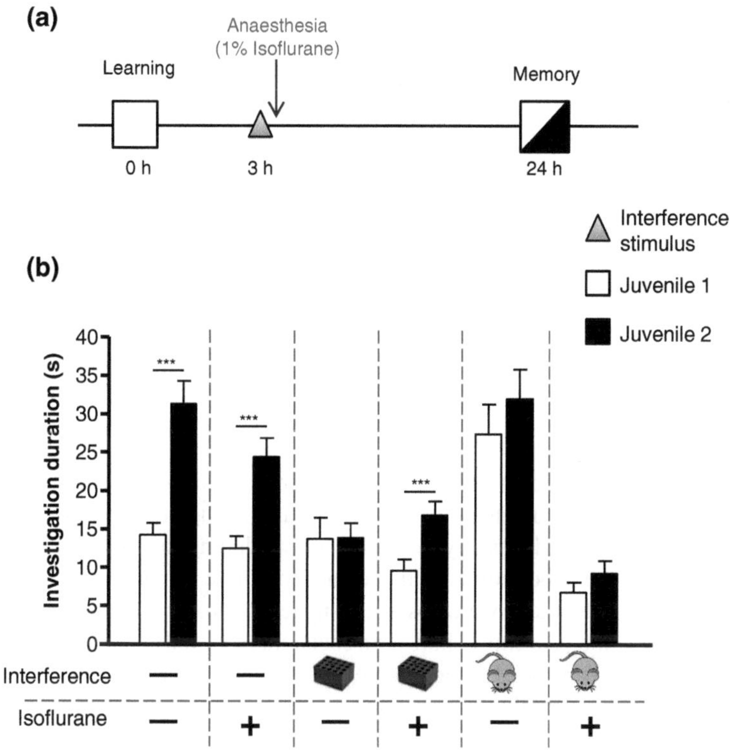

Fig. 3 Effects of short-lasting anaesthesia administered immediately after the presentation of a potential interference stimuli, on mice long-term social recognition memory. **a** Schematic drawing showing the experimental protocol of the experiments: interference stimuli (*toy brick*, previously not encountered "interference" juvenile) were presented 3 h after learning, followed by a short-lasting 1 % isoflurane anaesthesia. Memory was tested 24 h after the learning session by simultaneous presentation of the previously encountered stimulus juvenile 1 and a novel stimulus juvenile 2 to the experimental subject. **b** Data obtained with the protocol shown in (**a**): the Investigation durations of the experimental subject towards the two stimulus juveniles measured during the memory session under the different treatment conditions is represented only. Exposure to isoflurane 3 h after the learning session failed to abolish the significantly reduced investigation duration of the (familiar) juvenile 1 and, thus, did not affect the intact long-term social recognition memory. Anaesthesia blocked the interference effect of the toy brick. No impact of the anaesthesia was observed on the interference effect produced by the "interference" juvenile: in both cases, investigation durations towards the juveniles 1 and 2 were statistically not significantly different. For the pictograms: see legend of Fig. 2. Means + SEM; $n = 20$. $^{***}p < 0.01$; paired student's t test. Some of the data were obtained from Perna et al. (2015)

block the interference caused by the object but failed to block the interference induced by an unfamiliar conspecific (Fig. 3b). These findings shed first light on the processing of the interference-inducing stimuli interacting with ongoing memory consolidation and its timing. In the case of the interference produced by the encounter of the conspecific, obviously the processing of cues acquired via different sensory modalities during the interference session makes the poly-modal representations to compete and thus more difficult to be blocked. In contrast, the potential interference induced by stimuli that activate fewer sensory modalities is sensitive to the central nervous effects produced by the brief isoflurane anaesthesia. It is of note that beyond the missing nonvolatile odours and active movement of the interference stimuli, the main difference between conspecific and object relies on the presence or absence of the generation of social ultrasonic sounds. Sound presentation itself was found to produce interference 6 h after learning (Perna et al. 2015). Although previous studies showed no retrograde amnesia caused by isoflurane as tested in the Pavlovian fear conditioning, we observed that the defined anaesthesia is able to interfere with the processing of nonsocial stimuli and, thus, protects the ongoing consolidation of social memory.

4 Conclusions

Social behaviour of rodents relies primarily on the emission and detection of olfactory cues to discriminate between familiar and strange conspecifics. Different methods were developed to study this behaviour including the underlying memory performance in the laboratory. In the course of these experiments, differences between distinct rodent taxa were observed, suggesting under different experimental conditions that some taxa show short-term social memory only (rats), whereas other taxa display long-term social recognition memory (mice, prairie voles). This type of memory is highly susceptible to disruption during the consolidation. Recent studies focused on the phenomena associated with the presentation of stimuli that potentially interfere with social memory at selected time points during its consolidation. The data of these studies revealed the sensitivity of social memory to disruption by different stimuli. It seems clear that the processing of the olfactory signature from a conspecific is superior for the ability to consolidate social memory in mice. However, all sensory modalities activated by the social encounter are likely to contribute to a complex neuronal representation that is required for long-term memory. The processing of the information obtained from the sensory modalities activated by the presentation of an interference stimulus might compete and interfere with the pattern completion necessary to remember the original conspecific. However, once consolidated, social recognition memory is more difficult to disrupt. We now know that the ability to transfer social recognition from short-term into long-term memory involves that odour cues being initially processed by the olfactory system and later distributed to primary, secondary and tertiary processing brain areas in mice. This knowledge provides the basis for the analysis of the impact of other sensory inputs to generate long-term social recognition memory. Indeed, the contribution of ultrasonic vocalization in mice during testing has been widely overlooked so far and provides an interesting substrate for more detailed studies of how the whole processing network encodes and organizes social memories. Understanding the neurophysiological basis of social recognition memory offers the access into the analysis of the development and—possibly also the treatment—of abnormal social disorders in humans.

Acknowledgments We thank Rita Murau for expert technical assistance. J.C.P was supported by stipends from the Leibniz Graduate School and the Medical Faculty Otto-von-Guericke-Universität Magdeburg.

References

Alkire MT, Gorski LA (2004) Relative amnesic potency of five inhalational anesthetics follows the Meyer–Overton rule. Anesthesiology 101(2):417–429

Bannerman DM, Lemaire M, Beggs S, Rawlins JN, Iversen SD (2001) Cytotoxic lesions of the hippocampus increase social investigation but do not impair social-recognition memory. Exp Brain Res 138(1):100–109

Boillat M, Challet L, Rossier D, Kan C, Carleton A, Rodriguez I (2015) The vomeronasal system mediates sick conspecific avoidance. Curr Biol 25(2):251–255

Borelli KG, Blanchard DC, Javier LK, Defensor EB, Brandao ML, Blanchard RJ (2009) Neural correlates of scent marking behavior in C57BL/6J mice: detection and recognition of a social stimulus. Neuroscience 162(4):914–923

Burman OH, Mendl M (2000) Short-term social memory in the laboratory rat: its susceptibility to disturbance. Appl Anim Behav Sci. 67(3):241–254

Burman OH, Mendl M (2002) Recognition of conspecific odors by laboratory rats (*Rattus norvegicus*) does not show context specificity. J Comp Psychol 116(3):247–252

Burman O, Mendl M (2006) Long-term social memory in the laboratory rat (*Rattus norvegicus*) Anim Welf 15(4):379–382

Carter CS, DeVries AC, Getz LL (1995) Physiological substrates of mammalian monogamy: the prairie vole model. Neurosci Biobehav Rev 19(2):303–314

Curtis JT, Liu Y, Wang Z (2001) Lesions of the vomeronasal organ disrupt mating-induced pair bonding in female prairie voles (*Microtus ochrogaster*). Brain Res 901(1–2):167–174

Dantzer R, Bluthe RM, Koob GF, Le Moal M (1987) Modulation of social memory in male rats by neurohypophyseal peptides. Psychopharmacology 91(3):363–368

Dutton RC, Maurer AJ, Sonner JM, Fanselow MS, Laster MJ, Eger EI 2nd (2001) The concentration of isoflurane required to suppress learning depends on the type of learning. Anesthesiology 94(3):514–519

Eibl-Eibesfeldt I (1950) Beiträge zur Biologie der Haus—und Ährenmaus nebst einigen Beobachtungen an anderen Nagern. Z Tierpsychol 7:558–587

Eibl-Eibesfeldt I (1952) Ethologische Unterschiede zwischen Hausratte und Wanderratte. Verh Deutsch Zool Ges Freiburg, pp 169–180

Engelmann M (2009) Competition between two memory traces for long-term recognition memory. Neurobiol Learn Mem 91(1):58–65

Engelmann M, Wotjak CT, Landgraf R (1995) Social discrimination procedure: an alternative method to investigate juvenile recognition abilities in rats. Physiol Behav 58(2):315–321

Engelmann M, Ebner K, Wotjak CT, Landgraf R (1998) Endogenous oxytocin is involved in short-term olfactory memory in female rats. Behav Brain Res 90(1):89–94

Engelmann M, Hadicke J, Noack J (2011) Testing declarative memory in laboratory rats and mice using the nonconditioned social discrimination procedure. Nat Protoc 6(8):1152–1162

Feinberg LM, Allen TA, Ly D, Fortin NJ (2012) Recognition memory for social and non-social odors: differential effects of neurotoxic lesions to the hippocampus and perirhinal cortex. Neurobiol Learn Mem 97(1):7–16

Ferguson JN, Aldag JM, Insel TR, Young LJ (2001) Oxytocin in the medial amygdala is essential for social recognition in the mouse. J Neurosci 21(20):8278–8285

Ferguson JN, Young LJ, Insel TR (2002) The neuroendocrine basis of social recognition. Front Neuroendocrinol 23(2):200–224

Fidalgo AR, Cibelli M, White JP, Nagy I, Wan Y, Ma D (2012) Isoflurane causes neocortical but not hippocampal-dependent memory impairment in mice. Acta Anaesthesiol Scand 56 (8):1052–1057

Hammock EA (2015) Developmental perspectives on oxytocin and vasopressin. Neuropsychopharmacology 40(1):24–42

Haseneder R, Kratzer S, von Meyer L, Eder M, Kochs E, Rammes G (2009) Isoflurane and sevoflurane dose-dependently impair hippocampal long-term potentiation. Eur J Pharmacol 623(1–3):47–51

Heiss E, Natchev N, Rabanser A, Weisgram J, Hilgers H (2009) Three types of cutaneous glands in the skin of the salamandrid *Pleurodeles waltl*. A histological and ultrastructural study. J Morphol 270(7):892–902

Hitti FL, Siegelbaum SA (2014) The hippocampal CA2 region is essential for social memory. Nature 508(7494):88–92

Isles AR, Baum MJ, Ma D, Keverne EB, Allen ND (2001) Urinary odour preferences in mice. Nature 409(6822):783–784

Kaut KP, Bunsey MD (2001) The effects of lesions to the rat hippocampus or rhinal cortex on olfactory and spatial memory: retrograde and anterograde findings. Cogn Affect Behav Neurosci 1(3):270–286

Kogan JH, Frankland PW, Silva AJ (2000) Long-term memory underlying hippocampus-dependent social recognition in mice. Hippocampus 10(1):47–56

Martinez-Marcos A (2009) On the organization of olfactory and vomeronasal cortices. Prog Neurobiol 87(1):21–30

Mendl M (1999) Performing under pressure: stress and cognitive function. Appl Anim Behav Sci 65(3):221–244

Meredith M, Westberry JM (2004) Distinctive responses in the medial amygdala to same-species and different-species pheromones. J Neurosci 24(25):5719–5725

Moles A, Costantini F, Garbugino L, Zanettini C, D'Amato FR (2007) Ultrasonic vocalizations emitted during dyadic interactions in female mice: a possible index of sociability? Behav Brain Res 182(2):223–230

Moura PJ, Meirelles ST, Xavier GF (2010) Long-term social recognition memory in adult male rats: factor analysis of the social and non-social behaviors. Braz J Med Biol Res 43(7):663–676

Moura PJ, Venkitaramani DV, Tashev R, Lombroso PJ, Xavier GF (2011) Transport of animals between rooms: a little-noted aspect of laboratory procedure that may interfere with memory. Behav Process 88(1):12–19

Natynczuk SE, Macdonald DW (1994) Scent, sex, and the self-calibrating rat. J Chem Ecol 20 (8):1843–1857

Neumann ID, Landgraf R (2012) Balance of brain oxytocin and vasopressin: implications for anxiety, depression, and social behaviors. Trends Neurosci 35(11):649–659

Noack J, Richter K, Laube G, Haghgoo HA, Veh RW, Engelmann M (2010) Different importance of the volatile and non-volatile fractions of an olfactory signature for individual social recognition in rats versus mice and short-term versus long-term memory. Neurobiol Learn Mem 94(4):568–575

Noack J, Murau R, Engelmann M (2015) Consequences of temporary inhibition of the medial amygdala on social recognition memory performance in mice. Front Neurosci 9:152

Pena RR, Pereira-Caixeta AR, Moraes MF, Pereira GS (2014) Anisomycin administered in the olfactory bulb and dorsal hippocampus impaired social recognition memory consolidation in different time-points. Brain Res Bull 109:151–157
Perna JC, Wotjak CT, Stork O, Engelmann M (2015) Timing of presentation and nature of stimuli determine retroactive interference with social recognition memory in mice. Physiol Behav 143:10–14
Petrulis A, Eichenbaum H (2003) The perirhinal-entorhinal cortex, but not the hippocampus, is critical for expression of individual recognition in the context of the Coolidge effect. Neuroscience 122(3):599–607
Petrulis A, Alvarez P, Eichenbaum H (2005) Neural correlates of social odor recognition and the representation of individual distinctive social odors within entorhinal cortex and ventral subiculum. Neuroscience 130(1):259–274
Popik P, Vetulani J, Bisaga A, van Ree JM (1991) Recognition cue in the rat's social memory paradigm. J Basic Clin Physiol Pharmacol 2(4):315–327
Richter K, Wolf G, Engelmann M (2005a) Social recognition memory requires two stages of protein synthesis in mice. Learn Mem 12(4):407–413
Richter K, Wolf G, Engelmann M (2005b) Social recognition memory requires two stages of protein synthesis in mice. Learn Memory. 12(4):407–413
Samuelsen CL, Meredith M (2011) Oxytocin antagonist disrupts male mouse medial amygdala response to chemical-communication signals. Neuroscience 180:96–104
Sawyer TF, Hengehold AK, Perez WA (1984) Chemosensory and hormonal mediation of social memory in male rats. Behav Neurosci 98(5):908–913
Sekiguchi R, Wolterink G, van Ree JM (1991) Short duration of retroactive facilitation of social recognition in rats. Physiol Behav 50(6):1253–1256
Shahar-Gold H, Gur R, Wagner S (2013) Rapid and reversible impairments of short- and long-term social recognition memory are caused by acute isolation of adult rats via distinct mechanisms. PLoS ONE 8(5):e65085
Squires AS, Peddle R, Milway SJ, Harley CW (2006) Cytotoxic lesions of the hippocampus do not impair social recognition memory in socially housed rats. Neurobiol Learn Mem 85(1):95–101
Stranahan AM, Khalil D, Gould E (2006) Social isolation delays the positive effects of running on adult neurogenesis. Nat Neurosci 9(4):526–533
Thor DH, Holloway WR (1982) Social memory of the male laboratory rat. J Comp Physiol Psychol 96(6):1000–1006
Uekita T, Okanoya K (2011) Hippocampus lesions induced deficits in social and spatial recognition in *Octodon degus*. Behav Brain Res 219(2):302–309
van der Kooij MA, Sandi C (2012) Social memories in rodents: methods, mechanisms and modulation by stress. Neurosci Biobehav Rev 36(7):1763–1772
Wacker DW, Ludwig M (2012) Vasopressin, oxytocin, and social odor recognition. Horm Behav 61(3):259–265
Wanisch K, Wotjak CT, Engelmann M (2008) Long-lasting second stage of recognition memory consolidation in mice. Behav Brain Res 186(2):191–196
Witter MP, Groenewegen HJ, Lopes da Silva FH, Lohman AH (1989) Functional organization of the extrinsic and intrinsic circuitry of the parahippocampal region. Prog Neurobiol 33(3):161–253
Zhao X, Sun L, Jia H, Meng Q, Wu S, Li N et al (2009) Isolation rearing induces social and emotional function abnormalities and alters glutamate and neurodevelopment-related gene expression in rats. Prog Neuropsychopharmacol Biol Psychiatry 33(7):1173–1177
Zheng DJ, Foley L, Rehman A, Ophir AG (2013) Social recognition is context dependent in single male prairie voles. Anim Behav 86(5):1085–1095

Social Odors: Alarm Pheromones and Social Buffering

Yasushi Kiyokawa

Abstract In this chapter, I describe 2 types of olfactory communication in rats, which appear to arouse anxiety and relief, respectively. In alarm pheromonal communication, rats release 4-methylpentanal and hexanal from their perianal region when they are stressed. These molecules activate the anxiety circuit, including the bed nucleus of the stria terminalis, when 4-methylpentanal and hexanal are simultaneously detected by the vomeronasal system and the main olfactory system, respectively. Consequently, recipient rats show a variety of anxiety responses, depending on the threatening stimuli. In appeasing olfactory communication, non-stressed rats release an appeasing olfactory signal, which is detected by the main olfactory system of other rats. When detected, this olfactory signal suppresses activation of the basolateral complex of the amygdala and, as a result, ameliorates stress responses elicited by an auditory conditioned stimulus during social buffering phenomenon. Because social buffering appears to be based on affinity and attachment to accompanying animals, the appeasing olfactory signal may arouse relief in rats. A definition of social buffering is also proposed as we still have no set definition for the term social buffering yet.

Keywords Emotional communication · Definition of social buffering · Pheromone · Stress · Stress-related odor

Contents

1 Introduction 48
2 Definition of Pheromones 48
3 Discovery of the Alarm Pheromone 49
4 Stress-Related Odors and Possible Alarm Pheromones in Mammals 49
5 Stress-Related Odors in Rats 49

Y. Kiyokawa (✉)
Laboratory of Veterinary Ethology, The University of Tokyo, 1-1-1 Yayoi, Bunkyo-ku, Tokyo 113-8657, Japan
e-mail: akiyo@mail.ecc.u-tokyo.ac.jp

Curr Topics Behav Neurosci (2016) 30: 47–66
DOI 10.1007/7854_2015_406
Published Online: 20 November 2015

6 The Identification of Molecules Responsible for Anxiety Responses 50
7 Stress-Related Odor Is an Alarm Pheromone in Rats 53
8 An Accidental Encounter with Social Buffering 54
9 Olfactory Signaling Mediates Social Buffering of Conditioned Fear Responses in Male Rats 55
10 Social Buffering of Conditioned Hyperthermia 57
11 The Definition of Social Buffering 58
12 Social Buffering Appears to Be Based on Affinity and Attachment to Accompanying Animals 60
13 Possible Contribution of Emotional Olfactory Communications 61
14 Conclusion 61
References 62

1 Introduction

It is well known that many mammals have highly developed olfactory systems and are highly dependent on their olfactory system, to the same extent that humans depend on their visual system. With such a developed olfactory system, animals can communicate their physiological state, including their health status and reproductive availability, to other conspecifics using olfactory cues. In addition, research has suggested that olfactory communication modulates behavior by arousing emotions such as fear, anxiety, or relief. A subset of these olfactory signals is called "pheromones."

2 Definition of Pheromones

The term pheromone was coined and defined by Karlson and Luscher (1959) as "substances which are secreted to the outside by an individual and received by a second individual of the same species, in which they release a specific reaction, for example, a definite behavior or a developmental process. The principle of minute amounts being effective holds." This definition was based on findings in insects; therefore, it is debatable whether the original definition can be applied to mammals. Researchers have proposed revising the original definition by modifying it and/or specifying additional requirements (Meredith 1998; Otte 1974; Stern and McClintock 1998; Wyatt 2010). We have proposed the following definition for pheromones, which includes aspects of the original definition: "(i) substances that are secreted to the outside by an individual and received by a second individual of the same species, in which they cause a specific reaction; (ii) substances that are effective in minute amounts; (iii) substances that are released from living individuals; and (iv) substances that mediate communication for an evolutionarily adaptive function" (Inagaki et al. 2014).

3 Discovery of the Alarm Pheromone

In 1938, von Frisch discovered that an injured European minnow might release specific chemicals that evoked fright reactions in other minnows (von Frisch 1938). Such chemicals were named "alarm substances (Schreckstoff)" and defined as "substances that communicate the presence of danger, provided that they are produced by members of the same species" (Pfeiffer 1963; von Frisch 1941). Once the term pheromone had been coined, "alarm substances" became known as "alarm pheromones."

4 Stress-Related Odors and Possible Alarm Pheromones in Mammals

Rats avoid the blood and flesh of other rats and mice (Hornbuckle and Beall 1974), suggesting that alarm pheromones are present in these tissues. However, in mammals, an alarm pheromone usually refers to the odors that are actively released from distressed animals, rather than the odors passively released from their wounds. Such stress-related odors and potential alarm pheromones in mammals do not appear to induce a stereotypic avoidance response. Stress-related odors have been reported to induce a cautious behavior, rather than a stereotypical avoidance response, in deer (Muller-Schwarze et al. 1984), cattle (Boissy et al. 1998), and swine (Vieuille-Thomas and Signoret 1992). In humans, stress-related odors increased anxiety (Albrecht et al. 2011) and enhanced the acoustic startle reflex (ASR) (Prehn et al. 2006). Mice, however, appear to be an exception in mammals as stress-related odors evoked a stereotypic avoidance response and activated the hypothalamic–pituitary–adrenocortical (HPA) axis in recipient mice (Brechbuhl et al. 2013; Carr et al. 1970). This response is consistent with the notion that mice have simplicity as a survival trait, as opposed to the complexity in rats and other mammals (Whishaw 1999).

5 Stress-Related Odors in Rats

In 1968, the existence of stress-related odors was first suggested by the results of a study in which food-restricted rats underwent discriminative punishment training. When they pressed the bar in the presence of odors from unstressed rats, they received a sucrose solution reward. However, when they pressed the bar in the presence of odors from electrical shock-distressed rats, they received foot shocks. These food-restricted rats showed an elongated latency of bar pressing in the presence of odors from distressed rats (Valenta and Rigby 1968). These results suggest that rats are able to distinguish between stress-related odors and odors from unstressed rats. Although several research groups subsequently analyzed stress-related odors, they mostly only reported on the existence of stress-related odors in their experimental models, rather than giving a detailed analysis of these odors.

A series of studies conducted in the early 1990s characterized stress-related substances released from rats during the forced swim test. In this test, a subject rat is placed in a deep pool without any evacuation routes. The rat will vigorously swim at first and then gradually become immobile, eventually floating silently in the water. Such immobility dramatically decreased when a subject rat (Rat A) swam in water in which another rat (Rat B) had swum previously (Abel and Bilitzke 1990). Given that immobility was decreased by pre-exposure to stressors before the forced swim test (e.g., foot shocks or loud noise) (Abel and Bilitzke 1990), the former rat (Rat B) may have released stress-related substances into water, which then served as a stressor for the subsequent rat (Rat A). After observing this immobility response, the research group identified several characteristics of stress-related substances, including that these substances did not exist in feces or urine (Abel and Bilitzke 1990) and that rats responded to their own substances and those of other rats in a similar fashion (Abel and Bilitzke 1990). However, the details of the communication mediated by the stress-related chemicals were still unclear.

6 The Identification of Molecules Responsible for Anxiety Responses

The late Professor Yuji Mori and his colleague at Laboratory of Veterinary Ethology at The University of Tokyo started to investigate alarm pheromonal communication in rats. They first discovered that odor-recipient male rats showed a greater rise in body temperature when placed in a box in which 2 odor-donor male rats had recently received foot shocks than when they were placed in a box where 2 odor-donor rats had been kept without having received foot shocks (Kikusui et al. 2001). This phenomenon suggested that the 2 odor-donor rats had released stress-related odors after receiving foot shocks, which aggravated hyperthermia in recipient rats. Subsequently, we found that the ability to release a stress-related odor was not affected by testosterone because intact and castrated odor donors were able to similarly induce aggravated hyperthermia in recipient rats (Kiyokawa et al. 2004b). Next, we attempted to identify the release site for the stress-related odors. Previously, I had confirmed in a preliminary observation that feces and urine excreted during foot shocks did not contain stress-related odors. Therefore, it was likely that stress-related odors were released from the surface of the rat's body. In rats, the anal glands are located just inside the anal verge. Secretions from the anal glands are a well-known defensive reaction in animals confronted with stressful stimuli. Indeed, these defensive and/or alarming reactions are anecdotally associated with many species, such as the skunk or fox (Albone and Fox 1971; Blackman 1911). Therefore, we hypothesized that rats were releasing stress-related odors from their anal glands. To squeeze the odor out of the gland, we electrically contracted muscles around the anus in anesthetized odor-donor rats. We found that odors released from the perianal region, but not from other regions, induced aggravated hyperthermia in recipient rats, demonstrating that stress-related odors were being released from the perianal region (Kiyokawa et al. 2004a).

Because stress-related odors are released into the air, recipient rats and odor-donor rats need to share the same enclosed environment in order to be exposed to the stress-related odors. This was the biggest limitation in analyzing stress-related odors in more detail. To solve this problem, we needed a suitable carrier that could trap the stress-related odors, as well as transfer the odors to a recipient rat. In the above-mentioned experiments, I noticed that the data results were poor when all the water droplets were wiped out of the experimental box after washing. Additionally, stress-related substances in rats had been reported in the forced swim test (Abel and Bilitzke 1990). Therefore, it appeared reasonable to hypothesize that water was a suitable carrier for stress-related odors. To assess this hypothesis, stress-related odors were released from anesthetized odor-donor rats in a small box containing water droplets on the ceiling. The water droplets were then collected, so they could be exposed to recipient rats using filter paper. We found that the water droplets exposed to stress-related odors induced aggravated hyperthermia in recipient rats, suggesting that water could be used as a carrier for stress-related odors (Kiyokawa et al. 2005a). These results also demonstrated that stress-related odors are effective in minute amounts.

By establishing a method to trap the stress-related odors into a water sample, we were able to analyze stress-related odors using a variety of experimental models. We found that stress-related odors induced aggravated hyperthermia in response to home cage movement (Kiyokawa et al. 2005a), increased defensive and risk assessment behaviors without eliciting odor avoidance in a modified open-field test (Kiyokawa et al. 2006), deteriorated male sexual behavior (Kobayashi et al. 2011), and enhanced the ASR (Inagaki et al. 2008). Considering the wide range of stress-related odor effects, it seemed reasonable to hypothesize that the primary effect of a stress-related odor is to activate the neural circuit for anxiety, which can result in a wide range of secondary anxiety responses. In support of this hypothesis, stress-related odor responses in recipient rats were antagonized by a pretreatment with anxiolytics. For example, pretreatment with benzodiazepines (diazepam and midazolam), a non-selective monoamine oxidase inhibitor (phenelzine), a non-selective ß-adrenergic receptor antagonist (propranolol), an 2-adrenergic receptor agonist (clonidine), or a corticotropin-releasing hormone subtype 1 receptor antagonist (CP-154,526), but not a serotonin-1A receptor agonist (buspirone), blocked the stress-related odor effects on the ASR (Inagaki et al. 2008, 2010). In addition, a pretreatment with an opioid receptor antagonist (naloxone) or CP-154,526 suppressed the stress-related odor effects on male sexual behavior (Kobayashi et al. 2011, 2013b, 2015). In addition, exposure to a stress-related odor increased Fos expression in the bed nucleus of the stria terminalis (BNST) (Kiyokawa et al. 2005b; Kobayashi et al. 2013a, 2015), an important nucleus in the neural circuit for anxiety. The significant role of the BNST was further suggested by the finding that functional inactivation of the BNST blocked the stress-related odor effects on risk assessment behavior (Breitfeld et al. 2015). In addition, we demonstrated that the vomeronasal system was involved in stress-related odor detection because removal of the vomeronasal organ blocked the stress-related odor effects on hyperthermia, the defensive and risk assessment behaviors, and the ASR (Kiyokawa et al. 2007a, 2013a). By freezing the stress-related odor-containing water, we were able to distinguish between the collection time of the stress-related odor and its presentation to recipient rats. This enabled us to demonstrate that recipient rats showed similar enhancements of their ASR in response to their own stress-related odors and to the stress-related odors released from other rats (Inagaki et al. 2012). These results suggested that there were one or more specific molecules in stress-related odors that were responsible for eliciting anxiety responses.

To identify the specific molecules, we pooled the stress-related odors in Tenax, an absorbent that can trap a wide range of volatile molecules. When we prepared 3 fractions (Frac. 1, Frac. 2, and Frac. 3) from the adsorbent, only Frac. 1 enhanced the ASR in recipient rats, suggesting that the specific molecules were in Frac. 1. We further fractionated Frac. 1 into 2 fractions, Frac. 1-1 and Frac. 1-2. When we exposed these fractions to recipient rats, only Frac. 1-1 enhanced their ASR, suggesting that the specific molecules were within Frac. 1-1. Because the analyses of Frac. 1-1 revealed a large increase in the levels of hexanal and a presence of 4-methylpentanal compared to the control sample, we presented these molecules to recipient rats at the same ratio as that detected in the absorbent (13:87). The binary mixture, but not the individual molecules, enhanced the ASR when they were presented at a total concentration of 10^{-5} M or 10^{-6} M, but not 10^{-7} M. Pharmacological analyses confirmed the anxiogenic effects of the mixture because the effects of the mixture were blocked by a pretreatment with diazepam, but not buspirone. In addition, we found that the mixture evoked defensive and risk assessment behaviors in the modified open-field test, suggesting that the binary mixture effects were not specific to the ASR test. When we assessed the impact of 4-methylpentanal and hexanal on the brain, the presentation of 4-methylpentanal, but not hexanal, activated the accessory olfactory bulb where vomeronasal epithelium sensory neurons send their projections exclusively. The BNST was activated only when 4-methylpentanal and hexanal were presented simultaneously. These results indicated that the detection of 4-methylpentanal in the vomeronasal system required simultaneous detection of hexanal to activate the neural circuit for anxiety. These findings were consistent with our observations that the stress-related odor did not elicit anxiety responses when we removed the vomeronasal organ of recipient rats. The neural circuit for anxiety was not activated because recipient rats lacking the vomeronasal organ could not detect 4-methylpentanal, even though they could detect hexanal. Taken together, we concluded that the mixture of 4-methylpentanal and hexanal in stress-related odors was responsible for eliciting anxiety responses in rats (Inagaki et al. 2014) (Fig. 1).

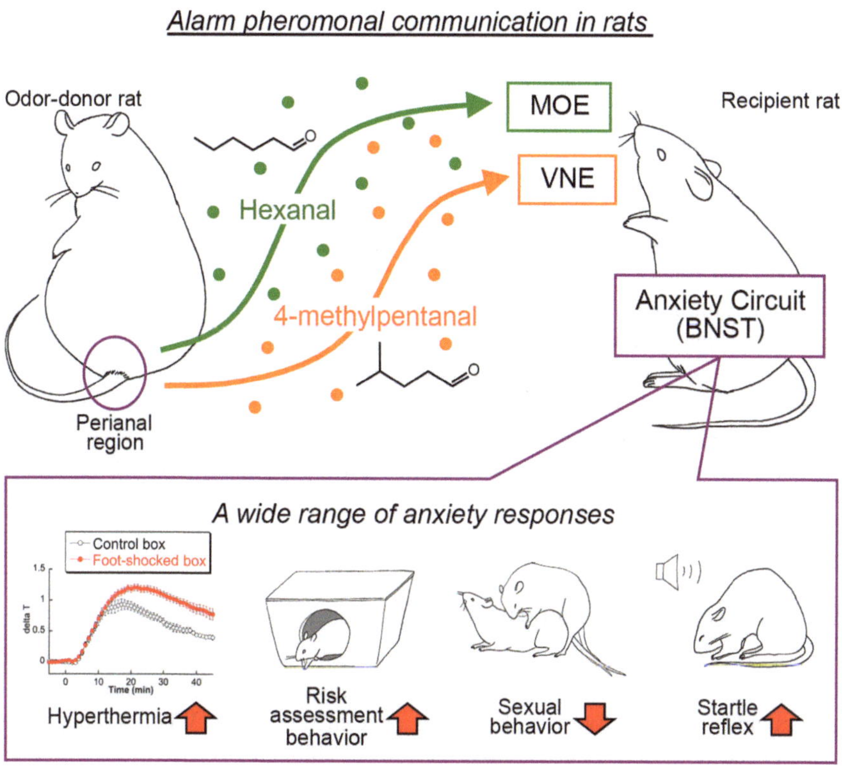

Fig. 1 Proposed alarm pheromonal communication in rats. Stressed rats release 4-methylpentanal and hexanal from their perianal regions. When simultaneously detected by the 2 olfactory systems, these molecules activate the anxiety circuit and evoke a variety of anxiety responses in other rats. *BNST* bed nucleus of the stria terminalis, *MOE* main olfactory epithelium, *VNE* vomeronasal epithelium

7 Stress-Related Odor Is an Alarm Pheromone in Rats

Stress-related odors meet our proposed definition of pheromones because they (i) are released from rats and elicit anxiety responses in other rats and (ii) are effective in minute amounts. In addition, (iii) stress-related odors are released by live rats. Furthermore, (iv) the communication mediated by a stress-related odor seems to have an evolutionarily adaptive function. The more complex the organism, the more situation-specific the response to the alarm might be. In addition, stress-related odors do not appear to provide specific information regarding the type of threatening stimulus. In such situations, activation of the anxiety circuit, rather than a specific stereotypical avoidance response, might be the most effective response for higher order mammals because it enables the recipient animal to show an appropriate response, depending on the situation and type of threatening stimulus. This may increase inclusive fitness. In summary, stress-related odors can be called alarm pheromones in rats.

In addition, we suggest that the mixture of 4-methylpentanal and hexanal could be the main component of alarm pheromone in rats. We revealed that this mixture shared many characteristics with alarm pheromones themselves. However, because our knowledge is limited, we still have many interesting questions. For example, it is unclear whether 4-methylpentanal and hexanal are effective if they are mixed in a different ratio. Because 4-methylpentanal is an easily oxidized molecule, it is unclear how long the mixture keeps its anxiogenic properties at room temperature and in the refrigerator. In addition, the intensity of the responses elicited by the mixture (Inagaki et al. 2014) seemed weak compared with those induced by native alarm pheromones in previous studies [ASR (Inagaki et al. 2008), defensive and risk assessment behaviors (Kiyokawa et al. 2006)]. Therefore, other molecules might contribute to the induction of anxiety response in recipient rats, even if the molecules themselves do not have anxiogenic properties. Another interesting question is whether the responses to alarm pheromones are affected by familiarity between the odor-donor and recipient rats. Further research is needed to address these questions.

8 An Accidental Encounter with Social Buffering

During our analyses of stress-related odors, rats were repeatedly used as odor donors and received foot shocks in an experimental box. When I placed the experienced odor donors in the box, these rats exhibited freezing to this environment, which is not surprising, as this procedure is the same one used in contextual fear conditioning. Thus, the experimental box served as the contextual conditioned stimulus (CS) and evoked freezing in the experienced odor-donor rats. During one experimental session, 2 odor-donor rats did not exhibit freezing at all. Although I thought that both rats were experienced odor donors, it turned out that one of these odor donors was a naïve rat. The fact that the naïve odor-donor rat was able to inhibit freezing of the experienced odor-donor rat was so surprising and fascinating that I decided to further examine this phenomenon. We reported that the presence of a non-conditioned rat suppressed freezing, hyperthermia, and HPA axis activation of a fear-conditioned subject rat in response to the contextual CS (Kiyokawa et al. 2004c), suggesting that the presence of a non-conditioned rat ameliorated stress of the subject rat. In the literature, such phenomenon has been called "social buffering" (Hennessy 2003). We also found that a non-conditioned rat had a more powerful effect than a conditioned rat, although both types of rats suppressed the above-mentioned responses in subject rats (Kiyokawa et al. 2004c).

9 Olfactory Signaling Mediates Social Buffering of Conditioned Fear Responses in Male Rats

After reporting that the presence of a non-conditioned rat ameliorated the stress responses of a subject rat to a contextual CS, I had an opportunity to launch a project on social buffering, which I owe to late Professor Mori. I decided to use an auditory CS rather than a contextual CS as a stressor in this project because an auditory CS evokes responses by a simpler neural pathway than the contextual CS. The hippocampus, BNST, and basolateral complex of the amygdala (BLA) participate in the responses to a contextual CS, whereas only the BLA contributes to the responses to an auditory CS (Kiyokawa et al. 2015). In addition, the presence of a conspecific animal does not affect the auditory CS per se. Indeed, in our previous study, we had to acknowledge the possibility that the presence of a non-conditioned rat disrupted the contextual CS, which may have caused the weakened stress responses in the subject rat.

In our model, the male subject rat was first fear-conditioned or not to an auditory CS. The next day, both types of subject rats were re-exposed to the auditory CS, either alone or with a non-conditioned, unfamiliar male rat ("an associate"). We found that the associate's presence completely blocked the freezing and HPA axis activation of the fear-conditioned subject rat in response to the auditory CS (Kiyokawa et al. 2007b) (Fig. 2). This social buffering effect was still observed even when the subject rat and associate were separated by a wire

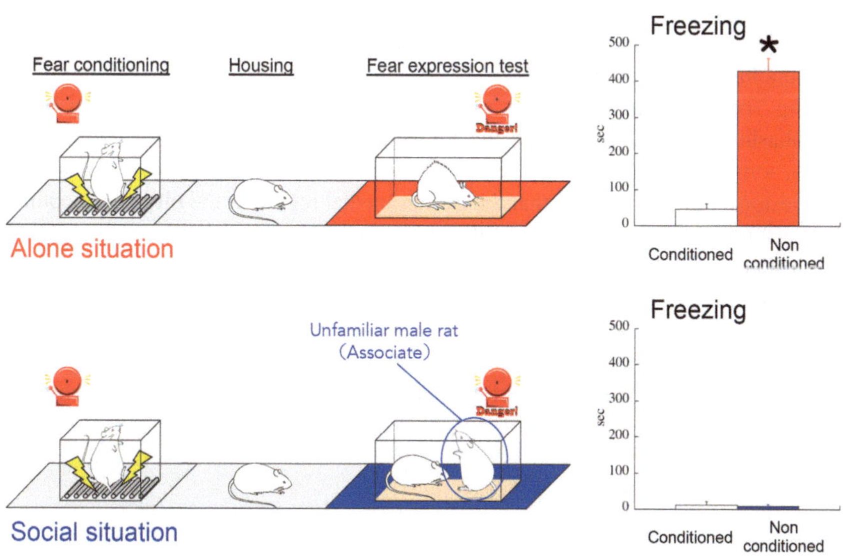

Fig. 2 Schematic diagram of the experiments assessing social buffering of conditioned fear responses. After being housed alone for 24 h after conditioning, fear-conditioned and non-conditioned subject rats were exposed to the conditioned stimulus with an associate

mesh partition or a double wire mesh partition separated by 5 cm (Kiyokawa et al. 2009, 2014a). We also confirmed that this phenomenon was not simply due to a distraction by the presence of another animal, because neither freezing nor HPA axis activation was suppressed by the presence of an unfamiliar male guinea pig separated by a wire mesh screen (Kiyokawa et al. 2009). Therefore, to induce social buffering, some communication must have existed between the subject rat and the associate. Because rats have a well-developed olfactory system, I focused on the role of olfactory communication in social buffering. We found that a lesion in the main olfactory epithelium (MOE) of the subject rat blocked social buffering (Kiyokawa et al. 2009). We also demonstrated that social buffering occurred when the subject rat was tested in a box that was previously odorized by an associate, i.e., only the olfactory signal from an associate was present in the box (Takahashi et al. 2013). Furthermore, this "appeasing olfactory signal" appeared to be volatile because social buffering was induced even if the subject rat was tested in an area that was separated from the associate's odorized area by a partition that allowed the penetration of only the volatile signals (Kiyokawa et al. 2014b). Together, these data indicate that olfactory communication induced social buffering in rats.

In parallel with these analyses, we investigated the neural pathway underlying this social buffering phenomenon. As mentioned above, social buffering was mediated by the detection of an appeasing olfactory signal by the MOE of the subject rat (Kiyokawa et al. 2009; Takahashi et al. 2013). After detection, the appeasing olfactory signal would be transmitted from the MOE to the main olfactory bulb (MOB), based on anatomical evidence that MOE sensory neurons only send their projections to the MOB. Because the auditory CS evokes stress responses by activating the BLA, this signal should be transmitted from the MOB to the BLA, thereby suppressing BLA activation. Indeed, electrophysiological and immunohistochemical analyses indicated that BLA activation was suppressed during social buffering (Fuzzo et al. 2015; Kiyokawa et al. 2007b, 2014b; Takahashi et al. 2013). Because the MOB does not have direct projections to the BLA, some nuclei must be intervening between the MOB and the BLA. Anatomical and lesion studies suggested that the posteromedial region of the olfactory peduncle (pmOP) relays the transmission of the appeasing olfactory signal from the MOB to the ipsilateral BLA (Kiyokawa et al. 2012). Further immunohistochemical analyses suggested that the posterior complex of the anterior olfactory nucleus (AOP) within the pmOP was primarily responsible (Takahashi et al. 2013). Taken together, we suggest that the appeasing olfactory signal is transmitted from the MOB to the pmOP, most likely the AOP region. The pmOP then suppresses BLA activation (Fig. 3).

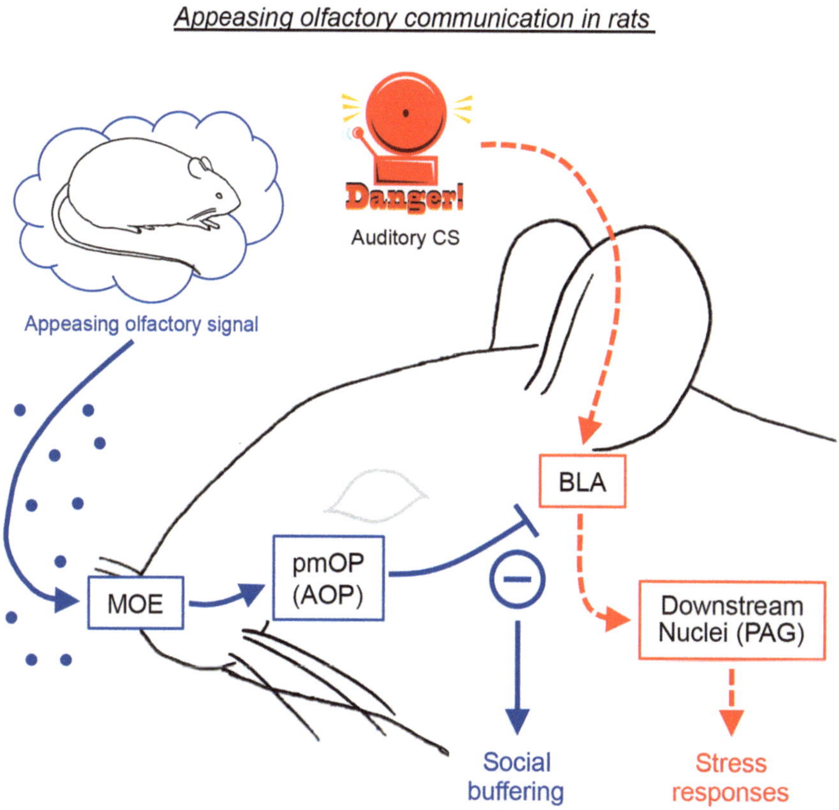

Fig. 3 Proposed neural circuit that induces social buffering of conditioned fear responses in rats. When the subject rat is exposed to an auditory conditioned stimulus (CS), the CS activates the basolateral complex of the amygdala (BLA) and elicits stress responses. The appeasing olfactory signal released from non-stressed rats suppresses activation of the BLA and, as a result, ameliorates these stress responses. *AOP* posterior complex of the anterior olfactory nucleus, *MOE* main olfactory epithelium, *PAG* periaqueductal gray, *pmOP* posteromedial region of the olfactory peduncle

10 Social Buffering of Conditioned Hyperthermia

When we started the above-mentioned experiments, I wondered if cohousing the subject rat with an associate from immediately after conditioning until auditory CS re-exposure would affect their stress responses. We decided to evaluate this idea, as it was supported by findings that cohousing with a conspecific rat after social defeat led to weight loss recovery and recovery from alteration in autonomic rhythms (de Jong et al. 2005). Male subject rats were first fear-conditioned or not to an auditory CS. Both types of subject rats were then housed either alone or with a non-conditioned associate. One day after the conditioning day, the subject rat alone was re-exposed to the auditory CS. In addition to freezing and HPA axis activation,

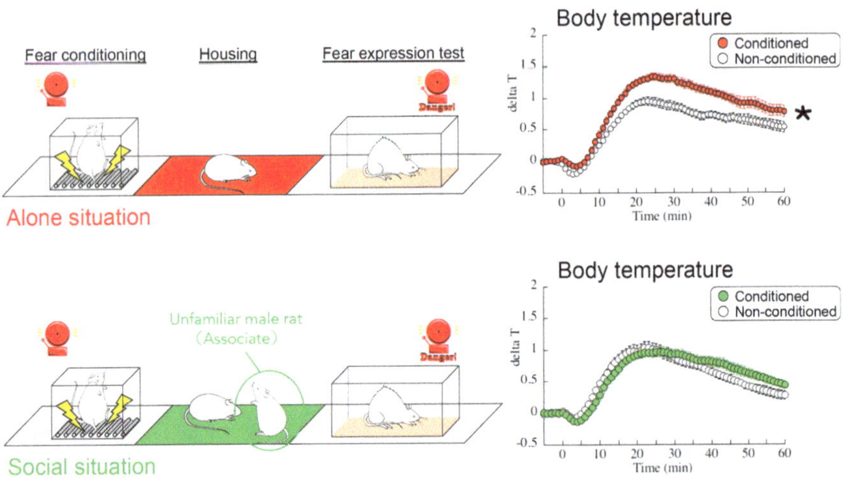

Fig. 4 Schematic diagram of the experiments assessing social buffering of conditioned hyperthermia. Immediately after the conditioning procedure, fear-conditioned and non-conditioned subject rats were cohoused with an associate for 24 h and then exposed to the conditioned stimulus alone

the fear-conditioned subjects that were housed alone immediately after conditioning until CS re-exposure exhibited hyperthermia in response to the CS. Cohousing with an associate during this period ameliorated the hyperthermia response, as well as HPA axis activation to re-exposure, even if only the subject rat was re-exposed to the CS (Kiyokawa et al. 2007b) (Fig. 4). Subsequent analyses of this social buffering effect revealed that 12 h of cohousing, but not 6 h, was sufficient to evoke social buffering (Kodama et al. 2011). In addition, this social buffering effect was still observed even if cohousing was started 24 h after conditioning (Kiyokawa et al. 2013b), or if the tactile signal between the rats was limited by a wire mesh screen during the 24 h of cohousing (Kiyokawa et al. 2013b).

11 The Definition of Social Buffering

When I initially launched this project, few research groups were analyzing social buffering. In parallel with an increase in the public's interest in social behavior, a wide range of social buffering phenomena were reported in many mammalian species (Hennessy et al. 2009; Rault 2012). However, several phenomena have been claimed to be social buffering even when stress of the subject animals seemed not to be ameliorated. Given that there is no set definition for the term "social buffering," we cannot truly evaluate whether all these phenomena can be called social buffering. Therefore, it is imperative that this term be defined.

The phenomena seem to be divided into 2 categories according to experimental sequences. In one sequence, the subject is exposed to distressing stimuli with a conspecific animal. In the other sequence, the subject alone is first exposed to distressing stimuli and then cohoused with a conspecific animal. These phenomena can be termed “exposure type” or “housing type,” respectively, based on the type of experimental sequence. Here, I propose a definition for exposure-type social buffering as “phenomena in which stress of the subject is ameliorated when the subject is exposed to distressing stimuli along with a conspecific animal(s).” Similarly, housing-type social buffering can be defined as “phenomena in which recovery from adverse alterations induced by previously distressing stimuli is led by subsequent cohousing with a conspecific animal(s).”

In exposure-type social buffering, stress of the subject should be ameliorated in addition to the reduction of specific responses. Although the definition of stress is still under discussion (Koolhaas et al. 2011), attenuation of HPA axis activation or sympathetic adrenomedullary system activation is the most widely accepted index of stress amelioration. When we assess social buffering in infants with an immature HPA axis, such as rat pups, their behavioral responses to isolation from their mother, such as distress calls, may be good measures. The reduction of behavioral measures may also be used as indices of stress amelioration in animals with a mature HPA axis, but only when responses to innately aversive stimuli are observed, such as avoidance of bright light, open-field, elevated platform, or predator cues. Care should be taken to distinguish social buffering from stress-induced analgesia when responses are mediated by pain. The use of changes in learned behaviors should be supported by the above-mentioned physiological measures, even if, for example, they are the fear responses or avoidance to an aversive CS. This is because the expression of learned behaviors can be decreased by simple distractions, such as the presence of a conspecific animal, even though stress in the subject animal is not ameliorated. In addition, “social facilitation” may be a more appropriate term when learned avoidance of the stimulus is decreased.

In housing-type social buffering, it is impossible to use acute responses as indices of stress amelioration because housing-type social buffering may require days to weeks of cohousing with a conspecific animal after a previously distressing stimulus to take effects. In contrast to the acute responses, there seems to be no long-term alteration that is widely accepted as an index of stress, which prevents us from standardizing the measures for housing-type social buffering. Nonetheless, several physiological measures may be good indices, based on research demonstrating that housing with a conspecific animal led to weight loss recovery and reduced alteration in autonomic rhythms, such as bradycardia and hypothermia during the dark phase, after a single social defeat in rats (de Jong et al. 2005). In the analyses of housing-type social buffering, an appropriate control subject that does not experience distressing stimuli and is housed alone for the same period should be employed. This control would allow for proper evaluation of the effects of housing conditions. It is possible that solitary housing induces adverse alterations, while housing with a conspecific animal has no effects, which may be incorrectly interpreted as housing-type social buffering leads to recovery from adverse alterations (de Jong et al. 2005; Ruis et al. 1999). When a previously distressing experience exacerbates subsequent responses to innately aversive stimuli, remission of this exacerbation may also be used as indices. In rats, a single social defeat shortened the time spent in the open arm of the elevated plus-maze test; this effect was ameliorated by housing with a conspecific animal after social defeat (Nakayasu and Ishii 2008). Similarly, cohousing with a conspecific animal after conditioning can alter the subsequent conditioned responses, even if the subject alone is re-exposed to the CS. Such phenomenon may be called social buffering only when the alteration in learned behaviors is supported by amelioration of physiological measures during cohousing or in response to the CS. Further analyses of housing-type social buffering could identify appropriate indices for this phenomenon.

While discussing social buffering, it is advisable to clarify the type of conspecific animal that induces the phenomenon. Research suggests that the neural mechanisms underlying social buffering differ depending on the type of conspecific animal. For example, we found that social buffering of conditioned fear responses in the presence of an unfamiliar male rat was induced by BLA suppression (Kiyokawa et al. 2012). However, in rat pups, the presence of a mother rat induced social buffering of HPA axis activation in response to foot shocks by suppressing norepinephrine release in the paraventricular nucleus of the hypothalamus of the pups (Shionoya et al. 2007). Moreover, pharmacological studies have suggested that dopamine D_1 and D_2 receptors are involved in the suppression of distress calls by a rat dam, but not by littermates (Shair et al. 2009). Therefore, ambiguity regarding the type of conspecific animal may result in confusions during scientific discussions of social buffering. The terms "maternal buffering," "mate buffering," or "conspecific buffering" may be useful for referring to social buffering induced by a mother, mate, or conspecifics without sexual relationships, respectively.

12 Social Buffering Appears to Be Based on Affinity and Attachment to Accompanying Animals

In social species, an animal is thought to have an affinity for other animals, even for unfamiliar same-sex or opposite-sex conspecific animals in a non-sexual relationship, because such affinity may help to form the basis of species' sociability. Therefore, it seems reasonable to assume that social buffering is based on an affinity and attachment to accompanying conspecific animals. In support of this notion, a familiar conspecific that had been housed with a subject rat for 3 weeks in the same cage was found to be more effective than an unfamiliar conspecific for exposure-type conspecific buffering in rats (Kiyokawa et al. 2014b). Furthermore, this may be one of the reasons why no clear conspecific buffering phenomenon has been observed in mice since mice are highly aggressive toward conspecifics.

Findings in maternal buffering further suggest that affinity and attachment to accompanying animals yield social buffering. Mammals depend on their mother during their postnatal period, which establishes an attachment to their mother. Based on this attachment, maternal buffering should be observed in all mammals. In rats, the presence of an anesthetized mother (more specifically, a dam of the same postpartum age as the pups' biological mother) suppressed HPA axis activation in response to foot shocks in the pups (Moriceau and Sullivan 2006). This maternal buffering seemed not to be induced when pups were close to weaning age (postnatal day 21–23). Because pups become less dependent on their mother by eating food independently around this age, the attachment to the mother may be decreased. The ontogeny of maternal buffering in the male guinea pig (Hennessy et al. 2006; Hennessy and Morris 2005) further supports the notion that maternal buffering is based on an attachment to the mother. When preweaning male pups (postnatal day 9–19) were isolated in a novel environment, they exhibited sickness behavior and HPA axis activation. These responses were ameliorated by the presence of a mother, but not by that of an unfamiliar female. After weaning, animals that induced social buffering changed from a mother to an unfamiliar female (postnatal day 49–61) and then to a favored female (postnatal day 270–330). These changes were consistent with the notion that male guinea pigs typically change the target of their attachment and affiliation from their mother to sexually matured, unfamiliar females, and then to a mating partner, depending on their sexual and social maturation.

13 Possible Contribution of Emotional Olfactory Communications

Emotional olfactory communications may mediate responses in other experimental models. For example, alarm pheromones in rats might enhance other types of alarm communications, such as the emission of ultrasonic alarm calls. Both alarm pheromones and appeasing olfactory signals might also mediate the observational learning of aversive stimuli or pro-social behaviors in rodents. From a practical point of view, emotional olfactory communication may increase the variation of the data in behavioral experiments. For example, stress-related odors or appeasing olfactory signals released from a precedent subject can enhance or suppress, respectively, the stress responses of the subsequent subjects. In addition, stress-related odors probably affect behavioral responses in swimming tests and water mazes. Therefore, possible olfactory communication between subjects should be taken into consideration in experimental designs and in the order of the experiments.

14 Conclusion

This chapter discussed 2 types of emotional olfactory communication in rats. Such communications may represent good experimental models for understanding emotions in animals. Because the emotion-evoking signal consisted of only one or a few odorant molecules, analyses of the neural mechanisms underlying the emotion will be simpler. Analyses of emotional communication may also further our understanding of animal's natural lives.

Acknowledgements I would like to thank Dr. Michael Hennessy, Akiko Ishii, and Kaori Mikami for their critical suggestion on the manuscript.

References

Abel EL, Bilitzke PJ (1990) A possible alarm substance in the forced swimming test. Physiol Behav 48(2):233–239. doi:10.1016/0031-9384(90)90306-O

Albone ES, Fox MW (1971) Anal gland secretion of the red fox. Nature 233:569–570. doi:10.1038/233569a0

Albrecht J, Demmel M, Schopf V, Kleemann AM, Kopietz R, May J, Schreder T, Zernecke R, Bruckmann H, Wiesmann M (2011) Smelling chemosensory signals of males in anxious versus nonanxious condition increases state anxiety of female subjects. Chem Senses 36(1):19–27. doi:10.1093/chemse/bjq087

Blackman MW (1911) The anal glands of Mephitus Mephitica. Anat Rec 5(11):491–515. doi:10.1002/ar.1090051102

Boissy A, Terlouw C, Le Neindre P (1998) Presence of cues from stressed conspecifics increases reactivity to aversive events in cattle: evidence for the existence of alarm substances in urine. Physiol Behav 63(4):489–495. doi:10.1016/S0031-9384(97)00466-6

Brechbuhl J, Moine F, Klaey M, Nenniger-Tosato M, Hurni N, Sporkert F, Giroud C, Broillet MC (2013) Mouse alarm pheromone shares structural similarity with predator scents. Proc Natl Acad Sci U S A 110(12):4762–4767. doi:10.1073/pnas.1214249110

Breitfeld T, Bruning J, Inagaki H, Takeuchi Y, Kiyokawa Y, Fendt M (2015) Temporary inactivation of the anterior part of the bed nucleus of the stria terminalis blocks alarm pheromone-induced defensive behavior in rats. Front Neurosci 9:321. doi:10.3389/fnins.2015.00321

Carr WJ, Martorano RD, Krames L (1970) Responses of mice to odors associated with stress. J Comp Physiol Psychol 71(2):223–228. doi:10.1037/h0029164

de Jong JG, van der Vegt BJ, Buwalda B, Koolhaas JM (2005) Social environment determines the long-term effects of social defeat. Physiol Behav 84(1):87–95. doi:10.1016/j.physbeh.2004.10.013

Fuzzo F, Matsumoto J, Kiyokawa Y, Takeuchi Y, Ono T, Nishijo H (2015) Social buffering suppresses fear-associated activation of the lateral amygdala in male rats: behavioral and neurophysiological evidence. Front Neurosci 9:99. doi:10.3389/fnins.2015.00099

Hennessy MB (2003) Enduring maternal influences in a precocial rodent. Dev Psychobiol 42 (3):225–236. doi:10.1002/dev.10095

Hennessy MB, Morris A (2005) Passive responses of young guinea pigs during exposure to a novel environment: influences of social partners and age. Dev Psychobiol 46(2):86–96. doi:10.1002/dev.20045

Hennessy MB, Hornschuh G, Kaiser S, Sachser N (2006) Cortisol responses and social buffering: a study throughout the life span. Horm Behav 49(3):383–390. doi:10.1016/j.yhbeh.2005.08.006

Hennessy MB, Kaiser S, Sachser N (2009) Social buffering of the stress response: diversity, mechanisms, and functions. Front Neuroendocrinol 30(4):470–482. doi:10.1016/j.yfrne.2009.06.001

Hornbuckle PA, Beall T (1974) Escape reactions to the blood of selected mammals by rats. Behav Biol 12(4):573–576. doi:10.1016/S0091-6773(74)92531-0

Inagaki H, Kiyokawa Y, Kikusui T, Takeuchi Y, Mori Y (2008) Enhancement of the acoustic startle reflex by an alarm pheromone in male rats. Physiol Behav 93(3):606–611. doi:10.1016/j.physbeh.2007.10.021

Inagaki H, Kiyokawa Y, Takeuchi Y, Mori Y (2010) The alarm pheromone in male rats as a unique anxiety model: psychopharmacological evidence using anxiolytics. Pharmacol Biochem Behav 94(4):575–579. doi:10.1016/j.pbb.2009.11.013

Inagaki H, Kiyokawa Y, Takeuchi Y, Mori Y (2012) Male rats respond to their own alarm pheromone. J Vet Med Sci 74(1):79–82. doi:10.1292/jvms.11-0225

Inagaki H, Kiyokawa Y, Tamogami S, Watanabe H, Takeuchi Y, Mori Y (2014) Identification of a pheromone that increases anxiety in rats. Proc Natl Acad Sci U S A 111(52):18751–18756. doi:10.1073/pnas.1414710112

Karlson P, Luscher M (1959) 'Pheromones': a new term for a class of biologically active substances. Nature 183(4653):55–56. doi:10.1038/183055a0

Kikusui T, Takigami S, Takeuchi Y, Mori Y (2001) Alarm pheromone enhances stress-induced hyperthermia in rats. Physiol Behav 72(1–2):45–50. doi:10.1016/S0031-9384(00)00370-X

Kiyokawa Y, Kikusui T, Takeuchi Y, Mori Y (2004a) Alarm pheromones with different functions are released from different regions of the body surface of male rats. Chem Senses 29(1):35–40. doi:10.1093/chemse/bjh004

Kiyokawa Y, Kikusui T, Takeuchi Y, Mori Y (2004b) Modulatory role of testosterone in alarm pheromone release by male rats. Horm Behav 45(2):122–127. doi:10.1016/j.yhbeh.2003.09.007

Kiyokawa Y, Kikusui T, Takeuchi Y, Mori Y (2004c) Partner's stress status influences social buffering effects in rats. Behav Neurosci 118(4):798–804. doi:10.1037/0735-7044.118.4.798

Kiyokawa Y, Kikusui T, Takeuchi Y, Mori Y (2005a) Alarm pheromone that aggravates stress-induced hyperthermia is soluble in water. Chem Senses 30(6):513–519. doi:10.1093/chemse/bji044

Kiyokawa Y, Kikusui T, Takeuchi Y, Mori Y (2005b) Mapping the neural circuit activated by alarm pheromone perception by c-Fos immunohistochemistry. Brain Res 1043(1–2):145–154. doi:10.1016/j.brainres.2005.02.061

Kiyokawa Y, Shimozuru M, Kikusui T, Takeuchi Y, Mori Y (2006) Alarm pheromone increases defensive and risk assessment behaviors in male rats. Physiol Behav 87(2):383–387. doi:10.1016/j.physbeh.2005.11.003

Kiyokawa Y, Kikusui T, Takeuchi Y, Mori Y (2007a) Removal of the vomeronasal organ blocks the stress-induced hyperthermia response to alarm pheromone in male rats. Chem Senses 32(1):57–64. doi:10.1093/chemse/bjl036

Kiyokawa Y, Takeuchi Y, Mori Y (2007b) Two types of social buffering differentially mitigate conditioned fear responses. Eur J Neurosci 26(12):3606–3613. doi:10.1111/j.1460-9568.2007.05969.x

Kiyokawa Y, Takeuchi Y, Nishihara M, Mori Y (2009) Main olfactory system mediates social buffering of conditioned fear responses in male rats. Eur J Neurosci 29(4):777–785. doi:10.1111/j.1460-9568.2009.06618.x

Kiyokawa Y, Wakabayashi Y, Takeuchi Y, Mori Y (2012) The neural pathway underlying social buffering of conditioned fear responses in male rats. Eur J Neurosci 36(10):3429–3437. doi:10.1111/j.1460-9568.2012.08257.x

Kiyokawa Y, Kodama Y, Kubota T, Takeuchi Y, Mori Y (2013a) Alarm pheromone is detected by the vomeronasal organ in male rats. Chem Senses 38(8):661–668. doi:10.1093/chemse/bjt030

Kiyokawa Y, Kodama Y, Takeuchi Y, Mori Y (2013b) Physical interaction is not necessary for the induction of housing-type social buffering of conditioned hyperthermia in male rats. Behav Brain Res 256:414–419. doi:10.1016/j.bbr.2013.08.037

Kiyokawa Y, Hiroshima S, Takeuchi Y, Mori Y (2014a) Social buffering reduces male rats' behavioral and corticosterone responses to a conditioned stimulus. Horm Behav 65(2):114–118. doi:10.1016/j.yhbeh.2013.12.005

Kiyokawa Y, Honda A, Takeuchi Y, Mori Y (2014b) A familiar conspecific is more effective than an unfamiliar conspecific for social buffering of conditioned fear responses in male rats. Behav Brain Res 267:189–193. doi:10.1016/j.bbr.2014.03.043

Kiyokawa Y, Mikami K, Mikamura Y, Ishii A, Takeuchi Y, Mori Y (2015) The 3-second auditory conditioned stimulus is a more effective stressor than the 20-second auditory conditioned stimulus in male rats. Neuroscience. doi:10.1016/j.neuroscience.2015.04.055

Kobayashi T, Kiyokawa Y, Takeuchi Y, Mori Y (2011) Pretreatment with CP-154526 blocks the modifying effects of alarm pheromone on components of sexual behavior in male, but not in female, rats. Chem Senses 36(7):623–632. doi:10.1093/chemse/bjr017

Kobayashi T, Kiyokawa Y, Arata S, Takeuchi Y, Mori Y (2013a) c-Fos expression during the modulation of sexual behavior by an alarm pheromone. Behav Brain Res 237:230–237. doi:10.1016/j.bbr.2012.09.042

Kobayashi T, Kiyokawa Y, Arata S, Takeuchi Y, Mori Y (2013b) Naloxone blocks the deterioration of male sexual behavior induced by alarm pheromone in rats. J Vet Med Sci 75 (6):761–765

Kobayashi T, Kiyokawa Y, Takeuchi Y, Mori Y (2015) Neural correlates underlying naloxone-induced amelioration of sexual behavior deterioration due to an alarm pheromone. Front Neurosci 9:52. doi:10.3389/fnins.2015.00052

Kodama Y, Kiyokawa Y, Takeuchi Y, Mori Y (2011) Twelve hours is sufficient for social buffering of conditioned hyperthermia. Physiol Behav 102(2):188–192. doi:10.1016/j.physbeh.2010.10.025

Koolhaas JM, Bartolomucci A, Buwalda B, de Boer SF, Flugge G, Korte SM, Meerlo P, Murison R, Olivier B, Palanza P, Richter-Levin G, Sgoifo A, Steimer T, Stiedl O, van Dijk G, Wohr M, Fuchs E (2011) Stress revisited: a critical evaluation of the stress concept. Neurosci Biobehav Rev 35(5):1291–1301. doi:10.1016/j.neubiorev.2011.02.003

Meredith M (1998) Vomeronasal, olfactory, hormonal convergence in the brain. Cooperation or coincidence? Ann N Y Acad Sci 855:349–361. doi:10.1111/j.1749-6632.1998.tb10593.x

Moriceau S, Sullivan RM (2006) Maternal presence serves as a switch between learning fear and attraction in infancy. Nat Neurosci 9(8):1004–1006. doi:10.1038/nn1733

Muller-Schwarze D, Altieri R, Porter N (1984) Alert odor from skin gland in deer. J Chem Ecol 10 (12):1707–1729. doi:10.1007/BF00987357

Nakayasu T, Ishii K (2008) Effects of pair-housing after social defeat experience on elevated plus-maze behavior in rats. Behav Processes 78(3):477–480. doi:10.1016/j.beproc.2008.02.007

Otte D (1974) Effects and funtions in the evolution of signaling systems. Annu Rev Ecol Evol Syst 5:385–417. doi:10.1146/annurev.es.05.110174.002125

Pfeiffer W (1963) Alarm substances. Experientia 19:113–123

Prehn A, Ohrt A, Sojka B, Ferstl R, Pause BM (2006) Chemosensory anxiety signals augment the startle reflex in humans. Neurosci Lett 394(2):127–130. doi:10.1016/j.neulet.2005.10.012

Rault JL (2012) Friends with benefits: social support and its relevance for farm animal welfare. Appl Anim Behav Sci 136(1):1–14. doi:10.1016/j.applanim.2011.10.002

Ruis MA, te Brake JH, Buwalda B, De Boer SF, Meerlo P, Korte SM, Blokhuis HJ, Koolhaas JM (1999) Housing familiar male wildtype rats together reduces the long-term adverse behavioural and physiological effects of social defeat. Psychoneuroendocrinology 24(3):285–300. doi:10.1016/S0306-4530(98)00050-X

Shair HN, Muller JM, Moore H (2009) Dopamine's role in social modulation of infant isolation-induced vocalization: I. Reunion responses to the dam, but not littermates, are dopamine dependent. Dev Psychobiol 51(2):131–146. doi:10.1002/dev.20353

Shionoya K, Moriceau S, Bradstock P, Sullivan RM (2007) Maternal attenuation of hypothalamic paraventricular nucleus norepinephrine switches avoidance learning to preference learning in preweanling rat pups. Horm Behav 52(3):391–400. doi:10.1016/j.yhbeh.2007.06.004

Stern K, McClintock MK (1998) Regulation of ovulation by human pheromones. Nature 392 (6672):177–179. doi:10.1038/32408

Takahashi Y, Kiyokawa Y, Kodama Y, Arata S, Takeuchi Y, Mori Y (2013) Olfactory signals mediate social buffering of conditioned fear responses in male rats. Behav Brain Res 240:46–51. doi:10.1016/j.bbr.2012.11.017

Valenta JG, Rigby MK (1968) Discrimination of the odor of stressed rats. Science 161(841):599–601. doi:10.1126/science.161.3841.599

Vieuille-Thomas C, Signoret JP (1992) Pheromonal transmission of an aversive experience in domestic pig. J Chem Ecol 18(9):1551–1557. doi:10.1007/BF00993228 von Frisch K (1938) Zur Psychologie des Fisch-Schwarmes. Naturwissenschaften 26(37):601–606. doi:10.1007/BF01590598

von Frisch K (1941) Über einen Schreckstoff der Fischhaut und seine biologische Bedeutung. Z Vgl Physiol 29:46–145. doi:10.1007/BF00304445

Whishaw IQ (1999) The laboratory rat, the Pied Piper of twentieth century neuroscience. Brain Res Bull 50(5–6):411. doi:10.1016/S0361-9230(99)00169-0

Wyatt TD (2010) Pheromones and signature mixtures: defining species-wide signals and variable cues for identity in both invertebrates and vertebrates. J Comp Physiol A Neuroethol Sens Neural Behav Physiol 196(10):685–700. doi:10.1007/s00359-010-0564-y

Acoustic Communication in Rats: Effects of Social Experiences on Ultrasonic Vocalizations as Socio-affective Signals

Markus Wöhr, K. Alexander Engelhardt, Dominik Seffer, A. Özge Sungur and Rainer K.W. Schwarting

Abstract Ultrasonic vocalizations (USV) serve important communicative functions as socio-affective signals in rats. In aversive situations, such as inter-male aggression and predator exposure, 22-kHz USV are emitted. They likely function as appeasement signals during fighting and/or as alarm calls to warn conspecifics. In appetitive situations, 50-kHz USV are uttered, most notably during social interactions, such as rough-and-tumble play and mating. It is believed that they fulfill an affiliative function as social contact calls. Social experiences or their lack, such as social isolation, can have profound impact on the emission of 22- and 50-kHz USV by the sender in later life, albeit direction and strength of observed effects vary, with time point of occurrence and duration being critical determinants. Little, however, is known about how social experiences affect the behavioral responses evoked by 22- and 50-kHz USV in the recipient. By means of our 50-kHz USV radial maze playback paradigm, we recently showed that the behavioral response elicited in the recipient is affected by post-weaning social isolation. Rats exposed to four weeks of isolation during the rough-and-tumble play period did not display social approach behavior toward 50-kHz USV but some signs of social avoidance. We further found that physical environmental enrichment providing minimal opportunities for social interactions has similar detrimental effects. Together, this indicates that social experiences can affect socio-affective communication in rodents, both at the level of sender and recipient. Deficits seen following post-weaning social isolation or physical environmental enrichment might be useful to model aspects of neurodevelopmental disorders characterized by social and communication deficits, such as autism and schizophrenia.

Keywords Ultrasonic communication · Social behavior · Social isolation · Animal model · Autism · Schizophrenia · Intense world syndrome

---

M. Wöhr (✉) · K.A. Engelhardt · D. Seffer · A.Ö. Sungur · R.K.W. Schwarting
Behavioral Neuroscience, Philipps-University of Marburg, Gutenbergstr. 18, 35032 Marburg, Germany
e-mail: markus.woehr@staff.uni-marburg.de

Curr Topics Behav Neurosci (2016) 30: 67–90
DOI 10.1007/7854_2015_410
Published Online: 16 November 2015

Contents

1 Introduction 68
2 Types of Ultrasonic Vocalizations in Rats 68
3 Alarming Function of 22-kHz USV 70
4 Affiliative Function of 50-kHz USV 72
5 Effects of Social Experiences on Ultrasonic Communication: 22-kHz USV 73
6 Effects of Social Experiences on Ultrasonic Communication: 50-kHz USV 76
7 The 50-kHz USV Radial Maze Playback Paradigm 79
8 Effects of Social Experiences on Ultrasonic Communication in the 50-kHz USV Radial Maze Playback Paradigm 80
9 Modeling the Intense World Syndrome in Rats: Effects of Environmental Enrichment 83
10 Conclusion 86
References 87

1 Introduction

Various rodent species, including mice and rats, developed a comparatively sophisticated acoustic communication system consisting of various distinct types of sound signals, such as alarm and social contact calls. The majority of such sound signals are not audible to humans as they are emitted in the ultrasonic range, i.e., above the human hearing threshold of ~20 kHz, typically referred to as ultrasonic vocalizations (USV). Such USV serve important communicative functions as socio-affective signals in mice (for a detailed overview, see Wöhr and Scattoni 2013) and rats (for a detailed overview, see Brudzynski 2013; Wöhr and Schwarting 2013).

2 Types of Ultrasonic Vocalizations in Rats

Typically, three main types of USV are distinguished in rats. The three main USV types can be differentiated based on a number of factors. Important ones include the rat's developmental stage, the emotional valence of the situation in which call production is elicited, and various acoustic call features, such as call duration and peak frequency. In infant rat pups, 40-kHz USV occur. Such 40-kHz USV are emitted in aversive situations, most notably social isolation, and serve an important communicative function in eliciting maternal caregiving behavior. They are mostly characterized by peak frequencies between 30 and 60 kHz and call durations of ~100 ms or higher.

In juvenile and adult rats, two main types of USV occur. Their occurrence mainly depends on the emotional valence of the situation (for a detailed overview, see Brudzynski 2013; Wöhr and Schwarting 2013). In aversive situations, 22-kHz USV are emitted (for an exemplary spectrogram, see Fig. 1). Emission is often restricted to a comparatively narrow peak frequency range between 18 and 24 kHz. They are further characterized by relatively long call durations of ~1000 ms, yet call duration is characterized by high variability. As compared to other USV types, however, 22-kHz USV are relatively stereotyped, with low variability in most other acoustic call features. Examples for situations in which 22-kHz USV are typically produced include inter-male aggression, predator exposure, and fear learning paradigms.

In appetitive situations, 50-kHz USV are uttered, which consist of many subtypes (for an exemplary spectrogram, see Fig. 1). The frequency range in which 50-kHz USV occur is very broad, ranging from 30 to 90 kHz with most of them occurring between 50 and 70 kHz. Call duration is comparatively short, and the

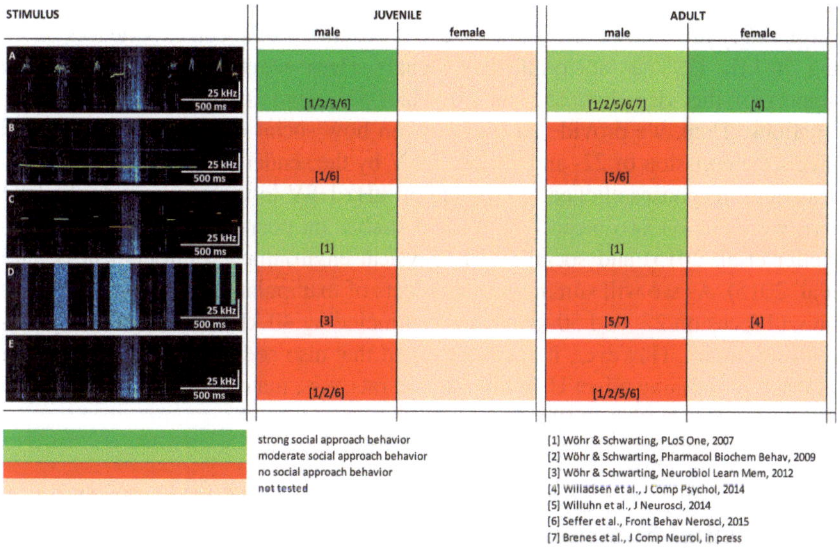

Fig. 1 Overview on the behavioral responses obtained in the 50-kHz USV radial maze playback paradigm [1/2/3/4/6/7] and in the Skinner box adaptation used for fast-scan cyclic voltammetry to measure dopamine release [5]. The following five acoustic stimuli were used: (*A*) natural 50-kHz USV (50-kHz USV), (*B*) natural 22-kHz USV (22-kHz USV), (*C*) artificial 50-kHz sine wave tones (SINE WAVE), (*D*) artificial time- and amplitude-matched white noise (WHITE NOISE), and (*E*) background noise (BACKGROUND NOISE). All stimuli were presented with a sampling rate of 192 kHz in 16-bit format at ~69 dB (measured from a distance of 40 cm), with the exception of background noise that was presented at ~50 dB (measured from a distance of 40 cm). Modified and updated version from Seffer et al. (2014)

great majority of 50-kHz USV are characterized by call durations clearly shorter than ~50 ms. In general, acoustic call features of 50-kHz USV are highly heterogeneous. This is reflected in a high level of variability in frequency modulation but also in the existence of various different call shapes. A number of methodological approaches were applied to define 50-kHz USV subtypes. Often, 50-kHz USV are categorized by means of call shape and complexity. The simplest approach is splitting 50-kHz USV into flat and frequency-modulated calls. Probably, the most commonly used approach is to further split frequency-modulated 50-kHz USV into frequency step calls and trills, resulting in three subtypes, namely flat, frequency step, and trill calls. However, there are much more complex classification systems with more than ten 50-kHz USV subtypes. The various different 50-kHz USV occur mainly during appetitive social interactions, such as rough-and-tumble play and mating. High rates of 50-kHz USV are also seen in response to drugs of abuse, such as amphetamine, and probably reflect the rewarding effects of the drugs.

Consistent with the emotional valence of the situation, 22-kHz USV are believed to serve an important communicative function as alarm calls, whereas 50-kHz USV appear to fulfill an affiliative communicative function as social contact calls. Surprisingly little, however, is known about the effects of social experiences on 22- and 50-kHz USV as socio-affective signals. Here, we first review evidence in support of the idea that 22- and 50-kHz USV serve important communicative functions. Then, we provide an overview on how social experiences or their lack affects the emission of 22- and 50-kHz USV by the sender, and how they affect the behavioral responses elicited by 22- and 50-kHz USV in the recipient. In the latter part, we focus on two recently published studies on post-weaning social isolation (Seffer et al. 2015) and social versus physical environmental enrichment (Brenes et al. 2015). As we will summarize the effects of manipulations following weaning, we will focus on 22- and 50-kHz USV, not including 40-kHz USV emitted by pups before weaning. However, it has to be noted that also social experiences and their lack before weaning affect USV emission in rats, e.g., maternal caregiving behavior (Wöhr and Schwarting 2008a).

3 Alarming Function of 22-kHz USV

While there are many hypotheses about the communicative function of 22-kHz USV, there is ample evidence supporting the ideas that 22-kHz USV serve as appeasement signals during inter-male fighting (Lore et al. 1976) and/or as alarm calls to warn conspecifics about danger (Blanchard et al. 1991). This latter idea was born out of a series of behavioral observations of a naturally established colony of rats living in a visible burrow system by Blanchard et al. (1991). When a rat living in the visible burrow system was exposed to a cat, emission of 22-kHz USV was facilitated by the presence of an audience, i.e., a group of familiar conspecifics living in the same environment. Importantly, 22-kHz USV emission by the sender then led to a rich set of defensive behaviors in the recipients, which had not been in contact with the cat themselves. Among others, 22-kHz USV emission by the sender induced 22-kHz USV in the recipients, indicating that the production of 22-kHz USV can be socially contagious.

The idea that 22-kHz USV serve an alarming function is further supported by a number of other studies showing that 22-kHz USV emission by the sender induces freezing behavior in the recipient, with some studies suggesting that 22-kHz USV emission is a key component of observational fear learning in rats. Importantly, however, the experiences recipients had made before exposure to 22-kHz USV appear to be highly relevant. For instance, it was shown that the production of 22-kHz USV by the sender leads to freezing behavior in recipients with prior aversive experience, i.e., which were exposed to a series of foot shocks the day before, but not in naïve recipients (Kim et al. 2010). In absence of 22-kHz USV, no freezing behavior in the recipients was seen even following aversive experiences. This indicates that 22-kHz USV emission leads to freezing behavior in experienced recipients, rather than sensitization. Consistently, disruption of the primary auditory pathway by lesioning the medial geniculate nucleus of the thalamus effectively blocked the freezing response in experienced recipients when exposed to 22-kHz USV. Other sensory stimuli, such as alarm pheromones, thus appear to be less relevant. The fact that rats displayed freezing behavior in response to 22-kHz USV only following aversive experiences indicates that the communicative significance of 22-kHz USV is learned. Since no other rat was present during the exposure to a series of foot shocks used to induce an aversive experience, its communicative significance is probably learned through auto-conditioning. In support of such an auto-conditioning hypothesis, it was found that recipients did not display freezing behavior in response to 22-kHz USV emitted by the sender following exposure to a series of foot shocks when their medial geniculate nucleus was temporally inactivated during the aversive stimulation. After such an inactivation of the primary auditory pathway, experienced recipients behaved like naïve rats, i.e., as if they were not exposed to the aversive stimulation. Together, this indicates that rats need to hear their own 22-kHz USV during aversive stimulation in order to form an association between 22-kHz USV and the aversive stimulation through auto-conditioning. Following auto-conditioning, recipients then display freezing behavior in response to a sender emitting 22-kHz USV.

The auto-conditioning hypothesis was later confirmed in other studies, e.g., in playback experiments (Parsana et al. 2012b). In such playback experiments, 22-kHz USV are presented through an ultrasonic loudspeaker, i.e., in the absence of a rat emitting 22-kHz USV. This allows to rule out other factors that typically accompany 22-kHz USV emission in rats, such as alarm pheromones. By this means, it was demonstrated that rats that had undergone an aversive experience show freezing behavior in response to playback of 22-kHz USV, while no freezing response was evident in rats that had not been exposed to such an aversive experience. Notably, the effects of an aversive experience were reported to be specific for 22-kHz USV because experienced rats did not respond to 50-kHz USV. Moreover, 22-kHz USV production during the aversive stimulation was positively correlated with and predicted subsequent freezing toward 22-kHz USV. It was therefore concluded that auto-conditioning "is sufficiently rapid, reliable, and stimulus-specific to serve an adaptive defensive function in rats" (Parsana et al. 2012b). The auto-conditioning phenomenon possibly explains some inconsistencies in the literature regarding the strength of the recipients' behavioral responses during playback of 22-kHz USV and shows that it is important to thoroughly control or experimentally manipulate the rats' affective experiences in studies focusing on the communicative functions of 22-kHz USV.

At the neuronal level, by means of immediate early gene activation patterns and electrophysiological recordings, it was shown that playback of 22-kHz USV results in increased neuronal activity in various brain areas, most notably amygdala and periaqueductal gray (Sadananda et al. 2008; Parsana et al. 2012a). Since both brain areas are strongly involved in emotion regulation and provide important components of the fear circuit, this activity pattern is in line with the idea that 22-kHz USV serve an alarming function.

4 Affiliative Function of 50-kHz USV

The idea that 50-kHz USV fulfill an affiliative communicative function as social contact calls is comparatively old, yet was almost exclusively studied in the sexual context, e.g., by means of devocalization and playback experiments. For instance, devocalization of males by cutting the laryngeal nerves was shown to result in a reduction in darting behavior displayed by females. They were also found more likely to move away while being mounted by a male unable to vocalize. Importantly, behavioral alterations were normalized via additional playback of tape-recorded male 50-kHz USV, and females were found to spend more time in front of an ultrasonic speaker emitting 50-kHz USV than in front of an inactive one (for a detailed overview, see Barfield and Thomas 1986).

More recently, however, strong evidence was provided that 50-kHz USV serve as social contact calls also during same-sex social interactions. For instance, juvenile rats emit high rates of 50-kHz USV during rough-and-tumble play (Knutson et al. 1998) and evidence was provided that such 50-kHz USV promote and maintain playful social interactions, while rough-and-tumble play behavior is reduced in pairs of devocalized rats (Kisko et al. 2015) and altered following deafening (Siviy and Panksepp 1987; for a detailed discussion, see Kisko et al. 2017). Moreover, it was found that rats spent more time with conspecifics emitting high rates of 50-kHz USV than with those emitting low levels (Panksepp et al. 2002). In an operant task, rats were found to even work for being exposed to playback of 50-kHz USV (Burgdorf et al. 2008). Also, 50-kHz USV production was shown to be driven by potential social contact and elicited by exposure to the odor of conspecifics, with 50-kHz USV rate being positively correlated with the number of rats that left their odor in the environment (Brudzynski and Pniak 2002). Likewise, separation from a conspecific results in a transient increase in 50-kHz

USV emission, most notably in flat calls, probably to reestablish social proximity (Wöhr et al. 2008). Beyond that, 50-kHz USV might further orchestrate complex social behaviors, such as cooperative behavior to obtain food rewards, as suggested by a positive correlation between cooperative actions and 50-kHz USV (Łopuch and Popik 2011). Finally, in a recent series of experiments employing a 50-kHz USV radial arm maze playback paradigm, it was consistently shown that 50-kHz USV induce social approach behavior, supporting the notion that they serve an affiliative function as social contact calls (Brenes et al. 2015; Seffer et al. 2015; Willadsen et al. 2014; Willuhn et al. 2014; Wöhr and Schwarting 2007, 2009, 2012). Social approach behavior is sometimes paralleled by USV emission in the sender, with both 22- and 50-kHz USV occurring (Willadsen et al. 2014; Willuhn et al. 2014; Wöhr and Schwarting 2007, 2009), the former possibly reflecting a state related to frustration, as highest levels were seen in rats characterized by strong social approach behavior (Wöhr and Schwarting 2009).

At the neuronal level, it was found by means of electrophysiological recordings and immunohistochemistry that playback of 50-kHz USV results in decreased firing rates and reduced c-fos expression in the amygdala as opposed to effects elicited by 22-kHz USV (Sadananda et al. 2008; Parsana et al. 2012a). This deactivation of the amygdala was paralleled by an activation of the nucleus accumbens, where immediate early gene expression was increased following 50-kHz USV playback (Sadananda et al. 2008). The nucleus accumbens is a key brain area implicated in reward processing, with dopamine being strongly involved. Consistently, fast-scan cyclic voltammetry recordings in the nucleus accumbens showed that dopamine is released during playback of 50-kHz USV but not in response to a number of acoustic control stimuli. Furthermore, dopamine release in the nucleus accumbens was positively correlated with social approach behavior in the Skinner box adaptation of the 50-kHz USV radial maze playback paradigm (Willuhn et al. 2014).

5 Effects of Social Experiences on Ultrasonic Communication: 22-kHz USV

While there are a large number of social experiences that likely affect ultrasonic communication in rats, post-weaning social isolation was most thoroughly studied. Long-lasting separation from conspecifics during adolescence is known to impair social behavior, and post-weaning social isolation in rats is a widely used animal model for neuropsychiatric disorders characterized by social and communication deficits, including neurodevelopmental disorders, such as autism and schizophrenia, but also affective disorders, most notably depression (Fone and Porkess 2008; Lapiz et al. 2003). Unfortunately, experimental protocols vary considerably with respect to time point of social isolation and its duration. It is likely that this variability at least partially explains some of the obtained inconsistencies.

In the seminal study by Blanchard et al. (1991) on the role of an audience for 22-kHz USV emission in response to a cat, which led to the idea that 22-kHz USV serve an alarming function, it was reported that the emission of 22-kHz USV is facilitated by the presence of a group of familiar conspecifics living in the same environment. In line with that view, no 22-kHz USV were detected when a rat was individually exposed to a cat. However, rats individually exposed to a cat were single-housed before testing. It appears therefore possible that social isolation before testing but not the absence of an audience during testing led to the lack of 22-kHz USV in response to a stressor. In fact, no evidence for an audience effect was reported by a more recent study (Wöhr and Schwarting 2008b). In this study, rats housed in groups were exposed to a stressor either alone or in the presence of a conspecific. Most of the rats vocalized, irrespective of the presence or absence of an audience.

In one of the earliest studies on the effects of social isolation on the emission of 22-kHz USV, which systematically compared single- and group-housed rats, Brudzynski and Ociepa (1992) exposed young adult male rats to gentle touch by a human hand in an unfamiliar environment. They reported that the number of rats uttering 22-kHz USV in response to gentle touch was markedly increased following social isolation for two weeks. Specifically, while only 60 % of the group-housed rats vocalized in response to the first gentle touch, 22-kHz USV were detected in 100 % of the single-housed rats. With repeated exposure to gentle touch over sessions, 22-kHz USV emission was found to be reduced to about 20 % in both groups and no group differences were detected anymore. Brudzynski and Ociepa (1992) hypothesized that "distress induced by isolation and hand touch were additive," with 22-kHz USV emission being caused by "potential danger or threat to the animal rather than by physical features" of the hand touch.

Inagaki et al. (2005) examined the effects of post-weaning social isolation during early development. They measured 22-kHz USV emission in male rats isolated on postnatal day (PND) 21 for a period for six months. Inagaki et al. (2005) found that rats exposed to post-weaning social isolation spent less time emitting 22-kHz USV in response to being gently nipped on the back with forceps than group-housed controls. While group-housed rats emitted a substantial amount of 22-kHz USV, rats exposed to social isolation from weaning on barely vocalized. Interestingly, freezing behavior and defecation were not affected by the experimental housing conditions, indicating that the stimulation with forceps was equally stressful in both conditions in terms of this classical visible measure. Of note, besides the marked difference between single- and group-housed rats, evidence for reduced 22-kHz USV emission in dominant as compared to subdominant rats was provided, further supporting the idea that social experiences have a prominent impact on 22-kHz USV emission. Inagaki et al. (2005) suggested that 22-kHz USV "developed through social interaction after weaning." As 22-kHz USV might serve to inhibit vigorous aggression (Lore et al. 1976) and the stimulation with forceps mimicked an attack of an opponent, they further speculated that the socially reared rats learned to emit 22-kHz USV as a component of the "submissive behavior repertoire during social interactions in pair housing" through conditioning. In a subsequent study,

Inagaki and Mori (2013) used air-puffs to induce 22-kHz USV and showed that male rats exposed to post-weaning social isolation vocalize less than group-housed controls after 16 weeks of isolation. Shorter periods of isolation of four, eight, and twelve weeks were less efficient in inducing a reduction in 22-kHz USV emission.

Similar findings were reported by Pohorecky (2008), also eliciting 22-kHz USV by means of air-puffs. Pohorecky (2008) found that social isolation for four weeks in adult male rats results in a prominent reduction in 22-kHz USV emission. This finding was replicated in an independent group of rats exposed to social isolation for eight weeks. While 22-kHz USV emission was reduced, anxiety-related behavior, as assessed on the elevated plus maze, was found to be enhanced. This shows that two important components of the rat's defensive behavior repertoire, namely 22-kHz USV and anxiety-related behavior, are differentially regulated by social isolation and that heightened anxiety not necessarily results in enhanced 22-kHz USV emission. Again, as suggested by the findings obtained by Inagaki et al. (2005) and Inagaki and Mori (2013), 22-kHz USV production appears to depend on social interactions in the weeks before testing. Also in line with the study by Inagaki et al. (2005), Pohorecky (2008) showed that dominant rats vocalize less than rats with a lower social rank. Specifically, highest 22-kHz USV emission rates were seen in subdominant rats, with subordinate rats being characterized by an intermediate 22-kHz USV level.

However, little is known about the effects of social isolation on the emission of 22-kHz USV in response to natural threats or during aggressive encounters. Von Frijtag et al. (2002) examined 22-kHz USV emission during aggressive encounters in male rats, which were isolated on PND21 for two weeks, and kept group-housed thereafter. In response to a periodic and unpredictable exposure to an aggressive resident, previously isolated rats responded with increased 22-kHz USV production, but suffered from more aggressive attacks, as compared to group-housed controls never exposed to social isolation. Therefore, Von Frijtag et al. (2002) suggested that 22-kHz USV production was dissociated from the sender's behavioral pattern and thus lost its aggression-reducing communicative function. In line with that idea, it was reported that previously isolated rats do not inhibit ongoing nonsocial and social activities in the presence of the aggressive resident.

Of note, there are also studies reporting social isolation effects on 22-kHz USV in response to mild stressors, such as novelty exposure and in the presence of an animal caretaker. For instance, Nunes Mamede Rosa et al. (2005), Tomazini et al. (2006), and Bassi et al. (2007) all reported reduced 22-kHz USV emission during novelty exposure following social isolation for ten or fourteen days in juvenile rats. Deficits were not rescued by one additional week of group housing (Bassi et al. 2007; Nunes Mamede Rosa et al. 2005). Social isolation for one day led to the opposite result pattern characterized by increased 22-kHz USV emission (Bassi et al. 2007; Tomazini et al. 2006). In these studies, on average, about 100 22-kHz USV were counted and the total time spent calling was reported to be around 10 s in group-housed controls. This means that on average, 22-kHz USV had call durations of $\sim$100 ms. Importantly, however, this appears very unlikely as 22-kHz USV are characterized by relatively long call durations of $\sim$1000 ms, indicating that the sounds recorded were possibly artifacts or very short and thus atypical 22-kHz USV. Likewise, Wen and Xu (2010) reported that social isolation in adult male rats for one day results in a reduction in the emission of 22-kHz USV during novelty exposure and that this reduction is even more prominent following two or seven days of social isolation. They further reported that prior sexual experience partially prevented this effect. In their study, Wen and Xu (2010) counted between 1000 and 1500 22-kHz USV in 5 min, meaning that three to five 22-kHz USV were emitted in one second, not including inter-call intervals. Again, this indicates that the sounds recorded were possibly artifacts or very short and thus atypical 22-kHz USV. In all four studies, no spectrographic information was provided. Moreover, Cloutier et al. (2013), using broadband high-frequency recordings, did not obtain evidence for differences in the emission of 22-kHz USV evoked by the presence of an animal caretaker between rats isolated on PND21 for four weeks and socially reared controls. Consistent with the fact that the presence of an animal caretaker is only a mild stressor, the numbers of 22-kHz USV emitted were relatively low.

Together, these studies show that social experiences or their lack, such as social isolation, can have a profound impact on the emission of 22-kHz USV by the sender in later life, albeit direction and strength of observed effects vary between studies and experimental protocols, with time point of occurrence and duration being critical determinants. Little is known about how social experiences affect the behavioral responses evoked by 22-kHz USV in the recipient.

6 Effects of Social Experiences on Ultrasonic Communication: 50-kHz USV

Already in the first study showing that rats emit 50-kHz USV as juveniles during rough-and-tumble play, it was expected that prior isolation for a few days might enhance play behavior and thus concomitant 50-kHz USV emission (Knutson et al. 1998). In this study, Knutson et al. (1998) weaned and individually housed rats at PND18 and compared them with rats that lived together with their mothers and littermates until testing at PND33. They found that rats exposed to social isolation emitted more than twice as many 50-kHz USV and that they played more vigorously than group-housed controls, possibly due to an increase in social motivation. This finding was shortly later confirmed in an extensive series of experiments conducted by Panksepp and Burgdorf (2000) in which rough-and-tumble play was mimicked through tickling by an experienced human experimenter. In their first experiment, Panksepp and Burgdorf (2000) measured tickling-induced 50-kHz USV in rats exposed to post-weaning social isolation from PND21. When being tickled at PND50, rats housed in isolation vocalized much more than group-housed controls. Importantly, when experimental housing conditions were switched a day after the second tickling exposure, 50-kHz USV emission patterns were reversed. Specifically, previously isolated rats almost completely stopped emitting 50-kHz USV, whereas previously group-housed rats suddenly emitted high rates of 50-kHz

USV comparable to the rats exposed to social isolation from weaning on. When switching back, group differences in 50-kHz USV emission rates were similar to the ones seen at PND50 at the start of the experiment. In subsequent experiments conducted at PND60, this difference in 50-kHz USV production was paralleled by elevated levels of instrumental approach for eliciting tickling and contextual conditioning of anticipatory 50-kHz USV to the environment in which they had been exposed to the tickling stimulation. In addition, rats exposed to post-weaning social isolation emitted more 50-kHz USV during the presentation of a conditioned stimulus associated with tickling through classical conditioning at PND30. During extinction, emission of 50-kHz USV gradually increased with extinction days, with rats exposed to social isolation consistently vocalizing more than group-housed controls. Also, play bites soliciting the tickling stimulation were more prominent in socially isolated rats and increased with extinction days, probably reflecting their enhanced need to re-establish social contact. Panksepp and Burgdorf (2000) emphasize that from a traditional conditioning theory, one might have expected that rats would have exhibited a rapid decline—and not an increase—in 50-kHz USV and play bites once the tickling stimulation was no longer available and state: "We take this to reflect the social communicative (i.e., 'come on and play') component of the vocalization." The finding that socially isolated rats vocalized more in response to and in anticipation of tickling than group-housed controls was confirmed by Burgdorf and Panksepp (2001) in an independent experiment. In this study, they further showed that post-weaning social isolation from PND21 modulated the effects of naloxone treatment on tickling-induced 50-kHz USV assessed about two weeks later. Naloxone is an opioid antagonist, and the opioid system is known to regulate social behavior. While naloxone reduced 50-kHz USV emission during tickling in socially isolated rats, it had opposite effects in group-housed rats and led to increased 50-kHz USV levels, particularly during repeated tickling exposures. Of note, increased 50-kHz USV emission during tickling following post-weaning social isolation was more recently confirmed by Burgdorf et al. (2008) and it was shown that particularly frequency-modulated but not flat calls were elevated by social isolation. However, no clear effects of post-weaning social isolation from PND21 for four weeks on 50-kHz USV in anticipation of tickling were found by Cloutier et al. (2013).

Effects of social isolation in adulthood were studied by Hamed et al. (2009; 2015), who housed adult male rats individually for three weeks and measured 50-kHz USV during social interactions. Pairs of rats exposed to social isolation or group-housed controls were compared. Hamed et al. (2009) showed that social isolation leads to an increase in 50-kHz USV emission and that this increase was more prominent in rats exposed to social isolation treated with the benzodiazepine diazepam but not the partial serotonin receptor 5-HT1A agonist buspirone. In a subsequent study, Hamed et al. (2015) confirmed enhanced 50-kHz USV emission in pairs of adult male rats exposed to social isolation for three weeks and showed that low doses of morphine, a μ-opioid agonist, reduces the augmentation in 50-kHz USV emission seen following social isolation. Also naltrexone, a μ- and κ-opioid antagonist, and U-50488, a κ-opioid agonist, reduced 50-kHz USV, while nor-binaltorphimine, a κ-opioid antagonist, further enhanced 50-kHz USV levels. The reduction seen in response to morphine was reversed by treatment with naltrexone and nor-binaltorphimine. In both studies, however, pharmacological manipulations were conducted in rats exposed to social isolation only. It is thus difficult to interpret the data, particularly as it is not clear whether the observed changes depend on social experience or not, i.e., simply reflect general drug effects on 50-kHz USV emission. Furthermore, Hamed et al. (2015) showed that social isolation leads to increased levels of dopamine and serotonin, together with its metabolites, in the nucleus accumbens immediately following social interactions. Related changes were seen in the ventral tegmental area. However, it is not clear whether these changes reflect long-term alterations in dopaminergic and serotoninergic neurotransmission or acute effects, since no controls not exposed to social interactions were included in the study.

Emission of 50-kHz USV during novelty exposure in adult male rats housed in social isolation for one, two, or seven days was measured by Wen and Xu (2010). They reported reduced 50-kHz USV emission following social isolation and that prior sexual experience partially prevented this effect. However, Cloutier et al. (2013) reported very low baseline 50-kHz USV levels, and no effect of post-weaning social isolation was detected.

Effects of social isolation after weaning and during adulthood on 50-kHz USV emission were compared by Willey and Spear (2013). They isolated rats on PND21 or PND63/64 for eleven/twelve days and assessed 50-kHz USV in a reward approach task toward either a food or a social stimulus, i.e., a confined conspecific. Rats exposed to social isolation emitted more 50-kHz USV in response to the social stimulus than group-housed controls in both age groups, with rats exposed to social isolation during adulthood emitting the highest 50-kHz USV rates. This response pattern was not seen when rats were exposed to the food reward, indicating that social isolation specifically affected the 50-kHz USV response toward a social stimulus. Notably, frequency-modulated calls were positively correlated with the time spent investigating the social stimulus but not the food reward. It has to be emphasized that social investigation behavior was not altered following social isolation, meaning that social isolation effects were detectable at the level of 50-kHz USV emission but not in visible behavioral responses.

While most studies focused on the effects of comparatively short-lasting social isolation exposures, long-term isolation effects were rarely assessed. In a recent study, however, Inagaki et al. (2013) demonstrated that rats isolated on PND21 for six months utter clearly less 50-kHz USV in response to both an anesthetized sexually receptive female and a non-receptive female as compared to group-housed controls. Whereas rats exposed to long-term post-weaning social isolation did not respond to their reproductive state and emitted similarly low amounts of 50-kHz USV in response to both stimuli, socially reared controls showed increased 50-kHz USV production during the presence of the receptive female as compared to the non-receptive female. Interestingly, both flat and frequency-modulated calls were reduced following long-term post-weaning social isolation. As flat calls have been more closely associated with an affiliative communicative function, while frequency-modulated calls appear to be more closely linked to reward and positive affect (Burgdorf et al. 2008; Wöhr et al. 2008), this indicates that long-term post-weaning social isolation results in social communication deficits possibly due to the rats' inability to recognize relevant appetitive social cues.

Together, several studies suggest that shorter periods of social isolation reinforce the incentive salience of social interaction which is then reflected in increased emission of 50-kHz USV during social encounters or simulated playful social interactions through tickling, irrespective of the time point of social isolation. Importantly, however, its duration appears to be critical in determining the outcome. Short-term social isolation was consistently shown to enhance 50-kHz USV emission, possibly due to an increase in social motivation. Conversely, long-term social isolation results in the opposite pattern. As there are no studies available on whether long-term isolation results in reduced 50-kHz USV emission irrespective of time point, the underlying mechanisms are unknown. It appears possible, however, that long-term post-weaning social isolation has more prominent effects than long-term isolation in adulthood, given that rats emit particularly high rates of 50-kHz USV in the rough-and-tumble play period following weaning (Knutson et al. 1998). As for 22-kHz USV, little is known about how social experiences affect the behavioral responses evoked by 50-kHz USV in the recipient.

7 The 50-kHz USV Radial Maze Playback Paradigm

A few years ago, we developed a radial maze playback paradigm to study the communicative functions of 50-kHz USV by assessing behavioral responses in the recipients. In this newly developed paradigm, a given rat is exposed to playback of 50-kHz USV and appropriate acoustic control stimuli in a counter-balanced manner on an elevated radial arm maze. For behavioral assessment, the radial arm maze is divided into three areas, namely proximal arms, i.e., the three arms close to the active ultrasonic speaker used for playback of 50-kHz USV, distal arms, i.e., the three arms opposite to the active ultrasonic speaker, and two neutral arms. Social approach behavior is assessed by comparing entries into proximal and distal arms and the time spent thereon. In addition, social exploratory behavior is assessed using locomotor activity during playback of 50-kHz USV.

Using this paradigm, we have repeatedly and consistently shown that 50-kHz USV lead to social approach behavior in juvenile and adult male rats (Brenes et al. 2015; Seffer et al. 2015; Willuhn et al. 2014; Wöhr and Schwarting 2007, 2009, 2012) as well as female rats (Willadsen et al. 2014). In addition, we have shown in these studies that the behavioral response pattern elicited in the recipients depends on certain acoustic features of 50-kHz USV by comparing it with the response patterns seen during the exposure to 22-kHz USV, 50-kHz sine wave tones, time- and amplitude-matched white noise, and background noise (Fig. 1). No evidence for social approach behavior was observed in response to 22-kHz USV, time- and amplitude-matched white noise, and background noise. Social approach was only seen in rats exposed to 50-kHz USV and 50-kHz USV sine wave tones. Specifically, juvenile male rats displayed strong social approach behavior in response to 50-kHz USV. A social approach response toward 50-kHz sine wave tones was also evident but less prominent. In adult male rats, weaker responses occurred and only moderate social approach behavior was seen in response to both 50-kHz USV and 50-kHz sine wave tones, whereas strong social approach behavior was seen in adult female rats exposed to 50-kHz USV. The effects of 50-kHz sine wave tones were not yet assessed in adult female rats. Similar levels of social approach behavior seen in response to playback of both 50-kHz USV and 50-kHz sine wave tones indicate that frequency modulation of 50-kHz USV is not needed to elicit social approach behavior. This is in line with the idea that flat calls mainly serve as social contact calls, as suggested by the transient increase in 50-kHz USV emission typically seen following separation from a conspecific (Wöhr et al. 2008), which is believed to reestablish social proximity under natural conditions. The 50-kHz sine wave tones used were identical to 50-kHz USV with respect to peak frequency, call length, and temporal patterning (for a detailed overview, see Seffer et al. 2014).

We also used this newly developed paradigm to study neurobiological processes involved in rodent acoustic communication by systematically manipulating relevant characteristics of the recipient. For instance, we showed that systemic application of low doses of the μ-opioid agonist morphine leads to enhanced social approach behavior in both juvenile and adult male rats, while the μ-opioid antagonist naloxone had opposite effects (Wöhr and Schwarting 2009). We further showed that memory consolidation of the 50-kHz USV experience depends on an intact acetylcholine system since it is blocked by the amnestic agent scopolamine, a muscarinic acetylcholine receptor antagonist (Wöhr and Schwarting 2012). Another neurotransmitter that appears to be strongly involved is dopamine, as indicated by our fast-scan cyclic voltammetry study (Willuhn et al. 2014). Very recently, we focused on the impact of social experiences on social approach behavior as assessed in the 50-kHz radial maze playback paradigm and studied the effects of post-weaning social isolation (Seffer et al. 2015) and social versus physical environmental enrichment (Brenes et al. 2015).

8 Effects of Social Experiences on Ultrasonic Communication in the 50-kHz USV Radial Maze Playback Paradigm

By means of our 50-kHz USV radial maze playback paradigm, we recently showed that not only 50-kHz USV production but also the behavioral response elicited in the recipient is affected by post-weaning social isolation (Seffer et al. 2015; Fig. 2). In this study, we compared three experimental housing conditions, namely group housing, short-term social isolation for one day, and long-term social isolation for four weeks (for an overview of the experimental design, see upper part of Fig. 2).

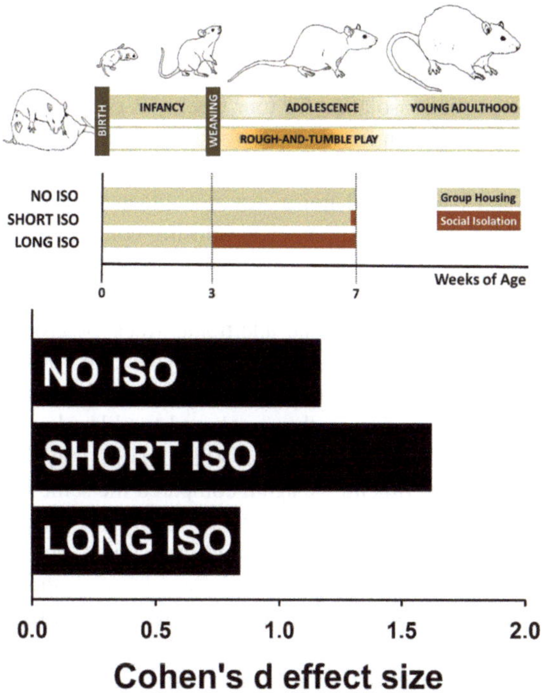

Fig. 2 Experimental design used to assess the impact of post-weaning social isolation on social approach behavior induced by pro-social 50-kHz USV (*left*). Rats (n = 12, per condition) were exposed to one of the following three experimental housing conditions for 4 weeks: (I) group housing (NO ISO), with rats being housed in groups of six; (II) short-term social isolation (SHORT ISO), with rats housed in groups of six, but isolated 24 h prior testing; or (III) long-term social isolation (LONG ISO), with rats being isolated for 28 days prior testing. Rats were exposed to the experimental housing conditions as weanlings at about 3 weeks of age. At about 7 weeks of age, they were exposed to playback of 50-kHz USV. Effect sizes (Cohen's d) are used as a quantitative measure of the strength of social approach behavior (time spent proximal to versus time spent distal from the active ultrasonic speaker) induced by pro-social 50-kHz USV (*right*). Often effect sizes of up to 0.3 are considered to represent "small" effects, effect sizes of up to 0.6 "medium" effects, and effect sizes higher than 0.9 "large" effects. Data from Seffer et al. (2015); modified graph

We then assessed the rats' behavioral responses during playback of 50-kHz USV. As acoustic controls, we used 22-kHz USV and background noise.

In the first experiment, we studied the effects of post-weaning social isolation and found that rats exposed to four weeks of isolation during the rough-and-tumble play period did not display social approach behavior toward 50-kHz USV. They even showed some signs of social avoidance and spent less time in proximity to the ultrasonic speaker during playback than before playback, i.e., they moved away from the ultrasonic speaker. In stark contrast, group-housed rats not exposed to post-weaning social isolation displayed social approach behavior, in line with previous studies. Social approach behavior in short-term isolated rats was even more prominent than in group-housed rats, probably because of an increase of social motivation due to social deprivation. This result pattern is also reflected in a reduced effect size as a quantitative measure of the strength of a phenomenon in rats exposed to long-term social isolation for four weeks, as compared to group housing and short-term social isolation for one day (for an overview of the effect sizes, see lower part of Fig. 2). Importantly, the behavioral response pattern during exposure of 22-kHz USV and background noise was not affected by post-weaning social isolation, with rats of all three experimental housing conditions displaying behavioral inhibition.

In the second experiment, we compared the same three experimental housing conditions, but exposed all rats to one additional week of group housing before assessing their responses toward 50-kHz USV playback. We found that one additional week of group housing is sufficient for reversing the social deficits induced by four weeks of isolation during the rough-and-tumble play period. All three groups displayed approach behavior toward 50-kHz USV playback.

Finally, in the third experiment, we again compared the same three experimental housing conditions, but housing conditions were implemented after the rough-and-tumble play period. As in the second experiment, no evidence for social deficits was obtained and all three groups displayed social approach behavior in response to playback of 50-kHz USV. Together, the three experiments show that post-weaning but not post-adolescent social isolation for four weeks leads to social deficits which can be rescued by resocialization. Although little is still known about the relevance of the duration of social isolation in modulating this effect, this series of experiments highlights the importance of social experience for affiliative behavior and suggests a critical involvement of the rough-and-tumble play period for responding appropriately to 50-kHz USV by displaying social approach behavior. It appears therefore possible that the lack of social approach behavior induced by post-weaning social isolation is due to deficits in interpreting 50-kHz USV as pro-social contact calls.

While the underlying neurobiological mechanisms are currently unknown, the fact that one week of group housing reverses social deficits seen following post-weaning social isolation indicates the involvement of a neurobiological mechanism that responds relatively fast to environmental changes. In line with that idea, we found that post-weaning social isolation leads to changes in microRNA expression regulating neuronal plasticity processes (Valluy et al. 2015). microRNAs are a large group of small noncoding RNAs known to act as important posttranscriptional regulators of gene expression in neurons. Specifically, we found that post-weaning social isolation results in an upregulation of the plasticity-regulating microRNA 134. In addition, Ube3a1, an alternative transcript of Ube3a, was upregulated, whose coding-independent function depends on an intact microRNA 134 pathway. Deletions or inactivation of *UBE3A* through genomic imprinting causes Angelman or Prader-Willi syndrome in humans, both characterized by a marked delay in development, often severe intellectual and/or learning disability, speech delay and/or impairment, and altered social behavior. Importantly, the social communication deficit observed in rats exposed to post-weaning social isolation mimics some of the symptoms (Seffer et al. 2015). Besides the lack of social approach behavior in response to 50-kHz USV, we found a severe impairment in novel object recognition reflecting memory deficits, which further supports the overlap in symptoms (Valluy et al. 2015). The two studies therefore suggest that our 50-kHz USV radial maze playback paradigm can help to detect social communication deficits in rat models with relevance to neurodevelopmental disorders, such as autism and related disorders (Silverman et al. 2010; Wöhr and Scattoni 2013). This will allow to test the hypotheses about neurodevelopmental disorders in rat models with face validity and help to identify relevant neurobiological mechanisms.

9 Modeling the Intense World Syndrome in Rats: Effects of Environmental Enrichment

A prominent hypothesis about neurodevelopmental disorders in general and autism in particular is the intense world syndrome/theory of autism by Markram and Markram (2010; see also: Markram et al. 2007), who hypothesized that the core neurophysiological pathology of autism is "excessive neuronal information processing and storage in local circuits of the brain, which gives rise to hyper-functioning of brain regions most affected. Such hyper-functioning in different brain regions is proposed to cause hyper-perception, hyper-attention, and hyper-memory that could potentially explain the full spectrum of symptoms in autism" (Markram et al. 2007). The intense world syndrome/theory of autism is grounded in experiments using valproate acid (VPA), an anticonvulsant and mood stabilizer, and the observation that VPA-treated rats exhibit amplified fear processing and memories. In humans, VPA intake during pregnancy has been associated with an increased risk of giving birth to a child with autism. Consistent with such an association, VPA was found to result in an autism-related behavioral phenotype in rats characterized by social and communication deficits, together with repetitive patterns of behavior (Schneider et al. 2006). Markram and Markram (2010) showed that VPA treatment in rats further leads to enhanced neuronal plasticity and reactivity. It was suggested that such enhanced neuronal processing in local neural microcircuits impairs long distance information processing between brain areas and thus leads to "obsessively detailed information processing of fragments of the world" (Markram and Markram 2010), which then results in social and communication deficits due to the complexity of social stimuli.

We hypothesized that besides VPA treatment, certain forms of environmental enrichment may allow modeling the intense world syndrome/theory of autism in rats (Brenes et al. 2015; Fig. 3): "Exposure to unpredictable environments and over-stimulation could, in fact, accelerate the development of autistic symptoms" (Favre et al. 2015). It is well known that environmental enrichment induces a hyperplastic state of the brain characterized by an increased level of adult hippocampal neurogenesis and enhanced neuronal connectivity due to dendritic and synaptic growth. Such neurobiological changes underlie the improved learning and

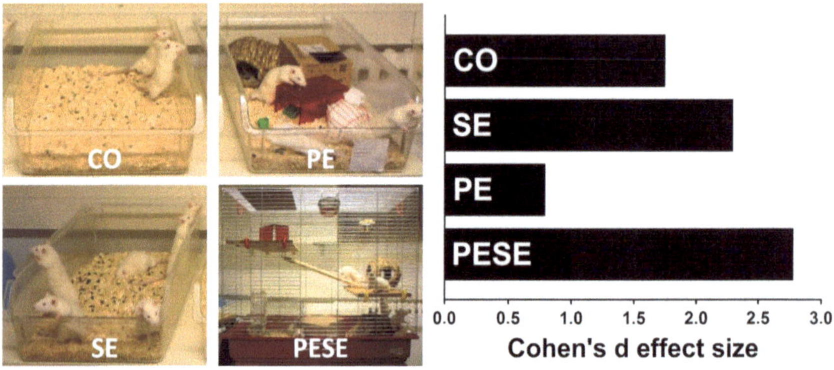

Fig. 3 Experimental design used to assess the impact of environmental enrichment on social approach behavior induced by pro-social 50-kHz USV (*left*). Rats ($n = 12$, per condition) were exposed to one of the following four experimental housing conditions for 5 weeks: (I) standard control (CO), with 2 rats in a non-enriched cage; (II) social enrichment (SE), with 6 rats in a non-enriched cage; (III) physical enrichment (PE), with 2 rats in an enriched cage; and (IV) physical plus social enrichment (PESE), with 6 rats in an enriched cage. Rats were exposed to the experimental housing conditions as weanlings at about 4 weeks of age. At about 9 weeks of age, they were exposed to playback of 50-kHz USV. Effect sizes (Cohen's d) are used as a quantitative measure of the strength of social approach behavior (time spent proximal to versus time spent distal from the active ultrasonic speaker) induced by pro-social 50-kHz USV (*right*). Often effect sizes of up to 0.3 are considered to represent "small" effects, effect sizes of up to 0.6 "medium" effects, and effect sizes higher than 0.9 "large" effects. Data from Brenes et al. (2015); modified graph

memory abilities, which might be viewed as "hyper-memory," typically seen in rats housed in environmental enrichment. However, it has to be noted that environmental enrichment is not well standardized experimentally and the various forms how environmental enrichment is studied in the laboratory substantially differ between studies. We compared four experimental housing conditions by using a 2×2 design with the factors social (2 vs. 6 rats per cage) and physical (enriched vs. non-enriched cages) environmental enrichment, with enrichment items being moved and replaced in an unpredictable manner (Brenes et al. 2015). The resulting four experimental housing conditions were standard control (CO: 2 rats in a non-enriched cage), social enrichment (SE: 6 rats in a non-enriched cage), physical enrichment (PE: 2 rats in an enriched cage), and physical plus social enrichment (PESE: 6 rats in an enriched cage; for an overview of the experimental design, see left part of Fig. 3). In line with the intense world syndrome/theory of autism, we found that physical environmental enrichment results in "hyperplasticity," as reflected in increased levels of adult hippocampal neurogenesis and the microRNA 132, a positive regulator of synaptic plasticity. In addition, evidence for "hyper-reactivity" was provided by means of increased c-fos and CREB1 expression levels. As expected, "hyperplasticity" and "hyper-reactivity" were accompanied by improved learning and memory, which might be viewed as "hyper-memory." Rats exposed to physical environmental enrichment displayed improved learning and memory during repeated open field exposures and the object recognition test. Most importantly, we further found that our form of unpredictable physical environmental enrichment providing minimal opportunities for social interactions leads to reduced social functioning, both at the level of sender and recipient. At the level of the sender, we found that this form of physical environmental enrichment leads to a profound reduction in the emission of 50-kHz USV. In particular, flat 50-kHz USV, most closely associated with an affiliative communicative function (Burgdorf et al. 2008; Wöhr et al. 2008), were reduced. By virtue of our 50-kHz USV radial maze playback paradigm, we further showed that this form of physical environmental enrichment also results in social deficits in the recipient. Unlike controls, rats exposed to physical environmental enrichment did not display social exploratory behavior when exposed to 50-kHz USV. They also did not show a preference for the speaker used for presenting 50-kHz USV. Such a preference was clearly evident in controls. This means that no evidence for social approach behavior in response to 50-kHz USV was obtained in rats housed under this certain type of environmental enrichment. Because rats exposed to physical environmental enrichment did not differ from controls in their behavioral response during exposure to an acoustic control stimulus, time- and amplitude-matched white noise, this indicates deficits in social exploratory and approach behavior following physical environmental enrichment despite intact acoustic information processing. Of particular relevance for treatment studies is our finding that deficits in social exploration and approach induced by exposure to physical environmental enrichment were prevented when exposing rats to social environmental enrichment in addition to physical environmental enrichment. This suggests that enhanced affiliative behavior provided by rearing rats in larger groups has positive effects on the rats' social development that counteract or compensate for deficits induced by physical environmental enrichment. Such result pattern is reflected in a clearly lower effect size in rats exposed to physical environmental enrichment as compared to controls, whereas the effects size in rats exposed to physical plus social environmental enrichment is similar to controls (for an overview of the effect sizes, see right part of Fig. 3). When comparing the effect sizes of rats exposed to long-term social isolation (Fig. 2) and physical environmental enrichment (Fig. 3), it becomes clear that the detrimental effects of our form of physical environmental enrichment were as strong as the ones seen following long-term social isolation. In our view, this is a remarkable finding as physical environmental enrichment is generally considered to be beneficial and to ameliorate deleterious effects of aversive environmental experiences, brain dysfunctions, and genetic alterations. Our findings of reduced social and communicative behavior suggest that the effects of physical environmental enrichment are not uniformly positive.

While our findings are in line with the intense world syndrome/theory of autism by Markram and Markram (2010; see also: Markram et al. 2007), they might appear surprising since environmental enrichment has successfully been used to reverse social deficits in animal models of autism, e.g., in the VPA rat model (Schneider et al. 2006). Schneider et al. (2006) even proposed environmental enrichment as an "important tool for the treatment of autism." Our results, however, suggest the exact opposite, namely that exposure to physical environmental enrichment can result in social deficits under certain circumstances. A likely explanation for this seeming discrepancy is that in the study by Schneider et al. (2006), reversal of VPA-induced autism-related behavioral deficits was obtained by exposing rats to physical and social environmental enrichment, and not just physical environmental enrichment. In fact, Schneider et al. (2006) reported that this combined approach results in more rough-and-tumble play behavior in juveniles and higher levels of social exploratory behavior in adulthood, suggesting that the combination of both factors has positive effects on social behavior. Moreover, other characteristics of the enriched environment also appear to be important. For instance, Favre et al. (2015) showed that a predictable enriched environment prevents the development of "hyper-emotionality" in VPA-treated rats—an effect that is not seen in an unpredictable enriched environment. In the predictable enriched environment, a constant set of enrichment items was used, whereas unpredictable enriched environment was defined by twice weekly exchange in part of stimuli identities, similar to our study (Brenes et al. 2015). Finally, in a study using an experimental design similar to ours, reduced sexual behavior was detected following physical environmental enrichment (Gruendel and Arnold 1974). Specifically, in this study, two factors were experimentally varied, namely social isolation and physical environmental enrichment, resulting in four experimental housing conditions: social isolation in a standard cage, social isolation in an enriched environment, group housing in a standard cage, and group housing in an enriched environment. As expected, rats exposed to social isolation displayed less sexual behavior. Likewise, rats exposed to environmental enrichment also displayed a marked reduction in sexual activity, with mounting and copulatory behavior being strongly reduced. Also, ejaculations were less frequent. The reduction of sexual behavior was very prominent, and in rats exposed to social isolation in an enriched environment sexual behavior was virtually absent and no ejaculations occurred. Of note, negative effects of physical environmental enrichment were also seen in fear learning and extinction paradigms, with rats exposed to physical environmental enrichment being characterized by impaired fear extinction (Hunter 2015).

10 Conclusion

Together, this evidence indicates that social experiences can affect socio-affective communication in rats, both at the level of sender and recipient. Deficits seen following post-weaning social isolation or physical environmental enrichment might be useful to model aspects of neurodevelopmental disorders characterized by social and communication deficits, such as autism and schizophrenia. For instance, exposure to physical environmental enrichment might be used to test hypotheses grounded in the intense world syndrome/theory of autism.

Acknowledgments The authors wish to thank J.C. Brenes for the pictures depicted in Fig. 3. M. W. is supported by the Deutsche Forschungsgemeinschaft (DFG WO 1732/4-1).

References

Barfield RJ, Thomas DA (1986) The role of ultrasonic vocalizations in the regulation of reproduction in rats. Ann N Y Acad Sci 474:33–43

Bassi SG, Nobre NJ, Carvalho MC, Brandao ML (2007) Substance P injected into the dorsal periaqueductal gray causes anxiogenic effects similar to long-term isolation as assessed by ultrasound vocalizations measurements. Behav Brain Res 182:301–307

Blanchard RJ, Blanchard DC, Agullana R, Weiss SM (1991) Twenty-two kHz alarm cries to presentation of a predator, by laboratory rats living in visible burrow systems. Physiol Behav 50:967–972

Brenes JC, Lackinger M, Höglinger GU, Schratt G, Schwarting RKW, Wöhr M (2015) Differential effects of social and physical environmental enrichment on brain plasticity, cognition, and ultrasonic communication in rats. J Comp Neurol (in press)

Brudzynski SM (2013) Ethotransmission: communication of emotional states through ultrasonic vocalization in rats. Curr Opin Neurobiol 23:310–317

Brudzynski SM, Ociepa D (1992) Ultrasonic vocalization of laboratory rats in response to handling and touch. Physiol Behav 52:655–660

Brudzynski SM, Pniak A (2002) Social contacts and production of 50-kHz short ultrasonic calls in adult rats. J Comp Psychol 116:73–82

Burgdorf J, Panksepp J (2001) Tickling induces reward in adolescent rats. Physiol Behav 72:167–173

Burgdorf J, Kroes RA, Moskal JR, Pfaus JG, Brudzynski SM, Panksepp J (2008) Ultrasonic vocalizations of rats (*Rattus norvegicus*) during mating, play, and aggression: behavioral concomitants, relationship to reward, and self-administration of playback. J Comp Psychol 122:357–367

Cloutier S, Baker C, Wahl K, Panksepp J, Newberry RC (2013) Playful handling as social enrichment for individually- and group-housed laboratory rats. Appl Anim Behav Sci 143:85–95

Favre MR, La Mendola D, Meystre J, Christodoulou D, Cochrane MJ, Markram H, Markram K (2015) Predictable enriched environment prevents development of hyper-emotionality in the VPA rat model of autism. Front Neurosci 9:127

Fone K, Porkess V (2008) Behavioral and neurochemical effects of post-weaning social isolation in rodents—Relevance to developmental neuropsychiatric disorders. Neurosci Biobehav Rev 32:1087–1102

Gruendel AD, Arnold WJ (1974) Influence of preadolescent experiential factors on the development of sexual behavior in albino rats. J Comp Physiol Psych 1:172–178

Hamed A, Jaroszewski T, Maciejak P, Szyndler J, Lehner M, Kamecka I, Olczak M, Kuzinska U, Taracha E, Płaźnik A (2009) The effects of buspirone and diazepam on aversive context- and social isolation-induced ultrasonic vocalization. Physiol Behav 98:474–480

Hamed A, Szyndler J, Taracha E, Turzyńska D, Sobolewska A, Lehner M, Krząścik P, Daszczuk P (2015) κ-opioid receptor as a key mediator in the regulation of appetitive 50-kHz ultrasonic vocalizations. Psychopharmacology 232:1941–1955

Hunter AS (2015) Impaired extinction of fear conditioning after REM deprivation is magnified by rearing in an enriched environment. Neurobiol Learn Mem 122:11–18

Inagaki H, Mori Y (2013) The critical point at which post-weaning individual housing conditions affect the emission of 22-kHz calls in male rats. J Vet Med Sci 75:527–529

Inagaki H, Kuwahara M, Kikusui T, Tsubone H (2005) The influence of social environmental condition on the production of stress-induced 22 kHz calls in adult male Wistar rats. Physiol Behav 84:17–22

Inagaki H, Kuwahara M, Tsubone H, Mori Y (2013) The effect of post-weaning individual housing on 50-kHz calls emitted from male rats to sexually receptive female rats. Physiol Behav 110–1:30–33

Kim EJ, Kim ES, Covey E, Kim JJ (2010) Social transmission of fear in rats: the role of 22-kHz ultrasonic distress vocalizations. PLoS ONE 5:e15077
Kisko TM, Himmler BT, Himmler SM, Euston DR, Pellis SM (2015) Are 50-kHz calls used as play signals in the playful interactions of rats? II. Evidence from the effects of devocalization. Behav Process 111:25–33
Kisko TM, Wöhr M, Pellis VC, Pellis SM (2017) From Play to Aggression: High-Frequency 50-kHz Ultrasonic Vocalizations as Play and Appeasement Signals in Rats. Curr Top Behav Neurosci 30:91–108
Knutson B, Burgdorf J, Panksepp J (1998) Anticipation of play elicits high-frequency ultrasonic vocalizations in young rats. J Comp Psychol 112:65–73
Lapiz MD, Fulford A, Muchimapura S, Mason R, Parker T, Marsden CA (2003) Influence of postweaning social isolation in the rat on brain development, conditioned behavior, and neurotransmission. Neurosci Behav Physiol 33:13–29
Łopuch S, Popik P (2011) Cooperative behavior of laboratory rats (*Rattus norvegicus*) in an instrumental task. J Comp Psychol 125:250–253
Lore R, Flannelly K, Farina P (1976) Ultrasounds produced by rats accompany decreases in intraspecific fighting. Aggress Behav 2:175–181
Markram K, Markram H (2010) The intense world theory—A unifying theory of the neurobiology of autism. Front Hum Neurosci 4:224
Markram H, Rinaldi T, Markram K (2007) The intense world syndrome–an alternative hypothesis for autism. Front Neurosci 1:77–96
Nunes Mamede Rosa ML, Nobre MJ, Ribeiro Oliveira A, Brandão ML (2005) Isolation-induced changes in ultrasonic vocalization, fear-potentiated startle and prepulse inhibition in rats. Neuropsychobiology 51:248–255
Panksepp J, Burgdorf J (2000) 50-kHz chirping (laughter?) in response to conditioned and unconditioned tickle-induced reward in rats: effects of social housing and genetic variables. Behav Brain Res 115:25–38
Panksepp J, Gordon N, Burgdorf J (2002) Empathy and the action-perception resonances of basic socio-emotional systems of the brain. Behav Brain Sci 25:43–44
Parsana AJ, Li N, Brown TH (2012a) Positive and negative ultrasonic social signals elicit opposing firing patterns in rat amygdala. Behav Brain Res 226:77–86
Parsana AJ, Moran EE, Brown TH (2012b) Rats learn to freeze to 22-kHz ultrasonic vocalizations through autoconditioning. Behav Brain Res 232:395–399
Pohorecky LA (2008) Psychosocial stress and chronic ethanol ingestion in male rats: Effects on elevated plus maze behavior and ultrasonic vocalizations. Physiol Behav 94:432–447
Sadananda M, Wöhr M, Schwarting RKW (2008) Playback of 22-kHz and 50-kHz ultrasonic vocalizations induces differential c-fos expression in rat brain. Neurosci Lett 435:17–23
Schneider T, Turczak J, Przewłocki R (2006) Environmental enrichment reverses behavioral alterations in rats prenatally exposed to valproic acid: issues for a therapeutic approach in autism. Neuropsychopharmacology 31:36–46
Seffer D, Rippberger H, Schwarting RKW, Wöhr M (2015) Pro-social 50-kHz ultrasonic communication in rats: post-weaning but not post-adolescent social isolation leads to social impairments—phenotypic rescue by re-socialization. Front Behav Neurosci 9:102
Seffer D, Schwarting RK, Wöhr M (2014) Pro-social ultrasonic communication in rats: insights from playback studies. J Neurosci Methods 234:73–81
Silverman JL, Yang M, Lord C, Crawley JN (2010) Behavioural phenotyping assays for mouse models of autism. Nat Rev Neurosci 11:490–502
Siviy SM, Panksepp J (1987) Sensory modulation of juvenile play in rats. Dev Psychobiol 20:39–55
Tomazini FM, Reimer A, Albrechet-Souza L, Brandao ML (2006) Opposite effects of short- and long-duration isolation on ultrasonic vocalization, startle and prepulse inhibition in rats. J Neurosci Meth 153:114–120
Valluy J, Bicker S, Aksoy-Aksel A, Lackinger M, Sumer S, Fiore R, Wüst T, Seffer D, Metge F, Dieterich C, Wöhr M, Schwarting R, Schratt G (2015) A coding-independent function of an alternative Ube3a transcript during neuronal development. Nat Neurosci 18:666–673

Von Frijtag JC, Schot M, van den Bos R, Spruijt BM (2002) Individual housing during the play period results in changed responses to and consequences of a psychosocial stress situation in rats. Dev Psychobiol 41:58–69
Wen F, Xu L (2010) Effects of isolation after sexual experience on anxiety-like, depressive-like behaviors and affective states in male rats. Chin Sci Bull 55:4136–4142
Willadsen M, Seffer D, Schwarting RKW, Wöhr M (2014) Rodent ultrasonic communication: male prosocial 50-kHz ultrasonic vocalizations elicit social approach behavior in female rats (*Rattus norvegicus*). J Comp Psychol 128:56–64
Willey AR, Spear LP (2013) The effects of pre-test social deprivation on a natural reward incentive test and concomitant 50 kHz ultrasonic vocalization production in adolescent and adult male Sprague-Dawley rats. Behav Brain Res 245:107–112
Willuhn I, Tose A, Wanat MJ, Hart AS, Hollon NG, Phillips PE, Schwarting RK, Wöhr M (2014) Phasic dopamine release in the nucleus accumbens in response to pro-social 50 kHz ultrasonic vocalizations in rats. J Neurosci 34:10616–10623
Wöhr M, Scattoni ML (2013) Behavioural methods used in rodent models of autism spectrum disorders: current standards and new developments. Behav Brain Res 251:5–17
Wöhr M, Schwarting RKW (2007) Ultrasonic communication in rats: can playback of 50-kHz calls induce approach behavior? PLoS ONE 2:e1365
Wöhr M, Schwarting RKW (2008a) Maternal care, isolation-induced infant ultrasonic calling, and their relations to adult anxiety-related behavior in the rat. Behav Neurosci 122:310–330
Wöhr M, Schwarting RKW (2008b) Ultrasonic calling during fear conditioning in the rat: no evidence for an audience effect. Anim Behav 76:749–760
Wöhr M, Schwarting RKW (2009) Ultrasonic communication in rats: effects of morphine and naloxone on vocal and behavioral responses to playback of 50-kHz vocalizations. Pharmacol Biochem Behav 94:285–295
Wöhr M, Schwarting RKW (2012) Testing social acoustic memory in rats: effects of stimulus configuration and long-term memory on the induction of social approach behavior by appetitive 50-kHz ultrasonic vocalizations. Neurobiol Learn Mem 98:154–164
Wöhr M, Houx B, Schwarting RKW, Spruijt B (2008) Effects of experience and context on 50-kHz vocalizations in rats. Physiol Behav 93:766–776
Wöhr M, Schwarting RKW (2013) Affective communication in rodents: ultrasonic vocalizations as a tool for research on emotion and motivation. Cell Tissue Res 354:81–97

From Play to Aggression: High-Frequency 50-kHz Ultrasonic Vocalizations as Play and Appeasement Signals in Rats

Theresa M. Kisko, Markus Wöhr, Vivien C. Pellis and Sergio M. Pellis

Abstract When rats engage in playful interactions, they emit appetitive 50-kHz ultrasonic vocalizations (USVs). We investigated the role of 50-kHz USVs in the playful behavior of both juvenile and adult rats. A cohort of juvenile rats was surgically devocalized and allowed to interact with either devocalized or intact partners as juveniles and again as adults. A substantial decrease in playful motivation was seen for pairs of devocalized rats, as well as all intact rats housed with devocalized ones. In pairs in which at least one partner could vocalize, there was no difference in the number of playful interactions as compared to controls. Further investigation revealed that, within the playful episode itself, 50-kHz USVs are more likely to appear before a playful attack is launched than after, regardless of the attacking partner's ability to vocalize, and when one partner is pinned on its back by another, it is the rat that is on top that is more likely to emit 50-kHz USVs. These findings suggest that, for juveniles, 50-kHz USVs may have a critical function in maintaining and facilitating playful motivation, but a more limited role in signaling playful actions. In adults, however, whatever the motivational role of such calling may be, the various kinds of USVs appear to serve critical communicatory functions. For instance, when pairs of adult males that are unfamiliar with one another encounter each other in a neutral arena, they play together, but if one partner is devocalized, there is a significantly higher likelihood that the interaction will escalate to become aggressive. While the relative roles of appetitive 50-kHz and aversive 22-kHz USVs in this context remain to be determined, our overall findings for play in both juveniles and adults suggest that 50-kHz USVs likely have multiple functions, with different functions being more prevalent at some ages and contexts than others.

T.M. Kisko (✉) · M. Wöhr
Behavioral Neuroscience, Philipps-University of Marburg, Gutenbergstr. 18, D-35032, Marburg, Germany
e-mail: theresa.kisko@gmail.com

V.C. Pellis · S.M. Pellis
Canadian Centre for Behavioural Neuroscience, Department of Neuroscience, University of Lethbridge, Lethbridge, AB T1K 3M4, Canada

Curr Topics Behav Neurosci (2016) 30: 91–108
DOI 10.1007/7854_2015_432
Published Online: 04 January 2016

Keywords Rough-and-tumble play · Rats · Play signal · Appeasement signal · Ultrasonic vocalizations

Contents

1 Introduction ... 92
2 Play Behavior in Rats ... 93
3 Play Signals ... 94
4 Ultrasonic Vocalizations in Rats ... 94
5 High-Frequency 50-kHz USV as Play Signals? ... 96
6 High-Frequency 50-kHz USVs as Appeasement Signals? ... 102
7 Conclusion ... 104
References ... 105

1 Introduction

One of the most challenging forms of play is rough-and-tumble play or play-fighting, in which pairs of animals typically compete for some advantage over one another (Aldis 1975). This advantage, often involving contacting some target on the body (Aldis 1975; Pellis 1988), does not involve the unrestrained competition often seen in serious fighting (Blanchard and Blanchard 1994), but rather involves some measure of restraint that leads to play-fighting being reciprocal. Such reciprocity has been characterized by the 50:50 rule, whereby each player wins about 50 % of its playful encounters (Altmann 1962). Subsequent game theory models have shown that, as win–loss ratios deviate from 50:50, play-fights become progressively less stable. This is not to say that in some cases, the win–loss ratio cannot deviate from 50:50 (for review see Pellis et al. 2010), but what it does suggest is that play-fighting will not remain playful if one partner attempts to dominate the encounters completely. Indeed, some empirical studies have shown that, when individuals do attempt to dominate playful interactions completely, their potential play partners ostracize them, reducing their ability to engage in further interactions (e.g., Suomi 2005).

To maintain the reciprocity needed for play-fights to remain playful, animals have to follow rules of restraint (Pellis et al. 2010), which requires them to monitor both their own actions and those of their partner, to evaluate any potential transgressions of the rules. To be precise, this requires that, while engaged in play, animals need to assess whether an inappropriate action by a partner is a one-off act of exuberance or a deliberate bending of the rules. The possibility of deliberate transgressions becomes particularly likely as animals become sexually mature and increasingly use play-fighting as a tool for social assessment and manipulation (Palagi 2011). Indeed, comparative data suggest that social play is a more demanding activity than nonsocial play. Comparative analyses in primates have shown that the size of socioemotional brain systems increases in species that engage in more play-fighting, but not in species that engage in more nonsocial play (i.e., locomotor-rotational play, object play) (Graham 2011). One of the mechanisms thought to be involved in maintaining play-fighting, not only in primates but also in other species, is the use of play signals to negotiate interactions (Palagi et al. 2015). Rats not only engage in complex patterns of play-fighting, but they also use signals that can potentially serve the negotiating functions needed to maintain playfulness (Pellis and Pellis 2009).

2 Play Behavior in Rats

Rats engage in a variety of forms of play, including playing with inanimate objects, solitary locomotor-rotational play, and play-fighting (Hole and Einon 1984). However, even though locomotor-rotational play and play-fighting in rats can be very complex, object play is limited, and by far, it is play-fighting that occupies the majority of their time when they are playing (see Pellis and Pellis 2009 for a review). Not surprisingly, the rat has been an important model species for the study of the behavioral, developmental, and neurobiological mechanisms underlying mammalian play-fighting (e.g., Siviy and Panksepp 2011; Vanderschuren and Trezza 2014).

In rats, play-fighting involves attack and defense of the nape of the neck, which is then nuzzled with the snout when contacted (Pellis and Pellis 1987; Siviy and Panksepp 1987). Such dorsal contact by one partner is defended against by the recipient by using one of two major classes of defensive tactics: (1) evasion, in which the rat turns to look away from the oncoming attacker and swerves, leaps or runs away, and (2) facing defense, in which the rat turns to face the attacker and uses a variety of movements to block access to its nape. Facing defense, in turn, can involve two different classes of tactical maneuvers: (1) rotation around a vertical axis, usually the mid-body or pelvis, thus maintaining a prone position, and (2) rotation around the longitudinal axis, with either the whole body rotating so that the defender lays supine on its back or with only the forequarters rotating so that the defender still maintains contact on the ground with at least one hind foot (Himmler et al. 2013). If successfully executed, the defender can then launch counterattacks of its own, which, if successful, can lead to a role reversal as the original attacker defends itself (Kisko et al. 2015a). Moreover, the attacker may execute movements that facilitate successful counterattacks by the defender (Pellis et al. 2005), thus ensuring reciprocity. Regardless of the pattern of defense used, the repeated attacks and defense often lead to one rat lying on its back and its partner standing over it in a pinning configuration (Panksepp 1981).

Due to the repeated cycles of attack, defense, and counterattack, play-fighting in rats is thought to be more complex than that reported in many other species of rodents (Pellis and Pellis 1998) and as complex as that reported in many species of primates and carnivores (Pellis and Pellis 2009). Consequently, play-fighting in rats, like in many primates (Palagi 2011), involves complex cognitive assessments and the regulation of emotions (Pellis et al. 2014). To maintain such complex processes and thus allow playful interactions to proceed, rats likely depend on the use of play signals.

3 Play Signals

Because the contact involved in play-fighting can be similar to that occurring in other social contexts, such as aggression and courtship, it has been hypothesized that animals can use play signals to inform their partners that the contact is playful (Fagen 1981). While play signals can be used to make amends if one animal is too rough in its actions (Aldis 1975), the traditional role of such play signals has been thought to be to inform a potential play partner that the imminent contact is playful (Bekoff 1975). Play signals can be produced in several sensory modalities, including olfactory (Wilson 1973) and auditory ones (Kipper and Todt 2002), but ones involving visual cues are the most widely reported (Palagi et al. 2015). Among canids and primates, facial gestures provide the richest source of signaling (Bekoff 1975; van Hoof 1967), but bodily movements and positions are also prevalent (Yanagi and Berman 2014). In rats, facial gestures are limited to basic ones exhibiting pleasure and revulsion (Berridge and Robinson 2003), and there is no evidence of olfactory signals being used in play (Hole and Einon 1984). Rats have a rich repertoire of jumps, twists, turns, and runs that are performed during playful interactions (Pellis and Pellis 1983), and these could potentially serve as play signals. Other rodents with complex playful wrestling, such as hamsters, do not have these bodily gyrations (Pellis and Pellis 1988), yet mice that do not engage in playful wrestling have a varied repertoire of jumps and rotations (van Oortmerssen 1971). Therefore, it is unlikely that *all* these bodily gestures function as play signals. Nonetheless, some of these jumps performed by rats are produced in contexts that are consistent with them being useful for facilitating play (Pellis and Pellis 1983); this suggests that some may serve as communicatory functions. More likely to function as play signals, however, are the rich diversity of ultrasonic vocalizations (USVs) that are emitted in a variety of prosocial contexts, including play-fighting (Burgdorf et al. 2008; Knutson et al. 1998; Wright et al. 2010).

4 Ultrasonic Vocalizations in Rats

Rats are able to emit sounds in the ultrasonic range, termed USVs. Typically, three main categories of USVs are distinguished, of which all serve distinct communicative functions as socioaffective signals (for a detailed overview, see Brudzynski 2013; Wöhr and Schwarting 2013). Infant rats emit 40-kHz USVs following separation from their mother and littermates. These 40-kHz USVs elicit maternal behaviors, most notably, search and retrieval behavior (Wöhr and Schwarting 2008). In juvenile and adult rats, two main USV types occur, with their occurrence strongly depending on the emotional valence of the situation. Low-frequency 22-kHz USVs occur in aversive situations and particularly high rates are observed during aggressive encounters of adult male rats (Lehman and Adams 1976; Lore et al. 1976; Sales 1972b; Sewell 1967). They likely reflect a negative affective state. In contrast, high-frequency 50-kHz USVs are observed in appetitive situations, most notably in juveniles both during rough-and-tumble play with peers (Burgdorf et al. 2008; Knutson et al. 1998; Wright et al. 2010) and when tickled by a human (Panksepp and Burgdorf 2000). However, some negative affective situations, such as resident–intruder tests, will also elicit 50-kHz USVs (Takahashi et al. 1983). In adulthood, 50-kHz USVs mainly occur during mating (Sales 1972a), but can also be seen in other rewarding situations, such as when given food (Burgdorf et al. 2000) and psychoactive drugs (Burgdorf et al. 2001, 2008). It is widely believed that they reflect a positive affective state. In the first study on 50-kHz USVs emitted during rough-and-tumble play, Knutson et al. (1998) showed that the emission of 50-kHz USVs is positively correlated with dorsal contacts during play and that 50-kHz USVs occur in anticipation of play. As described by Wöhr et al. (2017), they further found that rats exposed to a brief period of social isolation emitted more than twice as many 50-kHz USVs and that they played more vigorously than group-housed controls, possibly due to an increase in social motivation. In contrast, an aversive stimulus, such as a bright white light, led to a reduction in 50-kHz USV emission. In a subsequent study, Burgdorf et al. (2008) found that, of the many 50-kHz USV subtypes, the frequency-modulated (FM) 50-kHz USVs occur at particularly high rates during rough-and-tumble play. These subtypes are also greatly increased following a brief period of social isolation and are most closely associated with the occurrence of dorsal contacts during play, but are negatively correlated with pinning behavior. Interestingly, in rats selectively bred for low 50-kHz USV emission rates, rough-and-tumble play is altered and characterized by fewer dorsal contacts but more pinning behavior (Webber et al. 2012). Moreover, in rats selectively bred for low or high anxiety-related behavior, we found that highly anxious rats initiate less rough-and-tumble play and emit fewer 50-kHz USVs, possibly reflecting lack of positive affect (Lukas and Wöhr 2015). As the breeding lines differ in their hypothalamic vasopressin availability and vasopressin is strongly implicated in the regulation of social behavior, we further tested whether manipulating the vasopressin system alters the emission of 50-kHz USVs during rough-and-tumble play. While the administration of synthetic vasopressin did not alter rough-and-tumble play and the concomitant emission of 50-kHz USVs, blocking the central vasopressin system by means of a vasopressin 1a receptor antagonist resulted in lower levels of play behavior and fewer 50-kHz USVs (Lukas and Wöhr 2015). This indicates that the central vasopressin system is involved in the regulation of affiliative communication in rodents, which is of translational relevance because various findings repeatedly link alterations in the central vasopressin system to autism in humans. Recently, we further showed that rats exposed to valproic acid during pregnancy emit fewer 50-kHz USV during rough-and-tumble play (Raza et al. 2015). Exposure to valproic acid, which is a drug typically used to treat epilepsy and bipolar disorder, is one of the major environmental risk factors for developing autism in humans (Moore et al. 2000) and has been shown to induce autism-like phenotypes when administered to pregnant rats (Schneider and Przewłocki 2005).

5 High-Frequency 50-kHz USV as Play Signals?

The close relationship between the play behavior and the emission of 50-kHz USVs suggests that 50-kHz USVs might serve a communicative function as play signals. If 50-kHz USVs are being used as traditional play signals, signifying "I want to play with you" (Bekoff 1975), then the most important characteristic would be that they occur most frequently preceding playful attacks. In a recent study (Himmler et al. 2014), we provided support for such use of 50-kHz USVs in juvenile rats. We showed that there were significantly more 50-kHz USVs emitted preceding playful contact compared to when rats cease contact. We also showed that, consistent with other studies (Lukas and Wöhr 2015), 50-kHz USVs occur more often in males than in females during play-fighting. This sex difference may be associated with the play of males tending to be rougher (Pellis et al. 1997). Rougher play poses a bigger threat in escalating to serious aggression and so may be more reliant on play signals to avoid such escalation (Palagi et al. 2015). Furthermore, because of the variety of 50-kHz USVs, we also explored whether particular 50-kHz USV subtypes are associated with the onset of specific defense tactics. In the Himmler et al. (2014) study, the most frequently emitted 50-kHz USV subtype was the trill, but this subtype was not significantly associated with any specific defensive action. Short calls, although less frequent, mainly occurred when the defender used an evasive tactic. These findings, especially those showing the high frequency of calling preceding contact, provide compelling evidence supporting the traditional function of play signals, that of advertising imminent contact of one partner by another (Bekoff 1975). In this study, however, both rats could vocalize, so any particular call could not be empirically attributed to either partner. Therefore, it cannot be certain whether the rat launching the attack was in fact the one vocalizing prior to making contact.

A procedure to overcome this dilemma is surgical devocalization, which has been previously used to study the communicative function of USVs in adult rats (Lehman and Adams 1976; Takahashi et al. 1983; Takeuchi and Kawashima 1986; Thomas et al. 1983). Therefore, using pairs of juvenile rats in which one partner was vocal and the other devocalized, we examined which partner, prior to a playful attack, was vocalizing (Kisko et al. 2015a). It was predicted that, when a devocalized rat attacks a vocal partner, there should be very few, if any, 50-kHz USVs being emitted prior to that attack. However, this was not the case, and, in fact, the number of 50-kHz USVs emitted prior to an attack when the devocalized partner

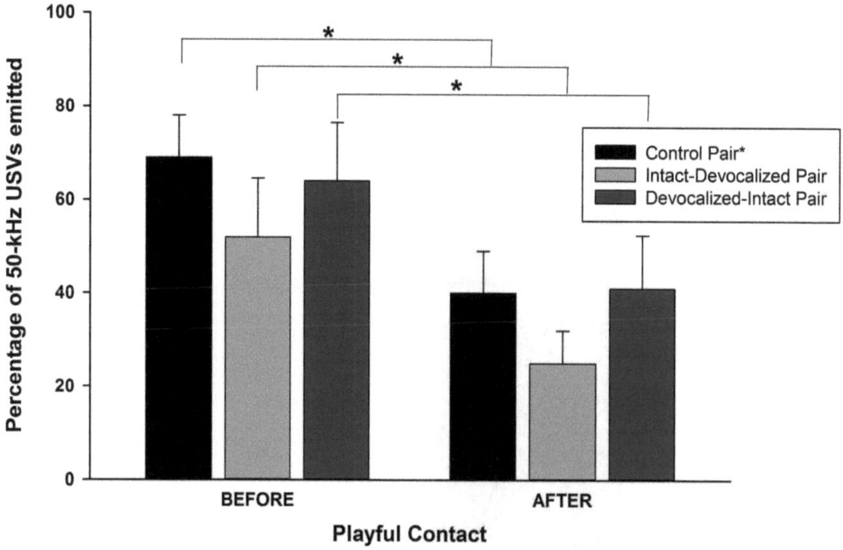

Fig. 1 Percentage (mean and SEM) of 50-kHz USVs emitted immediately before playful contact and immediately following the termination of contact. More 50-kHz USVs are emitted prior to contact whether the attacker is the one able to vocalize or not (*$p < 0.05$; the control pair is from Himmler et al. 2014; the graph is a combined data set from Himmler et al. 2014 and Kisko et al. 2015a)

attacked was comparable to the number of 50-kHz USVs emitted when a vocal rat was attacking. That is, the same pattern (Fig. 1) that was found whether both partners could vocalize (Himmler et al. 2014) or only one could do so (Kisko et al. 2015a). Moreover, we found no difference in the subtypes of 50-kHz USV emitted, irrespective of which partner was attacking (Kisko et al. 2015a). These findings suggest that the rats are not only using 50-kHz USVs to announce an attack but also to solicit playful contact from a partner.

Tickling juvenile rats by a human experimenter elicits high rates of 50-kHz USVs (Panksepp and Burgdorf 2000); this action is thought to mimic rough-and-tumble play between two rats. In particular, when tickled, rats roll over onto their backs, thus adopting a configuration similar to that of the pinning present in play-fighting. This suggests that rats produce many calls while on their backs. If this were the case, it would seem reasonable that many, if not the majority of 50-kHz USVs emitted during play-fights, should be emitted by the rat that is being pinned.

Contrary to expectation, data analyzed from our pairs of rats in which one partner was devocalized (Kisko et al. 2015a) revealed that more 50-kHz USVs occurred when the vocal rat was pinning the devocalized rat than when the devocalized rat was pinning the vocal rat (Fig. 2). However, given that the rate of pinning by devocalized rats was low, data based on six pairs of rats, even though significant, should be considered preliminary. If substantiated by further studies, these observations would suggest that it is not the tickling of the belly itself that

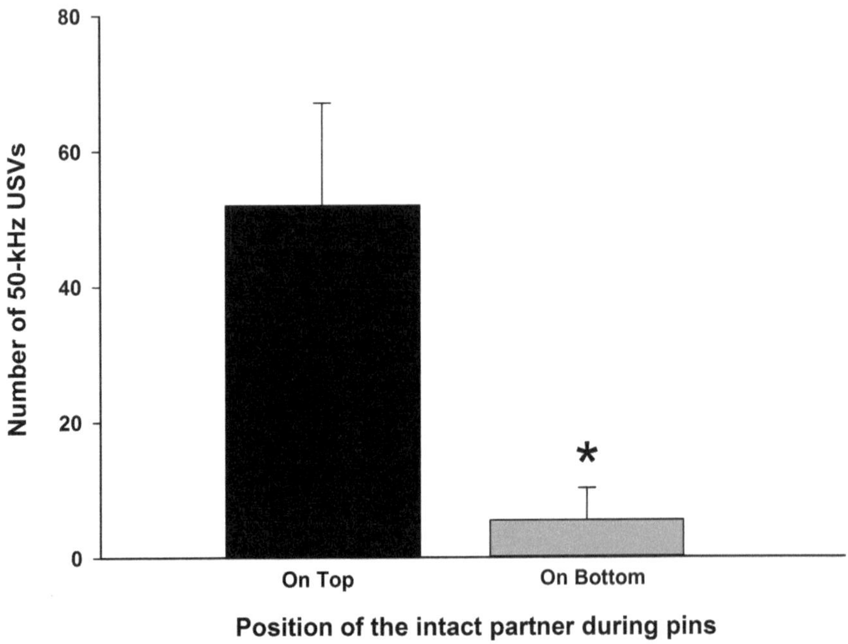

Fig. 2 Frequency of occurrence of 50-kHz USVs (mean and SEM) when rats are pinned during playful interactions. 50-kHz USVs are more frequent when the intact rat is on *top* than when it is on the *bottom* ($^{*}p < 0.05$)

elicits 50-kHz USVs during play-fighting, but rather, it is the rat on top—the "tickler"—that receives the most enjoyment, thus emitting more 50-kHz USVs. Some level of resistance by the partner being attacked seems to be critical in motivating playful attacks (Pellis and McKenna 1995), so that initiating, soliciting, and gaining contacts together ensure rewarding tactile experiences during play. The presence of 50-kHz USVs in all these phases of play-fighting provides support for the hypothesis that 50-kHz USVs express the rats' positive affective state and so function to maintain the animals' playful motivation.

In further support of the hypothesis that juvenile rats are using 50-kHz USVs to keep the mood playful and in doing so maintain playful interactions, we found that pairs of devocalized rats had a reduced frequency of playful interactions (Kisko et al. 2015a). When compared to pairs of vocal rats, devocalized pairs had almost 50 % fewer play-fights (Fig. 3). This suggests that 50-kHz USVs are being used to promote and maintain a playful mood and, in their absence, the rats are not nearly as motivated to engage in play. It is possible that this playful mood is linked to dopamine. Studies have shown that play-fighting is associated with the release of dopamine in the nucleus accumbens (Trezza et al. 2010) and that activation of the mesolimbic dopamine system induces the production of 50-kHz USVs (Burgdorf et al. 2001, 2007). Using the playback paradigm, we found that hearing 50-kHz

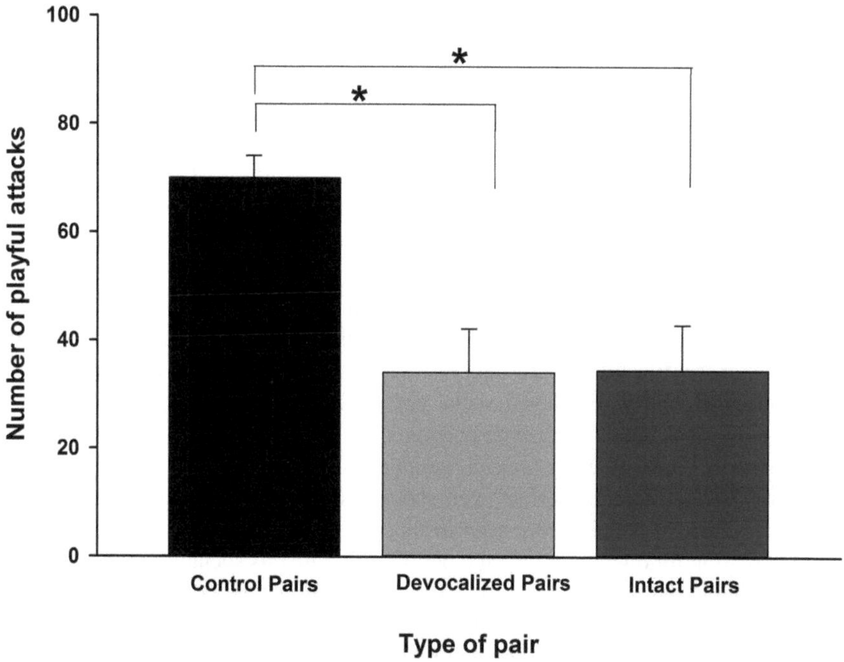

Fig. 3 Number of playful attacks (mean and SEM) initiated by pairs of intact rats reared with other intact rats (control pairs), by pairs of devocalized rats (devocalized pairs), and by pairs of intact rats reared with devocalized partners (intact pairs) in 10-min trials. Both the devocalized rats and the intact cage mates of devocalized rats exhibit a reduced motivation to engage in play (*$p < 0.05$)

USVs results in increased neuronal activity (Sadananda et al. 2008) and dopamine release (Willuhn et al. 2014) in the nucleus accumbens. This suggests that the release of dopamine in the nucleus accumbens is linked to both the production and the perception of 50-kHz USVs, possibly indicating that dopamine release in the nucleus accumbens functions as a translator of a motivational acoustic signal into a prosocial action. Such a perception-and-action loop is particularly relevant for appetitive social and reciprocal communicatory signals, with 50-kHz USVs reflecting a positive affective state in the sender and evoking a similar affective state in the receiver, thus promoting positive social interactions.

Interestingly, the playful mood can be reinstated to the typical control levels, seen in pairs of vocal rats, by pairing a devocalized rat with an unfamiliar vocal partner (Kisko et al. 2015b). This provides further support for the motivational role of 50-kHz USVs. The motivational role of 50-kHz USVs, however, may have a critical learning period. We observed that, in juveniles, the overall playful motivation was not only decreased in pairs of devocalized rats but was also significantly decreased in pairs of vocal rats that were housed with devocalized cage mates (Fig. 3). Juvenile cage mates often engage in playful interactions together, and it is possible that, in this critical learning period for juveniles, a vocal rat playing with a devocalized cage mate may not receive the necessary feedback from hearing 50-kHz USVs to learn about their contextual uses. That is not to say that the calls themselves are learned, but rather, that the proper context for their use in some situations could be learned through play (see, for further evidence, Wöhr et al. 2017). The animals in our study were housed in quads of two devocalized and two intact rats, so one would think that when the vocal cage mates played together it would be sufficient for them to learn the contextual cues, but this does not appear to be the case. In support of this critical learning period for juveniles, we have recently shown that prolonged social isolation in the four weeks after weaning, the juvenile period when play-fighting is most frequent results in a lack of appropriate behavioral responses toward 50-kHz USVs (Seffer et al. 2015). Specifically, while group-housed controls displayed social approach behavior in response to 50-kHz USVs, a response that is even more prominent in rats isolated for 24 h, rats exposed to long-term, post-weaning, social isolation did not display social approach behavior. Furthermore, these rats even showed some signs of social avoidance. In contrast, no social deficits were seen in rats given comparable levels of long-term social isolation following the juvenile period. Juvenile rats socially isolated for 24 h have an increased motivation to engage in playful interactions (Himmler et al. 2013); however, this increase can be curtailed by placing them with less playful partner. For example, a partner treated with scopolamine, a cholinergic antagonist, will explore the enclosure in which it is placed, but will not initiate playful attacks or respond to a playful attack (Pellis and McKenna 1995). Such a partner elicits playful attacks initially, but prolonged exposure to such a partner leads to reduced playful motivation in the un-drugged animal, as evidenced by a decrease in initiating playful attacks (Pellis and McKenna 1995). Furthermore, social play generally occurs only when a rat is free from physiological and social stress (Siviy et al. 2006). The decreased motivation to play that is seen in the devocalized cage mates could, in turn, negatively impact the playful motivation of the vocal cage mates. As a result, if a lack of playful motivation is consistent and prolonged, as would be the case for the vocal cage mates of the devocalized rats, the vocal rats may become depressed or stressed and thus much less motivated to play.

As well as regulating playful mood, 50-kHz USVs may also serve other important communicatory functions. For rats, pinning and being pinned during play-fighting appears to be highly rewarding and is thus a substantial component within their playful repertoire (Panksepp 1981). In a study by Siviy and Panksepp (1987), it was found that deafened rats pinned less, suggesting that not being able to hear 50-kHz USVs decreases the desire for close bodily contact in playful situations. Similarly, it was hypothesized that devocalized pairs would also show a reduction in playful pinning defenses, but the opposite turned out to be true (Kisko et al. 2015a). Pairs of devocalized rats had a higher frequency and preference for contact-promoting playful defenses than the intact control pairs. One hypothesis to explain these results could be that the 50-kHz USVs are acting as contact calls to help localize the partner within the play arena. Being nocturnal, the majority of playful interactions in rats take place in the dark, and so, being able to signal their location to their partner in a non-visual manner would be beneficial. If this were so, this could explain why devocalized pairs prefer to stay in close contact, in that it would avoid spending long amounts of time searching for one another in a test arena. Therefore, we predicted that, if a vocal rat were paired with a devocalized partner, the devocalized rat would adopt the typical playful defense tactics seen in control rat pairs, since calls from the vocal rat would provide the means to locate that partner. That is, by being able to hear their play partner's 50-kHz USVs and adopting the more typical tactics of defense, the devocalized rats would return to the pinning frequencies present in control pairs. However, even when paired with a vocal partner, the devocalized animals still appeared to prefer to use contact-promoting defense tactics significantly more often than evasive defense tactics. This suggests that the change in defensive actions by the devocalized rats is not to compensate for the absence of 50-kHz USVs. Therefore, at least within the confines of the test arena used in this study, the results do not support the contact call hypothesis.

Moreover, when given the choice of being presented in the same test arena, vocal rats are no more attractive as a play partner than are silent ones (Kisko et al. 2015a). Indeed, even when confronted with unfamiliar animals, the rats were just as likely to launch playful attacks on devocalized partners as they were on vocal ones (Kisko et al. 2015b). That is to say, among juveniles, there is little evidence that rats use 50-kHz USVs as traditionally conceived play signals (Bekoff 1975; Palagi et al. 2015)—they appear to be unnecessary for both initiating playful contact and in soliciting playful contact. That for juvenile rats 50-kHz USVs do not appear to provide rewarding social incentives (Willey and Spear 2012) is consistent with these findings (although see below). Rather, the role of 50-USVs seems more closely tied to regulating playful motivation and possibly in promoting the development of prosocial neural systems.

A commonly used measure of playful motivation is the frequency with which rats initiate playful contacts on the nape of their partner (Himmler et al. 2013). Such attacks are diminished when pairs of devocalized rats are tested together (Kisko et al. 2015a). Moreover, role reversals, in which the original defender launches a successful counterattack, forcing the original attacker to defend itself, are also reduced in such pairs (Kisko et al. 2015a). Given that the frequency of such counterattacks are decreased in tandem with initiating attacks (Pellis and Pellis 1990), the reduced frequency of role reversals is also likely to reflect a reduction in the motivation to play. That these reductions are, at least in part, due to an acute effect of the absence of 50-kHz USVs on playful motivation is suggested by the restoration of a high frequency of playful attacks when devocalized rats are tested with unfamiliar, vocal partners (Kisko et al. 2015b). However, that some of this effect is due to a more chronic influence of lack of exposure to normal levels of 50-kHz USVs over a prolonged period is shown by the finding that the vocal partners of devocalized cage mates also show a depressed level of initiating playful attacks (Fig. 3). In addition, the altered pattern of playful defense present in devocalized rats (Kisko et al. 2015a) is not ameliorated when playing with an unfamiliar, vocal partner (Kisko et al. 2015b), further suggesting deeper organizational changes in brain development due to a chronic lack of vocalizing. All our devocalized rats received their surgeries at around postnatal day 25, an age within a critical period for the development of several neurotransmitter and neuropeptide systems implicated in the regulation of social behavior (Trezza et al. 2010). The observation that the rats with sham surgeries did not display the same changes in play-fighting implicates the role of cutting laryngeal nerves, and the associated elimination of the ability to produce 50 or 22-kHz USVs, in these developmental disturbances.

Unlike the study by Wiley and Spear (2012), some playback studies have shown that 50-kHz USVs do appear to provide rewarding social incentives. For instance, Burgdorf et al. (2008) found that rats will nose-poke to elicit playback of 50-kHz USVs. Moreover, we showed that playback of 50-kHz USVs results in social approach behavior in the recipient (Wöhr and Schwarting 2007; Willuhn et al. 2014), and as described by Wöhr et al. (2017), this response is present in both juveniles and adults. However, as already mentioned, long-term, post-weaning social isolation results in a lack of social approach behavior in response to 50-kHz USVs (Seffer et al. 2015). These latter findings are consistent with the notion that the juvenile period is an important one for the development of the neural systems associated with 50-kHz USV production. Thus, given the possibility that the neural systems associated with the production of USVs and those associated with the regulation of social behavior overlap in their development, the changes in social play wrought by chronic devocalization in the early juvenile period that we have found (Kisko et al. 2015a, b) may not be coincidental. Such effects may be used as a vehicle for exploring how these neural systems may interact.

6 High-Frequency 50-kHz USVs as Appeasement Signals?

As noted above, for play-fighting to remain playful, the participants need to exercise some degree of reciprocity. Transgressions can lead to the partner escalating the encounter into serious aggression. Among juveniles, such escalation is rare, but not absent (Fagen 1981). It has been suggested that play signals can be used in such situations to de-escalate the encounter with the transgressor effectively using the signal to inform the partner that "it was only play" (Aldis 1975). That is, the signal can be used to appease the partner. In rats, play-fighting can also occasionally escalate into serious fighting, which can be unambiguously identified as when the rats stop attempting to nuzzle each others' napes and instead switch to bite the partner's lower flanks and rump (Pellis and Pellis 1987, 1990). If 50-kHz USVs are used as signals to de-escalate the risk of a playful encounter turning into aggression, then, in the absence of these calls, such escalation should be more likely. For none of our juvenile experimental animals were play-fights found to escalate into aggression—not when devocalized rats played together or when devocalized rats played with vocal partners (Kisko et al. 2015a). Even when tested with unfamiliar partners, so eliminating the possibility that rats with an established relationship can use other means to avoid escalation, there was no evidence that play-fights were more likely to escalate to aggression when one of the rats could not vocalize (Kisko et al. 2015b). The situation appears to be different when adult rats are involved.

In some species, adults also engage in play-fights, at which age it is likely to be used for social assessment and manipulation (Palagi 2011). Among adult male rats, dominance relationships can be negotiated with play-fights (Pellis and Pellis 2009). Within colonies of familiar rats, subordinate males will initiate and engage in a more gentle form of play with a dominant male. Furthermore, they will initiate less play with other subordinates, and when they do play together, it will be rougher. When unfamiliar adult rats encounter one another in a neutral arena, they can engage in a rough form of play-fighting which can lead to the establishment of a dominance relationship. When neither member of a pair adopts a submissive status, the encounter can escalate into serious fighting (reviewed in Pellis and Pellis 2009). It is hypothesized that, if 50-kHz USVs serve an important communicative function as appeasement signals, then this should become apparent when unfamiliar, adult males encounter one another in a neutral arena.

In pairs in which one play partner is devocalized, the risk of the interaction becoming aggressive is significantly higher than in pairs in which both rats can vocalize (Kisko et al. 2015b). In fact, in all pairs that included an unfamiliar devocalized partner, there were both agonistic displays, such as piloerection, lateral displays, and tail wiggles, and aggressive attacks, in which one partner directs bites at the flanks of the opponent. Such agonism was rare in the pairs in which both rats could vocalize, and their encounters never escalated to biting. This strongly indicates that, in potentially risky and ambiguous situations, adults may rely on 50-kHz USVs to modify each other's behavior tactically in a way that is not essential among juveniles. These findings are thus consistent with ones that show that 50-kHz USVs are used as signals in agonistic encounters in adult rats.

In resident–intruder tests, in which an unfamiliar adult male is placed in the home cage of a resident male, the resident typically attacks the intruder (Blanchard and Blanchard 1994), and in such encounters, 50-kHz USVs are frequently emitted (Sales 1972b; Sewell 1967). Moreover, rats are even found to emit 50-kHz USVs when entering an area associated with the potential presence of an aggressor, with the number of 50-kHz USVs emitted by the intruder being positively correlated with the number of aggressive encounters it has experienced in this enclosure (Tornatzky et al. 1994, 1995). Importantly, devocalization (Takahashi et al. 1983; Thomas et al. 1983) and pharmacological (Vivian and Miczek 1993) studies have implicated the intruder as the source of the 50-kHz USVs. Together, these findings indicate that 50-kHz USVs are emitted as a signal of appeasement, thus reducing the likelihood of being attacked by the resident.

It should be noted that devocalization abolishes not only the ability to produce 50-kHz USVs, but also 22-kHz USVs, and there is evidence for 22-kHz USVs being used as an appeasement signal, but results are conflicting. For instance, it was reported that in the resident–intruder paradigm, aggressive behavior is rarely observed following the emission of 22-kHz USVs (Lehman and Adams 1976;

Lore et al. 1976; Sales 1972b; Sewell 1967); yet devocalization experiments do not support the idea that 22-kHz USV emission modulates the aggressive behavior of the resident (Lehman and Adams 1976; Takeuchi and Kawashima 1986; Thomas et al. 1983). Thus, in the neutral test arena that we used (Kisko et al. 2015b), either 50-kHz USVs alone, 22-kHz USVs alone, or some combination of both may be used to diminish the likelihood of escalation from playful to serious fighting.

7 Conclusion

50-kHz USVs are emitted at a high frequency during play-fighting among juvenile rats (Burgdorf et al. 2008; Knutson et al. 1998; Wright et al. 2010). Examination of when these calls occur during play-fights shows that they are most likely to occur immediately prior to playful contact (Himmler et al. 2014), and this is true whether the attacker can vocalize or not (Kisko et al. 2015a). These findings suggest that the production of USVs is integral to play and that they may provide important communicatory functions. Such vocalizations may serve two kinds of communicatory functions that have been traditionally postulated for play signals (Bekoff 1975; Palagi et al. 2015): that of informing a potential recipient of a playful attack and that the imminent contact will be playful or that of a recipient soliciting such an attack from a nearby partner. However, our findings with devocalized rats indicate that play can occur in the absence of these presumed communicatory functions, at least among juveniles (Kisko et al. 2015a, b). More consistent with our data is the hypothesis advocated by Knutson et al. (1998) and supported by others that 50-kHz USVs are an expression of the positive affective state associated with play (Burgdorf et al. 2008). In this context, if these calls do serve a communicatory role, it is an indirect one, that of maintaining the playful mood of the producer and/or the receiver of the calls (Kisko et al. 2015a, b). From a developmental perspective, the production and perception of such calls may be important for the maturation of neural and behavioral systems that are associated with prosocial behavior. The case for a communicatory role of USVs is more compelling for adult males engaging in playful interactions with unfamiliar partners. In the absence of such calling, even when just one member of the pair cannot vocalize, there is a marked increase in the likelihood that play-fights escalate into aggression. Given the evidence from resident–intruder encounters (e.g., Lore et al. 1976; Sales 1972b), it seems highly likely that USVs, be they 50-kHz USVs, 22-kHz USVs, or both, are being used as appeasement signals to attenuate the risk of playful encounters escalating into aggression (Kisko et al. 2015b). Therefore, it seems that 50-kHz USVs may have multiple functions, and these may differ across different stages of development.

Further developmental studies are needed to understand the roles of maturation and learning in being able to emit 50-kHz USVs in contextually relevant ways during social interactions. Surgical devocalization provides a clear way to examine the role of such vocalizations and enables observers to identify, unambiguously, the rat that is vocalizing when only one member of a dyad is devocalized (Kisko et al.

2015a, b; Thomas et al. 1983). However, the technique is highly invasive and may have potential long-term side effects that have not yet been investigated. Therefore, other methods that do not require surgery to identify which member of the pair is calling would be helpful. One such technique, the use of multiple microphone arrays, has been successfully used to examine the emission of USVs during mating and other social interactions in mice (Neunuebel et al. 2015). However, play-fighting in rats is often very vigorous and fast-paced, with the rats alternating between wrestling and running. While worth trying, it seems unlikely that the multiple microphone array technique would be able to distinguish which rat is calling in all situations during play. Another new and potentially useful technique is one being used in songbirds, in which an ultraminiature backpack is used to record sound and acceleration in the bird carrying the pack (Anisimov et al. 2014). The small backpack, which weighs only 2.6 g, is harnessed on the bird's back. Moreover, the weight of the backpack can be further decreased, to 1.4 g, if necessary. This backpack monitoring system may be an ideal way to record the vocalizations of individual rats or mice. Nonetheless, the utility of this technique in recording vocalizations from individual pair mates during play-fights needs to be evaluated, as the presence of the backpacks may inhibit the play or modify the play that is performed. After all, as already noted above, rolling over on to their backs is an important part of playful wrestling, and the presence of a backpack may hamper such behavior. Also, it is possible that the vigorous nature of play may dislodge the device or obstruct the recording abilities of the microphone. Irrespective of these concerns, the value of gaining information on the vocalizations emitted by individuals during social encounters, and do so while avoiding the potential side effects of surgical manipulation, is so great that these techniques should be tested empirically as tools for studying the USVs used during play-fights.

Acknowledgements M.W. is supported by the Deutsche Forschungsgemeinschaft (DFG WO 1732/4-1). S.M.P. is supported by the National Sciences and Engineering Research Council of Canada (NSERC—4058).

References

Aldis O (1975) Play fighting. Academic Press, New York, NY

Altmann SA (1962) Social behavior of anthropoid primates: analysis of recent concepts. In: Bliss EL (ed) Roots of behavior. Harper, New York, pp 277–285

Anisimov VN, Herbst JA, Abramchuk AN, Latanov AV, Hahnloser RH, Vyssotski AL (2014) Reconstruction of vocal interactions in a group of small songbirds. Nat Methods 11:1135–1137

Bekoff M (1975) The communication of play intention: are play signals functional? Semiotica 15:231–239

Berridge KC, Robinson TE (2003) Parsing reward. Trends Neurosci 26:507–513

Blanchard RJ, Blanchard DC (1994) Environmental targets and sensorimotor systems in aggression and defense. In: Cooper SJ, Hendrie CA (eds) Ethology and psychopharmacology. Wiley, New York, NY, pp 133–157

Brudzynski SM (2013) Ethotransmission: communication of emotional states through ultrasonic vocalization in rats. Curr Opin Neurobiol 23:310–317

Burgdorf J, Knutson B, Panksepp J (2000) Anticipation of rewarding brain stimulation evokes ultrasonic vocalizations in rats. Behav Neurosci 97:320–327

Burgdorf J, Knutson B, Panksepp J, Ikemoto S (2001) Nucleus accumbens amphetamine microinjections unconditionally elicit 50-kHz ultrasonic vocalizations in rats. Behav Neurosci 115:940–944

Burgdorf J, Wood PL, Kroes RA, Moskal JR, Panksepp J (2007) Neurobiology of 50-kHz ultrasonic vocalizations in rats: electrode mapping, lesion, and pharmacology studies. Behav Brain Res 182:274–283

Burgdorf J, Kroes RA, Moskal JR, Pfaus JG, Brudzynski SM, Panksepp J (2008) Ultrasonic vocalizations of rats (*Rattus norvegicus*) during mating, play, and aggression: behavioral concomitants, relationship to reward, and self-administration of playback. J Comp Psychol 122:357–367

Fagen R (1981) Animal play behavior. Oxford University Press, New York

Graham KL (2011) Coevolutionary relationship between striatum size and social play in nonhuman primates. Am J Primat 73:314–322

Himmler BT, Pellis VC, Pellis SM (2013) Peering into the dynamics of social interactions: measuring play fighting in rats. J Vis Exp 71:e4288

Himmler BT, Kisko TM, Euston DR, Kolb B, Pellis SM (2014) Are 50-kHz calls used as play signals in the playful interactions of rats? I. Evidence from the timing and context of their use. Behav Process 106:60–66

Hole GJ, Einon DF (1984) Play in rodents. In: Smith PK (ed) Play in animals and children. Blackwell, Oxford, pp 95–117

Kipper S, Todt D (2002) The use of vocal signals in the social play of Barbary macaques. Primates 43:3–17

Kisko TM, Euston DR, Pellis SM (2015a) Are 50-kHz calls used as play signals in the playful interactions of rats? III. The effects of devocalization on play with unfamiliar partners as juveniles and as adults. Behav Process 113:113–121

Kisko TM, Himmler BT, Himmler SM, Euston DR, Pellis SM (2015b) Are 50-kHz calls used as play signals in the playful interactions of rats? II. Evidence from the effects of devocalization. Behav Process 111:25–33

Knutson B, Burgdorf J, Panksepp J (1998) Anticipation of play elicits high-frequency ultrasonic vocalizations in young rats. J Comp Psychol 112:65–73

Lehman MN, Adams DB (1976) A statistical and motivational analysis of the social behaviors of the male laboratory rat. Behaviour, LXI:3–4

Lore R, Flannelly K, Farina P (1976) Ultrasounds produced by rats accompany decreases in intraspecific fighting. Aggress Behav 2:175–181

Lukas M, Wöhr M (2015) Endogenous vasopressin, innate anxiety, and the emission of pro-social 50-kHz ultrasonic vocalizations during social play behavior in juvenile rats. Psychoneuroendocrinology 56:35–44

Moore SJ, Turnpenny P, Quinn A, Glover S, Lloyd DJ, Montgomery T, Dean JCS (2000) A clinical study of 57 children with fetal anticonvulsant syndromes. J Med Genet 37:489–497

Neunuebel JP, Taylor AL, Arthur BJ, Egnor SR (2015) Female mice ultrasonically interact with males during courtship displays. eLife 4:e06203

Palagi E (2011) Playing at every age: modalities and potential functions in non-human primates. In: Pellegrini AD (ed) Oxford handbook of the development of play. Oxford University Press, Oxford, pp 70–82

Palagi E, Burghardt GM, Smuts B, Cordoni G, Dall'Olio S, Fouts HN, Řeháková-Petrů M, Siviy SM, Pellis SM (2015). Rough-and-tumble play as a window on animal communication. Biol Rev, in press

Panksepp J (1981) The ontogeny of play in rats. Dev Psychobiol 14:327–332

Panksepp J, Burgdorf J (2000) 50-kHz chirping (laughter?) in response to conditioned and unconditioned tickle-induced reward in rats: effects of social housing and genetic variables. Behav Brain Res 115:25–38

Pellis SM (1988) Agonistic versus amicable targets of attack and defense: consequences for the origin, function and descriptive classification of play-fighting. Aggress Behav 14:85–104

Pellis SM, McKenna MM (1995) What do rats find rewarding in play fighting? An analysis using drug-induced non-playful partners. Behav Brain Res 68:65–73

Pellis SM, Pellis VC (1983) Locomotor-rotational movements in the ontogeny and play of the laboratory rat (*Rattus norvegicus*). Dev Psychobiol 16:269–286

Pellis SM, Pellis VC (1987) Play-fighting differs from serious fighting in both target of attack and tactics of fighting in the laboratory rat (*Rattus norvegicus*). Aggress Behav 13:227–242

Pellis SM, Pellis VC (1988) Play-fighting in the Syrian golden hamster (*Mesocricetus auratus* waterhouse), and its relationship to serious fighting during post-weaning development. Dev Psychobiol 21:323–337

Pellis SM, Pellis VC (1990) Differential rates of attack, defense and counterattack during the developmental decrease in play fighting by male and female rats. Dev Psychobiol 23:215–231

Pellis SM, Pellis VC (1998) The play fighting of rats in comparative perspective: a schema for neurobehavioral analyses. Neurosci Biobehav Rev 23:87–101

Pellis SM, Pellis VC (2009) The playful brain: venturing to the limits of neuroscience. Oneworld Press, Oxford

Pellis SM, Pellis VC, Himmler BT (2014) How play makes for a more adaptable brain: a comparative and neural perspective. Am J Play 7:73–98

Pellis SM, Pellis VC, Foroud A (2005) Play fighting: aggression, affiliation and the development of nuanced social skills. In: Tremblay R, Hartup WW, Archer J (eds) Developmental origins of aggression. Guilford Press, New York, NY, pp 47–62

Pellis SM, Pellis VC, Reinhart CJ (2010) The evolution of social play. In: Worthman C, Plotsky P, Schechter D, Cummings C (eds) Formative experiences: the interaction of caregiving, culture, and developmental psychobiology. Cambridge University Press, Cambridge, pp 404–431

Pellis SM, Field EF, Smith LK, Pellis VC (1997) Multiple differences in the play fighting of male and female rats. Implications for the causes and functions of play. Neurosci Biobehav Rev 21:105–120

Raza S, Himmler BT, Himmler SM, Harker A, Kolb B, Pellis SM, Gibb R (2015) Effects of prenatal exposure to valproic acid on the development of juvenile-typical social play in rats. Behav Pharmacol 26:707–719

Sadananda M, Wöhr M, Schwarting RKW (2008) Playback of 22-kHz and 50-kHz ultrasonic vocalizations induces differential c-fos expression in rat brain. Neurosci Lett 435:17–23

Sales GD (1972a) Ultrasound and mating behaviour in rodents with some observations on other behavioural situations. J Zool 168:149–164

Sales GD (1972b) Ultrasound and aggressive behaviour in rats and other small mammals. Anim Behav 20:88–100

Schneider T, Przewłocki R (2005) Behavioral alterations in rats prenatally exposed to valproic acid: animal model of autism. Neuropsychopharmacology 30:80–89

Seffer D, Rippberger H, Schwarting RKW, Wöhr M (2015) Pro-social 50-kHz ultrasonic communication in rats: post-weaning but not post-adolescent social isolation leads to social impairments—phenotypic rescue by re-socialization. Front Behav Neurosci 9:102

Sewell GD (1967) Ultrasound in adult rodents. Nature 215:512

Siviy SM, Panksepp J (1987) Sensory modulation of juvenile play in rats. Dev Psychobiol 20:39–55

Siviy SM, Panksepp J (2011) In search of the neurobiological substrates for social playfulness in mammalian brains. Neurosci Biobehav Rev 35:1821–1830

Siviy SM, Harrison KA, McGregor LS (2006) Fear, risk assessment, and playfulness in the juvenile rat. Behav Neurosci 120:49–59

Suomi SJ (2005) Genetic and environmental factors influencing the expression of impulsive aggression and serotonergic functioning in rhesus monkeys. In: Tremblay RE, Hartup WW, Archer J (eds) Developmental origins of aggression. Guilford Press, New York, pp 63–82

Takahashi LK, Thomas DA, Barfield RJ (1983) Analysis of ultrasonic vocalizations emitted by residents during aggressive encounters among rats (*Rattus norvegicus*). J Comp Psychol 97:207–212

Takeuchi H, Kawashima S (1986) Ultrasonic vocalizations and aggressive behavior in male rats. Physiol Behav 38:545–550

Thomas DA, Takahashi LK, Barfield RJ (1983) Analysis of ultrasonic vocalizations emitted by intruders during aggressive encounters among rats (*rattus norvegicus*). J Comp Psychol 97:201–206

Tornatzky W, Miczek KA (1994) Behavioural and autonomic responses to intermittent social stress: differential protection by clonidine and metoprolol. Psychopharmacol 116:346–356

Tornatzky W, Miczek KA (1995) Alcohol, anxiolytics and social stress in rats. Psychopharmacol 121:135–144

Trezza V, Baarendse PJ, Vanderschuren LJ (2010) The pleasures of play: pharmacological insights into social reward mechanisms. Trends Pharmacol Sci 31:463–469

Vanderschuren LJ, Trezza V (2014) What the laboratory rat has taught us about social play behavior: role in behavioral development and neural mechanisms. Curr Top Behav Neurosci 16:189–212

van Hooff J (1967) The facial displays of the catarrhine monkeys and apes. In: Morris D (ed) Primate Ethology. Weidenfeld & Nicolson, London, pp 9–88

van Oortmerssen GA (1971) Biological significance, genetics, and evolutionary origin of variability in behavior within and between inbred strains of mice (*Mus musculus*). Behav 38:1–91

Vivian JA, Miczek KA (1993) Diazepam and gepirone selectively attenuate either 20–32 or 32–64 kHz ultrasonic vocalizations during aggressive encounters. Psychopharmacol (Berl.) 112:66–73

Webber ES, Harmon KM, Beckwith TJ, Peña S, Burgdorf J, Panksepp J, Cromwell HC (2012) Selective breeding for 50 kHz ultrasonic vocalization emission produces alterations in the ontogeny and regulation of rough-and-tumble play. Behav Brain Res 229:138–144

Willey AR, Spear LP (2012) Development of anticipatory 50 kHz USV production to a social stimuli in adolescent and adult male Sprague-Dawley rats. Behav Brain Res 226:613–618

Willuhn I, Tose A, Wanat MJ, Hart AS, Hollon NG, Phillips PE, Schwarting RKW, Wöhr M (2014) Phasic dopamine release in the nucleus accumbens in response to pro-social 50 kHz ultrasonic vocalizations in rats. J Neurosci 34:10616–10623

Wilson SC (1973) The development of social behaviour in the vole (*Microtus agrestis*). Zool J Linn Soc 52:45–62

Wöhr M, Schwarting RKW (2007) Ultrasonic communication in rats: can playback of 50-kHz calls induce approach behavior? PLoS ONE 2:e1365

Wöhr M, Schwarting RKW (2008) Maternal care, isolation-induced infant ultrasonic calling, and their relations to adult anxiety-related behavior in the rat. Behav Neurosci 122:310

Wöhr M, Schwarting RKW (2013) Affective communication in rodents: ultrasonic vocalizations as a tool for research on emotion and motivation. Cell Tissue Res 354:81–97

Wöhr M, Engelhardt KA, Seffer D, Sungur AÖ, Schwarting RKW (2017) Acoustic communication in rats: effects of social experiences on ultrasonic vocalizations as socio-affective signals. Curr Top Behav Neurosci 30:67–89

Wright JM, Gourdon JM, Clarke PB (2010) Identification of multiple call categories within the rich repertoire of adult rat 50-kHz ultrasonic vocalizations: effects of amphetamine and social context. Psychopharmacol 211:1–13

Yanagi A, Berman CM (2014) Body signals during social play in free-ranging Rhesus macaques (*Macaca mulatta*): a systematic analysis. Am J Primatol 76:168–179

The Psycho-Neurology of Cross-Species Affective/Social Neuroscience: Understanding Animal Affective States as a Guide to Development of Novel Psychiatric Treatments

Jaak Panksepp

Abstract During the past half century of research with preclinical animal models, affective neuroscience has helped identify and illuminate the functional neuroanatomies and neurochemistries of seven *primary process*, i.e., genetically provided emotional systems of mammalian brains. All are subcortically localized, allowing animal models to guide the needed behavioral and neuroscientific analyses at levels of detail that cannot be achieved through human research, including modern brain imaging. They consist of the following neuronal processes: SEEKING/Enthusiasm, RAGE/Anger, FEAR/Anxiety, sexual LUST/Passion, maternal CARE/Nurturance, separation-distress PANIC/Grief and PLAY/Social Joy. Several of these systems figure heavily in social bonding. I will focus here especially on the genesis of depression. Its genesis is significantly influenced by (i) sustained overactivity of the separation-distress PANIC system reflecting severed social bonds and the excessive "psychological pain" of loneliness that can, if sustained, lead to a downward cascade known as psychological despair, and (ii) the despair phase that follows the acute PANIC response, which is characterized by abnormally low activity of the SEEKING, the so-called brain reward networks, leading to amotivational states that characterize depression. Depressive affect is promoted by such brain affective mechanisms of social attachments and social loss as well as diminished arousability of the SEEKING system, leading to chronic dysphoria. To understand why depression feels so bad, we must understand the neural mechanisms that mediate such social feelings.

J. Panksepp (✉)
Department of Integrative Physiology and Neuroscience, College of Veterinary Medicine, Washington State University, Pullman, WA 99164-6351, USA
e-mail: jpanksepp@vetmed.wsu.edu

Curr Topics Behav Neurosci (2016) 30: 109–126
DOI 10.1007/7854_2016_458
Published Online: 01 October 2016

Contents

1 Introduction: The Affective Neuroscience to Understanding the Neural Nature of Primal Emotional Feelings in Human Beings 110
2 The Overall Affective Neuroscience Strategy 112
3 Behavioral Models of Depression Contrasted with the Primal Affective Foundations of Depression 113
4 An Affective Neuroscientific Perspective on Why Depression Feel so Bad: Separation Distress/PANIC Is One Gateway to Depression 114
5 New PLAY Psycho-Chemotherapeutic Approaches: An Inroad to Antidepressant Development 118
6 Conclusions 119
References 121

Introductory Prelude: In September of 1965, as I initiated graduate work in clinical psychology at University of Massachusetts, Amherst, I was disappointed that there was no discussion of primal affective imbalances in various psychiatric disorders. That was the era of behavioral modification, where reinforcement contingencies were seen as the future of therapeutics. I promptly shifted to physiological psychology in order to study emotions in animals. Here is how my interest in emotions emerge: My last year of undergraduate work at University of Pittsburgh, I was working as an orderly in the back wards of a psychiatric hospital —a time when orderlies were allowed to read the case studies of all the patients in the nurses office in their spare time. As far as I could tell, I was the only orderly on the ward to pursue that self-education opportunity for understanding psychiatric issues and became fascinated by the nature of psychiatric problems, especially affective disorders. I realized promptly that the Darwinian view that all mammals share the same basic sets of emotional systems is true, and since then I remain convinced that the only way to credibly model raw human emotional feelings is to have animal neuroscience models where the basic science could be done. I could not have realized that at the end of my career this would remain as radical a position as it was at the beginning of my career, half a century ago.

1 Introduction: The Affective Neuroscience to Understanding the Neural Nature of Primal Emotional Feelings in Human Beings

The affective neuroscience approach to understanding the emotional mind of all mammals (Panksepp 1998, 2005a) is based on two premises that allow us to address the nature of valenced emotional experiences in non-human mammals. This provides model systems to address fundamental questions about the neural nature of the affective mind in basic animal research, which then should allow us to advance psychiatric clinical practice in both novel and more effective ways. First, brain emotional systems and their affective and behavioral manifestations evolved to *do* something specific, namely to encode survival values, in relation to behavioral and psychological strategies that routinely occur in diverse life-challenging situations. Emotions and their affective state are not mere epiphenomena. Secondly, the affective feelings of primary-process emotional arousals, namely diverse forms of valenced affects, appear to be built into the brain and shared homologously by all mammals.

The brain's affective action systems apparently motivate organisms to behave in ways that promote survival and hence maximize reproductive success. Positive emotional feelings (SEEKING, LUST, CARE and PLAY) automatically inform animals they are statistically on paths of survival. Negative affects (RAGE, FEAR and PANIC) inform animals they are in life-threatening situations. There are other categories of affects, for instance diverse *sensory* affects, from tastes and smells to orgasmic delights, as well as *homeostatic* affective systems, e.g., HUNGER and THIRST, which gauge bodily states needed for survival. In addition, there are mixed categories such as "disgust" that can be viewed as sensory (various bad tastes) or homeostatic (e.g., nausea) affects.

My goal here is to clarify how accruing knowledge about the *primal*, i.e., instinctual/unconditioned, emotional systems can promote better understanding of psychiatric disorders. Of course, these systems control a great deal of *secondary processes* that reflect learning and memory, much of which transpires in the basal ganglia (extremely well modeled by more strictly behavioral neuroscience models (LeDoux 2012; Rolls 2014), and higher cognitive *tertiary processes* heavily represented in neocortex, that are next to impossible to model in non-speaking animals. These systems can be conceptualized as "nested hierarchies" whereby the primary systems control learning, both of which are essential for tertiary, cognitive decision making. This nesting is not just evolutionary "layering" but the interpenetration of evolving systems that allow for organismic and psychological coherence. Obviously, evolution can continue to operate on many interactive details of the underlying brain systems that engender affective feelings.

The neurocausal manipulations that allow us to re-introduce emotional feelings (*affects*) of animal brains into the scientific conversations (Panksepp 1982, 1998) are the rewarding and punishing properties of deep-brain stimulations (DBS) that evoke distinct emotional behaviors. The key finding is that wherever in the brain we arouse emotional-instinctual behavior patterns, we can demonstrate that those brain arousals can serve as "rewards" and "punishments" in the control of learned behavior, i.e., self-stimulation of affectively positive emotional systems and escape from negative emotional states evoked by DBS.

The various primal affective states are presumably important in the genesis of various affective disorders, especially depression. Depression is promoted by diminished enthusiasm (underactivity of SEEKING) and reduced capacity to experience social joy (diminished PLAY arousal), as well as excessive psychological pain commonly promoted by social loss (too much PANIC arousal). Those primary processes can be directly studied in animals with DBS as well as instinctual indicators of emotional arousal. Obviously, animal brain behavior research cannot access higher-order cognitive changes that are common in humans, e.g., negative thoughts and ruminations that characterize human depression. Conversely, human research cannot readily access the subcortical emotional networks that critically important for human affective life.

2 The Overall Affective Neuroscience Strategy

Affective neuroscience has now outlined the general neuroanatomy (and some of the neurochemistries) of seven *primary processes*, i.e., genetically provided, emotional systems. All are subcortically concentrated (for extensive summaries, see Panksepp (1982, 1998, 2005a, b, c) please note that these systems largely survive neonatal decortication), allowing animal models to guide our understanding of homologous brain systems and functions in human beings. The emotional primes include at least the following seven fundamental neuropsychological processes: (i) SEEKING, (ii) RAGE, (iii) FEAR, (iv) sexual LUST, (v) maternal CARE, (vi) separation-distress PANIC/GRIEF (henceforth, simply PANIC) and (vii) joyful PLAY.

The capitalized terminology was chosen to highlight their primary-process nature and to cue readers that I am talking about *evolutionarily primal* or *basic* emotional processes that have dedicated brain systems that are increasingly recognized in modern psychiatry albeit with many different terminologies, i.e., they are official scientific names for brain systems that exist, with species typical variations not yet documented (Panksepp 2004, 2006). The terminology allows clearer discourse than the vernacular counterparts. The theoretical suggestion is that in humans, these valenced affects would, respectively, heavily influence psychological processes that in the vernacular are called: enthusiasm, anger, anxiety, passion, loving nurturance, psychological pain of loneliness, and social joy. Because of the importance of such psychological states for psychiatry, I devoted, together with my colleagues, considerable effort to develop personality tests specifically designed to evaluate the status of these affective states in humans (Davis et al. 2003; Davis and Panksepp 2011).

Considering depression as an example, I suggest that (i) sustained overactivity of the separation-distress PANIC can, if sustained, lead to an affectively negative brain cascade promoting depression with (ii) this sustained despair phase is characterized by diminished activity of the SEEKING (so-called brain reward networks), leading to depression. In other words, depression may reflect an emotional despair phase following protracted PANIC arousal. This neuropsychological point of view is completely concordant with the theoretical vision originally formulated by Bowlby (1960, 1980).

Specifically, I argue that the "primary-process" emotion systems that, in affective neuroscience terminology, are labeled as SEEKING and PANIC, are of direct importance understanding and treating the underlying affective dynamics that can lead to depression. The discussion of one of the underlying systems, namely the PANIC system, is framed within the seminal contributions of John Bowlby, who provided the clearest statement of how the genesis of depression is related intimately to the loss of supportive social bonds, such as the loss of parents in young children, and the death life-companions, now widely recognized as major antecedents to the psychological despair that can lead to depression upon which the present view is explicitly based (for a more detailed summary, see Watt and Panksepp (2009) and also see Panksepp and Watt (2011), Wright and Panksepp (2011, 2012) and Zellner et al. (2011) for neuropsychoanalytic perspectives).

These systems are "evolutionary memories," namely (i) affective psychobehavioral functions built into the brain and (ii) interacting with world events promote learning, especially via basal ganglia (e.g., amygdala, bed nucleus of the stria terminalis) to refine survival-promoting learned behavioral strategies. That acquired knowledge allows the top of the brain (neocortex) to generate adaptive decision making, namely to think. The most important practical contribution that a cross species affective neuroscience can provide is how subcortical brain systems can generate emotional feelings of critical importance for psychiatry, as well as psychology.

3 Behavioral Models of Depression Contrasted with the Primal Affective Foundations of Depression

Affective neuroscience offers testable hypotheses concerning the subcortical nature of affective states and imbalances, with implications for understanding human minds (Solms and Panksepp 2010; Zellner et al. 2011). A major gateway is sustained overactivity of the PANIC system, which was behaviorally modeled preclinically through the study of brief periods of separation-induced crying in the young of various species, especially dogs, guinea pigs and baby chickens (Panksepp et al. 1980, 1981, 1988, 2005b, c). Perhaps the negative affect generated by too much social separation diminishes arousability of the dopamine-energized, reward-SEEKING system, yielding eventually chronic anhedonic (despair) and other negative amotivational (depressive) states of mind, which may be treated, as a last resort, by direct DBS of the reward-SEEKING system (see Coenen et al. 2011; Schlaepfer et al. 2008). Thus, the animal emotion work has direct implications for understanding the sources of depressive "pain." In combination with associated neurochemical analyses, it can lead to the development of novel therapeutics (see Burgdorf et al. 2010, 2011; Moskal et al. 2011; Preskorn et al. 2015), something that has not yet been achieved by important generalized stress research, as pursued by many (e.g., de Kloet et al. 2005; McEwen 2007), including the various general brain biogenic amine changes (e.g., Schildkraut 1965; Harro and Oreland 2001).

Behavioral analyses have long focused on the capacity of very general brain neurochemical "state-control" systems, such as norepinephrine and serotonin, to regulate all emotional behaviors. Such very general overall state-control systems, which regulate everything animals do, are unlikely to *specifically* explain the morbid moods of depression (Delgado et al. 1990). Indeed, we now know that serotonin and norepinephrine regulate widespread but generalized brain arousal functions that influence all higher brain processes from cognitions to emotions to wakefulness. Thus, selective serotonin reuptake inhibitors, SSRIs, can provide only modest benefits for many psychiatric patients when evaluated across large cohorts of individuals, as highlighted by rather disappointing overall finding in the famous STAR*D studies (Rush et al. 2003; Rush 2007). Other recent preclinical work focusing on various neurotrophic factors (Koziek et al. 2008) and stress-induced hippocampal shrinkage and inflammation in the central nervous system (Miller et al. 2009) accompanied by diverse autonomic, psychophysiological problems (Levinson 2006) have also generated few new medicines. This may be because such strategies do not directly illuminate affective feelings that characterize depression.

In short, behavior-only preclinical models of depression, which are abundant (see the Special issue of *Neuroscience and Biobehavioral Reviews* vol. 30 (2006), have yielded very few new therapeutic breakthroughs). More affect-specific neuroscience approaches (e.g., as addressed by Burgdorf et al. 2011; Kroes et al. 2007; Watt and Panksepp 2009; Wright and Panksepp 2011, 2012) have the potential to do better.

4 An Affective Neuroscientific Perspective on Why Depression Feel so Bad: Separation Distress/PANIC Is One Gateway to Depression

Affective neuroscience has thus provided distinct brain networks that can help explain the psychological pain and general dysphoria of depression, especially as envisioned through the overactivity of brain separation-distress PANIC and the underactivity of the SEEKING system. The PANIC system has been proposed to be a primary emotional systems for social-loss-induced psychological pain (Panksepp 2003a, b, 2005b, c, 2010) and hence a conceptually clear foundation process for the biggest epidemiological stressor which most commonly leads to depression—social loss (Bowlby 1980; Heim and Nemeroff 1999).

The PANIC system probably evolved from general pain mechanisms (Panksepp 1981, 1998, 2003a, b) presumably well over a hundred million years ago or more, with birds possessing a homologous system. This critical opioid modulated system promotes social connection, helps forge social attachments and dependencies between infants and mothers, probably fortifies sexual relationships, and may ultimately be the foundation of group solidarity among group living species. As a reasonable index of its role in attachment, I would argue that when one misses someone with whom one is bonded, this system is aroused to some extent, underscoring its centrality in human social affairs. If someone is never missed, this suggests that one does not have an attachment to that individual. Indeed, when socially separated, the affective consequences of severed attachment bonds make individuals suffer in a distinct and powerfully aversive fashion. This type of psychological pain, which most humans will generally avoid at almost all costs, is apparently a gateway to major forms of depression.

The acute separation-distress response, although providing affective impact for depressive disorders, may not constitute a sustained depressive psychopathology on its own. For that, a set of neuroaffective changes are set in motion that promote feelings of lassitude and despair, and the neuroscience of these processes is not yet so complete. One line of research is suggesting that immune modulators, e.g., cytokines, such as interleukin 1, IL-6 and TNF-α, can instigate sickness-type affective states with relevance to the sustained despair of depression (Hennessy et al. 2001). An equally promising possibility is that sustained separation distress cascades into despair because of the ensuing diminution of SEEKING urges.

When protest fails to ensure social reunion, a gradual behavioral and psychological shutdown/depression emerges. At this critical transition from the protest to the despair phase of depression, a new form of sustained negative affect emerges, which is a fully developed depressive phenotype. The further elevation of negative affect contributed by "giving up" may yield a mixture of the sustained psychic pain of separation intermingled with the inability to recruit mental energies such as SEEKING-euphoria that characterizes a positive attitude to life. This end state is characterized by diminished engagements with the world and reduced pursuit of rewards, real or imagined.

This giving-up "despair" phase may need to be counteracted not only by brain chemistries that reduce the psychic pain of loss but also ways to elevate dopamine-driven SEEKING urges that characterize depressive despair. In a sense, opioid drugs can do both, yielding dopamine independent pleasures and promoting dopamine-SEEKING urges, especially at low doses. Thus, in the emergence of depressive affect, it is as important to emphasize the lassitude of diminished SEEKING as the lingering psychic pain and emptiness of separation distress. Indeed, ever since Anisman and Matheson's (2005) discovery that stressors that promote depressive profiles in animal models are accompanied by elevated thresholds in "brain reward" SEEKING arousal, there have been periodic reports of similar findings by others (Nestler and Carlezon 2006; Pereira Do Carmo et al. 2009). What causes this reduction in SEEKING urges is a central question for depression research. A key candidate is the gradually increasing influence of dynorphins, powerful and pervasive brain opioids that mediate a very distinct form of negative affect that is recruited by social loss, and demonstrably reduces the responsivity of the brain reward-SEEKING system (McLaughlin et al. 2006) both at synaptic terminals (Mu et al. 2011), and at closely related global neuropeptide modulators of positive arousal and affect such as orexin (Nocjar et al. 2012).

A solid neuroscience of such brain processes has been garnered through the mapping of the neuroanatomies and neurochemistries of separation distress, and they tell us something specific about social attachments and loss, as summarized in

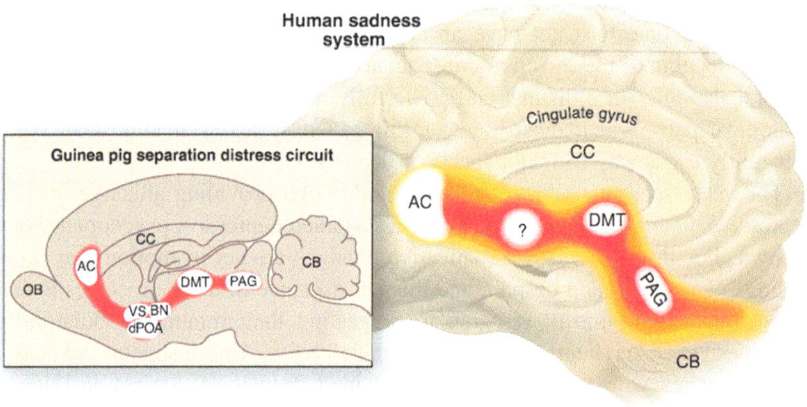

Fig. 1 Human and animal sadness of sadness and separation-distress systems. Animal data come from localized brain stimulation mapping of separation-distress circuits in guinea pigs (Herman and Panksepp 1981) and human data from Damasio et al. 2000; graph is taken from Panksepp 2003b

Fig. 1 (Panksepp 1998, 2003a, b). It is remarkable that these same systems mediate various addictive urges that can emerge from fundamental social bonding systems of the brain (Panksepp 1981). In fact, various addictive processes emerge from the capacity of some drugs such as opioids to modify affective states to artificially simulate social bonding feelings that are of critical importance for mental health.

The PANIC circuitry starts in midbrain central gray regions, currently commonly called the periaqueductal gray (PAG), and it ascends through medial diencephalic structures, especially the dorsomedial thalamus, and terminates in more ventral or subcallosal anterior cingulate forebrain regions. Inhibition of this system with DBS may have already figured positively in the direct neural systems modulation of treatment resistant depressions (Mayberg et al. 2005; Mayberg 2009).

The key neurochemistries that promote separation distress, and thus protest calls, are declining opioid and oxytocin chemistries and elevated corticotropin-releasing factor, combined with increased glutamatergic drive in PANIC circuits of the brain, with the neuropeptides being presumably more important than the excitatory amino acid in controlling the specific social-affective responses of the brain. Still, inhibition of both neuropeptide and excitatory amino acid (e.g., glutamatergic) promoters of PANIC (e.g., Normansell and Panksepp 2011; Panksepp et al. 1988) should help alleviate the bad feelings of depression, and recent work along these lines has been consistently promising (Holsboer 2000; Zarate et al. 2006).

The separation-distress mediating PANIC system is regulated by various prosocial neuropeptides that also promote CARE and PLAY behaviors, e.g., endogenous opioids, oxytocin and prolactin (Panksepp 1981, 1998). The ability of these systems to consolidate social bonds (Panksepp 1981, 1998) may also help explain why depression is almost twice as common in females than males. Female brains may be intrinsically more responsive to prosocial emotions than male brains (Swain et al. 2007). Each prosocial neuropeptide could be considered as a potential vector for beneficially countering the affective changes that promote depression. We should not forget that in primates social grooming releases brain opioids (Keverne et al. 1989, 1997) and human voices are, in part, a way for our species to groom each other indirectly. Of course, one would hesitate to use addictive opioids, which very effectively, but only temporarily, alleviate depression as routine treatments, because of their addictive liability. However, safe opioids such as ultra low-dose buprenorphine, which only stimulate opioid receptors at low doses and become antagonists at high doses, are very effective antidepressants for individuals where no other medications have provided sustained relief of depression (Bodkin et al. 1995; Yovell et al. 2016).

With regard to buprenorphine, prior to the modern era of psychopharmacology, psychiatrists only had opioids for treating mental suffering (Tenore 2008). Although very effective antidepressants in the short term, their addictive potential discouraged long-term use, even though one could obtain potent sustained effects with low prescription doses. Still, widespread addiction phobias have precluded full empirical evaluation of such ideas. The mixed mu-opioid receptor agonist/antagonist buprenorphine solves most of these problems, and open trials have highlighted the high and sustained efficacy of low doses in depressed clients who have had no relief from many accepted antidepressants (Bodkin et al. 1995). This "miracle drug" (long off patent) also has the uniquely desirable effect of blocking dynorphin receptors that are widespread in the brain, including suppressive effects on the euphoric potentials of the brain's reward-SEEKING system. Since high doses of buprenorphine actually block addictive mu-receptors, the drug has a fail-safe mechanism that limits addictive escalations and the ensuing abuse that characterizes pure opiate receptor stimulants, with their risk of respiratory arrest. One reason this medication has been badly neglected in research (few proper follow-up to Bodkin et al. (1995), provocative study in refractory depression but see the Yovell et al. (2016)) is its seriously diminished profit margin (it is off patent), as well as the resulting diminished financial investments available for conducting expensive clinical trials needed for medical approval.

Centrally administered oxytocin is also remarkably effective in alleviating separation distress and social bonding in animal models (Panksepp 1992; Nelson and Panksepp 1998; Uvnäs-Moberg 1998). Whether non-peptide oxytocinergic drugs that cross the blood–brain barrier, yet to be developed, can be harnessed to help re-establish affective homeostasis in an excessively aroused PANIC system in human beings remains to be seen, with psychological effects observed in intranasal studies being promising (e.g., Heinrichs and Domes 2008; Insel 2010; Nelson and Panksepp 1998). In this context, it is noteworthy that one of the rarely considered effects of this neuropeptide is its capacity to sustain the activity of endogenous opioid processes, thereby sustaining positive social feelings, by inhibiting the development of tolerance to opioids (Kovács et al. 1998), which is a pathway for better social-affect regulation (Panksepp 1981, 1998, 2005c).

In sum, although negative affective changes in the opioid- and oxytocin-driven attachment and affectional systems may be the pivotal precipitants of depressive cascades, it is the affective dysphoria of diminished SEEKING urge that puts "the nail in the coffin" so to speak. This scenario remains consistent with biogenic amine theories of depression, since those general features of brain-mind organization participate in the overall arousal level of every emotion animals exhibit. Because of the multi-dimensionality of depression, there are bound to be many variants on these basic themes among the many subtypes of depression. For example, the sustained affective separation-induced psychic-pain response may characterize some depressions more than others, while dynorphin-facilitated shutdown of dopamine-driven appetitive SEEKING (i.e., when some depressed persons "give up" in an almost illness-type of amotivational lassitude) may constitute another variant of depression (Nocjar et al. 2012).

Just as with psychotherapeutic disciplines, the goal of psychopharmacology is to counteract and reverse this downward cascade. In my estimation, new therapeutic approaches that take advantage of the positive hedonics of social CARE systems (the primal foundation of empathy) and PLAY systems (the primal source of social joy) may be especially important for better therapeutic outcomes. However, rather than develop psychotherapeutic concepts for treatment of acute depressive episodes by promoting social re-connection and re-attachment of depressed individuals, my remaining goal here will be to introduce new emerging concepts in chemotherapeutics of depression.

5 New PLAY Psycho-Chemotherapeutic Approaches: An Inroad to Antidepressant Development

During social PLAY, rats emit high-frequency ultrasonic vocalizations (Wöhr and Schwarting 2013), so-called 50-kHz USV (Burgdorf et al. 2008; Knutson et al. 1998; Lukas and Wöhr 2015). It is now widely believed that they reflect a positive affective state, and the term "rat laughter" was coined (Panksepp 2005d). Because it is difficult to experimentally control social PLAY behavior in rats, I developed, together with my colleague Jeffrey Burgdorf, an animal model of positive affect, namely systematic tickling of rats, as a mean to mimic social PLAY, which can bring hedonic 50-kHz USV, as during social PLAY, under experimental control (Panksepp and Burgdorf 2000, 2003). Systematic tickling can be used as a positive affect "assay," and it was demonstrated that antidepressant-type hippocampal neuronal proliferation can be promoted through tickling, i.e., by systematic playfulness that elevates such happy-playful 50-kHz USV in rats (Wöhr et al. 2009). Furthermore, these 50-kHz USV have been mapped within the brain to activation of the mesolimbic dopamine system providing a direct readout of the responsivity of this positive affective response (e.g., euphoric eagerness) that can help counteract depressive affect (Burgdorf et al. 2007). Importantly, those affectively positive

50-kHz USV, along with their counterpart, the negative 22-kHz USV, can therefore be used to index-specific affective shifts that can illuminate the underlying changes in affect that characterize depression and thus serve as a marker to assess treatment efficacy of novel drugs for treating depression (Kroes et al. 2007).

In this context, I would also note that the "power of PLAY" in psychotherapy remains largely completely untapped, at least in any systematic way. There are good reasons to believe that the long-term recruiting of such mental energies would be effective for the amelioration of various recalcitrant childhood problems, such as childhood impulsivity (Panksepp 2007), through the capacity of such prosocial activities to promote both socialization and brain maturation. For instance, play can "fertilize" the brain by promoting growth factors such as brain-derived neurotrophic factor (BDNF) gene expression within the brain (Gordon et al. 2003). BDNF is well known to promote antidepressant effects in the brain through various genetic cascades and thereby opposes the hippocampal dysgenesis that often accompanies depression (McEwen 2000, 2007).

In fact, the analysis of the genetic changes in animals undergoing abundant social PLAY, which operates in part through the mesolimbic dopamine-energized function of the medial forebrain bundle centered SEEKING system (Burgdorf et al. 2007; Panksepp and Moskal 2008), has yielded a variety of targets for new drug development (Moskal et al. 2011; also see Krishnan and Nestler (2008) for related work) and has led to the identification of other growth factors that may prove to be affectively positive adjuncts to playful psychotherapy. One of the biggest gene expression changes in the neocortex is in the elevated expression of insulin-like growth factor-1 (IGF-1; Burgdorf et al. 2010). When this growth factor was tested for functional changes in relevant social behaviors, using direct intracerebral injections of an IGF-1 receptor antagonist, as well as siRNA inhibition of IGF-1 brain activity, it yielded convergent evidence for the role of IGF-1 in promoting positive affect (Burgdorf et al. 2010). There are reasons to suspect that further research on the positive social-affect systems of the mammalian brain will yield new ways to promote feeling of secure affective well-being that can help counteract depressive cascades, but because of its capacity to promote tumor growth, further development of this medicinal concept was abandoned. The second top candidate was within the glutamate receptor-related family of regulatory systems and found to facilitate positive affect in preclinical models, i.e., it was demonstrated to have an antidepressive, pro-hedonic profile. One agent, GLYX-13, was taken through both animal and human toxicology, with no adverse effects observed (Burgdorf et al. 2011), and is currently in successful human clinical evaluation.

6 Conclusions

This brain analysis has built upon the psychological insights of Bowlby (1980), who originally conjectured that depression arises from sustained separation distress, which is followed, if sustained for too long, by chronic depressive despair. Based on affective neuroscience strategies (Panksepp 1998), we now have abundant data on the brain mechanisms of separation distress, and hence, the protest phase that leads to depression, offering ever better neuroscience views of how diminished SEEKING urges tend to promote a depressive phenotype when this system fails to sustain protest (e.g., sustained separation calls), but with diminished activity, leads to despair. The separation-distress mediating PANIC system is regulated by various prosocial neuropeptides that also promote CARE and PLAY behaviors, e.g., endogenous opioids, oxytocin and prolactin.

The pain of depression arising commonly from social loss and social defeat may be the price we mammals pay for the evolutionary advantages of social bonds that enormously enhance our survival and procreation, to say nothing of enriching our affective lives. Although animal research cannot inform us of the complex cognitive-affective amalgams (especially the ruminations and the "darkenings" of cognitions) that emerge in humans during depression, they can inform us of the evolutionarily conserved brain-mind affective mechanisms that lie at the very heart of depressive despair. From this point of view, the potential for depression is intimately linked to the pain of social loss and resulting diminution of engagement with the world that is an intrinsic vulnerability of highly prosocial brains.

The breadth and depth of our human cognitive consciousness have been widened enormously by the thought-filled intellectual potentials of our enlarged brains, and the resulting cultural supports that have been constructed historically. However, we remain inheritors of ancient biological values that constitute the very ground of meaning and existence within our minds. Although this affective ground of meaning is very hard to talk about clearly, it is from within our ancient animalian nature, full of primary-process affects that the subjectively experienced blessings and curses of our existence emerge. The primary-process emotion/affect generating systems are all situated in ancient medially situated subcortical brain regions that all mammals share because of their common ancestry, with large longitudinally coursing emotional systems. These powers of the mind get connected to many life experiences through learning, but their affective intensity is an evolutionary birthright of all mammalian minds. This makes the study of comparative neurophenomenology critically important for unraveling the affective processes that make depression, and many other emotional problems of the mind, affectively horrible. Affective neuroscience strategies have allowed us to envision how John Bowlby's seminal conceptual work on the genesis of depression can now be linked to specific affective networks that can be studied, in causal detail, in preclinical animal models.

In conclusion, why does depression feel bad? It feels bad, from my perspective, for two reasons, both related to diminished feelings of internal security: First, because of its intrinsic relationship to separation distress, the PANIC circuitry encourage us to form *and maintain* addictive attachments, particularly to early care-giving figures, but also with our adult companions and children, as well as extended social groups. Secondly, depression persuades us to give up hope if our attempts to re-unite with such figures or groups do not succeed within a limited timeframe, and thereby we become psychologically detached from the world. This sustained loss of affective "energy" which depletes cognitive "meaning" may be intimately linked to diminished SEEKING urges, which has long been known to be the epi-center of all major forms of drug addiction. In short, at the neurochemical level, we are addicted to the ones we love.

In light of the existence of brain structures that generate such feelings, it seems reasonable to hypothesize that the linchpin of at least one major form of depression rarely the things that have preoccupied contemporary psychiatric researchers over the past three decades. It is rather the evolutionarily conserved brain state that mediates the transition from “protest” to “despair” in the wake of social loss. I see this to be intimately linked to reduced arousal of the dopamine-energized SEEKING system. In other words, it seems reasonable to hypothesize that the core brain basis of depression revolves around the process by which separation distress is normally shut down (possibly by diminished dopamine arousal, declining mu and delta and increasing kappa-opioid activities, together with various inflammatory cytokines, which prompt animals and humans to “give up” when the affective mind is overfilled with distress). Affective neuroscience offers new strategies to counteract such degradations of mind, by analyzing the underlying details of the core affective processes that all mammals share.

In sum, as John Bowlby well-recognized, separation distress—the panicy “protest” that promptly follow social loss, especially in young animals—feels bad in a special way. It is a feeling of psychic pain. This knowledge can help clinicians craft new psychotherapeutic approaches, including the better utilization of positive affects that arise from SEEKING and PLAY systems, including new pharmacological and DBS treatments for depression. By taking the emotional feelings of animals seriously as targets for understanding human affective disorders, we are more likely to develop novel therapeutics than has been achieved with mere behavioral modeling (for additional perspectives on these issues, see Alcaro and Panksepp 2011; Panksepp et al. 1981, 2014; Panksepp 2015, 2016.

References

Alcaro A, Panksepp J (2011) The seeking mind: primal neuro-affective substrates for appetitive incentive states and their pathological dynamics in addictions and depression. Neurosci Biobehav Rev 35:1805–1820

Anisman H, Matheson K (2005) Stress, depression, and anhedonia: caveats concerning animal models. Neurosci Biobehav Rev 29:525–546

Bodkin JA, Zornberg GL, Lukas SE, Cole JO (1995) Buprenorphine treatment of refractory depression. J Clin Psychopharmacology 15:49–57

Bowlby J (1960) Separation anxiety. Int J Psychonanal 41:89–113

Bowlby J (1980) Attachment and loss, vol 3. Loss: sadness and depression. Basic Books, New York

Burgdorf J, Wood PL, Kroes RA, Moskal JR, Panksepp J (2007) Neurobiology of 50-kHz ultrasonic vocalizations in rats: electrode mapping, lesion, and pharmacology studies. Behav Brain Res 182:274–283

Burgdorf J, Kroes RA, Moskal JR, Pfaus JG, Brudzynski SM, Panksepp J (2008) Ultrasonic vocalizations of rats (Rattus norvegicus) during mating, play, and aggression: behavioral

concomitants, relationship to reward, and self-administration of playback. J Comp Psychol 122:357–367
Burgdorf J, Kroes RA, Beinfeld MC, Panksepp J, Moskal JR (2010) Uncovering the molecular basis of positive affect using rough-and-tumble play in rats: a role for insulin-like growth factor I. Neuroscience 163:769–777
Burgdorf J, Panksepp J, Moskal JR (2011) Frequency-modulated 50 kHz ultrasonic vocalizations: a tool for uncovering the molecular substrates of positive affect. Neurosci Biobehav Rev 35:1831–1836
Coenen VA, Schlaepfer TE, Maedler B, Panksepp J (2011) Cross-species affective functions of the medial forebrain bundle-Implications for the treatment of affective pain and depression in humans. Neurosci Biobehav Rev 35:1971–1981
Damasio AR, Grabowski TJ, Bechara A, Damasio H, Ponto LLB, Parvizi J, Hichwa RD (2000) Subcortical and cortical brain activity during the feeling of self-generated emotions. Nat Neurosci 3:1049–1056
Davis KL, Panksepp J (2011) The brain's emotional foundations of human personality and the affective neuroscience personality scales. Neurosci Biobehav Rev 35:1946–1958
Davis KL, Panksepp J, Normansell L (2003) The affective neuroscience personality scales: Normative data and implications. NeuroPsychoanalysis 5:21–29
De Kloet E, Joels M, Holsboer F (2005) Stress and the brain: from adaptation to disease. Nat Rev Neurosci 6:463–475
Delgado PL, Charney DS, Price LH, Aghajanian GK, Landis H, Heninger GR (1990) Serotonin function and the mechanism of antidepressant action: reversal of antidepressant-induced remission by rapid depletion of plasma tryptophan. Arch Gen Psychiatry 47:411–418
Gordon NS, Burke S, Akil H, Watson S, Panksepp J (2003) Socially-induced brain 'fertilization': play promotes brain derived neurotrophic factor transcription in the amygdala and dorsolateral frontal cortex in juvenile rats. Neurosci Lett 341:17–20
Harro J, Oreland L (2001) Depression as a spreading adjustment disorder of monoaminergic neurons: a case for primary implications of the locus coeruleus. Brain Resh Rev 38:79–128
Heim C, Nemeroff C (1999) The impact of early adverse experiences on brain systems involved in the pathophysiology of anxiety and affective disorders. Biol Psychiat 46:1509–1522
Heinrichs M, Domes G (2008) Neuropeptides and social behaviour: effects of oxytocin and vasopressin in humans. Prog Brain Res 170:337–350
Hennessy MB, Deak T, Schiml-Webb PA (2001) Stress-induced sickness behaviors: an alternative hypothesis for responses during maternal separation. Dev Psychobiol 39:76–83
Herman BH, Panksepp J (1981) Ascending endorphinergic inhibition of distress vocalization. Science 211:1060–1062
Holsboer F (2000) The corticosteroid receptor hypothesis of depression. Neuropsychopharmacology 23:477–501
Insel TR (2010) The challenge of translation in social neuroscience: a review of oxytocin, vasopressin, and affiliative behavior. Neuron 65:768–779
Keverne EB, Martensz N, Tuite B (1989) β-Endorphin concentrations in CSF of monkeys are influenced by grooming relationships. Psychoneuroendocrinology 14:155–161
Keverne EB, Nevison CM, Martel FL (1997) Early learning and the social bond. Ann N Y Acad Sci 807:329–339
Knutson B, Burgdorf J, Panksepp J (1998) Anticipation of play elicits high-frequency ultrasonic vocalizations in young rats. J Comp Psychol 112:65–73
Kovács GL, Sarnyai Z, Szabó G. (1998) Oxytocin and addiction: a review. Psychoneuroendocrino 23:945–962
Koziek M, Middlemas D, Bylund D (2008) Brain-derived neurotrophic factor and its receptor tropomyosin-related kinase B in the mechanism of action of antidepressant therapies. Phamacological Ther 117:30–51
Krishnan V, Nestler EJ (2008) The molecular neurobiology of depression. Nature 455:894–902

Kroes RA, Burgdorf J, Otto NJ, Panksepp J, Moskal JR (2007) Social defeat, a paradigm of depression in rats that elicits 22-kHz vocalizations, preferentially activates the cholinergic signaling pathway in the periaqueductal gray. Behav Brain Res 182:290–300

LeDoux J (2012) Rethinking the emotional brain. Neuron 73:653–676

Levinson DF (2006) The genetics of depression: a review. Biol Psychiatry 60:84–92

Lukas M, Wöhr M (2015) Endogenous vasopressin, innate anxiety, and the emission of pro-social 50-kHz ultrasonic vocalizations during social play behavior in juvenile rats. Psychoneuroendocrinology 56:35–44

Mayberg HS (2009) Targeted electrode-based modulation of neural circuits for depression. J Clin Invest 119:717–725

Mayberg H, Lozano A, Voon V, McNeely H, Seminowicz D, Hamani C, Schwalb J, Kennedy S (2005) Deep brain stimulation for treatment-resistant depression. Neuron 45:651–660

McEwen BS (2000) The neurobiology of stress: from serendipity to clinical relevance. Brain Res 886:172–189

McEwen BS (2007) Physiology and neurobiology of stress and adaptation: central role of the brain. Physiol Rev 87:873–904

McLaughlin JP, Li S, Valdez J, Chavkin TA, Chavkin C (2006) Social defeat stress-induced behavioral responses are mediated by the endogenous kappa opioid system. Neuropsychopharmacology 31:1241–1248

Miller AH, Maletic V, Raison CL (2009) Inflammation and its discontents: the role of cytokines in the pathophysiology of major depression. Biol Psychiatry 65:732–741

Moskal JR, Burgdorf J, Kroes RA, Brudzynski SM, Panksepp J (2011) A novel NMDA receptor glycine-site partial agonist, GLYX-13, has therapeutic potential for the treatment of autism. Neurosci Biobehav Rev 35:1982–1988

Mu P, Neumann PA, Panksepp J, Schlüter OM, Dong Y (2011) Exposure to cocaine alters dynorphin-mediated regulation of excitatory synaptic transmission in nucleus accumbens neurons. Biol Psychiatry 69:228–235

Nelson E, Panksepp J (1998) Brain substrates of infant-mother attachment, contributions of opioids, oxytocin, and norepinephrine. Neurosci Biobehav Rev 22:437–452

Nestler EJ, Carlezon WA Jr (2006) The mesolimbic dopamine reward circuit in depression. Biol Psychiatry 59:1151–1159

Nocjar C, Zhang J, Feng P, Panksepp J (2012) The rat social-defeat model of depression shows diminished levels of orexin in mesocortical regions of the brain reward seeking system, and of dynorphin and orexin in the hypothalamus. Neuroscience 218:138–153

Normansell L, Panksepp J (2011) Glutamatergic modulation of separation distress: profound emotional effects of excitatory amino acids in chicks. Neurosci Biobehav Rev 35:1890–1901

Panksepp J (1981) Brain opioids: a neurochemical substrate for narcotic and social dependence. In: Cooper S (ed) Progress in theory in psychopharmacology. Academic Press, London, pp 149–175

Panksepp J (1982) Toward a general psychobiological theory of emotions. Behav Brain Sci 5:407–467

Panksepp J (1992) Oxytocin effects on emotional processes: separation distress, social bonding, and relationships to psychiatric disorders. Ann N Y Acad Sci 652:243–252

Panksepp J (1998) Affective neuroscience: the foundations of human and animal emotion. Oxford University Press, New York

Panksepp J (2003a) Can anthropomorphic analyses of "separation cries" in other animals inform us about the emotional nature of social loss in humans? Psychol Rev 110:376–388

Panksepp J (2003b) Feeling the pain of social loss. Science 302:237–239

Panksepp J (2004) (ed) Textbook of biological psychiatry. Wiley-Liss, Inc., Hoboken

Panksepp J (2005a) Affective consciousness: core emotional feelings in animals and humans. Conscious Cogn 14:19–69

Panksepp J (2005b) Feelings of social loss: the evolution of pain and the ache of a broken heart. In: Ellis R, Newton N (eds) Consciousness & emotions, vol 1. John Benjamins, Amsterdam, pp 23–55

Panksepp J (2005c) Why does separation-distress hurt?: a comment on macdonald and leary. Psychol Bull 131:224–230
Panksepp J (2005d) Psychology. Beyond a joke: from animal laughter to human joy? Science 308:62–63
Panksepp J (2006) Emotional endophenotypes in evolutionary psychiatry. Prog Neuropsychopharmacol Biol Psychiat 30:774–784
Panksepp J (2007) Can PLAY diminish ADHD and facilitate the construction of the social brain. J Can Acad Child Adolesc Psychiat 10:57–66
Panksepp J (2010) The neurobiology of social loss in animals: some keys to the puzzle of psychic pain in humans. In: MacDonald G, Jensen-Campbell LA (eds) Social pain: neuropsychological and health implications of loss and exclusion. American Psychological Association, Washington, D.C., pp 11–52
Panksepp J (2015) Affective preclinical modeling of psychiatric disorders: taking imbalanced primal emotional feelings of animals seriously in our search for novel antidepressants. Dialog Clin Neurosci. 17:363–379
Panksepp J (2016) The cross mammalian neurophenomenology of primal emotional affects: From animal feelings to human therapeutics. J Comp Neurol 524:1624–1635
Panksepp J, Burgdorf J (2000) 50-kHz chirping (laughter?) in response to conditioned and unconditioned tickle-induced reward in rats: effects of social housing and genetic variables. Behav Brain Res 115:25–38
Panksepp J, Burgdorf J (2003) Laughing rats and the evolutionary antecedents of human joy? Physiol Behav 79:533–547
Panksepp J, Moskal J (2008) Dopamine and seeking: subcortical "reward" systems and appetitive urges. In: Elliot A (ed) Handbook of approach and avoidance motivation. Taylor & Francis Group LLC, New York, pp 67–87
Panksepp J, Watt J (2011) Why does depression hurt? Ancestral primary-process separation-distress (PANIC) and diminished brain reward (SEEKING) processes in the genesis of depressive affect. Psychiatry 74:5–14
Panksepp J, Herman BH, Vilberg T, Bishop P, DeEskinazi FG (1980) Endogenous opioids and social behavior. Neurosci Biobehav Rev 4:473–487
Panksepp J, Normansell LA, Herman B, Bishop P, Crepeau L (1988) Neural and neurochemical control of the separation distress call. In: Newman JD (ed) The physiological control of mammalian vocalizations. Plenum, New York, pp 263–300
Panksepp J, Wright JS, Döbrössy MD, Schlaepfer TE, Coenen VA (2014) Affective neuroscience strategies for understanding and treating depressions: from preclinical models to novel therapeutics. Clin Psychol Sci 2:472–494
Pereira Do Carmo G, Stevenson GW, Carlezon WA, Negus SS (2009) Effects of pain- and analgesia-related manipulations on intracranial self-stimulation in rats: further studies on pain-depressed behavior. Pain 144:170–177
Preskorn S, Macaluso M, Mehra V, Zammit G, Moskal JR, Burch RM (2015) Randomized proof of concept trial of GLYX-13, an N-Methyl-D-Aspartate receptor glycine site partial agonist, in major depressive disorder nonresponsive to a previous antidepressant agent. J Psychiat Pract 21:140–149
Rolls ET (2014) Emotion and decision making explained. Oxford University Press, Oxford
Rush AJ (2007) STAR*D: what have we learned? Am J Psychiatry 164:201–204
Rush AJ, Trivedi M, Fava M (2003) Depression IV: STAR*D treatment trial for depression. Am J Psychiatry 160:237
Schildkraut J (1965) The catecholamine hypothesis of affective disorders: a review of supportive evidence. Am J Psychiatry 122:509–522
Schlaepfer TE, Cohen MX, Frick C, Kosel M, Brodesser D, Axmacher N, Joe AJ, Kreft M, Lenartz D, Sturm V (2008) Deep brain stimulation to reward circuitry alleviates anhedonia in refractory major depression. Neuropsychopharmacology 33:368–377
Solms M, Panksepp J (2010) Why depression feels bad. In. Ashton CH, Perry EK (eds) The neuroscience of conscious/nonconscious interactions. John Benjamins, Amsterdam (in press)

Swain JE, Lorberbaum JP, Kose S, Strathearn L (2007) Brain basis of early parent-infant interactions: psychology, physiology, and in vivo functional neuroimaging studies. J Child Psychol Psychiat 48:262–287

Tenore PL (2008) Psychotherapeutic benefits of opioid agonist therapy. J Addict Dis 27:49–65

Uvnäs-Moberg K (1998) Oxytocin may mediate the benefits of positive social interaction and emotions. Psychoneuroendocrino 23:819–835

Watt D, Panksepp J (2009) Depression: an evolutionarily conserved mechanism to terminate separation distress? A review of aminergic, peptidergic, and neural network perspectives. (with commentaries) Neuropsychoanalysis. 11(1):5–51

Wöhr M, Schwarting RKW (2013) Affective communication in rodents: ultrasonic vocalizations as a tool for research on emotion and motivation. Cell Tissue Res 354:81–97

Wöhr M, Kehl M, Borta A, Schänzer A, Schwarting RKW, Höglinger GU (2009) New insights into the relationship of neurogenesis and affect: tickling induces hippocampal cell proliferation in rats emitting appetitive 50-kHz ultrasonic vocalizations. Neuroscience 163:1024–1030

Wright JS, Panksepp J (2011) Toward affective circuit-based preclinical models of depression: sensitizing dorsal PAG arousal leads to sustained suppression of positive affect in rats. Neurosci Biobehav Rev 35:1902–1915

Wright JS, Panksepp J (2012) An evolutionary framework to understand foraging, wanting, and desire: the neuropsychology of the seeking system. Neuropsychoanalysis 14:5–39

Yovell Y, Bar G, Mashiah M, Baruch Y, Briskman I, Asherov J, Panksepp J (2016) Ultra-low-dose buprenorphine as a time-limited treatment for severe suicidal ideation: a randomized controlled trial. Amer J Psychiat 173:491–498

Zarate CAJ, Singh JB, Carlson PJ, Brutsche NE, Ameli R, Luckenbaugh DA, Charney DS, Manji HK (2006) A randomized trial of an N-methyl-D-aspartate antagonist in treatment-resistant major depression. Arch Gen Psychiatry 63:856–864

Zellner MR, Watt DF, Solms M, Panksepp J (2011) Affective neuroscientific and neuropsychoanalytic approaches to two intractable psychiatric problems: why depression feels so bad and what addicts really want. Neurosci Biobehav Rev 35:2000–2008

Social Reward and Empathy as Proximal Contributions to Altruism: The Camaraderie Effect

Garet P. Lahvis

Abstract Natural selection favors individuals to act in their own interests, implying that wild animals experience a competitive psychology. Animals in the wild also express helping behaviors, presumably at their own expense and suggestive of a more compassionate psychology. This apparent paradox can be partially explained by ultimate mechanisms that include kin selection, reciprocity, and multilevel selection, yet some theorists argue such ultimate explanations may not be sufficient and that an additional "stake in others" is necessary for altruism's evolution. We suggest this stake is the "camaraderie effect," a by-product of two highly adaptive psychological experiences: social motivation and empathy. Rodents can derive pleasure from access to others and this appetite for social rewards motivates individuals to live together, a valuable psychology when group living is adaptive. Rodents can also experience empathy, the generation of an affective state more appropriate to the situation of another compared to one's own. Empathy is not a compassionate feeling but it has useful predictive value. For instance, empathy allows an individual to feel an unperceived danger from social cues. Empathy of another's stance toward one's self would predict either social acceptance or ostracism and amplify one's physiological sensitivity to social isolation, including impaired immune responses and delayed wound healing. By contrast, altruistic behaviors would promote well-being in others and feelings of camaraderie from others, thereby improving one's own physiological well-being. Together, these affective states engender a stake in others necessary for the expression of altruistic behavior.

G.P. Lahvis (✉)
Department of Behavioral Neuroscience, Oregon Health & Science University,
3181 S.W. Sam Jackson Park Road, Mail Code L-470, Portland, Oregon 97239, USA
e-mail: lahvisg@gmail.com; lahvisg@ohsu.edu

Curr Topics Behav Neurosci (2016) 30: 127–158
DOI 10.1007/7854_2016_449
Published Online: 01 September 2016

Contents

1 Introduction 128
2 Social Interaction 129
3 Social Reward 134
4 Empathy 137
5 Helping Behavior 142
6 Camaradarie Effect 144
References 149

1 Introduction

Since its inception, a challenge for evolutionary theory has been widespread evidence of altruistic behavior (Darwin 1888). Natural selection works at the level of the individual, not the species. When the environment favors the survival and reproduction of one individual over another, the frequency of that individual's alleles represents a larger proportion of the gene pool in the following generation. Inherent in this process is the requirement that individuals compete for limited resources. Yet wild animals also help one another, apparently at personal cost. Helping behaviors include biological altruism and cooperation. "Biological altruism" refers to instances when individuals seemingly pay a cost, at least in the short term, to benefit another individual. "Cooperation" refers to instances when individuals act together to acquire benefits for survival and reproduction that they might be unable to acquire by acting alone (Roberts 2005). Some altruistic behaviors are expressed by only a particular species; a vampire bat risks starvation when regurgitating a blood meal to share it with another (Wilkinson 1984). Other behaviors are widespread, such as participation in the mobbing of a predator, a risky behavior that can help protect other colony members (Graw and Manser 2007; Krams et al. 2006; Templeton et al. 2005). Efforts to reconcile altruistic behaviors within the context of natural selection have largely focused on the development of ultimate mechanisms, such as kin selection, various forms of reciprocity, and multilevel selection (Nowak 2006).

An implication of natural selection is that it favors a competitive and individualistic "natural" psychology; that individuals in the wild harbor affective experiences focused on acquiring resources and providing for progeny. But what do we actually know about animal psychology in the wild? Is a constant motivation for resource acquisition the "natural" affective state of wild non-human animals? If so, why do wild animals help each other when resources are restricted? Likewise, an implication of altruistic behavior is that it reflects an underlying psychological experience of compassion. In both of these instances, we confuse the expression of a behavior with affective experiences. This conflation of animal behavior with psychological experience is a persistent problem in behavioral science, sharing similarity with a much earlier problem regarding the psychology of an animal with its movement toward or away from an environmental stimulus.

We can dissociate behavior from the reading of intent into underlying affective states by employing operant and classical conditioning experiments that help us infer internal subjective responses to environmental rewards and punishments (Bardo and Bevins 2000). With classical conditioning experiments, we can infer affective experiences occurring within social interactions, whether a rodent can experience social reward and empathy, affective capacities that can be inherited and are thereby sensitive to natural selection.

How might social reward and empathy contribute to the expression of altruistic behavior in the wild? Social rewards can motivate individuals to shelter and forage together, behavioral strategies that can be adaptive. Empathy, the generation of an affective state more appropriate to the situation of another compared to one's own, has useful predictive value, allowing an individual to detect a threat from the behavioral cues of others. The "camaraderie effect" emerges from the motivation for social reward and the predictive sensitivity of empathy. A socially motivated individual can detect feelings held by others that may be interpersonal (e.g., affinity, indifference, and enmity) or may pertain to changes in the environment (e.g., calm and fear). Once established, the camaraderie effect gains its own survival value because the detection of amicable, dispassionate, or hostile feelings becomes predictive of well-being within the group versus ostracism and social isolation. As the psychological experience of social isolation can compromise immune responses and impair wound healing, one possibility is that the detection of social cues predicting social isolation, or serving as a surrogate for physical ostracism, would also compromise physiology. Likewise, an individual might be motivated to express altruistic behaviors to share in the positive affective experience of others, an experience that might promote one's physiological well-being. Components of the camaraderie effect have established neurobiological substrates (Dolen et al. 2013; Jeon et al. 2010) that are consistent with the theoretical underpinnings of affective neuroscience (Ekman and Davidson 1994; Panksepp 1998) and also meet the requirements for genetic self-interest imposed by natural selection.

Topics to be covered in this review include brief overviews of emotion and its expression in social interaction, social motivation, empathy, helping behavior, and the camaraderie effect. Rodent studies are a primary focus of this review because they underscore the widespread capacity for social reward and empathy in mammals and because rodent studies offer the most comprehensive neurobiological assessments underlying emotion expression and receptivity in non-human animals.

2 Social Interaction

There was speech in their dumbness,
language in their very gestures.

William Shakespeare

A meeting between individuals is comprised of a remarkably complex array of signals transmitted and received, spoken and heard, involving language responsive to how arousal moderates the acoustic parameters of speech (Scherer 1986) or existing as mere utterances: a laugh, a sigh, or a grunt. A visual cue might feature a redirection of eye gaze or an extra bounce in a step. An eye movement, a facial expression, or a gesture, each one is a social cue that occurs by volition or by reflex, each one having the capacity to pivot the gestalt and trajectory of a social interplay, shaping its direction, and turning it aggressive, compassionate, or disengaged.

Beneath this surface of signals are the emotions, moods, interpersonal stances, attitudes, and personality traits of the interacting participants—affective experiences and behavioral expressions that can be fleeting or long-standing, influencing what each participant expects from the other and what each participant detects or feels, correctly or incorrectly, in another's motives and emotions (Scherer 2003).

Emotion: A relatively brief episode of synchronized response to an external or internal event valued as being of major significance (anger, sadness, joy, fear, and desperation).

Mood: A diffuse affect state, most pronounced as change in subjective feeling, of low intensity but relatively long duration, often without apparent cause (cheerful, gloomy, and irritable).

Interpersonal stance: An affective stance taken toward another individual in a specific interaction, coloring the interpersonal exchange in that situation (cold, warm, supportive, and contemptuous).

Attitudes: Relatively enduring, affectively colored beliefs, references, and predispositions toward objects or persons (liking, loving, hating, and desiring).

Personality traits: Emotionally laden, stable dispositions and behavioral tendencies of an individual (nervous, anxious, hostile, and jealous).

Emotions, moods, interpersonal stances, attitudes, and personality traits are not directly measurable in humans and non-humans but these terms help us to conceptually dissociate affective states experienced during social interactions. For instance, dominant-subordinate relationships, which occur broadly in nature and in the laboratory (Blanchard et al. 1987), can be considered in the context of interpersonal stance. Adjustments in attitude might explain how Pinyon jays make transitive inferences about dominance hierarchies (Bond et al. 2004). Mice express variations in aggressive personality traits (Caramaschi et al. 2008). Such distinctions in affective states will also be useful for understanding wild animal social behavior.

Each of us might be aware of our own emotions, interpersonal stances, and attitudes, but these internal subjective experiences elude the scientific process. Science is a way-of-knowing that requires objectivity and a deliberate attempt to depart from subjective experience. As scientists, we measure physical entities, what can be poured into a beaker or measured in wavelengths, not sensory or emotional experiences.

When two people sit in the same room and look at a red apple, neither one can know the color perceived by the other (Russell 1912). My daughter might see magenta while I see crimson. We both call it "red." Our verbal report is unreliable, perhaps misleading, suggesting that our shared experiences, like the "red" in an apple, are identical. Our verbal report of affective states is even more likely misinterpreted. As a result, scientists often choose to avoid questions about subjective experience, more so when we lack a shared language with our non-human experimental subjects. Critically, to improve treatments for mental illness, which often features impaired social-emotional regulation, we are compelled to study the mechanisms underlying non-human animal affective experiences in a social context.

Studies of animal subjects focus on measurements of behavioral expressions that signify affective states. The Reciprocity Chain (Fig. 1) is a simple model that shows how affective experiences, in this case emotions, can be dissociated from the measurable behaviors expressed during a social interaction (Bishop and Lahvis 2011). This model is a simplified analogue of the Brunswikian lens model, a useful model of social-emotional interaction also employed in communication research (Scherer 2003). Measurable signals include the visual, auditory, tactile, and olfactory cues and their timing.

In *Expression of Emotion in Man and Animal*, Darwin argues that the expression of an emotional experience can occur reflexively, borrowed from a related experience (Darwin 1872). As depicted in the Reciprocity Chain, a minced facial expression (see Fig. 1b) responding to a bitter or repugnant thought or emotion (see Fig. 1a) taps into an analogous experience, perhaps the taste of a bitter berry or the smell of a putrid deposit. Even when involuntary, these expressions can engender emotional changes in those who perceive them that are, in turn, expressed. The Brunswikian lens model considers additional features, such as how a signal expressed might differ from the signal perceived; a minced look might be seen as a smile.

An expression might be intentionally directed toward another individual (Tomasello et al. 2005). Feeling frustrated after seeing a full-bellied raccoon waddle

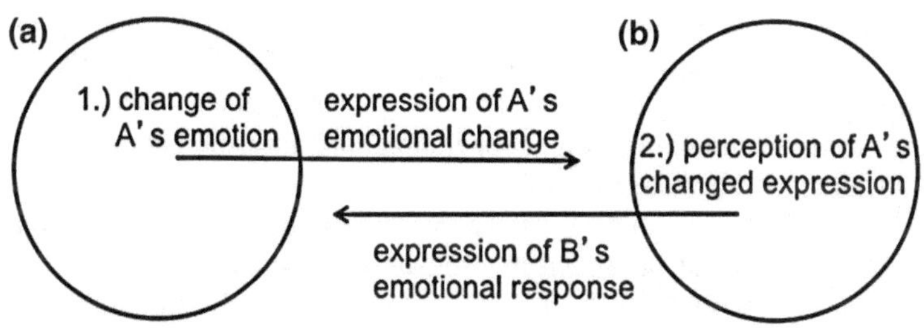

Fig. 1 The Reciprocity Chain represents emotions felt and expressed during a dyadic social interaction between Individual A (**a**) and Individual B (**b**). In this simple model, Individual A experiences a change in emotion while interacting with Individual B. This change in affect is expressed as a behavioral cue by Individual A that is detected by Individual B and in turn provokes an emotional change in Individual B, which is again expressed, and detected by Individual A. While emotions elude direct scientific observation, their expressions can be measured

from my grape trellis, I might keep this feeling to myself or point him out to my wife. I experience frustration either way but I'm intentionally sharing my feeling when I direct my wife's attention toward the raccoon. In this shared affective experience, she needs to "get it" and respond in turn, acknowledging my attempt to communicate. Perhaps she will look downward, sigh, or laugh. If her response is not perceptible to me, I'll remain unsure whether our experience was shared. As with reflexive signals, intentional signals connote emotion. A concerned facial expression and a softer voice help relate a sad story. A widening of one's eyes and staccato in one's voice help express surprise and excitement (see Markova and Legerstee 2006). Emotions signaled by intention can be different from emotions portrayed by actors (see Scherer 2003). The Reciprocity Chain and the Brunswikian lens models help us differentiate affective states from their expressions, but they may neglect the embodied cognition experienced in the first or second person, an emergent property experienced when one is engaged *within* a social interaction rather than just observing it (Schilbach et al. 2013).

Neuroscientists often study laboratory rodents to model social-emotional challenges: how adolescent social motivation responds to substance abuse and addiction, how vocalizations and social behaviors are integrated in a rodent model of autism, or how empathy can be re-instilled in an individual crippled by traumatic memories. We ask how social relationships form, what allows individuals to bond, how these relationships are maintained, where motivations for social bonding are orchestrated in the brain, and how these relationships fade. Laboratory rodents, such as mice and rats, are useful because their wild conspecifics live at a variety of densities (Fitzgerald et al. 1981) that require multiple layers of social competence: navigating complex and dynamic social hierarchies (Butler 1980; Pocock et al. 2005); assessing, accepting, and rejecting mating opportunities (Drickamer et al. 2000; Krackow and Matuschak 1991; Wolff 1985); escalating and resolving territorial disputes (Chambers et al. 2000); acquiring food preferences based upon social cues (Valsecchi et al. 1996); and even avoiding parasitized conspecifics (Kavaliers and Colwell 1995). Some rodent species, such as prairie voles, engage in social monogamy (McGraw and Young 2010). Others, such as members of the squirrel family, emit alarm calls (Baack and Switzer 2000; Blumstein et al. 1997; Mateo and Holmes 1999). By studying small rodents, we can collect large sample sizes that help us to elucidate the anatomy, physiology, and genetics underlying social bonding, social learning, empathy, cooperation, and altruistic behaviors.

In the laboratory, investigators often measure social approach (Lahvis and Black 2011). The two most common tests are the three-chambered social approach test (Nadler et al. 2004) and the social investigation (SI) test (Winslow and Camacho 1995). Both tests measure levels of *subject* approach toward a stimulus rodent, also called the *object*. In the SI test, both the test subject and the object are free to move inside a test structure. In the social approach test, the subject moves freely, while the object is confined to a small cage within the test structure. Social recognition tests (Ferguson et al. 2001) and social preference tests (Moy et al. 2004) are variants of these social approach measures. Social approach tests have been used to compare the responses of rodent sociability to genetic background (Moy et al. 2009;

Panksepp et al. 2007; Sankoorikal et al. 2006), variations in brain anatomy (Fairless et al. 2008), targeted alleles including knockout mice (Spencer et al. 2008), brain lesions (Yang et al. 2009), exposures to modulators of opiate and dopamine pathways (Benton et al. 1984; Gariepy et al. 1998; Kennedy et al. 2011), toxic chemicals (Belloni et al. 2011), and candidates for pharmacological treatments (Calamandrei et al. 2000; and for reviews, see Halladay et al. 2009; Moy et al. 2009; Silverman et al. 2010). Measures of social approach often tally how often the subject rodent brings itself into close proximity or physical contact with the object rodent. Strains that maintain close proximity are typically classified as more "social," whereas strains that approach less often or withdraw from the object rodent are often classified as "asocial."

Classifications of "sociality" based solely upon body movement imply that social interactions within a test structure are restricted to proximity, ignoring other dimensions of communication including vocal and olfactory signals. Rodents emit audible and ultrasonic vocalizations (USVs) that exceed the upper limit of human hearing, emitted at frequencies above 20 kHz (20,000 cycles per second). USVs can be transmitted across cage structures, can be associated with affiliative or aggressive social interactions, and can engender behavioral responses from others. Infant rodents emit wriggling calls ($\sim$35 kHz) to solicit maternal care (D'Amato et al. 2005; Ehret and Bernecker 1986) and distress calls ($\sim$90 kHz) soliciting return to the nest (Branchi et al. 1998). Vocalizations can also be associated with social approach. Adolescent mice that engage in more robust social approach vocalize more often (Panksepp et al. 2007). Laboratory rats emit 22- and 50-kHz calls to signal negative and positive affect (Burgdorf et al. 2005; Carden et al. 1993; Harmon et al. 2008) and female mice emit a 38-kHz calls to coordinate paternal pup retrieval (Liu et al. 2013). Like vocalizations, scent marking is used in social interactions, establishing territorial dominance among males (Hurst 1990) and attracting females (Roberts et al. 2014; Thonhauser et al. 2013). Assessments of scent marking, and of subject responses to maternal scent, have been used to phenotype mouse models of autism (Kane et al. 2012; Wöhr et al. 2011a, b).

Approach behaviors should not to be confused with the desire for a reward (Schneirla 1959). An amoeba, single-celled and brain-free, can move up a chemical gradient. While assessments of social approach, vocal, and scent marking behaviors give us a sense of the level of social interaction between experimental subjects, by themselves they offer no measurable insight to underlying affective experiences. Social approach behaviors are not necessarily an expressed indication of the affective experience of social seeking, a desire for a social reward. Similarly, social withdrawal behaviors do not necessarily indicate that the subject feels the social interaction is an aversive experience.

To elucidate a few of the affective experiences underlying social interaction, we can make use of classical conditioning. With conditioned place preference testing, we can infer that rodents find some experiences pleasurable, including both natural rewards, such as social interactions, and drug rewards, such as methamphetamine exposure. How these subjective experiences can be inferred is the topic of the next section.

3 Social Reward

Seeking and avoidance are of a higher evolutionary and developmental order than approach and withdrawal, and these terms should not be mismated.

Theodore C. Schneirla

Children and adults with developmental disabilities can be challenged by social interactions (Lord et al. 2001). For some, these social interactions may not feel rewarding or be sufficiently desirable for maintaining and enhancing relationships (Chevallier et al. 2012). Studying rodent models of autism, we can use Pavlovian conditioning approaches to determine whether subjects derive pleasure from social access. If a test subject finds a stimulus rewarding, it will return to an environment associated with that stimulus (Glickman and Schiff 1967; Schneirla 1959). The *social conditioned place preference (social CPP) test* is used to determine whether a mouse prefers housing with its peers versus housing in social isolation. Prior to testing, each subject is "conditioned," alternately placed within one of two housing environments specifically paired with the presence or absence of other mice. We hypothesized that social interactions would be most rewarding for mice under comfortable conditions because social interactions are generally most rewarding for humans in comfort. Thus, each environment contains novel bedding, such as wood chips or paper bedding, rather than the traditional steel bars and metal floors used for CPP tests in drug abuse experiments. PVC couplers are also added, either threaded or smooth, to add environmental complexity to the conditioning experience and testing conditions (Panksepp and Lahvis 2007). Each day, mice are transferred from one housing environment to the other, paired with the same social condition, in a group or alone. After conditioning, subjects are tested in a social CPP test structure with separate chambers offering rival beddings and their associated couplers. If a subject derives a rewarding experience from social housing, this feeling becomes paired with the bedding environment, and on test day, the subject chooses to spend more time in the bedding associated with access to that positive feeling (Panksepp and Lahvis 2007). Social CPP experiments allow us to determine whether the test subject seeks social reward, avoids social isolation, or is indifferent to social context.

Social CPP tests demonstrate that positive affective experiences occur during juvenile social interactions (Calcagnetti and Schechter 1992; Douglas et al. 2004), mating opportunities (Camacho et al. 2004; Jenkins and Becker 2003), access to offspring (Mattson et al. 2001), and even in response to aggressive social interactions (Martinez et al. 1995; Tzschentke 2007). CPP has demonstrated that laboratory rodents typically prefer environments associated with social access, a behavior driven by anticipation of a social reward (Calcagnetti and Schechter 1992; Douglas et al. 2004; Panksepp and Lahvis 2007). In a limited number of experiments, mouse strains expressing low levels of social CPP, or social indifference, also express minimal social approach (Panksepp and Lahvis 2007).

Social affect: The feeling experienced during a social encounter, such as pleasure from a social reward or fear from a social threat.

Social motivation: What drives an individual to engage in a social interaction, irrespective of the kind of affective experience or even the presence of affect.

Social reward: The positive affective experience associated with access to others can be inferred from a subject's behavioral response to the social conditioned place preference (CPP) test. Spending of more time exploring bedding experimentally paired with social housing versus bedding paired with isolation indicates that the subject derives pleasure from social access.

Social CPP reveals that social reward is mediated by coordinated activities of oxytocin and serotonin in the nucleus accumbens (Dolen et al. 2013). These neurological circuits, recruited during social interactions, also play critical roles in the rewarding experiences mediated by drugs of abuse. These findings confirm the prediction that natural reward systems of wanting and liking (Berridge 2004, 2007; Berridge and Robinson 2003; Smith and Berridge 2007) are engaged, systems that are co-opted by drugs of abuse (Kelley et al. 2005; Kelley and Berridge 2002; Schroeder et al. 2001) and serve as specific neurobiological substrates for the anticipating and rewarding experiences that motivate social interaction.

The design of the social CPP provides insight to how social reward contributes to group living in the wild. Social CPP identifies a motivation for individuals to interact with others *and* environments where they would encounter conspecifics. Outside the cage, individuals have opportunity to make choices about where to move based upon various features across a heterogeneous environment, including those stimuli associated with social access, such as shelters and common foraging areas. For many species, living within a group confers survival and reproductive benefits, either throughout their lives or during specific developmental life stages. Group living is not always adaptive, especially when local resources are depleted, when colony parasitism is high, or during adolescent maturation (Hoogland 1979), so we might expect that social reward is also adaptive only under specific environmental conditions.

While social CPP experiments show that laboratory rats and mice can derive pleasure from a social interaction, a possible explanation for social reward is that it emerged from breeding under highly constrained housing conditions that force individuals to live together. Domestication also relieves tame animals of the natural selective pressures to compete for limited resources, rendering them more docile (Nelson and Chiavegatto 2001). For instance, tame animals are less aggressive toward conspecifics (Boreman and Price 1972; Ebert 1976) and more readily engage in mating opportunities without requiring mate choice (Drickamer et al. 2000; Manning et al. 1992; Penn and Potts 1999).

To determine whether rodent social reward is an artifact of domestication, a social CPP test was conducted on undomesticated thirteen-lined ground squirrels (*Ictidomys tridecemlineatus*), a species considered asocial among ground squirrels

(Armitage 1981) because they appear to form colonies not out of attraction to one another but because they prefer living in a specific environment (McCarley 1966). However, in the social CPP test, captive juveniles that were second- and third-generation descendants of wild ground squirrels expressed a robust preference for environments paired with social access, indicating that social interactions can be rewarding for rodents with undomesticated genetic backgrounds (Lahvis et al. 2015). This finding suggests that wild squirrels, known for their diminished sociality, can derive pleasure from a social interaction. Additional comparisons of laboratory experiments with concurrent field experiments showed that while maturing wild juveniles gradually foraged at increasing distances from one another, a behavioral pattern that predicts dispersal, captive juveniles simultaneously expressed diminished social approach and increased play fighting (Lahvis et al. 2015). Taken together, this comparison supports the idea that the adolescent thirteen-lined ground squirrel can experience social reward and that social motivation diminishes as maturing adolescents begin to disperse, an idea akin to the "ontogenetic switch" (Holekamp 1984).

Social reward propels individuals into social proximity in the wild under conditions that are more expansive and more patchy than the laboratory cage. Social proximity facilitates opportunities for individuals to learn from others through interactions between demonstrator and observer, teacher and learner. Social reward also motivates communication with others, the expression of cues and receptivity to the cues of others that distinguishes social interaction from an encounter with a physical object. In a natural environment, individuals must be sufficiently motivated to attend to the alarm calls, tail flicks, abrupt stops in eating behaviors, and upright stances of their alert conspecifics.

Social motivation may also enhance the learning process. For instance, in a door opening experiment, observer mice must learn from a demonstrator mouse to swing a door to the left to obtain a food reward (Collins 1988). Adult males more readily learn to open the door from adult females than from other males. Perhaps males more closely attend to adult females and this added attention improves the learning process. In the wild, social learning can be an efficient alternative to untutored trial-and-error learning. For instance, naïve red squirrels must to learn to consume hickory nuts in a fashion that requires minimal investment of time and effort (i.e., energy). When introduced to hickory nuts in the presence of an experienced squirrel, naïve squirrels learn more efficient techniques for consuming the nuts, suggesting that red squirrels can learn by observation (Weigl and Hanson 1980).

In summary, the conditioned place preference test shows that rodents can derive pleasure from a social interaction, preferring to spend time in an environment that predicts access to peers versus an environment that predicts isolation. Experiences of social reward and illicit drug reward share common brain circuitry. In seeking the pleasure derived from social reward, or in avoiding the adversity of social isolation, individuals would be more likely to sleep and forage together and engage in social communication.

4 Empathy

Empathy is the generation of an affective state more appropriate to the situation of another compared to one's own (Hoffman 1975; Preston and de Waal 2002). In popular usage, empathy is jumbled with feelings of compassion or with behaviors often associated with compassion: acts of kindness, helpful acts, or displays of sorrow for a victim. Empathy is not a compassionate feeling nor is it a helping behavior. Empathy involves adopting the feelings of another: fear, joy, pain, anxiety, appreciation, or distain.

Empathy is not always a strictly affective experience and can include various levels of cognitive function. More extreme examples arise from abstractions, such as the experience you may feel upon learning that over two millennia ago, the armies of Alexander the Great conquered the city of Thebes, killed the majority of its inhabitants, sold the remaining 30,000 into slavery, and then burned the city to the ground. Even during this cognitive process, you might feel suffering, desperation, and anger.

Empathy: The generation of an affective state more appropriate to the situation of another compared to one's own. This change in affective state is not compassion and does not necessarily result in the expression of an altruistic act.

A term often associated with empathy is *emotional contagion*, which is *not* an affective experience. Rather, emotional contagion refers to an individual's spontaneous expression of a behavior that resembles the behavior expressed by another individual. Jointly expressed behaviors might include two people yawning together or several babies crying in unison. A classic rodent experiment that demonstrates emotional contagion involves different concentrations of acetic acid injected into two mice. When isolated, a mouse injected with a lower concentration of the irritant acetic acid writhes at a subdued level relative to an isolated mouse injected with a higher concentration of acetic acid. When these two mice are placed together, their responses change. The mouse exposed to the higher concentration of irritant writhes less. The mouse exposed to the lower concentration of irritant writhes more (Langford et al. 2006). This convergence of behavior also occurs when the paws of mice are injected with different concentrations of formalin. The frequency of paw licking is greater for a solitary mouse exposed to high concentrations of formalin, and diminishes if that mouse is placed next to a mouse exposed to lower concentrations of formalin (Langford et al. 2006).

As mentioned, emotional contagion suggests an affective experience but the term actually refers to a *behavior*, a tendency to automatically mimic and synchronize expressions, vocalizations, postures, and movements with those of another individual (Hatfield et al. 1994). The behavior suggests affective convergence but may include alternative explanations. For instance, in the Langford experiments, the mouse experiencing greater levels of abdominal pain may suppress its writhing frequency to match the behavior of its more comfortable partner, thereby masking its own expression of weakness. Likewise, social facilitation might serve as an alternative explanation for the increased writhing behavior of the more comfortable individual.

Emotional contagion: A reflexive behavioral change within the context of a motivationally salient event in which an individual spontaneously expresses a behavior that resembles the behavior expressed by another individual.

Irrespective of the emotions experienced by the two writhing mice, they appear to adjust their converging behaviors based upon visual cues expressed in the writhing of the nearby conspecific. Visual cues of emotions among rodents might also be signaled by facial expressions, which are responsive to painful stimuli (Langford et al. 2010). Facial expressions in rodents can indicate positive and negative affective states (Kelley and Berridge 2002). Olfactory cues, such as urine odors, and vocalizations also serve to signal emotional changes in rodents. When a mouse is offered two cotton balls, one soaked with urine from a recently shocked conspecific and the other soaked with the urine of an undisturbed conspecific, the mouse avoids the cotton ball soaked in urine of its shocked conspecific (Rottman and Snowdon 1972). More recent work shows that the urine of an alarmed mouse releases volatile molecules that evoke in others increased systemic corticosterone levels (Brechbühl et al. 2013).

Social CPP experiments help us to determine if a rodent derives pleasure from a social interaction by assessing whether its access to a social reward in a particular environment changes the affective salience of that paired environment. Vicarious fear learning experiments are used to infer empathy in a rodent, whether observation of a conspecific in pain in a particular environment changes the affective salience of the paired environment. In *context*-dependent vicarious fear learning experiments, demonstrator rodents are repeatedly exposed to a distressing stimulus (an electrical shock) within a chamber and they begin to freeze for short periods, nearly motionless and often trembling in place, and this response becomes more frequent as they are delivered more shocks. These responses reflect a change in the rodent's affective state because the rodent expresses freezing behavior primarily in the chamber where it experiences the shock, but not in other environments. In *context*-dependent vicarious fear learning experiments, rodents hear the vocalizations of conspecifics undergoing repeated shocks in a nearby chamber. When these rodents are in turn placed within the chamber, they express increased freezing behavior. The subjects acquire the affective fear experience of their distressed conspecifics to the context, the chamber, during the conditioning phase, which is subsequently expressed during the test phase of the experiment (Jeon and Shin 2011).

In *cue*-conditioned vicarious fear learning experiments, subjects become fearful of a repeated temporal cue, such as a tone, paired with an aversive stimulus, usually a shock, applied to the rodent in an adjacent chamber. In our experiments, a 30-s tone co-terminated with a 2-s electrical shock is applied to demonstrator mice inside the shock chamber, followed by 90s of silence. With each shock, the mice emit audible vocalizations that sound to the human ear like a squeak. These mouse vocalizations lack nuance, appearing on a spectrogram as undefined broadband noise (Chen et al. 2009), suggesting an expulsion of air, and lacking the audible nuance and precise overtone frequency modulation of referential alarm calls, such as those emitted by prairie dogs to reference predators (Kiriazis and Slobodchikoff 2006) or the 22-kHz vocalizations emitted by rats when they are shocked (Atsak et al. 2011). With repeated tone-shock pairings, mice inside the shock chamber freeze when they hear the tone. This change in behavior represents an affective experience analogous to a fear response in anticipation of an imminent shock.

After demonstrators are conditioned, subjects are placed inside the shock chamber. When they hear the tone, they freeze. This behavioral response suggests that the subject mice have learned from the conditioned demonstrators that the tone predicts an aversive experience. In other words, the affective experiences of the subject mice, garnered while observing the demonstrator mice undergo fear conditioning, alter their subsequent responses to the tone inside the shock chamber. The vocalizations emitted by the demonstrator mice when they are shocked are sufficient to engender fear in the subject mice. When experimenters replace demonstrators with playbacks of 2-second recordings of squeaks, these vocalizations alone are sufficient to induce a freezing response in mice (Chen et al. 2009). While playbacks of mouse vocalizations are sufficient for vicarious fear conditioning, playbacks of 22-kHz vocalizations emitted by rats during shock do not necessarily engender freezing behaviors by observer rats (Atsak et al. 2011), suggesting a role among rats for other signaling modalities during vicarious fear experiments.

An individual human more readily feels empathy for another after having experienced a similar pain in the past (Batson et al. 1996; Eklund et al. 2009; Preis and Kroener-Herwig 2012). Likewise, rodent subjects generally require one experience with the shock prior to demonstrator conditioning (Atsak et al. 2011; Chen et al. 2009; Sanders et al. 2013; but see Jeon et al. 2010). The observing subject rodent is delivered one shock *that is not paired with the tone or the unique context,* prior to observing the demonstrator undergo conditioning. If the subject is temporarily deafened during its single experience with the shock, so that it cannot hear its own vocalization in response to the shock, the subject then fails to vicariously learn from the vocalizations of demonstrators that the tone predicts a shock (Jeon et al. 2010). This result is also underscored by evidence that rats learn to freeze in response to 22 kHz USVs through autoconditioning, associating their own USVs with an internal fear state (Parsana et al. 2012). Experiments showing that subjects freeze in response to the tone only in the shock chamber, not the observation chamber, suggest that subjects have a sense of place, recognizing that the distress associated with shock is dependent upon their physical position within one chamber of the two-chambered test structure.

When demonstrator vocalizations are heard by subjects, they activate dopamine and serotonin circuits (Kim et al. 2014) within the anterior cingulate cortex (ACC) (Jeon et al. 2010), activity patterns that can be lateralized in rodents (Kim et al. 2012) and share similarity with the ACC activity in human subjects, as revealed by functional MRI (Singer et al. 2004, 2006).

One important caveat should be mentioned here. Experiments that use ethologically relevant threats may not require direct individual experience with the aversive stimulus. Mice bury themselves in cage bedding to escape from biting flies after they observe a single successful escape (Kavaliers et al. 2001). These subjects also express decreased pain sensitivity, suggesting they sense the pain experienced by the demonstrators.

Vicarious fear learning and social CPP tests share similarities. In both tests, the rodent encounters an affective experience under a specific set of temporal or contextual conditions. Both tests require classical conditioning; the subject must learn to associate a particular environmental cue or context with a specific affective experience, such as pain, reward, or fear. With repeated pairings of the cue or context with the affective experience, the subject learns that when it encounters the cue or context, it will also feel the reward, fear, or pain. The subject learns that certain affective experiences are contingent upon encountering specific environmental conditions. Since the test phase occurs at a time or place separate from the conditioning phase, the rewarding or aversive experience is carried as an affective memory, coloring the subject's affective response to the test conditions and moderating its behavioral response. Through conditioning experiments, we infer affective experience because, when properly controlled, variability in the rodent's behavioral response is dependent upon the retained affective experience.

Social reward brings individuals into proximity, catalyzing opportunities for interactions between demonstrator and observer. Again, social motivation likely includes attending to the behaviors of others, such as the tail flicks, standing upright, and abrupt stops in eating that might be expressed by conspecifics foraging nearby. In these situations, empathy also has a functional role in survival and reproduction. In response to the threat not directly perceived, empathy elicits arousal from these cues of others, engendering a more robust behavioral response to an urgent situation than what might be expressed with mere cognitive understanding.

In the wild, alarm signals can signify distress and referential information (Klump and Shalter 1984) and the extent that they represent differences in affective arousal versus cognitive referencing can be difficult to dissociate. For instance, under threat, chickadees emit characteristic "chicka-dee-dee" calls. More "dee" syllables correlate inversely with raptor wingspan and smaller raptors are more successful hunters, capable of pursuing their prey with a tighter turning radius (Templeton et al. 2005). It is unclear in this example whether chickadee alarm calls are proportional to their levels of arousal, but in other instances, "semantic" or referential information is communicated (Marler et al. 1992). Prairie dog alarm calls can be easily distinguished according to subtle variations in frequency modulation, differentiating raptors from humans, coyotes, and domesticated dogs, and each call elicits a predator-specific evasion technique (Fredericksen and Slobodchikoff 2007). Prairie dogs also appear to modulate their alarm calls for differences in the colored shirts worn by humans (Slobodchikoff et al. 2009), signaling differences that are less likely to be affected by variations in arousal.

Empathy might also be adaptive for juvenile play behavior, particularly rough-and-tumble play, or play fighting, which is common among juvenile and adolescent mammals. Rough-and-tumble play requires sensitivity to the moment-to-moment condition of the playmate: too hard a bite on the nape, too hard a flip, not a vigorous enough counter approach, too quick on the return from a tumble, not enough responsiveness to the moves of the playmate, and play fighting ends. Like vicarious fear learning, vocalizations play a prominent role in communicating between social participants. Emission of 50-kHz USV maintains playful contact among conspecifics in rats (Himmler et al. 2014; Kisko et al. 2015a, b). Play fighting is sensitive to opiate administration (Vanderschuren et al. 1997), suggesting hedonia is associated with this self-versus-other, back-and-forth activity (Panksepp 1998). Explanations for the energy demanding behavior among wild mammals include its role in improving emotional responsiveness to unexpected events (Nunes et al. 1999; Spinka et al. 2001), familiarizing participants with self-handicap and fair behavior (Bekoff 2004), improving abilities to cope with social challenges (van den Berg et al. 1999), establishing dominance relationships (Blumstein et al. 2013), and helping refine abilities to respond to subtle and ambiguous social signals (Pellis et al. 2010). The give-and-take of social play promotes normal brain development (Gordon et al. 2003; Pellis and Pellis 2007).

Empathy also supports nurturing behavior. USVs emitted by infant mice solicit maternal care and both pup generation of calls and maternal responses are sensitive to opiates, dopamine, and serotonin (D'Amato et al. 2005; Dastur et al. 1999; Moles et al. 2004), suggesting that USV emission and their behavioral responses require changes in affective state.

Empathy can be useful for social learning. For instance, juvenile black-tailed prairie dogs express appropriate levels of arousal and avoidance to predators after observing an experienced wild conspecific adult, and relative to unschooled controls, these captive trained juveniles are more likely to survive in the wild after their release (Shier and Owings 2007). Empathy might also aid in social learning by helping an individual feel into the intent of the individual observed. An early model of social learning described the process as "from an act witnessed, learn to perform that act" (Thorndike 1911), emphasizing observation of an action and not explicitly dissociating it from conceptualizing the goal (Thorndike 1911 and see Galef 2013 for thorough review). This process is called imitation, learning from observations how to perform the form of a novel behavior (production imitation) or a familiar act in an unfamiliar context (contextual imitation) (Janik and Slater 1997; Galef 2013). Rats can learn to push a joystick (Heyes et al. 1994) or swing open a door (Collins 1988) in a particular direction after watching a trained conspecific. In these kinds of experiments, the action and the goal seem one and the same.

Some now question imitation as a the sole means for social learning. By simply watching others, one cannot learn Tai Chi and downhill skiing (Galef 2013). An alternative idea is that social learning may inherently require trial-and-error learning and understanding of the goal to reproduce a behavior, a concept called "emulation" (Tomasello et al. 1993; Galef 2013; Zentall and Galef 1988). Experiments with puzzle boxes can help uncouple imitation from emulation by requiring learners to reproduce the goal of the behavior (Galef 2013). Puzzle boxes are intricate devices with solutions that include multiple behavioral steps, revealing several processes involved in social learning: trial and error, social exposure, and stimulus enhancement (Valsecchi et al. 2002). For example, mice can learn from others to manipulate a puzzle box, pushing a metal tab on its side with a paw to release food into a bin and then opening a drawer in front of the box to recover the food (Carlier and Jamon 2006; Valsecchi et al. 2002). To the extent that an observer experiences the intent of the demonstrator, empathy is involved in emulation.

Imitation: Learning from observation to perform the form of a novel behavior or a familiar behavior in an unfamiliar context.

Emulation: Learning from observation that a goal, the result of an observed action, is achievable, followed by trial-and-error learning to achieve the goal.

5 Helping Behavior

Since the late 1950s, experimental psychology studies have shown that laboratory rats help their conspecifics. George Rice and Priscilla Gainer built a structure that could hoist a rat in a harness, suspended so that its paws could not quite touch the floor, hung in a pendent condition that likely causes distress. A subject rat then had the opportunity to lower the suspended rat to the floor. Given a choice, the subject was more likely to press a lever to relieve its pendent and distressed conspecific versus an alternate lever to acquire a food reward, a chocolate chip (Rice and Gainer 1962).

In more recent experiments, rat subjects were given opportunity to release a conspecific held inside a small container. Faced with the dilemma of whether to free the constrained rat or gain access to chocolate, subjects were more likely to free their distressed conspecifics and share the chocolate (Bartal et al. 2011; Bartal et al. 2014). Authors state that these helping behaviors were "empathically motivated," a perhaps untenable inference because empathy does not by itself confer motivation to obtain a reward or to avoid an aversive experience. Consider empathy within the context of the conditioned place preference test, an experimental measure that allows us to draw inferences about the feelings that drive motivation. Empathic experience would not drive a rodent to predictably spend more time exploring one CPP-conditioned environment versus another. Rather, the direction of a rodent's preference for a particular environment would depend entirely upon the nature of the experience felt in that shared environment, whether the shared experience was inherently pleasurable or aversive, influenced by the valence of the conspecific's affective state.

The argument that empathy motivates the subject to free the rat distressed by confinement also requires evidence that restraint within the small container is aversive and that the subject is more likely to help if it previously experienced the same aversive condition (Greene 1969). When the experiments described above were repeated and elaborated by others (Silberberg et al. 2014), rats confined within the container and then freed and given choice to move inside and outside the container returned to move in and out. This finding is not consistent with the contention that constraint inside the container was an aversive experience (Silberberg et al. 2014). Critically, constrained rats in the original study emitted 22-kHz USVs and the frequency of these vocalizations strongly suggests that the constrained rats experienced an aversive condition. These experiments might be reconciled if we consider that prolonged and unelected restraint is aversive but when the subject has agency to move inside and outside the container, the rat no longer finds the context unpleasant. The rat might quickly realize the container door stays open. Further, if we assume for the sake of discussion that constraint was an aversive experience, a second important condition for empathy remains unclear: whether the subject would be more likely to free the constrained rat were it to experience the same aversive condition. Subjects in this experiment were not constrained prior to testing. An alternative explanation for these results is that the subject was motivated to release the conspecific to gain access to a social reward (Silberberg et al. 2014).

In a different helping experiment, a subject rat opened a door to allow a wet rat to escape from a pool, whereupon the subject often shared its food (Sato et al. 2015). In this experiment, helper rats were more likely to help if they had been soaked themselves, a response that suggests that empathy could play a role in this helping behavior. In other experiments, rat subjects gave their conspecifics access to food without any personal benefit (Hernandez-Lallement et al. 2014; Márquez et al. 2015).

Unlike their wild conspecifics, laboratory rodents do not experience resource restriction. In nature, resources fluctuate with years of bounty and others of paucity, unpredictably driving natural selection (Grant and Grant 2002). While ultimate explanations for altruistic behaviors in the wild attempt to reconcile helping behavior with natural selection (Graw and Manser 2007; Krams et al. 2006; Templeton et al. 2005; Wilkinson 1984), laboratory animals are afforded *ad libitum* access to food and shelter, so they do not face similar selective pressures. Instead, the standard laboratory animal cage imposes a different set of constraints upon the maturation of brain development and behavior, including an extreme poverty of temporal and spatial variation in food availability and quality, lack of social refuge, and invariant social structure (Lahvis 2016). Therefore, we must be cognizant of possible differences in the underlying affective conditions that engender expressions of helping behavior in the laboratory versus the wild.

6 Camaraderie Effect

Grief can take care of itself, but to get the full value of joy you must have somebody to divide it with.

Mark Twain

The popular phrase "survival of the fittest" is a tautology that literally means *survival of the one who survives the best* and, for some, conjures notions of wild animals motivated by an ongoing desire to acquire and defend food items, territory, and mating opportunities. Despite the pressure of natural selection, wild animals also express altruistic behaviors, actions that involve payment of a personal cost, at least in the short term, to benefit other individuals (Roberts 2005). To some, these helping behaviors suggest feelings of compassion. Conditioning experiments can help us dissociate competitive and helping behaviors from their underlying feelings, but we need to first step back again and consider explanations for behaviors expressed by wild animals.

Non-human animal behaviors in nature can be explained in ultimate and proximal terms, ranging from why a behavior might be adaptive for survival and reproduction (an ultimate explanation) to how a developmental or biological mechanism, such as a neuropeptide released in brain region, supports its expression (a proximal explanation) (Tinbergen 1963). Ultimate explanations for biological altruism include, but are not limited to, kin selection (Hamilton 1964), direct reciprocity (Trivers 1971), indirect reciprocity (Nowak and Sigmund 2005), handicap models (Griffin and West 2003), multilevel selection (Wilson 1997; Traulsen and Nowak 2006), and by-product benefits (Leimar and Connor 2003). Each of these models explains how the stability of an altruistic behavior is maintained under specific social conditions (Nowak 2006; West et al. 2007). For instance, kin selection requires that the altruistic individual be genetically related to a sufficient proportion of benefiting recipients that carry alleles for the altruistic behavior. Also helpful is the capacity of the individual to discriminate kin from non-kin (Griffin and West 2003; Hamilton 1964). Direct reciprocity requires opportunities for repeated social interactions so that recipients can reciprocate the help they receive (Trivers 1971). Reciprocity also requires that the recipient can associate the identity of the helping actor with the helping behavior and also remember either the act itself or, in light of the discussion above, associate the actor with a positive interpersonal affective experience. Multilevel selection requires that the altruistic individual helps members of its own group, which competes with other groups, to indirectly gaining access to resources at another group's expense (Traulsen and Nowak 2006). These social conditions vary with species identity, environmental harshness (Barash 1974), and social constraints and include the varied durations individuals share within a common environment, such as the temporary social arrangements of migrating birds (Wheatcroft and Price 2008).

Explanations for cooperation and altruism also include handicap and indirect reciprocity models that require a third party to witness whether the actor aids the recipient. Subsequent interactions between the actor and the witnesses are again sensitive to whether or not the actor initially provided help (Nowak and Sigmund 2005). Responses might include increased mating opportunities favoring an altruistic actor (Griffin and West 2003) and social punishment for an actor that did not help, including a decision not to reciprocate aid (Krams et al. 2008). When breeding pied flycatchers are experimentally restricted from helping in predator defense, they are not helped in return (Krams et al. 2006).

These ultimate models may not fully explain why individuals engage in altruistic behaviors—such as when they are sensitive to freeloaders that take advantage of altruistic behaviors and chose not to reciprocate (Roberts 2005). To ensure the stability of altruistic behavior, it would be ideal if an additional mechanism could buttress its adaptive value. Were the actor to have an additional "stake" in the well-being of the recipient, its long-term interests would be served as a secondary consequence even if the helping behavior is not reciprocated (Roberts 2005).

Could affective experience serve as an actor's stake in potential recipients?

Reciprocity and handicap models require that the recipient and third-party witnesses have a memory of the actor's identity and experience one of at least two mental processes: either a cognitive score sheet of the actor's provision or denial of help—or an affective experience associated with the actor's previous decision to offer or deny help. In models that require reciprocal action (help the actor that chooses to help, do not help the actor that chooses to deny help), these associated affective experiences might serve to motivate behaviors that favor or punish the actor.

If the actor were to have a capacity for empathy, would it be able to predict social threats by sensing the feelings of others? Could a difference in attitudes experienced and subtly expressed by the helped or unaided recipients and their witnesses serve as a palpable reward or punishment for the actor? CPP experiments show that laboratory rodents can express a strong preference for environments that predict future social interactions (Calcagnetti and Schechter 1992; Douglas et al. 2004) and avoid environments paired with social isolation (Panksepp and Lahvis 2007). In this regard, laboratory rodents may be similar to humans, who can have a basic need for social connections and find social separation or rejection psychologically painful (Eisenberger and Lieberman 2005). While domestication can engender prosocial behaviors that do not exist in the wild, experiments with captive ground squirrels show that undomesticated rodents can also derive pleasure from access to social interactions (Lahvis et al. 2015), indicating that social reward is a natural motivation.

For individuals living in groups, social ostracism may confer substantial costs to an outcast individual. Emmigration from a colony is associated with high mortality risks (Koopman et al. 2000; Pocock et al. 2005). Dispersal is most common for adolescents, particularly young males. Depending upon species, maturational age, and environment, adolescent dispersal may be explained both by diminished motivation of the adolescents for social interaction, indicated by an abrupt decrease in play behavior and increase in aggression (Festa-Bianchet and King 1984), and as a response to social ostracism. For instance, adult squirrels can act aggressively toward maturing adolescents, chasing them away (Festa-Bianchet and King 1984). Adults in marmot colonies can violently ostracize adolescents that fail to participate in a morning greeting (Barash 1974).

Ostracism can also extend to individuals or groups that fail to cooperate with the larger social structure. A wolf pack will reject an individual that mates with a subordinate wolf (Peterson 1979) and chimpanzees will shun individuals appearing abnormal (Goodall 1986). Social rejection is also found in rodent societies. Female Norway rats reject males that copulate with anestrous females (Galef et al. 2008).

How does an individual respond to the potential for ostracism when a motivation for social reward, for living with others, is coupled with empathy, the generation of an affective state more appropriate to the situation of another? Overt chasing behavior or violent aggression might not be required for an individual to feel social exclusion. Empathy would amplify an actor's own feelings of fear, pain, or distress when witnesses harbor enmity or indifference to the actor for past decisions, engendering painful experiences in the actor in anticipation of social rejection. Likewise, helping decisions that cement social acceptance would increase an actor's positive affective experience within the group. In rats, gene expression is enhanced in fear-related areas when rats hear 22-kHz vocalizations, USVs that are associated with pain and aggression (Sadananda et al. 2008). Empathy may also augment the positive affective experiences garnered when conspecifics are comforted by an altruistic act. When a rat hears affiliative 50-kHz USVs, gene expression is activated in reward-related areas of the brain (Sadananda et al. 2008) and dopamine is released in the nucleus accumbens (Willuhn et al. 2014), a physiological correlate of a reward experience (Kelley and Berridge 2002; Smith and Berridge 2007). These affective experiences might help build strong social bonds that confer a variety of reproductive benefits (Lahvis et al. 2015; Seyfarth and Cheney 2013).

Together, social motivation and empathy could be considered a *camaraderie effect*, a proximal explanation for biological altruism whereby an actor seeks social access and avoids social isolation, sensitive to the feelings of social acceptance or rejection that others hold for the actor (Lahvis et al. 2015). With social reward, an individual becomes susceptible to the actions of others; with empathy, these sensitivities expand to include the feelings of others, such as growing affinity or enmity, and social motivation engenders the drive to promote positive over negative feelings.

The camaraderie effect might have emerged in the evolution of mammalian psychology as a by-product of social reward and empathy. Both affective states are heritable (Chen et al. 2009; Panksepp and Lahvis 2007) and thus sensitive to natural selection. The camaraderie effect may have emerged from these related traits that offer considerable adaptive value, appearing as an evolutionary "spandrel."

The concept of a spandrel comes from a seminal essay that uses the ceiling supports of San Marco's Cathedral in Venice as a metaphor for the adaptive value of various biological traits (Gould and Lewontin 1979). The spandrels of San Marco's Cathedral are marvelous tapering triangles that are demarcated by the base

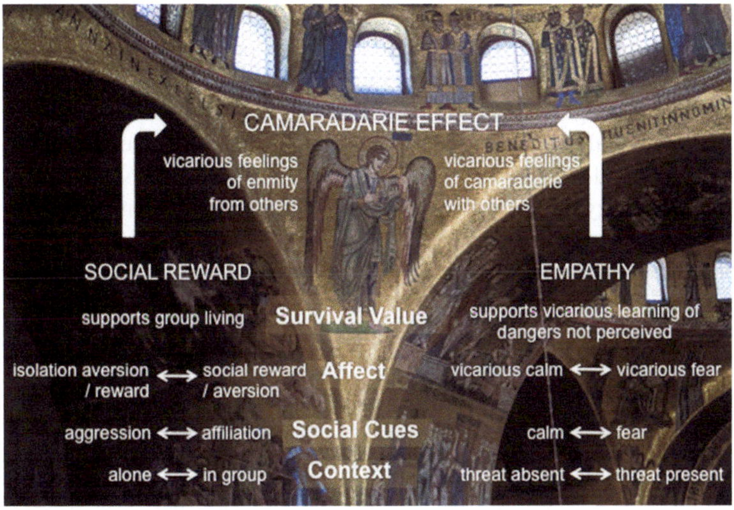

Fig. 2 The camaraderie effect may have evolved as a by-product, or a spandrel, of two adaptive psychological capacities, social reward and empathy. Social reward supports a motivation for living in groups and is responsive to aggressive and affiliative social cues. The affective states included in social motivation include isolation aversion and social reward. Social reward and isolation aversion are two of four affective states that can be inferred from social CPP testing. Other states include social aversion and isolation reward and may play a role in voluntary dispersal. The survival value of social reward is that it supports group living. Empathy is useful for predicting future events from social cues, such as the presence or absence of a threat in the environment from cues of calm or fear. In turn empathy engenders an affective state of vicarious calm or vicarious fear, so one can be alert to dangers not perceived directly or one can be calm under situations felt to be safe by others. In combination, social reward and empathy can generate the camaraderie effect, a process whereby vicarious feelings of others toward the actor can be either discomforting or can engender a sense of well-being. In turn, these psychological states may affect physiological health. These ideas are presented as text over a photograph of a spandrel in San Marco Cathedral, showing social reward and empathy as adaptive psychological states, represented by the arches holding up the dome with the camaraderie effect as a spandrel between these functional and highly adaptive arches

of the cathedral's dome and the curves of the supporting arches beneath the dome that are perpendicular to one another (see Fig. 2). Spandrels add esthetic value to the cathedral, beautiful in their subtle shape and in the images painted on their surfaces, but they serve no primary architectural function. Similarly, while social reward and empathy are adaptive experiences, the camaraderie effect may have emerged as a spandrel, arising only in consequence as self-sustaining. Among humans, social rejection may be unhealthy, increasing salivary cortisol levels (Blackhart et al. 2007). Social isolation also confers psychological and physiological costs to rodents. Social isolation impairs brain development (Black and Greenough 1998; Champagne and Curley 2005; Wiedenmayer 2009), immune reactivity (Boissy et al. 2007; Shanks et al. 1994; Tuchscherer et al. 2010), healing from burns and wounds (Detillion et al. 2004; İşeri et al. 2010), response to ischemia (Norman et al. 2010), recovery from social defeat (Ruis et al. 1999), resiliency to metastasis (Wu et al. 2000), social-emotional health (Seffer et al. 2015), and competence in social hierarchies (van den Berg et al. 1999). With empathy, psychological rejection or indifference could serve as a psychological surrogate for social isolation, resulting in physiological impairments similar to what might be expected from ostracism.

The psychology of the witnesses and their feelings of animosity toward the unhelpful actor, perhaps in anticipation of physical ostracism, might also incur a psychological (and hence physiological) cost for the ostracizers, as these norm enforcers also deprive themselves of camaraderie. Yet ostracizers would still benefit from camaraderie with other group members, whereas the ostracized actor, experiencing more universal rejection, would bear a larger burden of the costs.

By helping others, an individual sustains a feeling of camaraderie, a sense of well-being that augments one's own health and reproductive success. Perhaps this combination of social reward and empathy at times even obfuscates an individual's ability to distinguish self-interest from the interests of others. For instance, when a rat witnesses another rat consuming a highly palatable food reward, dopamine is released into the ventral striatum (Kashtelyan et al. 2014), suggesting a vicarious feeling of reward.

Taken together, these findings offer a proximal mechanism for maintaining altruism as a stable behavioral phenotype. This proximal stake in others, this interdependence, or camaraderie effect, offers a necessary sustenance for altruism genes within wild communities.

Camaraderie Effect: A proximal explanation for biological altruism whereby social reward and empathy (feeling what others feel) together promote feelings of well-being that result from helping others via the positive affect of the recipient and witnesses. This sense of well-being in turn improves one's own immunological responses to pathogens, burns, and wounds, and resiliency to ischemia, social defeat, and metastasis.

Summary A feeling of camaraderie exists when an individual derives pleasure from social interaction and can experience what others feel, including the positive feelings others may have toward the individual. Witnesses of a single altruistic or selfish act, or perhaps repeated actions, associate a favorable or unfavorable affective experience with the actor's identity, an affective experience that resurfaces when witnesses re-encounter the actor. With social affect, an individual becomes susceptible to the actions of others. We know that social isolation, even in rodents, can compromise immune responses and diminish wound repair. With empathy, the actor can predict future outcomes from the feelings of others, including the ability to predict social acceptance or physical rejection from the feelings of affinity or enmity harbored by group members. These changes in affective state, akin to interpersonal stances between witnesses and the actor, influence the nature of ongoing social interactions. In turn, the actor's affective state can bolster or compromise physiological resiliency, resulting in the sustainability of the camaraderie effect as a mechanism for altruistic behavior. The camaraderie effect may thus play a substantial role as an essential stake that an actor holds in the well-being and affinity of others.

References

Armitage KB (1981) Sociality as a life-history tactic of ground squirrels. Oecologia 48(1):36–49

Atsak P, Orre M, Bakker P, Cerliani L, Roozendaal B, Gazzola V, Keysers C (2011) Experience modulates vicarious freezing in rats: a model for empathy. Stress Cognit 6:17

Baack JK, Switzer PV (2000) Alarm calls affect foraging behavior in eastern chipmunks (Tamias striatus, Rodentia: Sciuridae). Ethology 106(12):1057–1066

Barash DP (1974) The evolution of marmot societies: a general theory. Science 185(4149):415–420

Bardo MT, Bevins RA (2000) Conditioned place preference: what does it add to our preclinical understanding of drug reward? Psychopharmacology 153(1):31–43

Bartal IB-A, Decety J, Mason P (2011) Empathy and pro-social behavior in rats. Science 334 (6061):1427–1430

Bartal IB-A, Rodgers DA, Sarria MSB, Decety J, Mason P (2014) Pro-social behavior in rats is modulated by social experience. Elife 3:e01385

Batson CD, Sympson SC, Hindman JL, Decruz P, Todd RM, Weeks JL, Burns CT (1996) "I've been there, too": effect on empathy of prior experience with a need. Pers Soc Psychol Bull 22 (5):474–482

Bekoff M (2004) Wild justice and fair play: cooperation, forgiveness, and morality in animals. Biol Philos 19(4):489–520

Belloni V, Dessi-Fulgheri F, Zaccaroni M, Di Consiglio E, De Angelis G, Testai E, Santucci D (2011) Early exposure to low doses of atrazine affects behavior in juvenile and adult CD1 mice. Toxicology 279(1–3):19–26

Benton D, Brain S, Brain PF (1984) Comparison of the influence of the opiate delta receptor antagonist, ICI 154,129, and naloxone on social interaction and behaviour in an open field. Neuropharmacology 23(1):13–17

Berridge KC (2004) Motivation concepts in behavioral neuroscience. Physiol Behav 81(2):179–209

Berridge KC (2007) The debate over dopamine's role in reward: the case for incentive salience. Psychopharmacology 191(3):391–431

Berridge KC, Robinson TE (2003) Parsing reward. Trends Neurosci 26(9):507–513

Bishop SL, Lahvis GP (2011) The autism diagnosis in translation: shared affect in children and mouse models of ASD. Autism Res 4(5):317–335

Black JE, Greenough WT (1998) Developmental approaches to the memory process. In: Martinez JL Jr, Kesner RP (eds) Neurobiology of learning and memory. Academic Press, San Diego, pp 55–88

Blackhart GC, Eckel LA, Tice DM (2007) Salivary cortisol in response to acute social rejection and acceptance by peers. Biol Psychol 75(3):267–276

Blanchard RJ, Hori K, Tom P, Blanchard DC (1987) Social structure and ethanol consumption in the laboratory rat. Pharmacol Biochem Behav 28(4):437–442

Blumstein DT, Steinmetz J, Armitage KB, Daniel JC (1997) Alarm calling in yellow-bellied marmots: II. The importance of direct fitness. Anim Behav 53(1):173–184

Blumstein DT, Chung LK, Smith JE (2013) Early play may predict later dominance relationships in yellow-bellied marmots (Marmota flaviventris). Proc R Soc B: Biol Sci 280 (1759):20130485

Boissy A, Manteuffel G, Jensen MB, Moe RO, Spruijt B, Keeling LJ, Aubert A (2007) Assessment of positive emotions in animals to improve their welfare. Physiol Behav 92(3):375–397

Bond AB, Kamil AC, Balda RP (2004) Pinyon jays use transitive inference to predict social dominance. Nature 430(7001):778–781

Boreman J, Price E (1972) Social dominance in wild and domestic Norway rats (Rattus norvegicus). Anim Behav 20(3):534–542

Branchi I, Santucci D, Vitale A, Alleva E (1998) Ultrasonic vocalizations by infant laboratory mice: a preliminary spectrographic characterization under different conditions. Dev Psychobiol 33(3):249–256

Brechbühl J, Moine F, Klaey M, Nenniger-Tosato M, Hurni N, Sporkert F, Broillet M-C (2013) Mouse alarm pheromone shares structural similarity with predator scents. Proc Natl Acad Sci 110(12):4762–4767

Burgdorf J, Panksepp J, Brudzynski SM, Kroes R, Moskal JR (2005) Breeding for 50-kHz positive affective vocalization in rats. Behav Genet 35(1):67–72

Butler RG (1980) Population size, social behaviour, and dispersal in house mice: a quantitative investigation. Anim Behav 28(1):78–85

Calamandrei G, Venerosi A, Branchi I, Valanzano A, Alleva E (2000) Prenatal exposure to anti-HIV drugs. Long-term neurobehavioral effects of lamivudine (3TC) in CD-1 mice. Neurotoxicol Teratol 22(3):369–379

Calcagnetti DJ, Schechter MD (1992) Place conditioning reveals the rewarding aspect of social interaction in juvenile rats. Physiol Behav 51(4):667–672

Camacho F, Sandoval C, Paredes R (2004) Sexual experience and conditioned place preference in male rats. Pharmacol Biochem Behav 78(3):419–425

Caramaschi D, de Boer SF, Koolhaas JM (2008) Is hyper-aggressiveness associated with physiological hypoarousal? A comparative study on mouse lines selected for high and low aggressiveness. Physiol Behav 95(4):591–598

Carden SE, Bortot AT, Hofer MA (1993) Ultrasonic vocalizations are elicited from rat pups in the home cage by pentylenetetrazol and U50,488, but not naltrexone. Behav Neurosci 107(5):851–859

Carlier P, Jamon M (2006) Observational learning in C57BL/6j mice. Behav Brain Res 174 (1):125–131

Chambers LK, Singleton GR, Krebs CJ (2000) Movements and social organization of wild house mice (Mus domesticus) in the wheatlands of northwestern Victoria, Australia. J Mammal 81 (1):59–69

Champagne FA, Curley JP (2005) How social experiences influence the brain. Curr Opin Neurobiol 15(6):704–709

Chen Q, Panksepp JB, Lahvis GP (2009) Empathy is moderated by genetic background in mice. PLoS ONE [Electronic Resource], 4(2)

Chevallier C, Kohls G, Troiani V, Brodkin ES, Schultz RT (2012) The social motivation theory of autism. Trends Cognit Sci 16(4):231–239

Collins RL (1988) Observational learning of a left-right behavioral asymmetry in mice (Mus musculus). J Comp Psychol 102(3):222–224

D'Amato FR, Scalera E, Sarli C, Moles A (2005) Pups call, mothers rush: does maternal responsiveness affect the amount of ultrasonic vocalizations in mouse pups? Behav Genet 35 (1):103–112

Darwin C (1872) The expression of the emotions in man and animals. Oxford University Press, New York

Darwin C (1888) The descent of man, and selection in relation to sex. Murray, John London

Dastur FN, McGregor IS, Brown RE (1999) Dopaminergic modulation of rat pup ultrasonic vocalizations. Eur J Pharmacol 382(2):53–67

Detillion CE, Craft TKS, Glasper ER, Prendergast BJ, DeVries AC (2004) Social facilitation of wound healing. Psychoneuroendocrinology 29(8):1004–1011

Dolen G, Darvishzadeh A, Huang KW, Malenka RC (2013) Social reward requires coordinated activity of nucleus accumbens oxytocin and serotonin. Nature 501(7466):179–184

Douglas LA, Varlinskaya EI, Spear LP (2004) Rewarding properties of social interactions in adolescent and adult male and female rats: impact of social versus isolate housing of subjects and partners. Dev Psychobiol 45(3):153–162

Drickamer LC, Gowaty PA, Holmes CM (2000) Free female mate choice in house mice affects reproductive success and offspring viability and performance. Anim Behav 59(2):371–378

Ebert PD (1976) Agonistic behavior in wild and inbred Mus musculus. Behavioral Biology 18 (2):291–294

Ehret G, Bernecker C (1986) Low-frequency sound communication by mouse pups (mus musculus): Wriggling calls release maternal behavior. Anim Behav 34(3):821–830

Eisenberger NI, Lieberman MD (2005) Why it hurts to be left out: the neurocognitive overlap between physical and social pain. The social outcast: Ostracism, social exclusion, rejection, and bullying, pp 109–130

Eklund J, Andersson-Stratberg T, Hansen EM (2009) "I've also experienced loss and fear": effects of prior similar experience on empathy. Scand J Psychol 50(1):65–69

Ekman PE, Davidson RJ (1994) The nature of emotion: fundamental questions: Oxford University Press, Oxford

Fairless AH, Dow HC, Toledo MM, Malkus KA, Edelmann M, Li H et al (2008). Low sociability is associated with reduced size of the corpus callosum in the BALB/cJ inbred mouse strain. Brain Res 16

Ferguson JN, Aldag JM, Insel TR, Young LJ (2001) Oxytocin in the medial amygdala is essential for social recognition in the mouse. J Neurosci 21(20):8278–8285

Festa-Bianchet M, King WJ (1984) Behavior and dispersal of yearling Columbian ground squirrels. Can J Zool 62(2):161–167

Fitzgerald B, Karl B, Moller H (1981) Spatial organization and ecology of a sparse population of house mice (Mus musculus) in a New Zealand forest. J Anim Ecol 489–518

Fredericksen J, Slobodchikoff C (2007) Referential specificity in the alarm calls of the black-tailed prairie dog. Ethol Ecol Evolut 19:87–99

Galef BG (2013) Imitation and local enhancement: detrimental effects of consensus definitions on analyses of social learning in animals. Behav Process 100:123–130

Galef BG Jr, Lim TC, Gilbert GS (2008) Evidence of mate choice copying in norway rats, Rattus norvegicus. Anim Behav 75(3):1117–1123

Gariepy JL, Gendreau PL, Cairns RB, Lewis MH (1998) D1 dopamine receptors and the reversal of isolation-induced behaviors in mice. Behav Brain Res 95(1):103–111

Glickman SE, Schiff BB (1967) A biological theory of reinforcement. Psychol Rev 74(2):81–109

Goodall J (1986) Social rejection, exclusion, and shunning among the Gombe chimpanzees. Ethol Sociobiol 7(3):227–236

Gordon NS, Burke S, Akil H, Watson SJ, Panksepp J (2003) Socially-induced brain 'fertilization': play promotes brain derived neurotrophic factor transcription in the amygdala and dorsolateral frontal cortex in juvenile rats. Neurosci Lett 341(1):17–20

Gould SJ, Lewontin RC (1979) The spandrels of San Marco and the Panglossian paradigm: a critique of the adaptationist programme. Proc R Soc Lond B: Biol Sci 205(1161):581–598

Grant PR, Grant BR (2002) Unpredictable evolution in a 30-year study of Darwin's finches. Science 296(5568):707–711

Graw B, Manser MB (2007) The function of mobbing in cooperative meerkats. Anim Behav 74 (3):507–517

Greene JT (1969) Altruistic behavior in the albino rat. Psychon Sci 14(1):47–48

Griffin AS, West SA (2003) Kin discrimination and the benefit of helping in cooperatively breeding vertebrates. Science 302(5645):634–636

Halladay AK, Amaral D, Aschner M, Bolivar VJ, Bowman A, DiCicco-Bloom E, Threadgill DW (2009) Animal models of autism spectrum disorders: information for neurotoxicologists. Neurotoxicology 5:811–821
Hamilton WD (1964) The genetical evolution of social behavior, I and II. J Theor Biol 7:1–52
Harmon KM, Cromwell HC, Burgdorf J, Moskal JR, Brudzynski SM, Kroes RA, Panksepp J (2008) Rats selectively bred for low levels of 50 kHz ultrasonic vocalizations exhibit alterations in early social motivation. Dev Psychobiol 50(4):322–331
Hatfield E, Cacioppo JT, Rapson RL (1994) Emotional contagion. Cambridge University Press, Cambridge
Hernandez-Lallement J, van Wingerden M, Marx C, Srejic M, Kalenscher T (2014) Rats prefer mutual rewards in a prosocial choice task. Front Neurosci 8
Heyes C, Jaldow E, Nokes T, Dawson G (1994) Imitation in rats (Rattus norvegicus): the role of demonstrator action. Behav Process 32(2):173–182
Himmler B, Kisko T, Euston D, Kolb B, Pellis S (2014) Are 50-kHz calls used as play signals in the playful interactions of rats? I. Evidence from the timing and context of their use. Behav Process 106:60–66
Hoffman ML (1975) Developmental synthesis of affect and cognition and its interplay for altruistic motivation. Dev Psychol 11:607–622
Holekamp K (1984) Dispersal in ground-dwelling sciurids. In: Murie JO, Michener GR (eds) Biology of ground-dwelling squirrels: annual cycles, behavioral ecology, and sociality. University of Nebraska Press, Lincoln, pp 297–320
Hoogland JL (1979) Aggression, ectoparasitism, and other possible costs of prairie dog (sciurudae, cyonmys spp.) coloniality. Behaviour 69(1/2):1–35
Hurst JL (1990) Urine marking in populations of wild house mice Mus domesticus Rutty. I. Communication between males. Anim Behav 40(2):209–222
İşeri SÖ, Düşünceli F, Erzik C, Uslu B, Arbak S, Yeğen BÇ (2010) Oxytocin or social housing alleviates local burn injury in rats. J Surg Res 162(1):122–131
Janik VM, Slater PJB (1997) Vocal learning in mammals. Adv Stud Behav 26:59–99
Jenkins WJ, Becker JB (2003) Female rats develop conditioned place preferences for sex at their preferred interval. Horm Behav 43(4):503–507
Jeon D, Shin HS (2011) A mouse model for observational fear learning and the empathetic response. Curr Protoc Neurosci 8–27
Jeon D, Kim S, Chetana M, Jo D, Ruley HE, Lin S-Y, Shin H-S (2010) Observational fear learning involves affective pain system and Cav1. 2 Ca2+ channels in ACC. Nat Neurosci 13(4):482–488
Kane MJ, Angoa-Peréz M, Briggs DI, Sykes CE, Francescutti DM, Rosenberg DR, Kuhn DM (2012) Mice genetically depleted of brain serotonin display social impairments, communication deficits and repetitive behaviors: possible relevance to autism. PLoS ONE 7(11):e48975
Kashtelyan V, Lichtenberg NT, Chen ML, Cheer JF, Roesch MR (2014) Observation of reward delivery to a conspecific modulates dopamine release in ventral striatum. Curr Biol 24 (21):2564–2568
Kavaliers M, Colwell DD (1995) Odours of parasitized males induce aversive responses in female mice. Anim Behav 50(5):1161–1169
Kavaliers M, Choleris E, Colwell DD (2001) Learning from others to cope with biting flies: social learning of fear-induced conditioned analgesia and active avoidance. Behav Neurosci 115 (3):661–674
Kelley AE, Berridge KC (2002) The neuroscience of natural rewards: relevance to addictive drugs. J Neurosci 22(9):3306–3311
Kelley AE, Baldo BA, Pratt WE, Will MJ (2005) Corticostriatal-hypothalamic circuitry and food motivation: integration of energy, action and reward. Physiol Behav 86(5):773–795
Kennedy BC, Panksepp JB, Wong JC, Krause EJ, Lahvis GP (2011) Age- and strain-dependent influences of morphine on mouse social investigation behavior. Behav Pharmacol 22(2):147–159

Kim S, Mátyás F, Lee S, Acsády L, Shin H-S (2012) Lateralization of observational fear learning at the cortical but not thalamic level in mice. Proc Natl Acad Sci 109(38):15497–15501

Kim BS, Lee J, Bang M, Am Seo B, Khalid A, Jung MW, Jeon D (2014) Differential regulation of observational fear and neural oscillations by serotonin and dopamine in the mouse anterior cingulate cortex. Psychopharmacology 231(22):4371–4381

Kiriazis J, Slobodchikoff C (2006) Perceptual specificity in the alarm calls of Gunnison's prairie dogs. Behav Process 73(1):29–35

Kisko TM, Euston DR, Pellis SM (2015a) Are 50-khz calls used as play signals in the playful interactions of rats? III. The effects of devocalization on play with unfamiliar partners as juveniles and as adults. Behav Process 113:113–121

Kisko TM, Himmler BT, Himmler SM, Euston DR, Pellis SM (2015b) Are 50-kHz calls used as play signals in the playful interactions of rats? II. Evidence from the effects of devocalization. Behav Process 111:25–33

Klump GM, Shalter MD (1984) Acoustic behaviour of birds and mammals in the predator context; I. factors affecting the structure of alarm signals. II. The functional significance and evolution of alarm signals. Zeitschrift fur Tierpsychologie 66(3):189–226. doi:10.1111/j.1439-0310.1984.tb01365.x

Koopman ME, Cypher BL, Scrivner JH (2000) Dispersal patterns of San Joaquin kit foxes (Vulpes macrotis mutica). J Mammal 81(1):213–222, 213

Krackow S, Matuschak B (1991) Mate choice for non-siblings in wild house mice: evidence from a choice test and a reproductive test. Ethology 88(2):99–108

Krams I, Krama T, Iguane K (2006) Mobbing behaviour: reciprocity-based co-operation in breeding Pied Flycatchers Ficedula hypoleuca. Ibis 148(1):50–54

Krams I, Krama T, Igaune K, Mand R (2008) Experimental evidence of reciprocal altruism in the pied flycatcher. Behav Ecol Sociobiol 62(4):599–605

Lahvis GP (2016) Rodent models of autism, epigenetics, and the inescapable problem of animal constraint. In: Kim Y-K, Gewirtz J (eds) Animal Models of Behavior Genetics (pp. 265–301). Springer Publishing Company, New York, p 395

Lahvis GP, Black LM (2011) Social interactions in the clinic and the cage: toward a more valid mouse model of autism animal models of behavioral analysis, vol 50, pp 153–192

Lahvis GP, Panksepp JB, Kennedy BC, Wilson CR, Merriman DK (2015) Social conditioned place preference in the captive ground squirrel (Ictidomys tridecemlineatus) social reward as a natural phenotype. J Comp Psychol 129(3):291

Langford DJ, Crager SE, Shehzad Z, Smith SB, Sotocinal SG, Levenstadt JS, Mogil JS (2006) Social modulation of pain as evidence for empathy in mice. Science 312(5782):1967–1970

Langford DJ, Bailey AL, Chanda ML, Clarke SE, Drummond TE, Echols S, Mogil JS (2010) Coding of facial expressions of pain in the laboratory mouse. Nat Methods 7(6):447–449

Leimar O, Connor RC (2003) By-product benefits, reciprocity, and pseudoreciprocity in mutualism. Genet Cult Evol Cooperation 203–222

Liu H-X, Lopatina O, Higashida C, Fujimoto H, Akther S, Inzhutova A, Yoshihara T (2013) Displays of paternal mouse pup retrieval following communicative interaction with maternal mates. Nat Commun 4:1346

Lord C, Leventhal BL, Cook EH Jr (2001) Quantifying the phenotype in autism spectrum disorders. Am J Med Genet 105(1):36–38

Manning CJ, Potts WK, Wakeland EK, Dewsbury DA (1992) What's wrong with MHC mate choice experiments? In: Doty RL, Muller-Schwarze D (eds) Chemical signals in vertebrates. Plenum, New York, pp 229–235

Markova G, Legerstee M (2006) Contingency, imitation, and affect sharing: Foundations of infants' social awareness. Dev Psychol 42(1):132

Marler P, Evans CS, Hauser MD (1992) Animal signals: motivational, referential, or both. Nonverbal vocal communication: comparative and developmental approaches, pp 66–86

Márquez C, Rennie SM, Costa DF, Moita MA (2015) Prosocial choice in rats depends on food-seeking behavior displayed by recipients. Curr Biol 25(13):1736–1745
Martinez M, Guillen-Salazar F, Salvador A, Simon VM (1995) Successful intermale aggression and conditioned place preference in mice. Physiol Behav 58(2):323–328
Mateo JM, Holmes WG (1999) Plasticity of alarm-call response development in Belding's ground squirrels (Spermophilus beldingi, Sciuridae). Ethology 105:193–206. doi:10.1046/j.1439-0310.1999.00389.x
Mattson BJ, Williams S, Rosenblatt JS, Morrell JI (2001) Comparison of two positive reinforcing stimuli: pups and cocaine throughout the postpartum period. Behav Neurosci 115(3):683–694
McCarley H (1966) Annual cycle, population dynamics and adaptive behavior of Citellus tridecemlineatus. J Mammal 47:294–316
McGraw LA, Young LJ (2010) The prairie vole: an emerging model organism for understanding the social brain. Trends Neurosci 33(2):103–109
Moles A, Kieffer BL, D'Amato FR (2004) Deficit in attachment behavior in mice lacking the mu-opioid receptor gene.[see comment]. Science 304(5679):1983–1986
Moy S, Nadler J, Perez A, Barbaro R, Johns J, Magnuson T, Crawley J (2004) Sociability and preference for social novelty in five inbred strains: An approach to assess autistic-like behavior in mice. Genes Brain Behav 3(5):287–302
Moy S, Nadler J, Young N, Nonneman R, Grossman A, Murphy D, Lauder J (2009) Social approach in genetically engineered mouse lines relevant to autism. Genes Brain Behavior 8 (2):129–142
Nadler J, Moy S, Dold G, Trang D, Simmons N, Perez A, Crawley J (2004) Automated apparatus for quantitation of social approach behaviors in mice. Genes Brain Behavior 3(5):303–314
Nelson RJ, Chiavegatto S (2001) Molecular basis of aggression. Trends Neurosci 24(12):713–719
Norman GJ, Zhang N, Morris JS, Karelina K, Berntson GG, DeVries AC (2010) Social interaction modulates autonomic, inflammatory, and depressive-like responses to cardiac arrest and cardiopulmonary resuscitation. Proc Natl Acad Sci 107(37):16342–16347
Nowak MA (2006) Five rules for the evolution of cooperation. Science 314(5805):1560–1563
Nowak MA, Sigmund K (2005) Evolution of indirect reciprocity. Nature 437(7063):1291–1298
Nunes S, Muecke E-M, Anthony JA, Batterbee AS (1999) Endocrine and energetic mediation of play behavior in free-living belding's ground squirrels. Horm Behav 36(2):153–165
Panksepp J (1998) Affective neuroscience. Oxford University Press, Oxford
Panksepp JB, Lahvis GP (2007) Social reward among juvenile mice. Genes Brain Behav 6 (7):661–671
Panksepp JB, Jochman K, Kim JU, Koy JJ, Wilson ED, Chen Q, Lahvis GP (2007) Affiliative behavior, ultrasonic communication and social reward are influenced by genetic variation in adolescent mice. PLoS ONE [Electronic Resource], 2(e351)
Parsana AJ, Moran EE, Brown TH (2012) Rats learn to freeze to 22-kHz ultrasonic vocalizations through autoconditioning. Behav Brain Res 232(2):395–399
Pellis SM, Pellis VC (2007) Rough-and-tumble play and the development of the social brain. Curr Dir Psychol Sci 16(2):95–98
Pellis SM, Pellis VC, Reinhart CJ (2010) The evolution of social play. In Wothman C, Plotsky P, Schecter D, Cummings CA (eds) Formative experiences: the interaction of caregiving, culture, and developmental psychobiology. Cambridge University Press, Cambridge, pp 404–431
Penn DJ, Potts WK (1999) The evolution of mating preferences and major histocompatibility complex genes. Am Nat 153(2):145–164
Peterson RO (1979) Social rejection following mating of a subordinate wolf. J Mammal 60 (1):219–221
Pocock MJO, Hauffe HC, Searle JB (2005) Dispersal in house mice. Linn Soc Biol J 84(3):565–583
Preis M, Kroener-Herwig B (2012) Empathy for pain: the effects of prior experience and sex. Eur J Pain 16(9):1311–1319

Preston SD, de Waal FB (2002) Empathy: its ultimate and proximate bases. Behav Brain Sci 25 (1):1–20

Rice GE, Gainer P (1962) "Altruism" in the albino rat. J Comp Physiol Psychol 55(1):123

Roberts G (2005) Cooperation through interdependence. Anim Behav 70:901–908. doi:10.1016/j.anbehav.2005.02.006

Roberts SA, Davidson AJ, Beynon RJ, Hurst JL (2014) Female attraction to male scent and associative learning: the house mouse as a mammalian model. Anim Behav 97:313–321

Rottman SJ, Snowdon CT (1972) Demonstration and analysis of an alarm pheromone in mice. J Comp Physiol Psychol 81(3):483–490

Ruis M, Te Brake J, Buwalda B, De Boer S, Meerlo P, Korte S, Koolhaas J (1999) Housing familiar male wildtype rats together reduces the long-term adverse behavioural and physiological effects of social defeat. Psychoneuroendocrinology 24(3):285–300

Russell B (1912) The problems of philosophy (1997 edn.). Oxford University Press, Oxford

Sadananda M, Wöhr M, Schwarting RK (2008) Playback of 22-kHz and 50-kHz ultrasonic vocalizations induces differential c-fos expression in rat brain. Neurosci Lett 435(1):17–23

Sanders J, Mayford M, Jeste D (2013) Empathic fear responses in mice are triggered by recognition of a shared experience. PLoS ONE 8(9):e74609

Sankoorikal GM, Kaercher KA, Boon CJ, Lee JK, Brodkin ES (2006) A mouse model system for genetic analysis of sociability: C57BL/6J versus BALB/cJ inbred mouse strains. Biol Psychiatry 59(5):415–423

Sato N, Tan L, Tate K, Okada M (2015) Rats demonstrate helping behavior toward a soaked conspecific. Anim Cogn 18(5):1039–1047

Scherer KR (1986) Vocal affect expression: a review and a model for future research. Psychol Bull 99(2):143

Scherer KR (2003) Vocal communication of emotion: a review of research paradigms. Speech Commun 40(1):227–256

Schilbach L, Timmermans B, Reddy V, Costall A, Bente G, Schlicht T, Vogeley K (2013) Toward a second-person neuroscience. Behav Brain Sci 36(04):393–414

Schneirla TC (ed) (1959) An evolutionary and developmental theory of biphasic processes underlying approach and withdrawal. Nebraska Press, Lincoln

Schroeder BE, Binzak JM, Kelley AE (2001) A common profile of prefrontal cortical activation following exposure to nicotine- or chocolate-associated contextual cues. Neuroscience 105 (3):535–545

Seffer D, Rippberger H, Schwarting RK, Wöhr M (2015) Pro-social 50-kHz ultrasonic communication in rats: post-weaning but not post-adolescent social isolation leads to social impairments—phenotypic rescue by re-socialization. Front Behav Neurosci 9

Seyfarth RM, Cheney DL (2013) Affiliation, empathy, and the origins of theory of mind. Proc Natl Acad Sci 110(Supplement 2):10349–10356

Shanks N, Renton C, Zalcman S, Anisman H (1994) Influence of change from grouped to individual housing on a T-cell-dependent immune response in mice: Antagonism by diazepam. Pharmacol Biochem Behav 47(3):497–502

Shier D, Owings D (2007) Effects of social learning on predator training and postrelease survival in juvenile black-tailed prairie dogs, Cynomys ludovicianus. Anim Behav 73(4):567–577

Silberberg A, Allouch C, Sandfort S, Kearns D, Karpel H, Slotnick B (2014) Desire for social contact, not empathy, may explain "rescue" behavior in rats. Anim Cogn 17(3):609–618

Silverman JL, Yang M, Lord C, Crawley JN (2010) Behavioural phenotyping assays for mouse models of autism. Nat Rev Neurosci 11(7):490–502

Singer T, Seymour B, O'Doherty J, Kaube H, Dolan RJ, Frith CD (2004) Empathy for pain involves the affective but not sensory components of pain. Science 303(5661):1157–1162

Singer T, Seymour B, O'Doherty JP, Stephan KE, Dolan RJ, Frith CD (2006) Empathic neural responses are modulated by the perceived fairness of others. Nature 439(7075):466–469

Slobodchikoff CN, Paseka A, Verdolin JL (2009) Prairie dog alarm calls encode labels about predator colors. Anim Cogn 12(3):1435–9448
Smith KS, Berridge KC (2007) Opioid limbic circuit for reward: Interaction between hedonic hotspots of nucleus accumbens and ventral pallidum. J Neurosci 27(7):1594–1605
Spencer CM, Graham DF, Yuva-Paylor LA, Nelson DL, Paylor R (2008) Social behavior in Fmr1 knockout mice carrying a human FMR1 transgene. Behav Neurosci 122(3):710–715
Spinka M, Newberry RC, Bekoff M (2001) Mammalian play: training for the unexpected. Q Rev Biol 76(2):143–168
Templeton CN, Greene E, Davis K (2005) Allometry of alarm calls: black-capped chickadees encode information about predator size. Science 308(5730):1934–1937
Thonhauser KE, Raveh S, Hettyey A, Beissmann H, Penn DJ (2013) Scent marking increases male reproductive success in wild house mice. Anim Behav 86(5):1013–1021
Thorndike EL (1911) Animal Intelligence. New York: Hafner
Tinbergen N (1963) On aims and methods of ethology. Zeitschrift fur Tierpsychologie 20:410–433
Tomasello M, Kruger AC, Ratner HH (1993) Cultural learning. Behav Brain Sci 16:495–552
Tomasello M, Carpenter M, Call J, Behne T, Moll H (2005) Understanding and sharing intentions: the origins of cultural cognition. Behav Brain Sci 28(05):675–691
Traulsen A, Nowak MA (2006) Evolution of cooperation by multilevel selection. Proc Natl Acad Sci 103(29):10952–10955
Trivers RL (1971) The evolution of reciprocal altruism. Q Rev Biol 46:189–226
Tuchscherer M, Kanitz E, Puppe B, Tuchscherer A (2010) Altered immunomodulation by glucocorticoids in neonatal pigs exposed to a psychosocial stressor. Pediatr Res 68(6):473–478
Tzschentke TM (2007) Measuring reward with the conditioned place preference (CPP) paradigm: update of the last decade. Addict Biol 12(3–4):227–462
Valsecchi P, Singleton GR, Price WJ (1996) Can social behaviour influence food preference of wild mice, Mus domesticus, in confined field populations? Austr J Zool 44(5):493–501
Valsecchi P, Bosellini I, Sabatini F, Mainardi M, Fiorito G (2002) Behavioral analysis of social effects on the problem-solving ability in the house mouse. Ethology 108(12):1115–1134
van den Berg CL, Hol T, Van Ree JM, Spruijt BM, Everts H, Koolhaas JM (1999) Play is indispensable for an adequate development of coping with social challenges in the rat. Dev Psychobiol 34(2):129–138. doi:10.1002/(sici)1098-2302(199903)34:2<129:aid-dev6>3.0.co;2-l
Vanderschuren LJ, Niesink RJ, Van Pee JM (1997) The neurobiology of social play behavior in rats. Neurosci Biobehav Rev 21(3):309–326
Weigl PD, Hanson EV (1980) Observational learning and the feeding behavior of the red squirrel Tamiasciurus hudsonicus: the ontogeny of optimization. Ecology 214–218
West SA, Griffin AS, Gardner A (2007) Evolutionary explanations for cooperation. Curr Biol 17 (16):R661–R672
Wheatcroft DJ, Price TD (2008) Reciprocal cooperation in avian mobbing: playing nice pays. Trends Ecol Evol 23(8):416–419
Wiedenmayer CP (2009) Plasticity of defensive behavior and fear in early development. Neurosci Biobehav Rev 33(3):432–441
Wilkinson GS (1984) Reciprocal food sharing in the vampire bat. Nature 308(5955):181–184
Willuhn I, Tose A, Wanat MJ, Hart AS, Hollon NG, Phillips PE, Wöhr M (2014) Phasic dopamine release in the nucleus accumbens in response to pro-social 50 kHz ultrasonic vocalizations in rats. J Neurosci 34(32):10616–10623
Wilson DS (1997) Altruism and organism: disentangling the themes of multilevel selection theory. Am Nat 150(S1):s122–S134. http://doi.org/10.1086/286053
Winslow JT, Camacho F (1995) Cholinergic modulation of a decrement in social investigation following repeated contacts between mice. Psychopharmacology 121(2):164–172
Wöhr M, Roullet FI, Crawley JN (2011a) Reduced scent marking and ultrasonic vocalizations in the BTBR T + tf/J mouse model of autism. Genes Brain Behav 10(1):35–43

Wöhr M, Roullet FI, Hung AY, Sheng M, Crawley JN (2011b) Communication impairments in mice lacking Shank1: reduced levels of ultrasonic vocalizations and scent marking behavior. PLoS ONE 6(6):e20631

Wolff RJ (1985) Mating behavior and female choice: their relation to social structure in wild caught house mice (mus musculus) housed in a semi-natural environment. J Zool Ser A 207 (1):43–51

Wu W, Yamaura T, Murakami K, Murata J, Matsumoto K, Watanabe H, Saiki I (2000) Social isolation stress enhanced liver metastasis of murine colon 26-L5 carcinoma cells by suppressing immune responses in mice. Life Sci 66(19):1827–1838

Yang M, Clarke AM, Crawley JN (2009) Postnatal lesion evidence against a primary role for the corpus callosum in mouse sociability. Eur J Neurosci 29(8):1663–1677

Zentall TR, Galef BG Jr (1988) Social learning: psychological and biological perspectives. Lawrence Erlbaum Associates Inc, Hillsdale

A Social Reinforcement Learning Hypothesis of Mutual Reward Preferences in Rats

Julen Hernandez-Lallement, Marijn van Wingerden, Sandra Schäble and Tobias Kalenscher

Abstract Although the use of neuroimaging techniques has revealed much about the neural correlates of social decision making (SDM) in humans, it remains poorly understood how social stimuli are represented, and how social decisions are implemented at the neural level in humans and in other species. To address this issue, the establishment of novel animal paradigms allowing a broad spectrum of neurobiological causal manipulations and neurophysiological recordings provides an exciting tool to investigate the neural implementation of social valuation in the brain. Here, we discuss the potential of a rodent model, Rattus norvegicus, for the understanding of SDM and its neural underpinnings. Particularly, we consider recent data collected in a rodent prosocial choice task within a social reinforcement framework and discuss factors that could drive SDM in rodents.

Keywords Prosocial · Rat · Social reinforcement learning · Amygdala · Mutual reward preference

Contents

1 Social Decision Making: From Humans to Animals 160
2 A Social Reinforcement Learning Framework for Understanding Social Decision Making in Animals 162
3 Individual Differences in Social Learning 166
4 Potential Mediators of Social Reinforcement 170
References 172

J. Hernandez-Lallement (✉) · M. van Wingerden · S. Schäble · T. Kalenscher
Department of Comparative Psychology, Institute of Experimental Psychology, Heinrich-Heine University Düsseldorf, Düsseldorf 40225, Germany
e-mail: julien.her@gmail.com

Curr Topics Behav Neurosci (2016) 30: 159–176
DOI 10.1007/7854_2016_436
Published Online: 30 April 2016

1 Social Decision Making: From Humans to Animals

Social decision making (SDM), i.e., choice behavior that directly or indirectly affects and/or is affected by others, is essential to navigate an ever more complex social environment. SDM is found in our desire to adopt a child and our decisions to give to charity or to punish social norm transgressions. A great amount of work has been devoted to outlining the behavioral dynamics underlying such decisions (Hastings et al. 2007; Wilson 2015), and a recent surge of interest has started to elucidate the general neural mechanisms underlying SDM in humans (Behrens et al. 2008; Izuma et al. 2008; Bhanji and Delgado 2013; Hernandez-Lallement et al. 2014; Ruff and Fehr 2014; Strombach et al. 2015).

Typically, studies exploring SDM in humans use noninvasive methods and correlative approaches (Rilling and Sanfey 2011; Crockett and Fehr 2014; Margittai et al. 2015). Although the use of such techniques has produced a formidable amount of data, the main limitation of these procedures is the lack of causal evidence for the contribution of a brain structure to SDM above and beyond the constraints of technologies available for human research (Knoch et al. 2006). As such, animal models of (social) decision making can complement human research at two different levels. First, they permit the use of neuroscientific methods that go beyond large-scale neural recording techniques in humans by providing direct access to neural activity with high temporal and spatial resolution, thus offering opportunities for causal interventions in the anatomy, activity, connectivity, genetics, and neurochemistry of the neural circuits implicated in SDM processes (Kalenscher and van Wingerden 2011). Second, through experimental analysis of behavior, such models provide a unique chance to compare the evolution of SDM across species (Crowley and Zentall 2013) and sample the spectrum of social behavior from markedly individualistic to highly social species. Therefore, animal models present essential tools to precisely delineate the neural pathways and mechanisms involved in SDM and provide a method to carry out between-species comparisons that are ultimately relevant for a better comprehension of human social cognition.

Up until now, the model of choice for investigating SDM is the nonhuman primate, representing a group of species closely related to humans (Silk and House 2011; Brosnan and de Waal 2014). Although such models are of great importance to study behavioral and evolutionary aspects of SDM, foremost ethical considerations limit their potential in neuroscientific research. Therefore, recent studies promote the use of rats (*Rattus norvegicus*) as an affordable, readily accessible, and standardized model to study SDM. There is a wealth of evidence suggesting that rodent decision making is often contingent on social contexts. For instance, social interaction modulates foraging behavior (Galef 1985; Galef and Whiskin 2008; Łopuch and Popik 2011) and motor learning (Zentall and Levine 1972), avoidance- (Masuda and Aou 2009) and fear-related behaviors (Kim et al. 2010; Atsak et al. 2011; Carrillo et al. 2015) as well as ultrasonic communication (Wöhr and Schwarting 2007; Wöhr et al. 2008; Łopuch and Popik 2011). Recent work demonstrates that rats reciprocate help to partners that previously helped them (direct reciprocity; Rutte and Taborsky 2007a) and show generalized helping behavior if they received assistance from others in the past (generalized reciprocity; Pfeiffer et al. 2005; Rutte and Taborsky 2007b). Helping behavior is modulated by social experience, that is, actor rats help partners they have previously been in contact with (Ben-Ami Bartal et al. 2014) and might depend on the current satiation state (Schneeberger et al. 2012), bodily mass (Hernandez-Lallement et al. 2015), and food-seeking behavior of conspecifics (Márquez et al. 2015). Finally, it has recently been shown that rats, tested in a prosocial choice task (PCT), prefer options that yield food for themselves as well as for other individuals over alternatives yielding reward only to themselves, suggesting that rats' choices are driven by social factors beyond their own-payoff (Marquez and Moita 2012; Hernandez-Lallement et al. 2015). The growing interest in rodents as a model for social neuroscience is illustrated by the steady increase in the proportion of publications on neuroscientific aspects of social behavior in rats over the last decades (Fig. 1a). Interestingly, other rodent species have also received much attention (Fig. 1b) and have been shown to exhibit similar social preferences (Panksepp and Lahvis 2011; Lahvis et al. 2015). This is in contrast to a rather limited increase in the ratio of publications on social neuroscience in nonhuman primates (Fig. 1b). Note that ethical restrictions and different experimental timescales might account for these discrepancies; nonetheless, the growing importance of rodent models in this field is undeniable.

The translation of social behavior from humans to animal models is a complicated matter, and any claim of human-animal translation of supposed social motives should be supported by rigorous controls that establish the true "social" component of the observed behavior. For example, cooperative behavior emerged in pairs of rats trained in a small chamber (Daniel 1942) but disappeared when the chamber length increased (Daniel 1943) or when physical contact became impossible (Marcuella and Owens 1975). Similarly, empirical evidence suggests that food deprivation levels can influence the establishment of social behavior in rats (Taylor

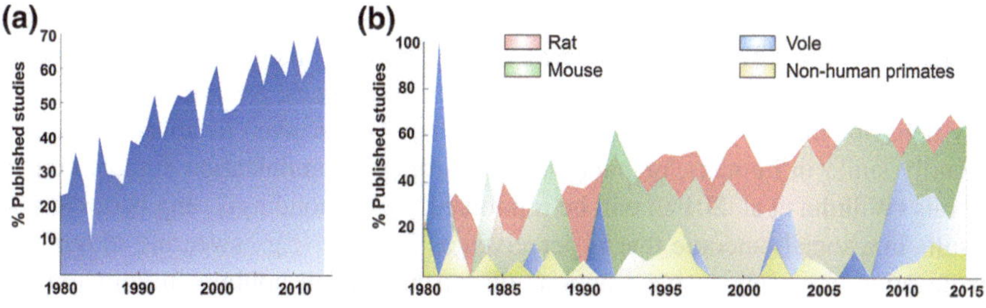

Fig. 1 Fraction of studies related to social behavior in different models indexed on Web of Science between 1980 and 2015. **a** Proportion of publications on the neuroscience of social behavior in rats in relation to the total number of studies on rat's social behavior (*Rat * Social * Neuroscience/Rat * Social*). **b**. Ratio between number of neuroscientific studies published on social behavior in a given animal model (rat, mouse, voles as rodents and macaques, chimpanzees, marmosets, cotton-top tamarin, and capuchin monkeys as nonhuman primate) and all studies using the same model published on social behavior (*Model * Social * Neuroscience/Model * Social*)

1975; Viana et al. 2010), higher deprivation levels resulting in decreased prosocial behavior. Finally, a decrease of social learning rate (Bunch and Zentall 1980) and cooperative moves (Gardner et al. 1984) is observed after abolishment of visual communication, thus suggesting that sensory and physical aspects of the experimental setup can have radical effects on putative social behavior. Thus, depending on task contingencies, experimental designs, physical setup and/or sensory communication possibilities, nonsocial cues might compete with social cues to affect decision making in social contexts. For instance, the establishment of cooperative coordination increases in pairs of rats when mediated by a nonsocial light cue, but not in the absence of the cue and furthermore disappears when reward is delivered for own instrumental behavior (Schuster 2002). These results emphasize the importance of including controls for nonsocial sources of behavioral reinforcement in experiments investigating social behavioral dynamics and mechanisms. Importantly, such nonsocial controls could reveal a relevant baseline behavioral (choice) pattern (not necessarily equivalent to economically defined indifference between the outcomes; Hernandez-Lallement et al. 2015) to which behavior in the social condition could be contrasted.

2 A Social Reinforcement Learning Framework for Understanding Social Decision Making in Animals

Animal choice behavior is often analyzed within a reinforcement learning framework (Schultz 2006). According to the most basic reinforcement learning principles, action–outcome contingencies are learned through positive reinforcement (i.e., the likelihood of an operant behavior increases if it is followed by a reward) and/or negative reinforcement (i.e., the likelihood decreases if it is followed by an aversive event, such as an electric shock; Niv and Montague 2008). SDM has been recently discussed in the light of a *social* reinforcement hypothesis (Chang et al. 2011; Hernandez-Lallement et al. 2015, 2016) which states that animals' choices in social contexts are also affected by a process that updates the likelihood of some actions over alternative courses of action based on *social* outcomes. According to this view, any behavior that results in a social outcome that is perceived as appetitive, e.g., a friendly smile in humans, or putatively rewarding communication signals emitted by rats (Willuhn et al. 2014), will be reinforced. Correspondingly, any behavior that results in a social outcome that is perceived as aversive (e.g., swearing in humans) or negative (e.g., aggressive reactions of conspecifics in nonhuman animals) will less likely be repeated in the future. In the social reinforcement framework, social reinforcers are thus social stimuli that carry positive or negative reinforcement properties. There is indirect evidence for this hypothesis in rats. For instance, putatively rewarding 50 kHz ultrasonic vocalizations (USVs; see below) emitted by a conspecific rat trigger dopamine release in an observer rat's nucleus accumbens (NAcc; Willuhn et al. 2014), a signal associated with reinforcement learning

(Schultz et al. 1997). Furthermore, witnessing a reward delivered to a conspecific rat elicits activation in an observer's NAcc—a possible mechanism for vicarious reinforcement (Kashtelyan et al. 2014).

Recently, we used this framework to discuss the dynamics of prosocial choice behavior in a rodent PCT (Fig. 2a; Hernandez-Lallement et al. 2015, 2016). In this task, pairs of rats (an actor and a partner) are trained in a double T-maze setting. Actors are the first movers and choose to enter one of two different compartments, either choice leading to an identical reward for themselves. However, entering one compartment triggers the delivery of an additional reward for the partner rat (both-reward, BR; Fig. 2a, upper panel), whereas entering the alternative compartment does not yield any additional reward to the partner (own reward, OR; lower panel). To control for nonsocial secondary reinforcement effects, actor rats

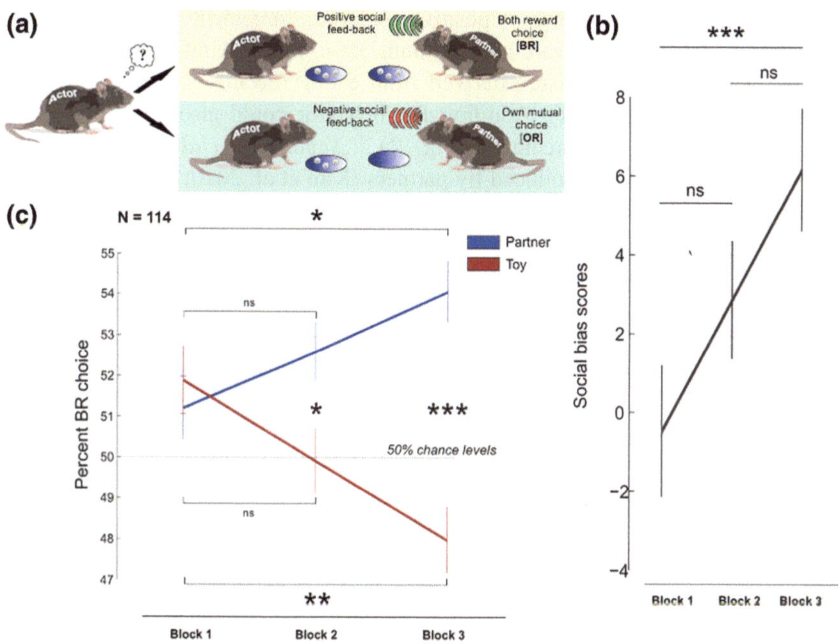

Fig. 2 Social reinforcement learning framework. **a** Putative reinforcement mechanisms in a prosocial choice task for rodents. An actor rat decides between rewarding (upper panel, *yellow background*) and not rewarding (lower panel, *mint background*) a partner rat at no cost to himself while being identically rewarded for both choices as well. The reinforcement learning hypothesis implies that both outcomes can lead to positive and negative social feedback from the partner in case it gets access to food (*upper panel*), or not (*lower panel*), respectively. **b** Social bias scores increased within sessions. Social bias score computed across 114 rats, eight sessions, and blocks of five trials. The distributions increased over blocks and became significantly different from the precedent block from block 2 onwards. **c** %BR preference increased and decreased in the partner and toy conditions, respectively. Preference for the BR alternative increased steadily across trials within sessions in the partner condition and decreased in the toy condition. Error bars are s.e.m. *$p < .05$; ***$p < .001$, *ns* not significant; Bonferroni corrected

are also tested in a nonsocial toy condition. In this control condition, the partner rat is replaced by a toy animal of similar shape and size, while keeping all other task parameters identical to the social condition, including the reward contingencies. Animals are allowed to sample both BR- and OR-outcomes for a certain number of forced trials (only one option available to the actor) followed by free choice trials (the actor can choose freely between OR- and BR-options) where their social preferences can be observed. Results show that rats prefer mutual rewards more in the partner condition than in the toy control condition (Hernandez-Lallement et al. 2015). We interpreted this behavior as evidence for prosocial preference in rats because the actors' inclination for providing food access to the partners was driven by social factors beyond their own-payoff.

The social reinforcement learning hypothesis provides a useful and parsimonious framework that equips us with conceptual tools to describe and predict the rats' behavior in the PCT task. As pointed out, an actor's prosocial choice could be driven by (i) the consequence of positive social reinforcement (Fig. 2a, "*Positive social feedback*"), e.g., rewarding communication signals emitted by the partner (Seffer et al. 2014) or pleasure derived from eating rewards in spatial proximity (Barnett and Spencer 1951). Additionally, behavior could also be reinforced by (ii) negative social reinforcement (Fig. 2a, "*Negative social feedback*"), e.g., potential distress signals produced by partners (Kim et al. 2010; Atsak et al. 2011) missing out on reward in OR choices. As previously noted (Hernandez-Lallement et al. 2015), positive and negative social reinforcement are not mutually exclusive, but could act in concert to reinforce prosocial choices. If the social reinforcement hypothesis accounts for the choice allocation observed in the PCT, one should be able to find signatures of social learning in the choice dynamics of actor rats. To search for signs of social learning, we exploited the reversal nature of the PCT task. Briefly, to control for side biases and habit formation, the compartments associated with BR- and OR-outcomes were pseudo-randomized across testing sessions and rats. Thus, on nearly every session, the OR/BR-compartment assignments were reversed with respect to the previous session, and animals had to re-learn the compartment-outcome contingencies anew. It is important to note again that the outcome for the actor was identical for both choices; OR- and BR-choices differed only in the outcome to the partner rat. Hence, flexible adaptation to the frequent contingency reversals could only be driven by the social reinforcing component of BR-outcomes, not by absolute differences in outcomes. Using a large data set of rats tested on the PCT (N = 114 rats; data taken from different, partly unpublished experiments), we divided the first eight sessions of testing (the number of training sessions differed across rats and experiments, but each animal in the data set was trained for at least 8 sessions per condition) in three blocks of five trials (each session consisted of 15 trials, which we subdivided into three blocks of five trials for analysis) and computed mean social bias scores across animals. Social bias scores quantify the normalized difference in mutual reward choices between partner and toy conditions, i.e., how much more (or less) an actor chooses the BR-option in the partner- compared to the overall BR preference levels. Social bias scores can be construed as the added social value of a conspecific's access to food

(See Hernandez-Lallement et al. 2015 for similar computation). The social bias score for rat i was computed with the following equation:

$$\mathrm{SB}_i = \left[\frac{\mathrm{BR(partner)}_i - \mathrm{BR(toy)}_i}{\mathrm{BR(partner)}_i + \mathrm{BR(toy)}_i}\right] * 100 \quad (1)$$

Note that in previous studies (Hernandez-Lallement et al. 2015, 2016), we used the BR preference in toy condition only in the denominator term of the social bias score equation which captured more directly the percent change from toy baseline levels. However, using only the percent change in the toy condition as normalization could potentially yield skewed distributions.[1] The formula used here, which produces strictly normalized values located between −100 and 100 %, yields qualitatively similar results while retaining a normal distribution of social bias scores at the population level. Accordingly, a positive social bias score for rat i (SB_i), i.e., higher BR preference in the partner than in the toy condition, reflects the added positive social value for a conspecific's access to food, whereas a negative social bias score can be interpreted as negative social value. Results are depicted in Fig. 2b. A repeated-measures ANOVA revealed a significant main effect of blocks on social bias scores ($F_{(2,226)} = 10.42$, $p < .001$, $\eta_p^2 = .08$), indicating that social preferences (re-)emerged across trials within sessions. Post hoc pairwise comparisons showed a significant increase in social bias scores between blocks 1 and 3 ($t_{(113)} = 4.45$, $p < .001$; $\mathrm{CI}_{99} = [-10.55, -2.73]$; Cohen's $d = .54$; Bonferroni corrected; $\alpha = .02$), whereas no significant difference was found between blocks 2 and 3 ($t_{(113)} = 2.26$, $p = .03$; $\mathrm{CI}_{99} = [-7.14, .54]$; $d = .28$) as well as between blocks 1 and 2 ($t_{(113)} = 2.38$, $p = .02$; $\mathrm{CI}_{99} = [-7.02,.34]$; $d = .28$).

Importantly, social bias scores quantify the normalized difference in BR preference between partner and toy condition (see Eq. 1). Therefore, to break down the processes underlying the increase of social bias scores previously reported, we computed the average fraction of BR-choices for the partner (blue) and toy (red) conditions, i.e., the percentage of BR-choices out of all choices (Fig. 2c). We found that rats were nearly indifferent between OR- and BR-alternatives at the beginning of a partner session, but their preferences for BR- over OR-options in the partner condition became increasingly pronounced as the session progressed. Surprisingly, this pattern was completely reversed in the toy condition, where animals decreased their preferences for BR over OR choices across trials within sessions. A repeated-measures ANOVA (with condition and block as within-subject factors) revealed a significant effect of condition on %BR-choices ($F_{(1,113)} = 13.23$, $p < .001$, $\eta_p^2 = .11$) and a significant condition * block interaction ($F_{(2,226)} = 10.62$, $p < .001$, $\eta_p^2 = .09$). Further post hoc pairwise comparisons revealed a significant difference in %BR-choices between partner and toy condition for block 2 as well as block 3 (paired-samples t test; block 2: $t_{(113)} = 2.54$, $p < .05$, $\mathrm{CI}_{99} = [-.09, 5.62]$; $d = .34$; block 3: $t_{(113)} = 5.53$, $p < .001$, $\mathrm{CI}_{99} = [3.23, 9.05]$; $d = .74$, Bonferroni

[1]This was not the case in previous studies from Hernandez-Lallement et al. 2015, 2016.

corrected), but not in block 1 ($t_{(113)} = -.51$, $p = .58$, $CI_{99} = [-3.78: 2.55]$; $d = -.07$). Moreover, %BR-choices were significantly different between blocks 1 and 3 in both partner ($t_{(113)} = 3.00$, $p < .05$, $CI_{99} = [-.37, -5.40]$; $d = .35$) and toy conditions ($t_{(113)} = 3.61$, $p < .01$, $CI_{99} = [6.68, 1.06]$; $d = .44$). No additional significant differences were found between the blocks.

These findings have two important implications. First, they show that preference for the BR-option increased across trials in the partner sessions, a process which might reflect the updating of the social value of the choice outcomes. Second, we observed a within-session decrease of BR preference in the toy condition which suggests that animals developed an aversion against additional rewards delivered to the opposite compartment in a nonsocial context, possibly reflecting frustration effects related to rats' inability to access uneaten rewards in the opposite compartment. This bifurcating pattern implicates that "baseline" preference levels in the PCT are dynamic; the actual preference for social outcomes should, therefore, not be compared to indifference levels (50 %), but rather to the BR-choice levels observed in the nonsocial context control condition. This is precisely why social bias scores, i.e., the percent change of BR-choice between partner and toy condition, in our opinion is a better estimate of mutual reward preferences than comparison of BR-choices against chance. Overall, these data are consistent with the idea that the emergence of rats' social preferences reflects social learning.

3 Individual Differences in Social Learning

An identical social context might affect individual animals in different ways. For instance, social behavior in rats seems to be differentially influenced by group hierarchy (Baenninger 1966) or social experience (Ben-Ami Bartal et al. 2014). Such inter-individual differences in social behavior should be prominent in PCT performance, too. To characterize individual differences in social preferences, we compared individual social bias scores to a bootstrapped reference distribution obtained through random permutation. Briefly (See Hernandez-Lallement et al. 2015 for exact procedure), we generated a distribution of permuted social bias scores, computed by drawing scores (with replacement, $N = 5000$ times) from sessions in both partner and toy conditions while shuffling the session labels. We then compared actual social bias scores to the 95 % confidence interval on this simulated distribution of social bias scores (Fig. 3a; confidence interval limits: [−2.66; 2.66]). Animals with social bias scores exceeding the upper limit of the confidence interval were categorized as "*prosocial*" ($n = 55$; 48 % of all animals), whereas all remaining animals were categorized as "*nonsocial*" ($n = 59$; 52 %). Strikingly, in comparison with baseline levels (toy condition), prosocial animals had between 2 to nearly 21 more BR-choices in the partner compared to the toy condition, illustrating that social preference levels varied substantially, even within the category of rats classified as prosocial. Additionally, animals classified as nonsocial included those that showed rather indifferent choice allocations across

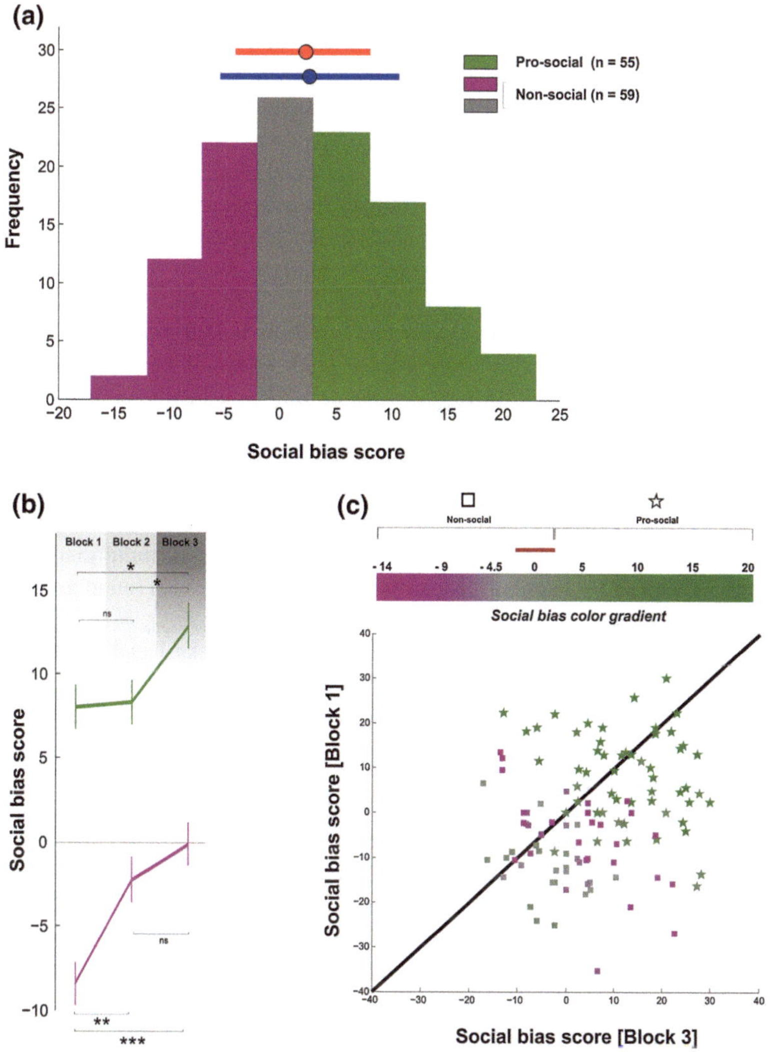

Fig. 3 Individual differences in prosocial choice. **a** Individual differences in social learning. Histogram of social bias scores. Social bias scores exceeding the upper limit of confidence interval (upper limit: 5.47) were categorized as "prosocial" (*green*; $n = 55$; 48 % of all rats) and remaining animals were categorized as "nonsocial" (*violet/gray*; $n = 59$, 52 % of all rats). The *gray bar* represents animals from the nonsocial group located within the 95 % confidence interval. *Blue dot* and *line* are the mean and standard deviation of the social bias score distribution, respectively. *Red dot* and *line* are the distribution's median and the 25 and 75 % percentile values, respectively. **b** Average social bias scores across blocks for prosocial (*green*) and nonsocial groups (*violet*). Both groups showed significant increase in social bias score across blocks. **c** Increasing social bias scores from block 1 to block 3. Scatter plot of individual social bias scores levels in block 1 (*y*-axis) and block 3 (*x*-axis) for prosocial (*star*) and nonsocial animals (*squares*). Data points under the *diagonal* represent animals that had an increase in social bias score from block 1 to block 3. *Color gradient* inform on overall social bias score values (See panel **a**). The *red horizontal line* represents the 95 % confidence interval. Error bars are s.e. m. $*p < .05$; $**p < .01$; $***p < .001$, *ns* not significant; Bonferroni corrected

conditions (SB within the bootstrapped confidence interval; Fig. 3a, gray bar) and others that even showed "*antisocial*[2]" behavior, i.e., negative social bias scores reflecting lower BR preferences for a conspecific than for inanimate toys. Note that negative social bias scores reached only modest levels compared to the positive social bias scores of the prosocial group.

In order to further investigate whether nonsocial animals truly showed overall indifference and/or aversion toward mutual rewards across trials, we computed social bias scores in each block of trials for both prosocial and nonsocial groups. We hypothesized that, contrary to prosocial animals, rats in the nonsocial group would not show significant change in social bias scores across blocks (Fig. 3b). A repeated-measures ANOVA (blocks and group as within- and between-subject factors, respectively) revealed a significant main effect of block on social bias scores ($F_{(2,224)} = 10.25$, $p < .001$, $\eta_p^2 = .08$), as well as a significant block * group interaction ($F_{(2,224)} = 4.07$, $p < .05$, $\eta_p^2 = .04$). Moreover, there was a significant difference in social bias scores between blocks 1 and 3 ($t_{(54)} = 2.21$, $p < .05$, $CI_{99} = [-10.72, 1.01]$; $d = .47$) as well as 2 and 3 in the prosocial group ($t_{(54)} = 2.07$, $p < .05$, $CI_{99} = [-10.44, 1.32]$; $d = -.44$), but not between blocks 1 and 2 ($t_{(54)} = .15$, $p = 1.00$, $CI_{99} = [-5.65, 5.06]$; $d = .03$), confirming that prosocial animals showed social learning. However, and crucially, we also found a significant difference in social bias scores between blocks 1 and 2 ($t_{(58)} = 3.17$, $p < .01$, $CI_{99} = [-11.27, -1.09]$; $d = -.63$) as well as between blocks 1 and 3 ($t_{(58)} = 4.00$, $p < .001$, $CI_{99} = [-12.70, -2.91]$; $d = -.87$) in the nonsocial group, although no difference was found between blocks 2 and 3 ($t_{(58)} = -.95$, $p = .74$, $CI_{99} = [-7.33, 3.09]$; $d = -.22$). These results suggest that animals initially classified as nonsocial also showed social learning. While 64 % of prosocial animals ($n = 35$) increased their social bias scores from blocks 1–3 (Fig. 3c; stars under the diagonal), 70 % of nonsocial animals ($n = 41$) showed a similar increase (squares under the diagonal), adding further support to the notion that nonsocial animals showed social learning, too. Therefore, although overall mean social bias scores differed between groups, the social learning rate might have been comparable across animals in both groups. To address this possibility, we computed the absolute difference in social bias scores between blocks 1 and 3 for every animal in each group. Direct comparison showed that rats in both groups showed comparable increases in social bias scores from block 1 to block 3 (Fig. 4a; $t_{(112)} = -1.16$, $p = .25$; $CI_{95} = [-9.35, 2.47]$, $d = -.21$). Overall, this analysis suggests that animals classified as prosocial or nonsocial differed predominantly in their baseline social preference levels rather than in social learning capabilities, which were robust across the whole population.

While the increase in social bias scores across blocks was comparable between groups, it is conceivable that prosocial and nonsocial animals differed in their social learning rate *within* the partner and the toy conditions. To address this possibility,

[2]The term "antisocial" needs to be interpreted with caution, because rats' choices may have been motivated by nonsocial factors that were unrelated to malicious, egocentric, or other "antisocial" motives. We use the term "antisocial" agnostically to describe the negative effect of social context on social preferences.

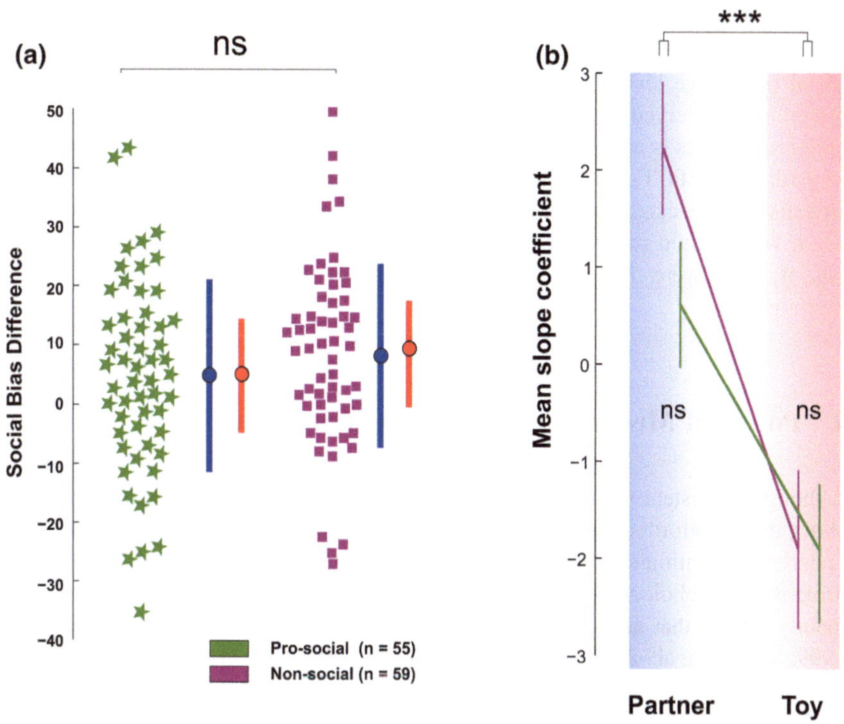

Fig. 4 Individual differences in prosocial choice. **a** Magnitude of change in social bias score between the blocks. There was no significant difference in social bias score difference (block 3–block 1) between prosocial (*green*) and nonsocial groups (*violet*). *Blue dot* and *line* are the distribution's mean and standard deviation, respectively. *Red dot* and *line* are the distribution's median and the 25 and 75 % percentile values, respectively. **b** Slope coefficient for %BR across blocks per group in the partner (*blue background*) and toy conditions (*red background*). While both groups showed higher slope coefficients in the partner than in the toy condition (main effect of condition), there was no difference between groups in either condition. Error bars are s.e.m. ***$p < .001$, *ns* not significant

we regressed, for each condition separately, the rats' individual %BR-choices against block and extracted the individual regression coefficients as estimates of the steepness of the slopes across blocks as a proxy of the rats' learning rates (linear fit of the %BR in each block, per animal; steeper slopes indicate higher learning rates). A mixed ANOVA revealed a significant effect of condition ($F_{(1,112)} = 20.01$, $p < .001$, $\eta_p^2 = .15$), but not in the condition * group interaction ($F_{(1,112)} = 1.23$, $p = .27$, $\eta_p^2 = .01$). While both groups showed significantly higher slope values in the partner than in the toy condition (prosocial: $t_{(112)} = 2.28$, $p < .05$; $CI_{99} = [-.43, 5.47]$, $d = .46$; nonsocial: $t_{(112)} = 4.12$, $p < .01$; $CI_{99} = [1.47, 6.88]$, $d = .77$), slope coefficients did not differ between groups in either condition (partner: $t_{(112)} = 1.69$, $p = .10$; $CI_{99} = [-.89, 4.11]$, $d = -.31$; toy: $t_{(112)} = -.05$, $p = .96$; $CI_{99} = [-2.88, 2.77]$, $d = .01$). Thus, this analysis also confirms that both prosocial and nonsocial animals showed comparable social learning in each condition.

Altogether, these results show that prosocial and nonsocial rats show comparable social reinforcement learning capabilities and that individual differences in initial social preference levels between animals can account for differences in prosocial preferences observed at the group level. Thus, considering learning rates next to preference levels is advisable when investigating social choice behavior in rodents. Regarding the PCT, one challenge for future research is to determine whether animals initially classified as nonsocial, i.e., rats that had lower social bias scores to begin with, would reach similar levels of mutual reward preferences as prosocial rats if they were trained more extensively. This possibility remains to be investigated.

4 Potential Mediators of Social Reinforcement

Although consistent with the social reinforcement hypothesis, the results presented above do not inform on what kind of social reinforcement, negative and/or positive, underlies the within-session increase of BR-preference. Several social stimuli could drive the rats' choice allocation in the PCT. Prime candidates are auditory stimuli, mainly USVs, that are known to carry affective state information (Knutson et al. 1999; Litvin et al. 2007) not only in rodents (Burgdorf et al. 2008; Wöhr and Schwarting 2008; Seffer et al. 2014) but also in other species (Sharp et al. 2005; Gadziola et al. 2012a). Notably, substantial evidence obtained in big brown bats suggest that amygdala neurons discriminate between different social USVs (Naumann and Kanwal 2011; Gadziola et al. 2012b; Peterson and Wenstrup 2012; Grimsley et al. 2013). Similar results were obtained in rats showing that USVs reflecting negative (22 kHz) and positive (50 kHz) affective state can modulate approach behavior (Wöhr et al. 2008) and are coupled to tonic increase and decrease of amygdala neuron firing rates, respectively (Parsana et al. 2012). Finally, the fact that 50 kHz USVs elicit phasic dopamine release in the nucleus accumbens (Willuhn et al. 2014), as mentioned above, is consistent with the idea that USVs have social significance and qualify as social reinforcers. Other stimuli, such as odors (Wang et al. 2006; Wesson 2013) might also carry reinforcing properties for rats. However, the idea that olfaction would drive prosocial choice allocation in the PCT would require highly dynamic chemical processes, which we believe unlikely given the trial-based design. Assessing the influence of several putative social signals in transmitting partner feedback and their effect on SDM remains an unresolved issue for now.

Recent evidence showing that lesion to the basolateral amygdala (BLA) abolishes mutual reward preferences in rats (Fig. 5; Hernandez-Lallement et al. 2016) sheds light on the potential neural bases of mutual reward preferences. The BLA, a neuronal cluster located in the temporal lobe involved in associative (social) learning (Adolphs 2009), receives strong innervations from visual, auditory, and somatosensory tracts, as well as from olfactory and vomeronasal pathways in rodents, and is therefore often considered as the amygdala sensory interface

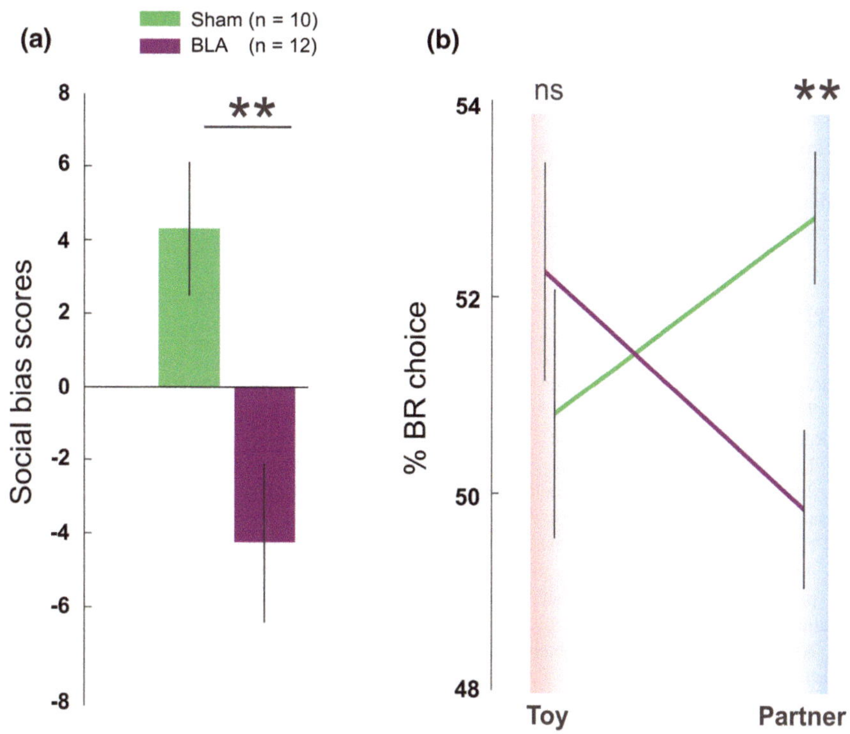

Fig. 5 Basolateral amygdala lesions impair mutual reward preferences in rats. **a** Social bias scores in sham-operated (*green*) and BLA-lesioned groups (*violet*). BLA-lesioned animals had significantly lower social bias scores than sham-operated animals. **b** Percentage of mutual reward choices for sham (*green*) and BLA group (*purple*). In comparison with sham-operated animals, BLA-lesioned rats made significantly less mutual reward choices in the partner but not the toy condition. *Shading* blue partner; red toy condition. Error bars represent the standard error of the mean, s.e.m. $**p < .01$, independent-samples t test; Bonferroni corrected; *ns* not significant. All panels were adapted with permission from Hernandez-Lallement et al. 2016

(Phelps and LeDoux 2005; Brennan and Kendrick 2006). Particularly, it has been proposed that the BLA may act as a vigilance device important for linking the incentive properties and outcome values of rewards and punishments to predictive sensory cues, and enhancing their affective salience (Schoenbaum et al. 1999, 2003). Accordingly, the BLA might sensitize individuals to the emotional value of social information and thus contribute to social learning (Adolphs 2009). Thus, BLA lesion-related impairments in the establishment of mutual reward preferences could reflect a deficit in rodent decision making in the social domain, reminiscent of similar deficits in human populations with impaired amygdala function (Adolphs 2010; Decety et al. 2013).

In conclusion, we believe that the emergence of rodent models of SDM within a social reinforcement learning framework provides exciting opportunities to study social choice using the full range of the neurobiological toolbox. Novel behavioral paradigms such as the PCT and others (Márquez et al. 2015) pave the way toward a mechanistic model of social preferences and therefore contribute to a better understanding of the neural circuits involved in nonhuman and human SDM.

Acknowledgments This work was supported by a grant from Deutsche Forschungsgemeinschaft (DFG Grant No. KA 2675/6-1). MvW was supported by the Volkswagen Stiftung "Freigeist" fellowship, AZ 88216.

References

Adolphs R (2009) The social brain: neural basis of social knowledge. Annu Rev Psychol 60:693–716

Adolphs R (2010) What does the amygdala contribute to social cognition? Ann N Y Acad Sci 1191:42–61. Available at: http://www.pubmedcentral.nih.gov/articlerender.fcgi?artid=2871162&tool=pmcentrez&rendertype=abstract. Accessed 10 Dec 2014

Atsak P, Orre M, Bakker P, Cerliani L, Roozendaal B, Gazzola V, Moita M, Keysers C (2011) Experience modulates vicarious freezing in rats: a model for empathy. PloS one 6:e21855. Available at: http://www.pubmedcentral.nih.gov/articlerender.fcgi?artid=3135600&tool=pmcentrez&rendertype=abstract. Accessed 29 May 2013

Baenninger LP (1966) The reliability of dominance orders in rats. Anim Behav 14:367–371. Available at: http://linkinghub.elsevier.com/retrieve/pii/S0003347266800994

Barnett SA, Spencer MM (1951) Feeding, social behaviour and interspecific competition in wild rats. Behaviour 229–242. Available at: http://psycnet.apa.org/psycinfo/1952-01970-001. Accessed 27 Mar 2012

Behrens TEJ, Hunt LT, Woolrich MW, Rushworth MFS (2008) Associative learning of social value. Nature 456:245–249

Bartal IBA, Rodgers DA, Sarria MSB, Decety J, Mason P (2014) Pro-social behavior in rats is modulated by social experience. Elife 3:e01385. Available at: http://www.pubmedcentral.nih.gov/articlerender.fcgi?artid=3884117&tool=pmcentrez&rendertype=abstract. Accessed January 20, 2014

Bhanji JP, Delgado MR (2013) The social brain and reward: social information processing in the human striatum. WIRE Cogn Sci

Brennan PA, Kendrick KM (2006) Mammalian social odours: attraction and individual recognition. Philos Transac Royal Soc Lond Ser B, Biol Sci 361:2061–2078

Brosnan SF, de Waal FBM (2014) Evolution of responses to (un)fairness. Science. Published online before print. Available at: http://www.sciencemag.org/cgi/doi/10.1126/science.1251776 . Accessed 16 Oct 2014

Bunch GB, Zentall TR (1980) Imitation of a passive avoidance response in the rat. Bull Psychon Soc 15:73–75. Available at: http://link.springer.com/10.3758/BF03334469

Burgdorf J, Kroes RA, Moskal JR, Pfaus JG, Brudzynski SM, Panksepp J (2008) Ultrasonic vocalizations of rats (Rattus norvegicus) during mating, play, and aggression: behavioral concomitants, relationship to reward, and self-administration of playback. J Comp Psychol 122:357–367. Available at: http://doi.apa.org/getdoi.cfm?doi=10.1037/a0012889

Carrillo M, Migliorati F, Bruls R, Han Y, Heinemans M, Pruis I, Gazzola V, Keysers C (2015) Repeated Witnessing of Conspecifics in Pain: effects on emotional contagion. Plos One 10: e0136979. Available at: http://dx.plos.org/10.1371/journal.pone.0136979

Chang SWC, Winecoff AA, Platt ML (2011) Vicarious reinforcement in rhesus macaques (Macaca Mulatta). Front Neurosci 5:10

Crockett MJ, Fehr E (2014) Social brains on drugs: tools for neuromodulation in social neuroscience. Soc Cogn Affect Neurosci 9:250–254

Crowley PH, Zentall TR (eds) (2013) Comparative decision making, 1 edition. Oxford University Press, New York, 27 Feb 2013

Daniel WJ (1942) Cooperative problem solving in rats. J Comp Psychol 34:361–368. Available at: http://content.apa.org/journals/com/34/3/361

Daniel WJ (1943) Higher order cooperative problem solving in rats. J Comp Psychol 35:297–305. Available at: http://content.apa.org/journals/com/35/3/297

Decety J, Chen C, Harenski C, Kiehl KA (2013) An fMRI study of affective perspective taking in individuals with psychopathy: imagining another in pain does not evoke empathy. Front Human Neurosci 7:489. Available at: http://www.pubmedcentral.nih.gov/articlerender.fcgi?artid=3782696&tool=pmcentrez&rendertype=abstract

Gadziola MA, Grimsley JMS, Faure PA, Wenstrup JJ (2012a) Social vocalizations of big brown bats vary with behavioral context. PLoS ONE 7

Gadziola MA, Grimsley JMS, Shanbhag SJ, Wenstrup JJ (2012b) A novel coding mechanism for social vocalizations in the lateral amygdala. J Neurophysiol 107:1047–1057

Galef B, Whiskin E (2008) "Conformity" in Norway rats? Animal Behav 75:2035–2039. Available at: http://linkinghub.elsevier.com/retrieve/pii/S0003347208001000. Accessed 21 July 2011

Galef BG (1985) Direct and indirect behavioral pathways to the social transmission of food avoidance. Ann N Y Acad Sci 443:203–215. Available at: http://www.ncbi.nlm.nih.gov/pubmed/2990282

Gardner R, Corbin TL, Beltramo J, Nickell G (1984) The prisoner's dilemma game and cooperation in the rat. Psychol Rep 55:687–696

Grimsley JMS, Hazlett EG, Wenstrup JJ (2013) Coding the meaning of sounds: contextual modulation of auditory responses in the basolateral amygdala. J Neurosci 33:17538–17548. Available at: http://www.pubmedcentral.nih.gov/articlerender.fcgi?artid=3812514&tool=pmcentrez&rendertype=abstract

Hastings PD, Utendale WT, Sullivan C (2007) The socialization of prosocial development. In: Grusec JE, Hastings PD (eds) Handbook of socialization: theory and research, the Guilfo. Guilford Publications, New York, pp 638–664

Hernandez-Lallement J, Kuss K, Trautner P, Weber B, Falk A, Fliessbach K (2014) Effort increases sensitivity to reward and loss magnitude in the human brain. Social cogn Affect Neurosci. Available at: http://www.ncbi.nlm.nih.gov/pubmed/23202663

Hernandez-Lallement J, van Wingerden M, Marx C, Srejic M, Kalenscher T (2015) Rats prefer mutual rewards in a prosocial choice task. Front Neurosci 8:1–9. Available at: http://www.frontiersin.org/Decision_Neuroscience/10.3389/fnins.2014.00443/abstract. Accessed 16 Jan 2015

Hernandez-Lallement J, van Wingerden M, Schäble S, Kalenscher T (2016) Basolateral amygdala lesions abolish mutual reward preferences in rats. Neurobiol Learn Mem 127:1–9

Izuma K, Saito DN, Sadato N (2008) Processing of social and monetary rewards in the human striatum. Neuron 58:284–294

Kalenscher T, van Wingerden M (2011) Why we should use animals to study economic decision making—a perspective. Fron Neurosci 5:82. Available at: http://www.pubmedcentral.nih.gov/articlerender.fcgi?artid=3118901&tool=pmcentrez&rendertype=abstract. Accessed 21 July 2011

Kashtelyan V, Lichtenberg NT, Chen ML, Cheer JF, Roesch MR (2014) Observation of reward delivery to a conspecific modulates dopamine release in ventral striatum. Curr Biol 24:2564–2568. Available at: http://linkinghub.elsevier.com/retrieve/pii/S0960982214011361. Accessed 24 Oct 2014

Kim EJ, Kim ES, Covey E, Kim JJ (2010) Social transmission of fear in rats: the role of 22-kHz ultrasonic distress vocalization. PloS one 5:e15077. Available at: http://www.pubmedcentral.nih.gov/articlerender.fcgi?artid=2995742&tool=pmcentrez&rendertype=abstract. Accessed 14 June 2011

Knoch D, Pascual-Leone A, Meyer K, Treyer V, Fehr E (2006) Diminishing reciprocal fairness by disrupting the right prefrontal cortex. Science (New York, NY) 314:829–832. Available at: http://www.ncbi.nlm.nih.gov/pubmed/17023614. Accessed 19 July 2011

Knutson B, Burgdorf J, Panksepp J (1999) High-frequency ultrasonic vocalizations index conditioned pharmacological reward in rats. Physiol Behav 66:639–643. Available at: http://www.ncbi.nlm.nih.gov/pubmed/10386908

Lahvis GP, Panksepp JB, Kennedy BC, Wilson CR, Merriman DK (2015) Social conditioned place preference in the captive ground squirrel Ictidomys tridecemlineatus: social reward as a natural phenotype. J Comp Psychol 129:291–303

Litvin Y, Blanchard DC, Blanchard RJ (2007) Rat 22 kHz ultrasonic vocalizations as alarm cries. Behav Brain Res 182:166–172. Available at: http://www.ncbi.nlm.nih.gov/pubmed/17173984. Accessed 5 Nov 2011

Łopuch S, Popik P (2011) Cooperative behavior of laboratory rats (Rattus norvegicus) in an instrumental task. J Comp Psychol (Washington, DC: 1983) 125:250–253. Available at: http://www.ncbi.nlm.nih.gov/pubmed/21341907. Accessed 6 Mar 2013

Marcuella H, Owens K (1975) Cooperative problem solving by albino rats: a re-evaluation. Psychol Rep 37:591–598

Margittai Z, Strombach T, van Wingerden M, Joëls M, Schwabe L, Kalenscher T (2015) A friend in need: time-dependent effects of stress on social discounting in men. Horm Behav 73:75–82. Available at: http://www.scopus.com/inward/record.url?eid=2-s2.0-84935493433&partnerID=40&md5=a43a37fa689897eaac7e52a6fcef99e9

Marquez C, Moita M (2012) Rats cooperate in the absence of direct benefit: development of a new behavioural task. In: Society for neuroscience, pp 14–15. New Orleans

Márquez C, Rennie SM, Costa DF, Moita MA (2015) Prosocial choice in rats depends on food-seeking behavior displayed by recipients. Curr Biol 1–10. Available at: http://linkinghub.elsevier.com/retrieve/pii/S0960982215005928

Masuda A, Aou S (2009) Social transmission of avoidance behavior under situational change in learned and unlearned rats. PloS one 4:e6794. Available at: http://www.pubmedcentral.nih.gov/articlerender.fcgi?artid=2728840&tool=pmcentrez&rendertype=abstract. Accessed 11 Nov 2013

Naumann RT, Kanwal JS (2011) Basolateral amygdala responds robustly to social calls: spiking characteristics of single unit activity. J Neurophysiol 105:2389–2404

Niv Y, Montague PR (2008) Theoretical and empirical studies of learning. Learning 329–350

Panksepp JB, Lahvis GP (2011) Rodent empathy and affective neuroscience. Neurosci Biobehav Rev 35:1864–1875. Available at: http://www.pubmedcentral.nih.gov/articlerender.fcgi?artid=3183383&tool=pmcentrez&rendertype=abstract. Accessed 26 Sept 2013

Parsana AJ, Li N, Brown TH (2012) Positive and negative ultrasonic social signals elicit opposing firing patterns in rat amygdala. Behav Brain Res 226:77–86. Available at: http://dx.doi.org/10.1016/j.bbr.2011.08.040

Peterson DC, Wenstrup JJ (2012) Selectivity and persistent firing responses to social vocalizations in the basolateral amygdala. Neuroscience 217:154–171. Available at: http://dx.doi.org/10.1016/j.neuroscience.2012.04.069

Pfeiffer T, Rutte C, Killingback T, Taborsky M, Bonhoeffer S (2005) Evolution of cooperation by generalized reciprocity. Proc Biol Sci Royal Soc 272:1115–1120. Available at: http://www.pubmedcentral.nih.gov/articlerender.fcgi?artid=1559812&tool=pmcentrez&rendertype=abstract. Accessed 26 July 2011

Phelps EA, LeDoux JE (2005) Contributions of the amygdala to emotion processing: from animal models to human behavior. Neuron 48:175–187. Available at: http://www.ncbi.nlm.nih.gov/pubmed/16242399. Accessed 10 July 2014

Rilling JK, Sanfey AG (2011) The neuroscience of social decision-making. Ann Rev Psychol 62:23–48. Available at: http://www.ncbi.nlm.nih.gov/pubmed/20822437. Accessed 11 June 2011

Ruff CC, Fehr E (2014) The neurobiology of rewards and values in social decision making. Nat Rev Neurosci 15:549–562. Available at: http://www.ncbi.nlm.nih.gov/pubmed/24986556. Accessed 9 July 2014

Rutte C, Taborsky M (2007a) The influence of social experience on cooperative behaviour of rats (Rattus norvegicus): direct vs generalised reciprocity. Behav Ecol Sociobiol 62:499–505. Available at: http://www.springerlink.com/index/10.1007/s00265-007-0474-3. Accessed 5 July 2011

Rutte C, Taborsky M (2007b) Generalized reciprocity in rats. PLoS Biol 5:e196. Available at: http://www.pubmedcentral.nih.gov/articlerender.fcgi?artid=1914408&tool=pmcentrez&rendertype=abstract. Accessed 14 June 2011

Schneeberger K, Dietz M, Taborsky M (2012) Reciprocal cooperation between unrelated rats depends on cost to donor and benefit to recipient. BMC Evol Biol 12:41. Available at: http://www.pubmedcentral.nih.gov/articlerender.fcgi?artid=3340299&tool=pmcentrez&rendertype=abstract. Accessed 13 Nov 2012

Schoenbaum G, Chiba AA, Gallagher M (1999) Neural encoding in orbitofrontal cortex and basolateral amygdala during olfactory discrimination learning. The J Neurosci Official J Soc Neurosci 19:1876–1884

Schoenbaum G, Setlow B, Saddoris MP, Gallagher M (2003) Encoding predicted outcome and acquired value in orbitofrontal cortex during cue sampling depends upon input from basolateral amygdala. Neuron 39:855–867

Schultz W (2006) Behavioral theories and the neurophysiology of reward. Ann Rev Psychol 57:87–115. Available at: http://www.ncbi.nlm.nih.gov/pubmed/16318590. Accessed 8 Mar 2012

Schultz W, Dayan P, Montague PR (1997) A neural substrate of prediction and reward. Science 275:1593–1599. Available at: http://www.sciencemag.org/cgi/doi/10.1126/science.275.5306.1593. Accessed 4 Oct 2012

Schuster R (2002) Cooperative coordination as a social behavior. Human Nat 13:47–83

Seffer D, Schwarting RKW, Wöhr M (2014) Pro-social ultrasonic communication in rats: Insights from playback studies. J Neurosci Methods 1–9. Available at: http://www.ncbi.nlm.nih.gov/pubmed/24508146. Accessed 3 April 2014

Sharp SP, McGowan A, Wood MJ, Hatchwell BJ (2005) Learned kin recognition cues in a social bird. Nature 434:1127–1130. Available at: http://www.ncbi.nlm.nih.gov/pubmed/15858573

Silk JB, House BR (2011) Evolutionary foundations of human prosocial sentiments. Proc National Acad Sci USA 108 Suppl:10910–10917. Available at: http://www.pubmedcentral.nih.gov/articlerender.fcgi?artid=3131813&tool=pmcentrez&rendertype=abstract. Accessed 19 March 2014

Strombach T, Weber B, Hangebrauk Z, Kenning P, Karipidis II, Tobler PN, Kalenscher T (2015) Social discounting involves modulation of neural value signals by temporoparietal junction. Proc National Acad Sci 112:1619–1624. Available at: http://www.pnas.org/lookup/doi/10.1073/pnas.1414715112

Taylor CJ (1975) Study of altruism in rats in an appetitive situation. Psychological reports 36:571–574. Available at: http://www.amsciepub.com/doi/abs/10.2466/pr0.1975.36.2.571

Viana DS, Gordo I, Sucena E, Moita MAP (2010) Cognitive and motivational requirements for the emergence of cooperation in a rat social game. PloS one 5:e8483. Available at: http://www.pubmedcentral.nih.gov/articlerender.fcgi?artid=2799661&tool=pmcentrez&rendertype=abstract. Accessed 14 Oct 2011

Wang Y, Fontanini A, Katz DB (2006) Temporary basolateral amygdala lesions disrupt acquisition of socially transmitted food preferences in rats. Learn Mem 13:794–800

Wesson DW (2013) Sniffing behavior communicates social hierarchy. Curr Biol 23:575–580. Available at: http://dx.doi.org/10.1016/j.cub.2013.02.012

Willuhn I, Tose A, Wanat MJ, Hart AS, Hollon NG, Phillips PEM, Schwarting RKW, Wöhr M (2014) Phasic dopamine release in the nucleus accumbens in response to pro-social 50 kHz ultrasonic vocalizations in rats. J Neurosci Official J Soc Neurosci 34:10616–10623. Available at: http://www.ncbi.nlm.nih.gov/pubmed/25100595. Accessed 7 Nov 2014

Wilson DS (2015) Altruism in everyday life. In: Does Altruism Exist?: culture, genes, and the welfare of others, pp 117–129. Yale University Press, January 13 2015

Wöhr M, Houx B, Schwarting RKW, Spruijt B (2008) Effects of experience and context on 50-kHz vocalizations in rats. Physiol Behav 93:766–776. Available at: http://www.ncbi.nlm.nih.gov/pubmed/18191963. Accessed 7 Nov 2013

Wöhr M, Schwarting RKW (2007) Ultrasonic communication in rats: can playback of 50 kHz calls induce approach behavior? PLoS ONE 2:e1365. Available at: http://dx.plos.org/10.1371/journal.pone.0001365

Wöhr M, Schwarting RKW (2008) Ultrasonic calling during fear conditioning in the rat: no evidence for an audience effect. Animal Behav 76:749–760. Available at: http://linkinghub.elsevier.com/retrieve/pii/S0003347208002108. Accessed 11 Dec 2011

Zentall T, Levine J (1972) Observational learning and social facilitation in the rat. Science (New York, NY) 178:1220–1221

Part II
Social Behavior in Humans

A Plea for Cross-species Social Neuroscience

Christian Keysers and Valeria Gazzola

Abstract Over the past two decades, the question of how our brain makes us sensitive to the state of conspecifics and how that affects our behaviour has undergone a profound change. Twenty years ago what would now be called social neuroscience was focused on the visual processing of facial expressions and body movements in temporal lobe structures of primates (Puce and Perrett 2003). With the discovery of mirror neurons, this changed rapidly towards the modern field of social neuroscience, in which high-level vision is but one of many focuses of interest. In this essay, we will argue that for the further progress of the field, the integration of animal neuroscience and human neuroscience is paramount. We will do so, by focusing on the field of embodied social cognition. We will first show how the combination of animal and human neuroscience was critical in how the discovery of mirror neurons placed the motor system on the map of social cognition. We will then argue why an integrated cross-species approach will be pivotal to our understanding of the neural basis of emotional empathy and its link to prosocial behaviour.

Keywords Empathy · fMRI · Single cell · Animal physiology · Neuroimaging · Emotional contagion · Mirror neuron

Contents

1 The Importance of Animal Neuroscience for Our Understanding of Action Observation 180

C. Keysers (✉) · V. Gazzola
Netherlands Institute for Neuroscience, A Research Institute of the Royal Netherlands Academy of Arts and Sciences, Meibergdreef 47, 1105BA Amsterdam, The Netherlands
e-mail: c.keysers@nin.knaw.nl

V. Gazzola
e-mail: v.gazzola@nin.knaw.nl

C. Keysers · V. Gazzola
Department of Psychology, University of Amsterdam, Amsterdam, The Netherlands

Curr Topics Behav Neurosci (2016) 30: 179–192
DOI 10.1007/7854_2016_439
Published Online: 01 March 2016

2 The Discovery of Vicarious Emotional Activations and the Limits of Human Neuroscience 182
3 Behavioural Evidence for Affect Sharing in Rodents 183
4 Anatomical Considerations Towards the Validity of a Rodent Model 185
5 Why We Should Use Rodents to Advance Our Understanding of Empathy 186
6 Summary 188
References 188

1 The Importance of Animal Neuroscience for Our Understanding of Action Observation

The discovery of mirror neurons in monkeys (di Pellegrino et al. 1992; Gallese et al. 1996; Keysers et al. 2003; Kohler et al. 2002) was a game-changer for how neuroscientists conceived of social perception (Keysers 2009). That neurons in the motor system could contain highly reliable information about complex visual and auditory stimuli such as the sight and sound of someone performing an action was unorthodox. It forced the field to rethink how semantic meaning is associated with visual and auditory input, by introducing the possibility of embodied cognition.

Embodied cognition is a term that has been used with slightly different meanings, but here we will use it to refer to information that is not in an abstract format, or in that of the input modality (e.g. visual in the case of visual stimuli), but that involves representations that are specific to the body of the perceiver (Goldman and de Vignemont 2009). Core examples are motor formats, i.e. commands that normally aim to move the body, and somatosensory formats, i.e. normally triggered by afferent input from the somatosensory receptors.

What mirror neurons showed was that witnessing the actions of others triggered in addition to the well-known high-level visual representations of regions in the temporal lobe (Puce and Perrett 2003), also embodied information in premotor cortices, that would permit the observer to replicate the observed action. The observer mirrored the observed action in its own motor cortex, hence the name 'mirror neuron' (Keysers 2009).

The acceptance that social perception involved embodied cognition was a true paradigm shift that only occurred because single cell recordings demonstrating that the very same neurons are involved in performing an action and in perceiving it (via vision or sound) were *direct* evidence that neural substrates involved in motor planning are recruited during perception. Had the data permitted any alternative explanation, it is unlikely that this paradigm shift would have occurred. That one could classify what action the animal itself had done, or what action someone else had performed, from the activity of the same neuron with >90 % accuracy was important to determine just how reliable embodied information can be (Keysers et al. 2003). These data were available from invasive single cell recordings in monkeys. In addition, a key element of the animal literature on mirror neurons was the description of what mirror neurons responded to exactly (Gallese et al. 1996;

Keysers et al. 2003; Kohler et al. 2002; Umilta et al. 2001). By comparing the execution and observation of numerous actions, and by exploiting the very high signal to noise of single cell measurements, it became clear that individual neurons sometimes have very sharp preferences for a particular action (e.g. tearing apart) that is shared between action observation and execution in what is called strictly congruent mirror neurons (Gallese et al. 1996). However, neurons with different preferences (e.g. grasping) are often found in the exact same penetration, less than 100 μm away.

Human neuroscience was used rapidly to verify whether a similar system might exist in humans. Initially, positron emission tomography confirmed that regions in the frontal lobe associated with action execution were recruited during action observation in humans too (Rizzolatti et al. 1996). Later, more refined fMRI studies showed that the exact same voxels are involved in performing, hearing and seeing actions (Gazzola et al. 2006; Gazzola and Keysers 2009) and that pattern classifiers could discriminate which of two actions were performed across action execution and perception (Etzel et al. 2008). We can vouch from our own experience that sceptics consistently argue that finding that a voxel is involved in action observation and execution is not strong evidence for the involvement of mirror neurons (Gazzola and Keysers 2009). Some argue that a voxel contains so many neurons that it can be activated in both cases without any of the neurons being involved in both cases. Others argue that even if neurons are involved in both cases, it does not show that they encode action-specific information: action observation and execution both require attention and executive resources that can be the source of the common activation. An additional constraint of neuroimaging in this field is the limited ability to unravel what neurons actually represent in these brain regions. Neurons in the premotor cortex, as mentioned above, have exquisitely specific response profiles, responding for instance to precision but not whole hand prehension during both observation and execution (Gallese et al. 1996), which support classification accuracies above 90 % (Keysers et al. 2003) between different hand actions. However, neurons with different selectivity occur so close to one another that averaging signals over the volume of a typical fMRI voxel would lead to very unspecific signals. Indeed, pattern classification from fMRI BOLD signals drops to levels typically below 60 % even when hand and mouth actions are compared (Etzel et al. 2008), which is thus a gross underestimate of the selectivity of the actual neurons. Such unspecific signals make it difficult to unravel what information such processes could contribute to perception.

It is fair to say that social neuroscience would never have embraced embodied cognition if neuroimaging evidence of overlapping voxels during action observation and execution would have been the only source of information. The animal single cell data were absolutely instrumental in shaping our understanding of this new component of social perception. Neuroimaging on the other hand has also been important. The whole-brain field of view of typical fMRI studies is in stark contrast to the very focal field of view of single cell recordings that is typically confined to a very specific cortical area. While the original mirror neuron work was confined to the ventral premotor cortex (Gallese et al. 1996), neuroimaging naturally expanded the search to the entire brain, with careful studies revealing that inferior parietal, somatosensory, dorsal premotor, supplementary motor and cerebellar cortices all consistently were activated both during the observation and execution of actions (Caspers et al. 2010; Gazzola and Keysers 2009; Keysers and Gazzola 2009). This has guided single cell recordings in animals and humans towards similar brain regions, and there is now evidence from single cell recordings for the existence of mirror neurons in most of these regions (Cisek and Kalaska 2004; Hihara et al. 2015; Rozzi et al. 2008; Mukamel et al. 2010).

In summary, the realization that the motor cortices are part of the network of brain regions involved in action observation has significantly reshaped social neuroscience by introducing embodied cognition into the mainstream. A combination of invasive animal physiology and non-invasive human neuroimaging has been key to the success of this endeavour.

2 The Discovery of Vicarious Emotional Activations and the Limits of Human Neuroscience

A decade after the discovery of mirror neurons in the motor domain, the time was right to explore whether the same basic principle, namely using embodied neural substrates involved in representing states of the self are also used to represent similar states in others, would also apply outside of the motor system (Gallese et al. 2004). In 2003 and 2004, two independent groups showed that experiencing an emotion such as disgust (Wicker et al. 2003) or pain (Singer et al. 2004) triggered activity in the anterior insula (aIns) and in the rostral cingulate cortex (rCC) that overlapped with the activity triggered while experiencing disgust and pain. Simply seeing someone else being touched on a body part also triggered activity in the secondary somatosensory cortex in a region overlapping with activity triggered by the experience of a similar touch (Keysers et al. 2004). Somehow, we vicariously activate our aIns, rCC and somatosensory cortices when we witness the emotions and sensations of others (Keysers and Gazzola 2006). These neuroimaging findings have since been replicated numerous times, with meta-analyses confirming how reliably these regions are recruited during both the experience and perception of emotional and tactile states (Keysers et al. 2010; Lamm et al. 2011).

Individual differences between people who report experiencing more or less empathy have been leveraged to explore the functional significance of these activations (Jabbi et al. 2007; Singer et al. 2004). This correlational approach has been expanded to explore individuals on the extreme of the empathic continuum. So for instance, psychopathic criminals recruit these regions less when witnessing the pain of others (Meffert et al. 2013). It has also been expanded to see whether people more willing to help have stronger activations in these regions (Hein et al. 2010). This has led many to propose that vicarious activations in these regions, the aIns and rCC in particular, are what causes us to share the emotions of others (Engen and Singer 2013). More or less implicit to this line of thinking is the notion that the same neural substrate within the voxels common to observing and experiencing emotions and sensations is recruited in the two conditions. If it were different neurons triggered by the observation and experience of an emotion, the notion that we experience our own emotions in response to the emotions of others makes little sense.

Since 2003, hundreds of studies have investigated vicarious activations in the emotional domain using fMRI, and even just the initial 3 studies have attracted over 2500 citations. However, the core hypotheses of the initial studies, namely that vicarious activations cause us to share the emotions of others and that the same neurons are involved in the experience and observation of emotions, remain untested. This is because these experiments are conducted in humans, in which it is difficult to test either of them. The aIns and rCC are relatively deep structures. To test whether they cause us to share the emotions of others, one would need to modulate their activity and show that this would modulate our ability to share the emotions of others. This is difficult to do, because state-of-the-art methods for non-invasive brain manipulation in humans, including TMS or tDCS, are currently unable to modulate brain activity so deep without having larger effects on the cortices that are closer to the surface. Also, the assumption that the same neurons are active during the observation and experience of emotions in these regions cannot be systematically tested in humans. Although surgical procedures for the treatment of epilepsy offer occasional opportunities to record from single neurons in these regions in humans (Hutchison et al. 1999), they seldom offer the opportunity to carefully characterize the response pattern of neurons over multiple hours of testing, making it difficult to ascertain that neurons truly selectively represent a particular emotional experience (e.g. pain) during the observation and experience of an emotion. It is thus unlikely that human neuroscience will be able, any time soon, to test the hypotheses that seminal neuroimaging studies have helped generate.

3 Behavioural Evidence for Affect Sharing in Rodents

We therefore argue that it is essential that the field now starts to supplement the non-invasive human work with invasive animal neuroscience. Rodents seem a natural model to undertake that work. In this section, we will review behavioural evidence that rodents show affective responses to the distress of others. In the next section, we will review evidence that the neural substrates for pain experience and for affect sharing might be similar enough to humans to make them a powerful model.

In 1959, Church showed that a rat shows signs of distress when witnessing another receive shocks (Church 1959). In 1962, Rice and Gainer showed that a rat would work to help free another rat from an uncomfortable situation (Rice and Gainer 1962), and in 1969, Greene showed that most rats would forgo an easy reward if that meant delivering a shock to another rat (Greene 1969). This initial evidence that rodents find the pain of others aversive and show signs of a motivation to avoid pain to others been supplemented by a second, more recent, series of studies. Langford et al. (2006) injected acetic acid into the abdomen of mice to induce abdominal writhing. Writhing was increased in the mice if they witnessed another suffering similar pain—as if they had shared the pain of the other. Knapska et al. (2010) have shown that rats will start licking rats that had previously been electroshocked, showing that they sense and care about the distress of other rats. Finally, several teams, including us, showed that rats and mice can show 'vicarious freezing' (Atsak et al. 2011; Kim et al. 2010; Jeon et al. 2010; Wohr and Schwarting 2008) in response to the distress of others. In our paradigm, two rats were separated only by a perforated Plexiglas divider. One of the rats was exposed to a series of moderate but startling electroshocks. The other, that either had or had not experienced electroshocks in the past, was made to witness this event through the divider. We found that shock-experienced, but not shock-naïve, rats freeze in response to the other's pain as if they had been shocked themselves, a phenomenon we call 'vicarious freezing' (Carrillo et al. 2015; Atsak et al. 2011). Playing back the sound of the interaction also caused freezing (only in shock-experienced listeners), but playing back the 22 kHz ultrasonic vocalizations alone did not, suggesting that the effect was not mediated by species-specific vocalizations alone but at least partially by a recognition of the situation. The silence caused by the freezing of the demonstrator appears to play an important role in this process (Pereira et al. 2012). Other teams have observed that the amount of 22 kHz vocalizations emitted by the demonstrators can predict the degree of vicarious freezing in the observers (Wohr and Schwarting 2008), a finding compatible with the fact that listening to 22 kHz vocalizations can trigger activity in limbic structures of the listener (Wohr and Schwarting 2010). This suggests that multiple channels of auditory communication can be at play during social fear transmission. Kim and colleagues (Kim et al. 2010) used a fear conditioning paradigm to trigger freezing in the demonstrator rat during the social exchange and arrived at similar conclusions: experienced but not naïve witnesses show vicarious freezing to the distress of a fellow rat. Finally, Jeon et al. (2010) conducted an experiment similar to ours in mice and also observed vicarious freezing in witness mice. However, they also varied how long the demonstrator and witness mice had been housed together previously and found that animals that had spent more time together before the experiment showed more vicarious freezing—an effect reminiscent of the stronger vicarious emotional activations in humans to the pain of in-group members (Avenanti et al. 2010; Hein et al. 2010; Martin et al. 2015; Xu et al. 2009).

Together, these independent rodent experiments converge to show two important findings. First, rats and mice evidence (through vicarious freezing) signs of distress to the distress of others, suggesting that they may provide a valuable animal model for empathy. Second, vicarious freezing is modulated by social factors (familiarity with the demonstrator) and is therefore potentially a valid model for similar factors in humans. It should be noted that in humans, different forms of reactions to another's distress have been distinguished (Preston and de Waal 2002). *Emotional contagion* occurs when the witness experiences an emotion similar to that of the demonstrator. *Empathy* occurs if the witness is additionally aware of the fact that the vicarious emotional state is not his/her own, but a reflection of the demonstrators' state. *Sympathy*, finally, occurs when the witness no longer only experiences vicarious pain in response to the pain of the demonstrator, but transforms this into a motivation to help the other. It is generally assumed that emotional contagion is a prerequisite for empathy, which is in turn a prerequisite for sympathy. How far along this hierarchy of empathy rodents might be remains entirely unclear, but even if vicarious freezing were only a reflection of emotional contagion, understanding its neural basis would illuminate the neural basis of a constituent part of empathy and sympathy (Panksepp and Panksepp 2013).

An additional feature of social fear transmission in rodents is that it shows interesting individual variability associated with genetic background (Chen et al. 2009) and chemically induced disorders (Jung et al. 2013).

4 Anatomical Considerations Towards the Validity of a Rodent Model

For rodents to enable us to attain a better understanding of the neural basis of affect sharing, it is important to explore whether the neural basis of the behavioural signs of affect sharing is homologous to that in humans. As in humans, affect sharing has been linked to the vicarious activation of regions involved in emotional experiences, and we will start by exploring whether the neural basis of emotional experiences is similar across rodents and primates. We will do so in particular, for the regions involved in the experience of pain. We will then review the limited evidence for the role of these regions in affect sharing.

The central representation of pain in rodents is not identical to that in humans (Tracey 2008), but similar enough to be considered a valuable model (Mogil 2009). Like humans, their pain system is composed of a lateral and a medial pain system (Gauriau and Bernard 2002). In the lateral pain system, nociceptive information is sent via lateral nuclei of the thalamus (VPL, VPM, Po) to nociceptive neurons in SI and SII. In the medial pain system, nociceptive information travels through parabrachial and medial nuclei of the thalamus (MD in particular) to the rCC and the rostral agranular insula (Gauriau and Bernard 2002; Iwata et al. 2011). Stimulation of nociceptive fibres in the rats paw triggers activity in $\sim 30\ \%$ of neurons in the rCC and SI, with stronger stimulation triggering more spikes per neuron and activating more neurons in both structures (Zhang et al. 2011). fMRI and deoxyglucose mapping in rats have confirmed that nociceptive stimulation consistently triggers activation in SI, SII and rCC (Shih et al. 2008; Zhao et al. 2011; Porro et al. 1999), much as it does in human (Lanz et al. 2011). While the rCC is therefore very similar in rodents and humans, the similarity of the aIns is notable but potentially less profound. First, as in humans, the insula of rodents receives nociceptive information as revealed by anatomical (Gauriau and Bernard 2002), electrical (Ito 1998; Rodgers et al. 2008), fMRI (Shih et al. 2008) (but see Zhao et al. 2011) and deoxyglucose mapping (Porro et al. 1999) data. However, it receives that input via a thalamic nucleus (PoT) differs from that in primates (VMpo). Second, the human insula has a primary pain representation in the posterior insula and a higher-level representation in the anterior insula, with the latter related to feelings (Craig 2009). This meta-representation may be lacking in rats (Craig 2009). Overall, the organization of the pain matrix in mice is less studied but generally considered similar to that in rats (Mogil 2009).

At present, relatively little is still known about which regions of the pain matrix in rodents show vicarious activation that could trigger vicarious freezing. However, there is increasing evidence that the rCC is involved—a region central to vicarious emotional for pain in humans as we have seen above. Chemical deactivation of the rCC reduces the social transmission of fear in mice (Jeon et al. 2010; Kim et al. 2012) and the electrical stimulation of this region increases social transmission of fear (Kim et al. 2012). Both of these effects are stronger in the right hemisphere (Kim et al. 2012). Also, oscillations in the theta band are increased in this region, while observers witness the distress of another (Jeon et al. 2010; Kim et al. 2012). Much as in humans, this finding does not show that the rCC contains single cells involved in both the experience and observation of pain, but suggests that the rodent rCC may be a good model of the human rCC in relation to empathy for pain.

In summary, rodents show behaviour compatible with vicarious emotions and possess a central pain representation resembling that of humans in many ways, and one of the regions involved in human vicarious emotions (rCCs) is vicariously activated in rodents; its deactivation reduces and its activation increases vicarious freezing.

5 Why We Should Use Rodents to Advance Our Understanding of Empathy

Human neuroscience has powerful methods to map what brain regions are associated with a particular task. This mapping enterprise has very successfully identified that the aIns and rCC are hot spots in the human brain when it comes to sharing the affect of others. Whether this activity causes us to share the affect of others, however, remains difficult to test. It also remains unclear what it is that activity in these brain regions represents (and hence could make us share) about the state of the self and of the other. Both regions are activated by a wide array of tasks, ranging from the experience of negative and positive emotions to a variety of other cognitive tasks (Yarkoni et al. 2011). This makes it impossible to determine with any level of certainty, what neurons in these regions really encode about emotions of the self and other. Do they represent individual emotions, such as disgust or pain, and do they represent emotions in more dimensional fashion (e.g. its level of arousal or valence). Or do they represent even more abstract properties such as salience or uncertainty? It is difficult to imagine how this question can be rigorously addressed without systematically measuring the response of individual neurons over a wide range of emotions and visual stimuli.

What makes rodents a particularly powerful model to shed more light on the neural basis of empathy is the rapidly expanding pallet of techniques neuroscientists have at their disposal to address the very questions that are most difficult to address in humans. This involves both sophisticated ways to modulate and to measure brain activity at the neural level. In terms of modulation, neuroscientists have tools at their disposal that vary from relatively classic chemical deactivations using local anaesthetics such as lidocaine or GABA agonists such as muscimol that powerfully deactivate most neurons in a well-controlled volume (Martin 1991), to highly sophisticated genetically encoded methods, most notably optogenetics, that permit the silencing or stimulation of neurons in a time-resolved fashion (Bernstein and Boyden 2011). These methods have already shown that chemically deactivating the rCC in mice reduces and activating the rCC electrically increases signs of sharing the distress of others (Jeon et al. 2010; Kim et al. 2012), offering what might be argued to be the first powerful experimental demonstration of a link between vicarious rCC activation and affect sharing. More selective modulations of activity in specific cell types, in the aIns, and time-resolved modulations promise to provide valuable insights into this system in the near future.

Methods to measure brain activity in rodents have also made tremendous progress. With the development of silicon-based probes (Buzsaki et al. 2015) that track the activity of dozens of neurons at a time in freely moving animals over days, we have the potential to characterize neurons, while rodents undergo a large pallet of experiences, thereby providing us with the opportunity to characterize the code through which neurons encode emotional states with unprecedented precision. With the development of hybrid devices that combine silicon probe electrodes for recording with miniature LED to stimulate neurons that have photosensitive ion channels will make it possible to combine the identification of local neural codes with the selective activation of these codes to study their impact on the affective state of the animal and behaviour. The possibilities for brain activity recording have been further expanded using genetically encoded calcium indicators (Scanziani and Hausser 2009). This involves the expression of proteins that alter their fluorescence when neurons are active [e.g. GCaMP6 (Chen et al. 2013)], which can then be visualized using two-photon microscopy in vivo. Using such methods, scientists can monitor the activity of hundreds of neurons at a time, while rodents undergo behavioural paradigms. Most importantly, the method can be combined with the staining of different cell types in different colours to enable experimenters to identify which cell types are involved in particular tasks. For the case of mirror neurons for actions, only about 10 % of neurons in the premotor cortex have the ability to respond to the observation of the actions of others. To this day, we do not know what makes these 10 % of mirror neurons any different from the 90 % that do not have mirror properties. Methods available in rodents, such as calcium imaging, would enable us to tackle such questions for the very first time and bring our level of understanding of emotion sharing to an entirely new, cellular level.

6 Summary

Over the past decades, social neuroscience has made great advances. With regard to our understanding of how we perceive the actions of others, the discovery of mirror neurons, and the combination of animal and human neuroscience, has been seminal in revealing the presence of embodied cognition. Animal neuroscience contributed the proof of the overlap at single cell levels of actions of the self and others and a thorough understanding of what it is about actions that is represented in these neurons. Human neuroimaging has contributed a system-level understanding of the brain regions involved and has confirmed striking homologies between the macaque and human brain with regard to action processing. With regard to the neural basis of empathy, we are still in the beginning of our understanding. Human neuroscience has so far not been seconded by systematic animal neuroscience experiments. We have evidenced that regions involved in our own emotions are recruited while witnessing the emotions of others, and correlational evidence suggests a link between these vicarious activations and empathy. However, we still lack an understanding of the properties of neurons in these regions and the causal link between activity of these neurons and sharing the emotions of others. We argue that rodents show signs of emotional contagion and afford a powerful pallet of methods to measure and alter brain activity and that the neuroscience of empathy would now be well advised to combine human experiments with rodent experiments to bring our understanding of empathy to a cellular and causal level that will otherwise remain evasive.

Acknowledgements We thank Sören Krach and Markus Wöhr for their comments on an early version of the manuscript. The work was supported by the NARSAD (Young Investigator Grant 22453 to V.G.), N.W.O. (VIDI grant 452-14-015 to V.G.) and the European Research Council of the European Commission (StG grant 312511 to C.K.).

References

Atsak P, Orre M, Bakker P, Cerliani L, Roozendaal B, Gazzola V, Moita M, Keysers C (2011) Experience modulates vicarious freezing in rats: a model for empathy. PLoS ONE 6:e21855

Avenanti A, Sirigu A, Aglioti SM (2010) Racial bias reduces empathic sensorimotor resonance with other-race pain. Curr Biol 20:1018–1022

Bernstein JG, Boyden ES (2011) Optogenetic tools for analyzing the neural circuits of behavior. Trends Cogn Sci 15:592–600

Buzsaki G, Stark E, Berenyi A, Khodagholy D, Kipke DR, Yoon E, Wise KD (2015) Tools for probing local circuits: high-density silicon probes combined with optogenetics. Neuron 86:92–105

Carrillo M, Migliorati F, Bruls R, Han Y, Heinemans M, Pruis I, Gazzola V, Keysers C (2015) Repeated witnessing of conspecifics in pain: effects on emotional contagion. PLoS ONE 10: e0136979

Caspers S, Zilles K, Laird AR, Eickhoff SB (2010) ALE meta-analysis of action observation and imitation in the human brain. Neuroimage 50:1148–1167

Chen Q, Panksepp JB, Lahvis GP (2009) Empathy is moderated by genetic background in mice. PLoS ONE 4:e4387

Chen TW, Wardill TJ, Sun Y, Pulver SR, Renninger SL, Baohan A, Schreiter ER, Kerr RA, Orger MB, Jayaraman V, Looger LL, Svoboda K, Kim DS (2013) Ultrasensitive fluorescent proteins for imaging neuronal activity. Nature 499:295–300

Church RM (1959) Emotional reactions of rats to the pain of others. J Comp Physiol Psychol 52:132–134

Cisek P, Kalaska JF (2004) Neural correlates of mental rehearsal in dorsal premotor cortex. Nature 431:993–996

Craig AD (2009) A rat is not a monkey is not a human: comment on Mogil (Nature Rev Neurosci 10:283–294 (2009)). Nat Rev Neurosci 10:466–466

di Pellegrino G, Fadiga L, Fogassi L, Gallese V, Rizzolatti G (1992) Understanding motor events: a neurophysiological study. Exp Brain Res 91:176–180

Engen HG, Singer T (2013) Empathy circuits. Curr Opin Neurobiol 23:275–282

Etzel JA, Gazzola V, Keysers C (2008) Testing simulation theory with cross-modal multivariate classification of fMRI data. PLoS ONE 3:e3690

Gallese V, Fadiga L, Fogassi L, Rizzolatti G (1996) Action recognition in the premotor cortex. Brain 119(Pt 2):593–609

Gallese V, Keysers C, Rizzolatti G (2004) A unifying view of the basis of social cognition. Trends Cogn Sci 8:396–403

Gauriau C, Bernard JF (2002) Pain pathways and parabrachial circuits in the rat. Exp Physiol 87:251–258

Gazzola V, Keysers C (2009) The observation and execution of actions share motor and somatosensory voxels in all tested subjects: single-subject analyses of unsmoothed fMRI data. Cereb Cortex 19:1239–1255

Gazzola V, Aziz-Zadeh L, Keysers C (2006) Empathy and the somatotopic auditory mirror system in humans. Curr Biol 16:1824–1829

Goldman A, de Vignemont F (2009) Is social cognition embodied? Trends Cogn Sci 13:154–159

Greene JT (1969) Altruistic behavior in albino rat. Psychon Sci 14:47

Hein G, Silani G, Preuschoff K, Batson CD, Singer T (2010) Neural responses to ingroup and outgroup members' suffering predict individual differences in costly helping. Neuron 68:149–160

Hihara S, Taoka M, Tanaka M, Iriki A (2015) Visual responsiveness of neurons in the secondary somatosensory area and its surrounding parietal operculum regions in awake macaque monkeys. Cereb Cortex 25:4535–4550

Hutchison WD, Davis KD, Lozano AM, Tasker RR, Dostrovsky JO (1999) Pain-related neurons in the human cingulate cortex. Nat Neurosci 2:403–405

Ito SI (1998) Possible representation of somatic pain in the rat insular visceral sensory cortex: a field potential study. Neurosci Lett 241:171–174

Iwata K, Miyachi S, Imanishi M, Tsuboi Y, Kitagawa J, Teramoto K, Hitomi S, Shinoda M, Kondo M, Takada M (2011) Ascending multisynaptic pathways from the trigeminal ganglion to the anterior cingulate cortex. Exp Neurol 227:69–78

Jabbi M, Swart M, Keysers C (2007) Empathy for positive and negative emotions in the gustatory cortex. Neuroimage 34:1744–1753

Jeon D, Kim S, Chetana M, Jo D, Ruley HE, Lin SY, Rabah D, Kinet JP, Shin HS (2010) Observational fear learning involves affective pain system and $Ca_v1.2$ Ca^{2+} channels in ACC. Nat Neurosci 13:482–488

Jung S, Seo JS, Kim BS, Lee D, Jung KH, Chu K, Lee SK, Jeon D (2013) Social deficits in the AY-9944 mouse model of atypical absence epilepsy. Behav Brain Res 236:23–29

Keysers C (2009) Mirror neurons. Curr Biol 19:R971–R973

Keysers C, Gazzola V (2006) Towards a unifying neural theory of social cognition. Prog Brain Res 156:379–401

Keysers C, Gazzola V (2009) Expanding the mirror: vicarious activity for actions, emotions, and sensations. Curr Opin Neurobiol 19:666–671

Keysers C, Kohler E, Umilta MA, Nanetti L, Fogassi L, Gallese V (2003) Audiovisual mirror neurons and action recognition. Exp Brain Res 153:628–636
Keysers C, Wicker B, Gazzola V, Anton JL, Fogassi L, Gallese V (2004) A touching sight: SII/PV activation during the observation and experience of touch. Neuron 42:335–346
Keysers C, Kaas JH, Gazzola V (2010) Somatosensation in social perception. Nat Rev Neurosci 11:417–428
Kim EJ, Kim ES, Covey E, Kim JJ (2010) Social transmission of fear in rats: the role of 22-kHz ultrasonic distress vocalization. PLoS ONE 5:e15077
Kim S, Matyas F, Lee S, Acsady L, Shin HS (2012) Lateralization of observational fear learning at the cortical but not thalamic level in mice. Proc Natl Acad Sci USA 109:15497–15501
Knapska E, Mikosz M, Werka T, Maren S (2010) Social modulation of learning in rats. Learn Mem 17:35–42
Kohler E, Keysers C, Umilta MA, Fogassi L, Gallese V, Rizzolatti G (2002) Hearing sounds, understanding actions: action representation in mirror neurons. Science 297:846–848
Lamm C, Decety J, Singer T (2011) Meta-analytic evidence for common and distinct neural networks associated with directly experienced pain and empathy for pain. Neuroimage 54:2492–2502
Langford DJ, Crager SE, Shehzad Z, Smith SB, Sotocinal SG, Levenstadt JS, Chanda ML, Levitin DJ, Mogil JS (2006) Social modulation of pain as evidence for empathy in mice. Science 312:1967–1970
Lanz S, Seifert F, Maihofner C (2011) Brain activity associated with pain, hyperalgesia and allodynia: an ALE meta-analysis. J Neural Transm 118:1139–1154
Martin JH (1991) Autoradiographic estimation of the extent of reversible inactivation produced by microinjection of lidocaine and muscimol in the rat. Neurosci Lett 127:160–164
Martin LJ, Hathaway G, Isbester K, Mirali S, Acland EL, Niederstrasser N, Slepian PM, Trost Z, Bartz JA, Sapolsky RM, Sternberg WF, Levitin DJ, Mogil JS (2015) Reducing social stress elicits emotional contagion of pain in mouse and human strangers. Curr Biol 25:326–332
Meffert H, Gazzola V, den Boer JA, Bartels AA, Keysers C (2013) Reduced spontaneous but relatively normal deliberate vicarious representations in psychopathy. Brain 136:2550–2562
Mogil JS (2009) Animal models of pain: progress and challenges. Nat Rev Neurosci 10:283–294
Mukamel R, Ekstrom AD, Kaplan J, Iacoboni M, Fried I (2010) Single-neuron responses in humans during execution and observation of actions. Curr Biol 20:750–756
Panksepp J, Panksepp JB (2013) Toward a cross-species understanding of empathy. Trends Neurosci 36:489–496
Pereira AG, Cruz A, Lima SQ, Moita MA (2012) Silence resulting from the cessation of movement signals danger. Curr Biol 22:R627–R628
Porro CA, Cavazzuti M, Baraldi P, Giuliani D, Panerai AE, Corazza R (1999) CNS pattern of metabolic activity during tonic pain: evidence for modulation by beta-endorphin. Eur J Neurosci 11:874–888
Preston SD, De Waal FB (2002) Empathy: its ultimate and proximate bases. Behav Brain Sci 25:1–20; discussion 20–71
Puce A, Perrett D (2003) Electrophysiology and brain imaging of biological motion. Philos Trans R Soc Lond B Biol Sci 358:435–445
Rice GE, Gainer P (1962) "Altruism" in the albino rat. J Comp Physiol Psychol 55:123–125
Rizzolatti G, Fadiga L, Matelli M, Bettinardi V, Paulesu E, Perani D, Fazio F (1996) Localization of grasp representations in humans by PET: 1. Observation versus execution. Exp Brain Res 111:246–252
Rodgers KM, Benison AM, Klein A, Barth DS (2008) Auditory, somatosensory, and multisensory insular cortex in the rat. Cereb Cortex 18:2941–2951
Rozzi S, Ferrari PF, Bonini L, Rizzolatti G, Fogassi L (2008) Functional organization of inferior parietal lobule convexity in the macaque monkey: electrophysiological characterization of motor, sensory and mirror responses and their correlation with cytoarchitectonic areas. Eur J Neurosci 28:1569–1588
Scanziani M, Hausser M (2009) Electrophysiology in the age of light. Nature 461:930–939

Shih YY, Chen YY, Chen CC, Chen JC, Chang C, Jaw FS (2008) Whole-brain functional magnetic resonance imaging mapping of acute nociceptive responses induced by formalin in rats using atlas registration-based event-related analysis. J Neurosci Res 86:1801–1811
Singer T, Seymour B, O'Doherty J, Kaube H, Dolan RJ, Frith CD (2004) Empathy for pain involves the affective but not sensory components of pain. Science 303:1157–1162
Tracey I (2008) Imaging pain. Br J Anaesth 101:32–39
Umilta MA, Kohler E, Gallese V, Fogassi L, Fadiga L, Keysers C, Rizzolatti G (2001) I know what you are doing. A neurophysiological study. Neuron 31:155–165
Wicker B, Keysers C, Plailly J, Royet JP, Gallese V, Rizzolatti G (2003) Both of us disgusted in My insula: the common neural basis of seeing and feeling disgust. Neuron 40:655–664
Wohr M, Schwarting RKW (2008) Ultrasonic calling during fear conditioning in the rat: no evidence for an audience effect. Anim Behav 76:749–760
Wohr M, Schwarting RKW (2010) Activation of limbic system structures by replay of ultrasonic vocalization in rats. In: Brudzynski SM (ed) Handbook of Mammalian vocalization. An integrative neuroscience approach. Elsevier, Amsterdam
Xu X, Zuo X, Wang X, Han S (2009) Do you feel my pain? Racial group membership modulates empathic neural responses. J Neurosci 29:8525–8529
Yarkoni T, Poldrack RA, Nichols TE, van Essen DC, Wager TD (2011) Large-scale automated synthesis of human functional neuroimaging data. Nat Methods 8:665–670
Zhang Y, Wang N, Wang JY, Chang JY, Woodward DJ, Luo F (2011) Ensemble encoding of nociceptive stimulus intensity in the rat medial and lateral pain systems. Mol Pain 7:64
Zhao F, Welsh D, Williams M, Coimbra A, Urban MO, Hargreaves R, Evelhoch J, Williams DS (2011) fMRI of pain processing in the brain: a within-animal comparative study of BOLD vs. CBV and noxious electrical vs. noxious mechanical stimulation in rat. Neuroimage

Models, Mechanisms and Moderators Dissociating Empathy and Theory of Mind

Philipp Kanske, Anne Böckler and Tania Singer

Abstract Most instances of social interaction provide a wealth of information about the states of other people, be it sensations, feelings, thoughts, or convictions. How we represent these states has been a major question in social neuroscience, leading to the identification of two routes to understanding others: an affective route for the direct sharing of others' emotions (empathy) that involves, among others, anterior insula and middle anterior cingulate cortex and a cognitive route for representing and reasoning about others' states (Theory of Mind) that entails, among others, ventral temporoparietal junction and anterior and posterior midline regions. Additionally, research has revealed a number of situational and personal factors that shape the functioning of empathy and Theory of Mind. Concerning situational modulators, it has been shown, for instance, that ingroup membership enhances empathic responding and that Theory of Mind performance seems to be susceptible to stress. Personal modulators include psychopathological conditions, for which alterations in empathy and mentalizing have consistently been demonstrated; people on the autism spectrum, for instance, are impaired specifically in mentalizing, while spontaneous empathic responding seems selectively reduced in psychopathy. Given the multifaceted evidence for separability of the two routes, current research endeavors aiming at fostering interpersonal cooperation explore the differential malleability of affective and cognitive understanding of others.

Philipp Kanske and Anne Böckler are joint first authors/contributed equally.

P. Kanske (✉) · A. Böckler (✉) · T. Singer
Department of Social Neuroscience, Max Planck Institute for Human Cognitive and Brain Sciences, Stephanstr. 1a, 04103 Leipzig, Germany
e-mail: kanske@cbs.mpg.de

A. Böckler
Department of Psychology, Julius-Maximilians-Universität Würzburg, Röntgenring 11, 97070 Würzburg, Germany
e-mail: aboeckler@cbs.mpg.de; anne.boeckler@uni-wuerzburg.de

Curr Topics Behav Neurosci (2016) 30: 193–206
DOI 10.1007/7854_2015_412
Published Online: 20 November 2015

Keywords Empathy • Theory of Mind • Social cognition • Temporoparietal junction • Anterior insula

Contents

1 Introduction 194
2 Empathy 195
3 Theory of Mind 196
4 Dissociating Empathy and Theory of Mind 197
5 Situational Influences on Empathy and Theory of Mind 200
6 Personal Influences on Empathy and Theory of Mind 201
7 What Now? Conclusions and Future Directions 202
References 203

1 Introduction

Our everyday lives embed us in various social networks, from breakfast at the family table to formal meetings with colleagues, in which we constantly and flexibly need to adjust our actions to those of others. A crucial step in order to successfully master these different kinds of social encounters is the accurate understanding of others' emotional and mental states. This includes sensing another's mood, inferring how to interpret another's (ironic?) utterance, or identifying our interaction partner's expectations. How do we achieve such understanding of others and what are the underlying mechanisms? Social neuroscience has addressed this question during the last decades and identified two routes allowing for representing others' states (Frith and Frith 2005; Singer 2006): The affective route entails directly sharing an observed person's emotions (empathy) (de Vignemont and Singer 2006), while the cognitive route enables inferences of other's thoughts, goals, or intentions (Theory of Mind, mentalizing) (Frith and Frith 2005; Premack and Woodruff 1978).

This chapter will first introduce these routes and their neural underpinnings in more detail and then outline a recent approach to dissociate empathy and mentalizing on a behavioral and brain level. Subsequently, situational and personal modulators of empathic responding and Theory of Mind capacities will be presented and a final outlook sketches current investigations of the malleability of the two routes to understanding others.

2 Empathy

The affective route to understanding others has mainly been studied under the term empathy. The concept was introduced by Robert Vischer and Theodor Lipps, who emphasized the role of "inner imitation" what they called "Einfühlung" (see Carr et al. 2003). The current literature defines empathy as an affective state, that is, (1) isomorphic to another person's affective state, (2) elicited by observing or imagining another person's affective state, and (3) includes knowing that the observed person's state is the source of one's own affect (de Vignemont and Singer 2006). Isomorphism refers to the idea that empathizing with somebody means directly sharing, "feeling with," his or her emotions, such as pain, sadness, or joy. Complementarily, observing another in an emotional state may, of course, also elicit non-isomorphic emotions such as envy, schadenfreude, or compassion. Crucially, even though these are all vicarious affective states (Paulus et al. 2013), empathy is per definition shared rather than complementary. Another important distinction of empathy concerns emotional contagion, the reaction in which one shares an emotion with another person *without* realizing that the other's emotion was the trigger. An example is the observation that infants start crying when they hear other infants cry, even before they presumably develop a clear sense of being an independent agent (Singer and Lamm 2009).

In order to investigate the neural underpinnings of empathy, psychologists and neuroscientists have used paradigms in which both the participant and an observed other received painful stimulation. Critically, this allowed for a direct comparison of neural activity elicited by being the recipient and by being a mere observer of pain (e.g., Singer et al. 2004). Clear overlap between these two conditions was found in a network of areas including bilateral anterior insula and middle anterior cingulate cortex. Therefore, "shared" brain networks have been proposed as an underlying mechanism for our ability to empathize (Decety 2010). Such shared neuronal representations subserving both the first-hand as well as the vicarious experience of emotions have been reported in different domains ranging from disgust (e.g., Wicker et al. 2003), touch (e.g., Keysers et al. 2004 and facial emotional expressions (e.g., Carr et al. 2003) to higher-order emotions such as social exclusion (Masten et al. 2011) and embarrassment (Muller-Pinzler et al. 2015).

Recent meta-analyses corroborate the involvement of anterior insula and middle anterior cingulate cortex in empathy for pain (Lamm et al. 2011). When including different types of empathy paradigms and emotional stimuli, a more extensive neural network was identified, including anterior insula and anterior cingulate cortex, but also more dorsal regions of the medial frontal cortex, inferior frontal gyri and anterior/dorsal parts of the temporoparietal junction (Bzdok et al. 2012). Interestingly, this network seems independent of whether spontaneous or instructed empathizing is tested (Fan et al. 2011).

As pointed out earlier, a distinction between isomorphic and complementary emotions has been suggested in the conceptual specification of empathy (de Vignemont and Singer 2006). For instance, imagine running into an angry boss at the non-functioning copy machine. While your boss' anger may elicit anger in you too, you may also be intimidated and feel anxious. Empirical support for the distinction of isomorphic and complementary emotions in response to the same situation comes from recent neuroimaging studies on isomorphic empathic responding ("feeling with") and complementary compassion ("feeling for"), which can both be induced by the suffering of others. Compassion, the feeling of warmth, care, and affiliation, does not rely on "shared" neural networks but rather activates regions commonly associated with positive affect and reward such as ventral striatum and medial orbitofrontal cortex (Klimecki et al. 2013). It has been hypothesized that empathy and compassion also differ with regard to the roles they play in interpersonal behavior. While empathic responding to another's suffering does not per se entail a prosocial motivation (i.e., the wish to alleviate the other's suffering) and might even lead to withdrawal from the other due to empathic distress, compassion toward others includes the motivation to enhance their well-being. In line with this hypothesis, recent studies showed increased helping behavior related to compassion trainings (Leiberg et al. 2011).

3 Theory of Mind

The cognitive route to understanding others has been investigated under the terms perspective taking, mentalizing, or Theory of Mind. The latter term was originally coined by Premack and Woodruff (1978), discussing whether chimpanzees can represent others' mental states such as desires, intentions, and beliefs. Similarly, developmental research has been dedicated to the question at what age children "have" a Theory of Mind. This conceptual definition of the term has been closely linked to the usage of so-called false-belief tasks that measure whether participants can correctly predict the actions of somebody who holds a false belief that is different from the participants' true belief. More recent investigations approach the concept by focusing on the cognitive operations that are involved in reasoning about another's (different) mental states (see Apperly 2012) or perceptions (Bockler and Zwickel 2013). In conclusion, research defines Theory of Mind as the process of inferring and reasoning about the perceptions, beliefs, thoughts, or emotions of others (Frith and Frith 2005).

In order to investigate the neural underpinnings of this cognitive route to understanding others, paradigms require participants to judge (mis-)assumptions or (false) beliefs of others that differ from their own, requiring the inhibition of the own mental state (Wimmer and Perner 1983). For example, participants are presented with a short story about a boy who believes that his bike is in the garden, while in fact his mother had taken it into the garage. Contrasting such stories with true belief conditions or with stories about changes in the physical world (e.g., an outdated map) yields activation in temporoparietal junction, superior temporal sulcus, temporal poles, and anterior and posterior midline regions (Dodell-Feder et al. 2011).

Recent meta-analyses convincingly demonstrate the involvement of the described neural network when incorporating a wide variety of different Theory of Mind tasks (Bzdok et al. 2012). Testing more stringently for regions that are consistently activated in most of these different tasks reveals a core network comprising temporoparietal junction and medial prefrontal cortex (Schurz et al. 2014). Comparing the different types of Theory of Mind tasks shows a more differentiated pattern within the overall network, with specific activation clusters for (among others) false-belief tasks (e.g., Aichhorn et al. 2009), social animations (e.g., Blakemore et al. 2003), and rational action judgment tasks (e.g., Walter et al. 2004). While judging others' mental states in realistically complex social situations has been shown to lead to more widespread activation within the Theory of Mind network (Wolf et al. 2010), the above described separations may prove helpful for delineating the specific processes that are represented in different parts of the network and that contribute to full-blown Theory of Mind.

Regarding the role of Theory of Mind in interpersonal behavior, it can be speculated that the capacity to understand other's mental states allows for more prosocial decision making. Possible mechanisms could include a better understanding of others' needs and circumstances or, based on the capacity to flexibly inhibit one's own mental states, an enhanced ability to inhibit one's immediate selfish needs (Batson 2011). In fact, recent evidence in children shows that Theory of Mind performance predicts enhanced prosocial behavior and thereby leads to better peer group integration (Caputi et al. 2012).

4 Dissociating Empathy and Theory of Mind

As described in the previous sections, empathy and Theory of Mind have inspired different research traditions that have employed different types of experimental paradigms and identified largely distinct neural networks underlying the two functions. Spelling out the differentiation on a conceptual level, empathy denotes the embodied sharing of a sensory or affective state, while Theory of Mind refers to the propositional knowledge about the state of another (including others' affective states). Beyond the differences between the two routes to understanding others, however, both functions yield access to another person's inner state. Coming back to the initial examples of breakfast at the family table or meetings among colleagues, both empathy and Theory of Mind are necessary to adequately react to your stressed husband's ironic remark about the timeless beauty of your vintage morning gown or to your new colleague's mixed feelings about his first group presentation. It is therefore plausible to assume that both functions and the related neural networks are concurrently required and active in almost all everyday social interactions. Indeed, meta-analytical evidence suggests that the temporoparietal junction, which is a core region of the Theory of Mind network (Schurz et al. 2014), is also activated in empathy studies (Bzdok et al. 2012). In particular, a recent meta-analysis gives some indication that empathizing with another's pain activates not only the typical empathy network (anterior insula, middle anterior cingulate cortex), but also Theory of Mind-related areas (temporoparietal junction, temporal poles, superior temporal sulcus, anterior and posterior midline regions) in situations where the other's state is not immediately visible and needs to be inferred (Lamm et al. 2011). Research in social neuroscience, therefore, needs to investigate the specifics and the separability of the neural networks underlying empathy and Theory of Mind, especially in ecologically valid instances of realistically complex social understanding that require both routes.

While some studies have compared cognitive and affective aspects of Theory of Mind (i.e., mentalizing on others' cognitive or affective states; (e.g., Bruneau et al. 2012; Shamay-Tsoory and Aharon-Peretz 2007) or have studied empathy and Theory of Mind in separation (Dziobek et al. 2011), the simultaneous manipulation and assessment of empathy and Theory of Mind within individuals has only recently tested (Kanske et al. 2015). Of course, empathy and mentalizing processes may co-occur in many experimental depictions of social situations. For instance, when manipulating empathic requirements by means of observed social exclusion, participants likely also take the (cognitive) perspective of the displayed agents, which is reflected in concurrent activation of both empathy- and Theory of Mind-related networks (Masten et al. 2011). Critically, because the two processes are not manipulated independently, clear-cut attribution of empathy and mentalizing functions to the specific neural activations are impossible. In the following paragraphs, we will outline a paradigm that explicitly applied an orthogonal manipulation of empathy and Theory of Mind and delineate the results in greater detail.

The task, henceforth termed EmpaToM, confronts participants with brief video clips of a person reporting on an autobiographic episode (see Fig. 1a). These episodes vary in emotionality (entailing neutral events such as the preparation of a meal or negative events such as loss of a loved one) and in Theory of Mind demands (necessitating inferences about mental states such as a planned deception or about physical states such as weather conditions). As a behavioral measure of empathic responding, participants rate the valence of their own affective state after each video. A subsequent rating asks for participants' compassion for the person in the video. Theory of Mind performance is assessed with multiple choice questions that ask either about the mental states of the person in the video or about factual relations in the story (control condition). This design orthogonally manipulates empathy and Theory of Mind requirements, hence including a condition in which both processes are elicited.

Using the EmpaToM, the typical neural networks for empathy and Theory of Mind could be identified, including anterior insula, anterior cingulate/medial frontal cortex for empathy and temporoparietal junction, temporal poles, and anterior and posterior midline regions for Theory of Mind. The revealed activation overlapped with meta-analytical masks (Bzdok et al. 2012) and with established empathy and Theory of Mind paradigms assessed in the same individuals (Dodell-Feder et al. 2011; Klimecki et al. 2013); behavioral parameters of empathizing and mentalizing could also be validated with existing tasks. Hence, even in complex situations of

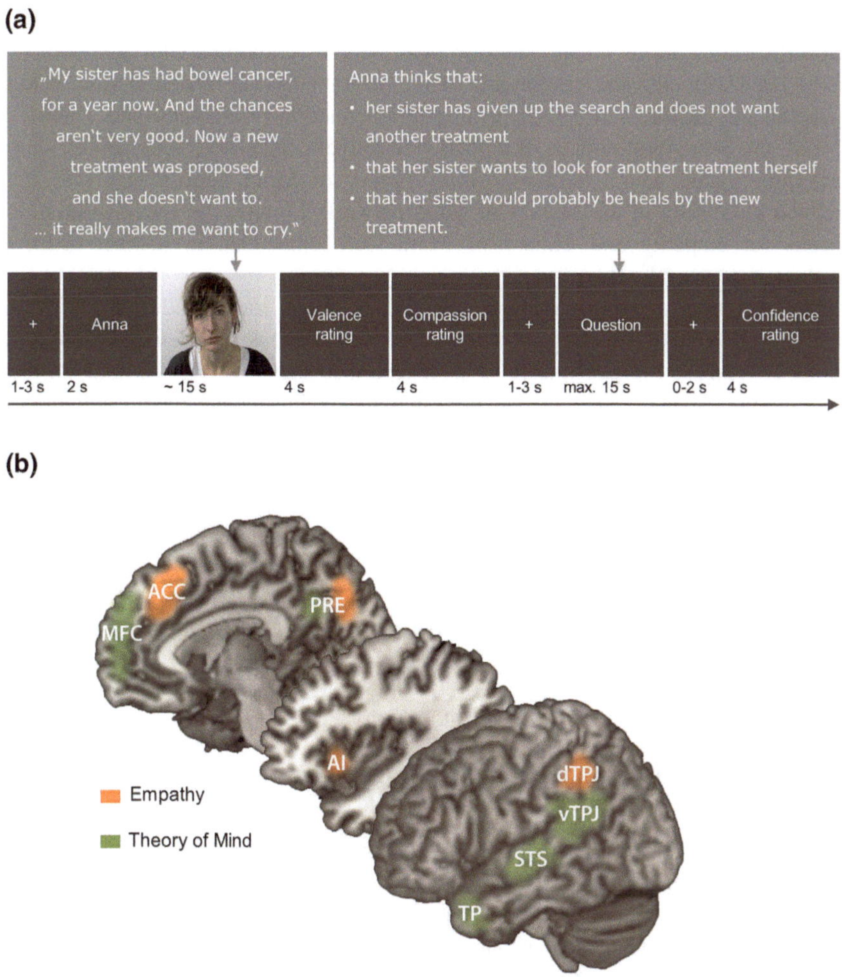

Fig. 1 **a** Overview of the EmpaToM task with an example video narration and corresponding question of the negative emotion/Theory of Mind condition. **b** Empathy and Theory of Mind networks (schematically depicted in the *left hemisphere*). Anterior cingulate cortex (ACC), anterior insula (AI), dorsal/ventral temporo-parietal junction (d/vTPJ), medial frontal cortex (MFC), precuneus/posterior cingulate cortex (PRE), superior temporal sulcus (STS), and temporal poles (TP). Adapted from Kanske et al. (2015)

social understanding that simultaneously require affect sharing and inferring others' mental states, the two networks can be clearly identified and mirror the results of studies investigating the functions separately (Lamm et al. 2011; Schurz et al. 2014). Assessing both functions simultaneously in the same individuals allows further investigation of the specifics by directly contrasting the activation related to empathy and Theory of Mind (see Fig. 1b for a schematic illustration). These specific contrasts replicated the typical networks, but also allowed differentiation of neighboring but distinct activations, for example, in temporoparietal junction. Specifically, a more dorsal anterior peak in the temporoparietal junction was associated with empathic responding, while mentalizing was related to more ventral parts of this region. Further supporting the distinction of empathy and Theory of Mind, the behavioral indices of empathy selectively correlated with activity in the empathy-related neural network, while Theory of Mind performance was only related to activity in the Theory of Mind network. Finally, the specific activation peaks for empathy and Theory of Mind were embedded in distinct task-free resting-state neural networks.

In sum, there is strong evidence for distinct neural routes underlying affective and cognitive understanding of others, even in situations that demand both functions concurrently. Open questions with regard to these two routes concern their interrelation, that is, whether and how activity in the distinct neural networks is orchestrated during online social understanding. For instance, some situations may call for a prioritization of one function over the other as has been demonstrated for cognitive control in emotional situations (Kanske et al. 2013), such as a physician who needs to infer whether a patient conceals a previous valium addiction, while treating the patient's painful injury. Another question on the interrelation of empathy and Theory of Mind is how the respective capacities are distributed inter-individually, that is, whether people with a particular proficiency in mentalizing are also strong empathizers.

5 Situational Influences on Empathy and Theory of Mind

In recent years, numerous situational modulators of empathic responding have been identified. It seems to be crucial for affect sharing, for instance, that attention is directed toward the painful aspects of a situation (Gu and Han 2007). When judging the painful consequences of situations, participants showed higher activation in empathy-related regions than when focusing attention on neutral aspects of the same stimulus (e.g., counting objects). Also, instructing participants to imagine being in the painful situation themselves clearly increased empathic responding on a neural level (Avenanti et al. 2005) and, accordingly, better perspective taking predicted enhanced early emotion detection (Kanske et al. 2013). A further top-down modulation concerns prior knowledge about the pain sensitivity of an observed person: Enhanced empathic responding was found when participants believed the other to actually feel a painful stimulation as compared to when they believed the other to be anesthetized (Lamm et al. 2007).

Interestingly, the social relation to the observed other also plays an important role for the strength of an empathic response. In particular, ingroup versus outgroup membership has been demonstrated to moderate empathic responding (Hein et al. 2010; Xu et al. 2009). Neural activation in anterior insula while observing others' pain has been found to be increased for racial ingroup as compared to outgroup members (Xu et al. 2009). Inviting fans of rivaling soccer teams, Hein et al. (2010) additionally showed that the anterior insula response predicted subsequent helping behavior for fans of the same soccer team (ingroup members). Besides group membership, it is also the sympathy toward another that shapes affect sharing. In a game show paradigm, participants showed activation of the ventral striatum in response to own gains, but also to gains of sympathetic versus unsympathetic others (representing an example of empathy for others' positive emotions) (Mobbs et al. 2009). An indication for the influence of interpersonal behavior on subsequent empathizing stems from evidence linking the fairness of an observed other in a previous interaction to the strength of empathic anterior insula activity (Singer et al. 2006).

A recent set of studies has probed how empathic responding is influenced by the observer's own emotional state (Silani et al. 2013). Incongruent affective stimulation of observer and observed (e.g., one receiving pleasant and the other receiving unpleasant touch) biases the empathic judgments of the observer toward the valence of his own stimulation (emotional egocentricity bias). Overcoming this bias is related to activity in an anterior/dorsal part of the temporoparietal junction that closely corresponds with the specific empathy peak that was identified in the previously described EmpaToM task (Kanske et al. 2015). Hence, the anterior/dorsal temporoparietal junction may play a role in separating one's own emotional state from the emotional state of the other, a function that seems crucial for many situations (and paradigms) involving empathic responding.

While the situational moderators of empathy have been studied in different domains, less is known about how the propensity and capacity to take another's perspective is shaped by contextual modulators. One critical factor for successful Theory of Mind performance seems to be the current psychosocial stress levels. Smeets et al. (2009) demonstrated that higher individual cortisol responses to the Trier Social Stress Test impair mentalizing in women, but enhance mentalizing in men. In developmental research, studies have isolated a number of sociodemographic and educational factors that predict Theory of Mind performance in children. In addition to low socioeconomic status, a parenting style relying strongly on power assertion (e.g., physical punishment) predicts worse performance in false-belief tasks, while communication focused parenting styles enhances childrens' capacity to understand others' perspectives (Cutting and Dunn 1999; Pears and Moses 2003).

6 Personal Influences on Empathy and Theory of Mind

In addition to situational influences on empathy and Theory of Mind, both functions are shaped by inter-individual differences in the personality and clinical domain. Gender differences are especially prominent in the literature, suggesting that women are more empathic and more prone to mentalize than men. The fact that gender effects have been mainly observed in self-reports (Rueckert and Naybar 2008) has been taken as an indication for demand characteristics in a sense that women feel they are expected to be more socially sensitive (Rueckert and Naybar 2008).

However, some studies have also reported behavior-based and neural activation differences, showing for instance that empathy in women is less susceptible to previous unfairness of the observed other (Singer et al. 2006) and that emotion recognition activates mirror regions to a stronger degree in women (Schulte-Ruther et al. 2008).

While deficits in both empathy and Theory of Mind have been described for a number of different mental disorders including depression (Wolkenstein et al. 2011) and schizophrenia (Mohnke et al. 2014), claims for selective deficits in one or the other function have also been made (Blair 2008). Theory of Mind deficits are a core component of autism spectrum disorder. Already during development, children on the autism spectrum pass classical false-belief tasks later than healthy children or than clinical control groups, even when controlling for language abilities (Baron-Cohen et al. 1985). Using ecologically valid and complex assessments of Theory of Mind, more recent studies show deficits also in high-functioning adult patients with autism spectrum disorder (Dziobek et al. 2006). Suggesting compensatory processes in high-functioning individuals, neuroimaging studies demonstrated hyperactivation during mentalizing in Theory of Mind-related regions (Mason et al. 2008). Interestingly, empathic responding seems preserved in autism when controlling for alexithymia (a trait characterized by difficulties in describing own emotional states (Bird et al. 2010). Further supporting this dissociation, the cortical structures subserving empathy do not differ between neurotypical and autistic individuals, whereas the structural network subserving Theory of Mind is hampered in autism (Bernhardt et al. 2014). The opposite pattern of empathy and Theory of Mind functioning has been described in psychopathy. While performance on cognitive and affective aspects of Theory of Mind seem to be intact in psychopaths (Blair et al. 1996), spontaneous empathic responding is largely reduced (Meffert et al. 2013).

7 What Now? Conclusions and Future Directions

The reviewed evidence strongly suggests the existence of two separable routes to understanding others. Based on the conceptualizations and underlying neural networks as well as on situational and personal moderators, empathy and Theory of Mind can be distinguished. In particular with regard to specific impairments of the functions in psychopathology, the question arises how (differentially) malleable the capacities to empathize and mentalize are. Therefore, one of the core tasks for future research will be to establish and evaluate targeted interventions aiming, for instance, at enhancing perspective taking skills in autism spectrum disorder and at improving the ability to share others' feelings in psychopathy.

Recent research has provided first indications for plasticity of empathy and compassion capabilities in healthy individuals (Klimecki et al. 2014). Participants underwent two one-week meditation-based trainings emphasizing either empathic or compassionate responding to other people's suffering. Neuroimaging results demonstrated a specific activation increase of anterior insula and middle anterior cingulate cortex after empathy training and of ventral striatum and medial orbitofrontal cortex after training compassion. This finding was mirrored by behavioral results pointing toward higher sharing of negative affect after empathy training and enhanced reports of positive affect after the compassion training. Plasticity of social understanding has also been revealed in the domain of Theory of Mind. For example, children with autism spectrum disorder showed specifically improved mentalizing performance after a 16-week Theory of Mind training (Begeer et al. 2011).

A hitherto unanswered question concerns the specificity of such interventions when applied in the same population. Hence, do trainings of the affective route to understanding others selectively enhance behavioral and neural markers of empathy while not influencing Theory of Mind performance, and vice versa? Or can shared mechanisms of social cognition be identified that enhance empathic responding even after specific trainings of mentalizing abilities? This and similar questions could be addressed with paradigms such as the EmpaToM task described earlier in this chapter since they manipulate and assess both functions of social cognition simultaneously.

We believe that the evidence outlined in this chapter demonstrates that research on the mechanisms of social understanding has moved beyond its infancy, revealing insight into the processes that are at work, for instance, during our family breakfasts and while we collaborate with our colleagues. The next steps must entail assessing the interplay of these affective and cognitive processes in controlled as well as in more complex interactive situations and probing their respective contributions to different types of social encounters, ranging from simple action coordination to large-scale cooperation.

References

Aichhorn M, Perner J, Weiss B, Kronbichler M, Staffen W, Ladurner G (2009) Temporo-parietal junction activity in theory-of-mind tasks: falseness, beliefs, or attention. J Cogn Neurosci 21 (6):1179–1192. doi:10.1162/jocn.2009.21082

Apperly IA (2012) What is "theory of mind"? Concepts, cognitive processes and individual differences. Q J Exp Psychol (Hove) 65(5):825–839. doi:10.1080/17470218.2012.676055

Avenanti A, Bueti D, Galati G, Aglioti SM (2005) Transcranial magnetic stimulation highlights the sensorimotor side of empathy for pain. Nat Neurosci 8(7):955–960. doi:10.1038/nn1481

Baron-Cohen S, Leslie AM, Frith U (1985) Does the autistic-child have a theory of mind. Cognition 21(1):37–46. doi:10.1016/0010-0277(85)90022-8

Batson CD (2011) Altruism in humans. Oxford University Press, New York

Begeer S, Gevers C, Clifford P, Verhoeve M, Kat K, Hoddenbach E, Boer F (2011) Theory of mind training in children with autism: a randomized controlled trial. J Autism Dev Disord 41 (8):997–1006. doi:10.1007/s10803-010-1121-9

Bernhardt BC, Valk SL, Silani G, Bird G, Frith U, Singer T (2014) Selective disruption of sociocognitive structural brain networks in autism and alexithymia. Cereb Cortex 24(12):3258–3267. doi:10.1093/cercor/bht182

Bird G, Silani G, Brindley R, White S, Frith U, Singer T (2010) Empathic brain responses in insula are modulated by levels of alexithymia but not autism. Brain 133(Pt 5):1515–1525. doi:10.1093/brain/awq060

Blair RJ (2008) Fine cuts of empathy and the amygdala: dissociable deficits in psychopathy and autism. Q J Exp Psychol (Hove) 61(1):157–170. doi:10.1080/17470210701508855

Blair RJ, Sellars C, Strickland I, Clark F, Williams A, Smith M, Jones L (1996) Theory of mind in the psychopath. J Forensic Psychiatry 7(1):15–25. doi:10.1080/09585189608409914

Blakemore SJ, Sarfati Y, Bazin N, Decety J (2003) The detection of intentional contingencies in simple animations in patients with delusions of persecution. Psychol Med 33(8):1433–1441

Bockler A, Zwickel J (2013) Influences of spontaneous perspective taking on spatial and identity processing of faces. Soc Cogn Affect Neurosci 8(7):735–740. doi:10.1093/scan/nss061

Bruneau EG, Pluta A, Saxe R (2012) Distinct roles of the 'shared pain' and 'theory of mind' networks in processing others' emotional suffering. Neuropsychologia 50(2):219–231. doi:10.1016/j.neuropsychologia.2011.11.008

Bzdok D, Schilbach L, Vogeley K, Schneider K, Laird AR, Langner R, Eickhoff SB (2012) Parsing the neural correlates of moral cognition: ALE meta-analysis on morality, theory of mind, and empathy. Brain Struct Funct 217(4):783–796. doi:10.1007/s00429-012-0380-y

Caputi M, Lecce S, Pagnin A, Banerjee R (2012) Longitudinal effects of theory of mind on later peer relations: the role of prosocial behavior. Dev Psychol 48(1):257–270. doi:10.1037/a0025402

Carr L, Iacoboni M, Dubeau MC, Mazziotta JC, Lenzi GL (2003) Neural mechanisms of empathy in humans: a relay from neural systems for imitation to limbic areas. Proc Natl Acad Sci USA 100(9):5497–5502. doi:10.1073/pnas.0935845100

Cutting AL, Dunn J (1999) Theory of mind, emotion understanding, language, and family background: Individual differences and interrelations. Child Dev 70(4):853–865. doi:10.1111/1467-8624.00061

de Vignemont F, Singer T (2006) The empathic brain: how, when and why? Trends Cogn Sci 10 (10):435–441. doi:10.1016/j.tics.2006.08.008

Decety J (2010) To what extent is the experience of empathy mediated by shared neural circuits? Emot Rev 2(3):204–207. doi:10.1177/1754073910361981

Dodell-Feder D, Koster-Hale J, Bedny M, Saxe R (2011) fMRI item analysis in a theory of mind task. Neuroimage 55(2):705–712. doi:10.1016/j.neuroimage.2010.12.040

Dziobek I, Fleck S, Kalbe E, Rogers K, Hassenstab J, Brand M, Convit A (2006) Introducing MASC: a movie for the assessment of social cognition. J Autism Dev Disord 36(5):623–636. doi:10.1007/s10803-006-0107-0

Dziobek I, Preissler S, Grozdanovic Z, Heuser I, Heekeren HR, Roepke S (2011) Neuronal correlates of altered empathy and social cognition in borderline personality disorder. Neuroimage 57(2):539–548. doi:10.1016/j.neuroimage.2011.05.005

Fan Y, Duncan NW, de Greck M, Northoff G (2011) Is there a core neural network in empathy? An fMRI based quantitative meta-analysis. Neurosci Biobehav Rev 35(3):903–911. doi:10.1016/j.neubiorev.2010.10.009

Frith C, Frith U (2005) Theory of mind. Curr Biol 15(17):R644–R646. doi:10.1016/j.cub.2005.08.041

Gu X, Han S (2007) Attention and reality constraints on the neural processes of empathy for pain. Neuroimage 36(1):256–267. doi:10.1016/j.neuroimage.2007.02.025

Hein G, Silani G, Preuschoff K, Batson CD, Singer T (2010) Neural responses to ingroup and outgroup members' suffering predict individual differences in costly helping. Neuron 68 (1):149–160

Kanske P, Heissler J, Schonfelder S, Forneck J, Wessa M (2013a) Neural correlates of emotional distractibility in bipolar disorder patients, unaffected relatives, and individuals with hypomanic personality. Am J Psychiatry 170(12):1487–1496. doi:10.1176/appi.ajp.2013.12081044

Kanske P, Schonfelder S, Wessa M (2013b) Emotional modulation of the attentional blink and the relation to interpersonal reactivity. Front Hum Neurosci 7:641. doi:10.3389/fnhum.2013.00641

Kanske P, Böckler A, Trautwein FM, Singer T (2015) Dissecting the social brain: introducing the EmpaToM to separate empathy and theory of mind in brain and behavior. Neuroimage 122:6–19

Keysers C, Wicker B, Gazzola V, Anton JL, Fogassi L, Gallese V (2004) A touching sight: SII/PV activation during the observation and experience of touch. Neuron 42(2):335–346

Klimecki OM, Leiberg S, Lamm C, Singer T (2013) Functional neural plasticity and associated changes in positive affect after compassion training. Cereb Cortex 23(7):1552–1561. doi:10.1093/cercor/bhs142

Klimecki OM, Leiberg S, Ricard M, Singer T (2014) Differential pattern of functional brain plasticity after compassion and empathy training. Soc Cogn Affect Neurosci 9(6):873–879. doi:10.1093/scan/nst060

Lamm C, Nusbaum HC, Meltzoff AN, Decety J (2007) What are you feeling? Using functional magnetic resonance imaging to assess the modulation of sensory and affective responses during empathy for pain. PLoS ONE 2(12):e1292. doi:10.1371/journal.pone.0001292

Lamm C, Decety J, Singer T (2011) Meta-analytic evidence for common and distinct neural networks associated with directly experienced pain and empathy for pain. Neuroimage 54 (3):2492–2502. doi:10.1016/j.neuroimage.2010.10.014

Leiberg S, Klimecki O, Singer T (2011) Short-term compassion training increases prosocial behavior in a newly developed prosocial game. PLoS ONE 6(3):e17798. doi:10.1371/journal.pone.0017798

Mason RA, Williams DL, Kana RK, Minshew N, Just MA (2008) Theory of mind disruption and recruitment of the right hemisphere during narrative comprehension in autism. Neuropsychologia 46(1):269–280. doi:10.1016/j.neuropsychologia.2007.07.018

Masten CL, Morelli SA, Eisenberger NI (2011) An fMRI investigation of empathy for 'social pain' and subsequent prosocial behavior. Neuroimage 55(1):381–388. doi:10.1016/j.neuroimage.2010.11.060

Meffert H, Gazzola V, den Boer JA, Bartels AAJ, Keysers C (2013) Reduced spontaneous but relatively normal deliberate vicarious representations in psychopathy, vol 136

Mobbs D, Yu R, Meyer M, Passamonti L, Seymour B, Calder AJ, Dalgleish T (2009) A key role for similarity in vicarious reward. Science 324(5929):900. doi:10.1126/science.1170539

Mohnke S, Erk S, Schnell K, Schutz C, Romanczuk-Seiferth N, Grimm O, Walter H (2014) Further evidence for the impact of a genome-wide-supported psychosis risk variant in ZNF804A on the theory of mind network. Neuropsychopharmacology 39(5):1196–1205. doi:10.1038/npp.2013.321

Muller-Pinzler L, Gazzola V, Keysers C, Sommer J, Jansen A, Frassle S, Krach S (2015) Neural pathways of embarrassment and their modulation by social anxiety. Neuroimage 119:252–261. doi:10.1016/j.neuroimage.2015.06.036

Paulus FM, Muller-Pinzler L, Westermann S, Krach S (2013) On the distinction of empathic and vicarious emotions. Front Hum Neurosci 7:196. doi:10.3389/fnhum.2013.00196

Pears KC, Moses LJ (2003) Demographics, parenting, and theory of mind in preschool children. Soc Dev 12(1):1–20. doi:10.1111/1467-9507.00219

Premack D, Woodruff G (1978) Does the chimpanzee have a theory of mind? Behav Brain Sci 515–526

Rueckert L, Naybar N (2008) Gender differences in empathy: the role of the right hemisphere. Brain Cogn 67(2):162–167. doi:10.1016/j.bandc.2008.01.002

Schulte-Ruther M, Markowitsch HJ, Shah NJ, Fink GR, Piefke M (2008) Gender differences in brain networks supporting empathy. Neuroimage 42(1):393–403. doi:10.1016/j.neuroimage.2008.04.180

Schurz M, Radua J, Aichhorn M, Richlan F, Perner J (2014) Fractionating theory of mind: a meta-analysis of functional brain imaging studies. Neurosci Biobehav Rev 42C:9–34. doi:10.1016/j.neubiorev.2014.01.009

Shamay-Tsoory SG, Aharon-Peretz J (2007) Dissociable prefrontal networks for cognitive and affective theory of mind: a lesion study. Neuropsychologia 45(13):3054–3067. doi:10.1016/j.neuropsychologia.2007.05.021

Silani G, Lamm C, Ruff CC, Singer T (2013) Right supramarginal gyrus is crucial to overcome emotional egocentricity bias in social judgments. J Neurosci 33(39):15466–15476. doi:10.1523/JNEUROSCI.1488-13.2013

Singer T (2006) The neuronal basis and ontogeny of empathy and mind reading: review of literature and implications for future research. Neurosci Biobehav Rev 30(6):855–863. doi:10.1016/j.neubiorev.2006.06.011

Singer T, Lamm C (2009) The social neuroscience of empathy. Ann NY Acad Sci 1156:81–96. doi:10.1111/j.1749-6632.2009.04418.x

Singer T, Seymour B, O'Doherty J, Kaube H, Dolan RJ, Frith CD (2004) Empathy for pain involves the affective but not sensory components of pain. Science 303(5661):1157–1162. doi:10.1126/science.1093535

Singer T, Seymour B, O'Doherty JP, Stephan KE, Dolan RJ, Frith CD (2006) Empathic neural responses are modulated by the perceived fairness of others. Nature 439(7075):466–469. doi:10.1038/nature04271

Smeets T, Dziobek I, Wolf OT (2009) Social cognition under stress: differential effects of stress-induced cortisol elevations in healthy young men and women. Horm Behav 55(4):507–513. doi:10.1016/j.yhbeh.2009.01.011

Walter H, Adenzato M, Ciaramidaro A, Enrici I, Pia L, Bara BG (2004) Understanding intentions in social interaction: the role of the anterior paracingulate cortex. J Cogn Neurosci 16 (10):1854–1863. doi:10.1162/0898929042947838

Wicker B, Keysers C, Plailly J, Royet JP, Gallese V, Rizzolatti G (2003) Both of us disgusted in My insula: the common neural basis of seeing and feeling disgust. Neuron 40(3):655–664

Wimmer H, Perner J (1983) Beliefs about beliefs—representation and constraining function of wrong beliefs in young childrens understanding of deception. Cognition 13(1):103–128. doi:10.1016/0010-0277(83)90004-5

Wolf I, Dziobek I, Heekeren HR (2010) Neural correlates of social cognition in naturalistic settings: a model-free analysis approach. Neuroimage 49(1):894–904

Wolkenstein L, Schönenberg M, Schirm E, Hautzinger M (2011) I can see what you feel, but I can't deal with it: impaired theory of mind in depression. J Affect Disord 132(1–2):104–111. doi:10.1016/j.jad.2011.02.010

Xu X, Zuo X, Wang X, Han S (2009) Do you feel my pain? Racial group membership modulates empathic neural responses. J Neurosci 29(26):8525–8529. doi:10.1523/JNEUROSCI.2418-09.2009

Reward: From Basic Reinforcers to Anticipation of Social Cues

Lena Rademacher, Martin Schulte-Rüther, Bernd Hanewald and Sarah Lammertz

Abstract Reward processing plays a major role in goal-directed behavior and motivation. On the neural level, it is mediated by a complex network of brain structures called the dopaminergic reward system. In the last decade, neuroscientific researchers have become increasingly interested in aspects of social interaction that are experienced as rewarding. Recent neuroimaging studies have provided evidence that the reward system mediates the processing of social stimuli in a manner analogous to nonsocial rewards and thus motivates social behavior. In this context, the neuropeptide oxytocin is assumed to play a key role by activating dopaminergic reward pathways in response to social cues, inducing the rewarding quality of social interactions. Alterations in the dopaminergic reward system have been found in several psychiatric disorders that are accompanied by social interaction and motivation problems, for example autism, attention deficit/hyperactivity disorder, addiction disorders, and schizophrenia.

Keywords Reward system · Anticipation · Social reward · Dopamine · Nucleus accumbens · Oxytocin · Autism · ADHD · Addiction · Schizophrenia

---

L. Rademacher (✉)
Department of Psychiatry and Psychotherapy, Social Neuroscience Lab,
University of Lübeck, Ratzeburger Allee 160, 23538 Lübeck, Germany
e-mail: rademacher@snl.uni-luebeck.de

M. Schulte-Rüther
Department of Child and Adolescent Psychiatry and Psychotherapy,
RWTH Aachen University, Neuenhofer Weg 21, 52074 Aachen, Germany
e-mail: mschulte@ukaachen.de

B. Hanewald
Department of Psychiatry and Psychotherapy, Justus Liebig University Gießen,
Am Steg 22, 35385 Gießen, Germany
e-mail: bernd.hanewald@psychiat.med.uni-giessen.de

S. Lammertz
Department of Psychiatry, Psychotherapy and Psychosomatics,
RWTH Aachen University, Pauwelsstraße 30, 52074 Aachen, Germany
e-mail: slammertz@ukaachen.de

Curr Topics Behav Neurosci (2016) 30: 207–222
DOI 10.1007/7854_2015_429
Published Online: 30 January 2016

Contents

1 Introduction 208
1.1 The Dopaminergic Reward System 208
1.2 fMRI Studies on Reward Processing 210
2 Social Reward 211
2.1 fMRI Studies on Social Reward 211
2.2 Oxytocin, Dopamine, and Social Behavior 212
3 Clinical Relevance 213
3.1 Autism 213
3.2 ADHD 215
3.3 Addiction 215
3.4 Schizophrenia 216
4 Summary and Outlook 217
References 218

1 Introduction

Rewards play a crucial part in almost all aspects of life and form the basis for goal-directed behavior and motivation. For example, the rewarding nature of food and sex assures survival and reproduction, the expectation of money and status drives many humans to work hard, and the desire for social relatedness and affiliation guides our behaviors during social interactions. Rewards such as food are called 'primary reinforcers' as they satisfy basic needs and do not require learning of the reinforcing value. In contrast, 'secondary reinforcers' such as money acquire their rewarding value by learned associations with primary reinforcers. Generally, two temporally distinct phases of reward processing need to be considered: an anticipatory phase and a consummatory phase. During reward anticipation, incentive salience is attributed to reward-predictive cues, rendering the stimulus a desirable goal and inducing approach behavior, whereas reward consumption is characterized by hedonic reactions to the pleasure gained by the reward (Berridge and Robinson 1998).

1.1 The Dopaminergic Reward System

On the neural level, reward processing is mediated by a number of different brain regions. At the heart of this complex network is a cortical–basal ganglia circuit comprising the ventral striatum (in particular the nucleus accumbens), orbital prefrontal cortex, anterior cingulate cortex, ventral pallidum, and the midbrain dopamine neurons in the ventral tegmental area (VTA) and substantia nigra (SN) (Haber and Knutson 2010). Dopamine projections from the VTA to the nucleus accumbens (NAcc, see Fig. 1), the so-called mesolimbic pathway, are considered the most important pathway for reward learning, anticipation, and approach behavior.

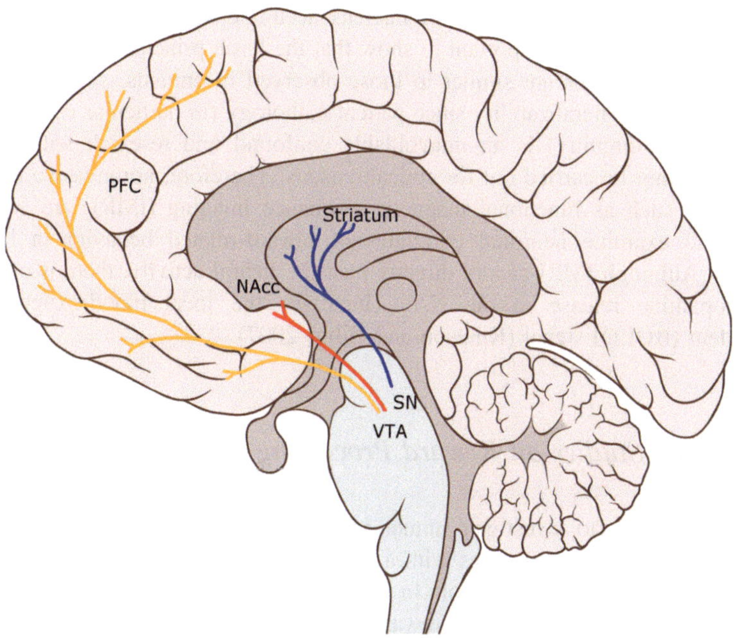

Fig. 1 Dopaminergic pathways in the human brain. The mesolimbic (*red*) and mesocortical (*orange*) pathways begin in the ventral tegmental area (VTA) and project to the nucleus accumbens (NAcc) and prefrontal cortex (PFC). The nigrostriatal pathway (*blue*) transmits dopamine from the substantia nigra (SN) to the dorsal striatum. See URL: https://commons.wikimedia.org/wiki/File:Brain_human_sagittal_section.svg

Electrophysiological studies in monkeys and other mammals (Mirenowicz and Schultz 1996) have demonstrated that midbrain dopamine neurons in the VTA and SN (as well as neurons in their projection sites at the striatum, frontal cortex, and amygdala) encode reward prediction error signals. That is, they show phasic activations in response to unpredicted and novel rewards. However, if the reward is contingently preceded by a predictive cue, the firing responses to reward consumption fade away and shift toward the predictive cue, reflecting reward anticipation. If an expected reward is omitted, neuronal firing is further suppressed, encoding a negative prediction error. Further research showed that such dopamine-related firing in the midbrain can also encode magnitude, probability, and temporal delay of rewards (Schultz 2013).

Most of the knowledge about dopaminergic reward processing is based on animal research, but recently there have been a few pioneering studies exploring reward responses at the single neuron level in humans (e.g., Zaghloul et al. 2009; Lega et al. 2011). For example, neuronal activity was recorded during reward-related tasks in the SN of Parkinson patients and in the NAcc of individuals with major depression and obsessive–compulsive disorder who underwent deep brain stimulation surgery. In line with the animal data, these studies showed reward-responsive spike activity of nucleus accumbens and midbrain dopamine neurons. Although it is important to show that the basic principles guiding reward processing in humans are similar to those observed in animals, such studies are limited in their generalizability since patient pathology (in particular related to the dopaminergic circuitry) is an unavoidable confound and research with healthy subjects cannot be carried out for ethical reasons. Therefore, noninvasive imaging techniques such as functional magnetic resonance imaging (fMRI) are typically applied to examine neuronal correlates of reward-related behavior in healthy humans. Although fMRI cannot directly measure neural activity, there is evidence that dopamine release in the NAcc increases the local blood-oxygen-level-dependent (BOLD) signal (Knutson and Gibbs 2007).

1.2 fMRI Studies on Reward Processing

The simplest way to investigate human reward processing is to use experimental designs, which contrast pictures with a positive valuation to those with a neutral, negative, or less positive valuation. In this manner, fMRI studies have used a wide range of primary and secondary rewards such as food, erotic pictures, attractive faces, sports cars, funny cartoons, images of a romantic partner, and visual art (e.g., Stark et al. 2005; Beaver et al. 2006). Other studies use learning paradigms during fMRI to map reward prediction error signals and the acquisition of predictive value of stimuli to their respective brain areas (Daniel and Pollmann 2014).

Another common experimental paradigm to study reward processing using fMRI is the incentive delay task. This paradigm includes explicit cues indicating whether and which reward can be expected, if the participant performs a task correctly (e.g., hitting a button as fast as possible upon the appearance of a target). The incentive delay paradigm is particularly useful to investigate reward anticipation in comparison with reward consumption. In recent years, it has been applied to examine the anticipation of various primary and secondary rewards, for example money, pleasant taste stimuli, smiling faces, erotic pictures, and professional success (e.g., Knutson et al. 2001; Paulus 2015; Spreckelmeyer et al. 2009).

In line with the electrophysiological results, fMRI studies have reported activations in dopamine-innervated brain regions during reinforcement learning and reward processing, in particular the NAcc and orbital prefrontal cortex (see (Liu et al. 2011) for meta-analysis). More specifically, imaging results confirm both the encoding of reward signals *and* reward expectation/anticipation in the human striatum, in prefrontal cortex, and in midbrain nuclei, which is strikingly similar to the firing pattern of dopamine neurons that are observed in animal studies. These findings support the crucial role of the mesocorticolimbic reward circuit for incentive-based learning and reward anticipation. For reward anticipation, nucleus accumbens, medial prefrontal, and orbitofrontal activities have been suggested to represent a 'common currency' for the valuation of different reward types, ranging from primary rewards to more complex and abstract reinforcers. Thus, mesocorticolimbic activity not only increases proportionally to the magnitude of an expected reward but also represents the relative personal reward value (e.g., Gross et al. 2014; Sescousse et al. 2014).

2 Social Reward

In social psychology, the quest for social acceptance and belonging is considered a basic motive of humans: We have a strong motivation to seek social relationships and enjoy friendship, mutual support, and understanding. On the other hand, social belongingness makes us care for the needs of others and motivates us to act pro-socially. Thus, it seems reasonable to assume that social connectedness and pro-social behavior have a rewarding quality mediated by the mesolimbic reward circuit. In the last decade, neuroscientific researchers have become increasingly interested in the rewarding nature of social interaction (Krach et al. 2010). In this context, the term 'social reward' has been employed for a broad variety of positive social stimuli and experiences, ranging from static pictures of faces to complex social experiences such as cooperation.

2.1 *fMRI Studies on Social Reward*

Genetic and pharmacological studies have shown that the dopamine system is crucial for social interaction. For example, genes involved in striatal dopamine transmission modulate social approach behavior (Enter et al. 2012) and a pharmacological increase of the dopamine concentration improves learning about a partner's pro-social preferences (Eisenegger et al. 2013). In line with these findings, several functional neuroimaging studies have demonstrated that structures of the dopaminergic reward circuit are activated by social stimuli and experiences.

For instance, data obtained in neuroimaging studies support the view that social belonging, acceptance, and support function as 'social rewards,' confirming social psychological theories: The ventral striatum was found to be activated when subjects feel understood (Morelli et al. 2014) or receive positive feedback about themselves (Izuma et al. 2008) but deactivated during a conflict with the prevailing group opinion (Klucharev et al. 2009). Also, the mere expectation of positive social feedback (e.g., in terms of a smiling face, verbal praise, or liking statements by others) was shown to elicit ventral striatal activity (e.g., Kirsch et al. 2003; Rademacher et al. 2010), reflecting a motivational drive for social approval.

Another type of social interaction experienced as rewarding is mutual support in terms of collaboration. Even toddlers were shown to have a preference to access a reward collaboratively instead of individually (Rekers et al. 2011). Accordingly, fMRI studies in adults have demonstrated that cooperation with an unknown person is associated with activations of the reward network (e.g., the NAcc, ACC, orbitofrontal and ventromedial prefrontal cortex). However, ventral striatal and ACC (but not OFC) activity is specific to human social interaction and not elicited when cooperating for instance with a computer partner (Rilling et al. 2002). These findings have been interpreted to reflect reward learning in terms of who can be trusted to cooperate, but also the rewarding nature of cooperation and reciprocated favors, which motivates us to act pro-socially and to withstand the temptation of being selfish (Strang and Park 2017).

The assumption that pro-social behavior has an intrinsic motivational value is evidenced by situations where a rewarding experience occurs in the absence of any direct benefit for oneself. For example, imaging studies have shown that donating is associated with activations in structures of the mesolimbic network (ventral striatum, VTA) (e.g., Moll et al. 2006). Moreover, monetary payoffs to charities activate the mesolimbic network in similar ways as monetary rewards to oneself, suggesting that the common neural currency of reward also relates to more complex incentives like the 'joy of giving' and vicarious pleasure (Müller-Pinzler et al. 2017), which make pro-social behavior inherently rewarding. This assumption receives further support from a study that used a gambling task, in which subjects could win rewards either for themselves or for another person (Braams et al. 2014): Anticipatory ventral striatal activity was equally strong when subjects played for the best friend and when they played for their own benefit. However, the social relationship with the others plays an important role: Anticipatory activity was found to be decreased when playing for an unknown or a disliked other person (e.g., Braams et al. 2014). Hence, the intrinsic motivational value of pro-social behavior and vicarious reward anticipation vary with social distance. This is in line with data showing that subjective feelings of excitement and ventral striatal activity when receiving rewards together with another person depend on the social distance of this person (Fareri et al. 2012).

These findings point to a complex involvement of the dopaminergic reward system in social interaction, which is presumably modulated by other neurotransmitters involved in social behavior, for example the neuropeptide oxytocin.

2.2 *Oxytocin, Dopamine, and Social Behavior*

Oxytocin is known to influence social behavior. It has been found to play a role in pair bonding, sexual behavior, and parenting in different species—also in humans. An increasing number of studies in humans also report an involvement of oxytocin in higher-order social behavior. For example, there is evidence that oxytocin increases emotion recognition, morality, altruism, trust, and generosity, but also ethnocentrism (preferring to help and support the in-group) (see Evans et al. (2013) for review). The underlying mechanisms of these effects of oxytocin are not yet fully understood, but there is some evidence indicating that the dopaminergic reward system plays an important role. A first important hint is the high density of oxytocin receptors in brain regions that are also rich in dopamine receptors like the amygdala, the NAcc, and the VTA (Skuse and Gallagher 2009). Second, an interaction of these two neurotransmitters has been observed. The effect of intranasally administered oxytocin on the amygdala's response to social stimuli has been reported to be modulated by the availability of dopamine (Sauer et al. 2013). Furthermore, animal studies have demonstrated that oxytocinergic circuits and the mesocorticolimbic dopamine pathway are directly connected and the strength of this connection is related to maternal caregiving behavior. Infusions of oxytocin in the VTA result in increased dopamine release within the ventral striatum along with increased maternal behaviors, whereas the injection of an oxytocin antagonist results in the opposite behavior of maternal neglect (Strathearn 2011).

These findings have led to the assumption that an interaction of oxytocin with dopamine regulates socio-affiliative behaviors. Social interaction may be rewarding because oxytocin activates the dopaminergic reward system in response to social cues. There are a few fMRI studies in humans that support this assumption. An intranasal application of oxytocin was found to increase the BOLD response of the VTA to cues signaling social reward or social punishment, which might indicate that oxytocin modulates social behavior by attaching motivational salience to socially relevant cues through mesolimbic dopamine projections (Groppe et al. 2013). Further studies could show that oxytocin facilitates learning with social (rewarding and punishing) feedback and increases the VTA response of female subjects to infant and sexual images as well as VTA and NAcc responses of male subjects to their female partners (e.g., Scheele et al. 2013; Gregory et al. 2015).

So far, these studies support the hypothesis of an interaction of oxytocin with dopamine. More research is needed to shed more light on this interaction and to clarify whether it is the underlying mechanism that makes social interaction rewarding and worth to seek. Other studies, however, have linked oxytocin administration to different emotion processing systems. For example, oxytocin modulates amygdala reactivity in response to threatening stimuli which may result in anxiolytic effects. Thus, the effect of oxytocin is likely not restricted to the dopaminergic system. Future studies will need to address in more detail to what extent the modulation of the dopaminergic system may contribute to the effects of oxytocin for social behavior.

3 Clinical Relevance

3.1 Autism

Autism spectrum disorder (ASD) is characterized by substantial impairments in social interaction and communication, as well as repetitive behaviors and restricted interests. Several recent theorists have focused on reduced social motivation for reciprocal social behaviors to explain deficits in the domain of social interaction and communication. An early lack of interest and pleasure with respect to social encounters may result in cascading negative consequences such as decreased expertise in faces, voices, and ultimately social cognition and behavior (e.g.,

Chevallier et al. 2012). Consequently, research into the behavioral and neural consequences of altered reward processing in ASD has received growing interest in recent years. The results from behavioral studies targeting the social motivation hypothesis are mixed. Several studies confirm a general decrease in sensitivity to social reinforcement in comparison with other reinforcers, or a failure to improve performance under social motivation conditions. Other studies revealed no differences or very subtle effects between social and nonsocial conditions. On the neural level, several studies have demonstrated a domain-general dysfunction of the reward system in ASD that does not seem to be restricted to social stimuli (e.g., Kohls et al. 2013), comprising diverse areas typically associated with reward processing (including the striatum). Other studies have found differences associated with the type of reward, for example more pronounced hypoactivations for social reward conditions (Scott-Van Zeeland et al. 2010) in comparison with monetary reward or differences only for social but not for nonsocial conditions (Delmonte et al. 2012). Interestingly, reward stimuli that are associated with specific restricted interests reliably engage reward circuitry also in ASD (Dichter and Adolphs 2012). To summarize, these results suggest that the brain circuitry mediating dopaminergic reward processing and reinforcement learning may remain functional in ASD, but engagement of the system is highly dependent on the individual motivational value of the employed reward stimuli.

This conclusion is well in accordance with clinical observations. The most effective early behavioral interventions in autism typically use reinforcement-based learning to shape behavior (e.g., applied behavior analysis (ABA)), and an essential part of such interventions is giving salient, frequent positive feedback in response to desired behaviors. Interestingly, this also includes social feedback (such as hugs, smiling, 'thumbs up,' verbal praise), but in a very salient and explicit way. Additional ways of increasing saliency and motivational value of social stimuli thus are a very promising approach for future interventions in ASD. Oxytocin has been shown to modulate the processing of social stimuli (in particular during social reward) and therefore might be a candidate pharmacological agent in this regard (see Sect. 2.2 oxytocin, dopamine and social reward and Poustka and Kamp-Becker 2017). However, the first clinical trials testing the effect of chronic oxytocin administration on the symptomatology of ASD have been somewhat disappointing (see Guastella and Hickie (2015) for review). Future trials are needed to show whether oxytocin proves effective in more targeted approaches—e.g., in combination with specific behavioral strategies or in specific subgroups of patients.

Taken together, it is not clear which aspects of reward processing are particularly impaired in ASD or how these relate to deficits in social motivation and restricted interests, and how they could be most effectively targeted in specific interventions. Results from previous experimental studies are mixed, due to diverse differences with respect to type of employed reward stimuli and behavioral paradigms. In future studies, it would be useful to make use of computational models of reinforcement learning that provide the possibility to tear apart different aspects of dysfunctional reward processing, i.e., impaired reinforcement learning per se, failure to attach motivational value during learning, or diminished reward value of primary and secondary reinforcers. The few studies employing such a strategy indicate that learning per se might be slower in ASD and particularly impaired for social stimuli, whereas reward value is enhanced for items related to ASD-specific interests. Establishing stable associations via reinforcement learning seems to be more difficult for individuals with ASD who rely more on trial-by-trial updates (Solomon et al. 2015).

3.2 ADHD

Attention deficit/hyperactivity disorder (ADHD) is closely related to dysfunctions of the dopaminergic system. It has been suggested that altered reinforcement mechanisms due to changes in dopamine signaling result in stronger immediate behavioral sensitivity to rewards but at the same time impairments in reward anticipation (Tripp and Wickens 2008). These changes might contribute to poor behavioral control and increased impulsivity typically observed in patients with ADHD. Several behavioral studies have confirmed atypical reward processing in ADHD, including preference for immediate as compared to delayed rewards, enhanced seeking of large but risky rewards, and less behavioral adaptation in response to reward (see Luman et al. 2010 for review). On the neural level, reward anticipation is associated with decreased activation of the ventral striatum in ADHD (Scheres et al. 2007) for both immediate and delayed rewards, whereas ventral striatal and orbitofrontal responses appear to be enhanced during reward outcome (von Rhein et al. 2015). It should be highlighted that the vast majority of studies has focused on monetary reward conditions, but only few studies have investigated social reward in ADHD (e.g., Kohls et al. 2014). These studies suggest a hyper-responsivity when task performance is coupled with social reward in comparison with monetary reward and an associated increase in striatal and medial frontal brain areas. However, more research is needed to elucidate atypical sensitivity for social rewards in ADHD and whether these are evident during processing of reward outcome or reward anticipation.

3.3 Addiction

The dopaminergic reward system plays a crucial role in addictive disorders. All known drugs of abuse acutely increase dopamine release in the nucleus accumbens, but animal studies have shown that dopamine is already released in response to drug-associated cues before the drug administration. Therefore, a sensitization of the reward system is assumed (Robinson and Berridge 2000): Stimuli that were learned to be associated with drug consumption attract attention, become desirable, and motivate drug-taking behavior. In line with this theory, several fMRI studies demonstrated increased striatal activity in smokers and alcoholics during the presentation of drug-associated pictures (e.g., Wrase et al. 2007). In contrast, other studies which used stimuli not associated with the drug (e.g., monetary rewards) reported decreased ventral striatal activation during reward anticipation (Wrase et al. 2007; Peters et al. 2011). However, not all studies could replicate this finding.

So far, there has been little research on social reward in addiction. There are several observations that suggest an interaction of social reward processing with addictive behavior. First, patients with antisocial behavior disorders have an increased risk for addiction and show an earlier start of consumption and more severe abuse than other substance abusers. Second, social experiences during development strongly affect drug use in later life, social bonds have a protective effect, and social emotions can often be a driver for recovery. In this context, oxytocin appears to have an important role: Exogenous oxytocin administration was found to prevent development of tolerance to several drugs, to inhibit self-administration, and to reduce withdrawal symptoms (McGregor and Bowen 2012). Third, addiction alters social behaviors and is often associated with social dysfunction and isolation. It is assumed that the increased motivation for drugs leads to decreased striving for naturally rewarding stimuli such as social interaction. In line with this suggestion, a recent study using a social interaction paradigm (Preller et al. 2014) demonstrated diminished emotional engagement and blunted reward responses in the medial orbitofrontal cortex of cocaine users that were related to real-life social network size. These results highlight the importance of social reward in the treatment of drug addiction and point to the potential of oxytocin-based therapeutics.

3.4 Schizophrenia

Reward processing in patients suffering from schizophrenia represents an area of particular clinical interest since motivational deficits often affect patients' quality of life and the common drug treatment is of limited effect. Not only motivational deficits can be seen as a result of anhedonia, but also a dissociation between the joyful reaction to a rewarding stimulus and the motivational behavior could be detected: Patients with schizophrenia show relatively intact consummatory pleasure, but a limited motivation to achieve a reward.

Irregularities in dopamine transmission are an important part of the pathophysiology of schizophrenia. Motivational deficits in patients suffering from schizophrenia may be related to orbitofrontal cortex-driven value representation deficits as well as to deficits in 'effort–cost' computation (Strauss et al. 2014). In the latter case, an overexpression of postsynaptic D2 receptors rather than reduced striatal dopamine release as well as cingulate dysfunction might contribute to aberrant effort–value computations in patients suffering from schizophrenia. Moreover, in previous studies, impairments in reward-related learning in people with schizophrenia were found, which were associated with disturbed ventral striatal activation: For example, individuals suffering from schizophrenia showed reduced BOLD activation in the ventral striatum, including the nucleus accumbens, in response to prediction errors (Schlagenhauf et al. 2014). Furthermore, patients showed inappropriately strong ventral striatal activations in response to neutral stimuli as compared to healthy controls (e.g., Diaconescu et al. 2011). These findings suggest that alterations in the mesolimbic dopamine system underlie deficits in reward-based learning. As a result, the discrimination between important and unimportant environmental stimuli appears to be more difficult for patients suffering from schizophrenia. Consequently, 'aberrant salience' to nonrelevant stimuli may underlie some psychotic symptoms (Mier and Kirsch 2017), whereas decreased salience attribution to reward-predicting cues may be associated with negative symptoms like motivational deficits and avolition.

Several studies that examined monetary reward anticipation in schizophrenic patients are consistent with this interpretation. fMRI studies showed a significantly reduced activation in the ventral striatum in unmedicated and drug-naïve patients suffering from schizophrenia during the anticipation of monetary gains, indicating a dysfunction of the reward system in schizophrenic patients (e.g., Juckel et al. 2006a). In unmedicated patients, the ventral striatal hypoactivity is correlated with negative symptoms, while positive symptoms can be seen as a hyper-responsiveness of the reward system. Studies in treated patients showed a less pronounced hypofunction in the ventral striatum during reward anticipation when patients were treated with atypical, but not with typical antipsychotics—this can be explained by a stronger blockade of D2 receptors by typical antipsychotics (e.g., Juckel et al. 2006b).

So far, all studies on reward anticipation in schizophrenia have used monetary rewards. However, there is evidence that disturbed social reward processing may play a particular role in schizophrenia. Patients with schizophrenia have severe impairments in social functioning and show reduced engagement in social interactions—it seems plausible to assume that there is a disturbed sensitivity to social rewards. A recent fMRI study using a trust game supports this hypothesis: Gromann et al. (2013) found decreased striatal activity in patients with schizophrenia, which was interpreted as a diminished sensitivity to rewarding aspects of social interaction, resulting in reduced motivation to seek interaction. In addition, a behavioral trust game study (Fett et al. 2012) demonstrated that patients with psychosis not only showed lower initial levels of trust, but also were unable to change their trusting behavior according to the interaction partner's behavior—this might reflect a prediction problem and insensitivity to social reward.

4 Summary and Outlook

To conclude, there is evidence that the dopaminergic reward network is of crucial importance for social interaction. Mesocorticolimbic activity is suggested to represent a 'common currency' for the personal reward value of social and nonsocial rewards and to motivate social behavior. However, there is also evidence for differences between social and nonsocial reward processing. For example, the neuropeptide oxytocin is suggested to be involved in the activation of the dopaminergic system in response to social stimuli. Although there is still no clear picture, several psychiatric disorders have been reported to be associated with disturbed social reward processing. Future research will need to clarify, which aspects of reward processing might be dysfunctional in these disorders and whether for example diminished personal reward values of social rewards can be increased by therapeutic interventions.

References

Beaver JD, Lawrence AD, van Ditzhuijzen J et al (2006) Individual differences in reward drive predict neural responses to images of food. J Neurosci 26:5160–5166. doi:10.1523/JNEUROSCI.0350-06.2006

Berridge KC, Robinson TE (1998) What is the role of dopamine in reward: hedonics, learning, or incentive salience? Brain Res Rev 28:308–367

Braams BR, Güroğlu B, De Water E et al (2014) Reward-related neural responses are dependent on the beneficiary. Soc Cogn Affect Neurosci 9:1030–1037. doi:10.1093/scan/nst077

Chevallier C, Kohls G, Troiani V et al (2012) The social motivation theory of autism. Trends Cogn Sci 16:231–238

Daniel R, Pollmann S (2014) A universal role of the ventral striatum in reward-based learning: evidence from human studies. Neurobiol Learn Mem 114:90–100. doi:10.1016/j.nlm.2014.05.002

Delmonte S, Balsters JH, McGrath J et al (2012) Social and monetary reward processing in autism spectrum disorders. Mol Autism 3:7. doi:10.1186/2040-2392-3-7

Diaconescu AO, Jensen J, Wang H et al (2011) Aberrant effective connectivity in schizophrenia patients during appetitive conditioning. Front Hum Neurosci 4:239. doi:10.3389/fnhum.2010.00239

Dichter G, Adolphs R (2012) Reward processing in autism: a thematic series. J Neurodev Disord 4:20. doi:10.1186/1866-1955-4-20

Eisenegger C, Pedroni A, Rieskamp J et al (2013) DAT1 polymorphism determines L-DOPA effects on learning about others' prosociality. PLoS ONE. doi:10.1371/journal.pone.0067820

Enter D, Colzato LS, Roelofs K (2012) Dopamine transporter polymorphisms affect social approach-avoidance tendencies. Genes, Brain Behav 11:671–676. doi:10.1111/j.1601-183X.2012.00791.x

Evans SL, Monte OD, Noble P, Averbeck BB (2013) Intranasal oxytocin effects on social cognition: a critique. Brain Res 1580:69–77. doi:10.1016/j.brainres.2013.11.008

Fareri DS, Niznikiewicz MA, Lee VK, Delgado MR (2012) Social network modulation of reward-related signals. J Neurosci 32:9045–9052. doi:10.1523/JNEUROSCI.0610-12.2012

Fett AKJ, Shergill SS, Joyce DW et al (2012) To trust or not to trust: the dynamics of social interaction in psychosis. Brain 135:976–984. doi:10.1093/brain/awr359

Gregory R, Cheng H, Rupp HA et al (2015) Oxytocin increases VTA activation to infant and sexual stimuli in nulliparous and postpartum women. Horm Behav 69:82–88. doi:10.1016/j.yhbeh.2014.12.009

Gromann PM, Heslenfeld DJ, Fett AK et al (2013) Trust versus paranoia: abnormal response to social reward in psychotic illness. Brain 136:1968–1975. doi:10.1093/brain/awt076

Groppe SE, Gossen A, Rademacher L et al (2013) Oxytocin influences processing of socially relevant cues in the ventral tegmental area of the human brain. Biol Psychiatry 74:172–179

Gross J, Woelbert E, Zimmermann J et al (2014) Value signals in the prefrontal cortex predict individual preferences across reward categories. J Neurosci 34:7580–7586. doi:10.1523/JNEUROSCI.5082-13.2014

Guastella A, Hickie I (2015) Oxytocin treatment, circuitry and autism: a critical review of the literature placing oxytocin into the autism context. Biol Psychiatry. doi:10.1016/j.biopsych.2015.06.028

Haber SN, Knutson B (2010) The reward circuit: linking primate anatomy and human imaging. Neuropsychopharmacology 35:4–26. doi:10.1038/npp.2009.129

Izuma K, Saito DN, Sadato N (2008) Processing of social and monetary rewards in the human striatum. Neuron 58:284–294. doi:10.1016/j.neuron.2008.03.020

Juckel G, Schlagenhauf F, Koslowski M et al (2006a) Dysfunction of ventral striatal reward prediction in schizophrenia. Neuroimage 29:409–416. doi:10.1016/j.neuroimage.2005.07.051

Juckel G, Schlagenhauf F, Koslowski M et al (2006b) Dysfunction of ventral striatal reward prediction in schizophrenic patients treated with typical, not atypical, neuroleptics. Psychopharmacology 187:222–228. doi:10.1007/s00213-006-0405-4

Kirsch P, Schienle A, Stark R et al (2003) Anticipation of reward in a nonaversive differential conditioning paradigm and the brain reward system: an event-related fMRI study. Neuroimage 20:1086–1095. doi:10.1016/S1053-8119(03)00381-1

Klucharev V, Hytönen K, Rijpkema M et al (2009) Reinforcement learning signal predicts social conformity. Neuron 61:140–151. doi:10.1016/j.neuron.2008.11.027

Knutson B, Adams CM, Fong GW, Hommer D (2001) Anticipation of increasing monetary reward selectively recruits nucleus accumbens. J Neurosci 21:RC159

Knutson B, Gibbs SEB (2007) Linking nucleus accumbens dopamine and blood oxygenation. Psychopharmacology 191:813–822. doi:10.1007/s00213-006-0686-7

Kohls G, Schulte-Rüther M, Nehrkorn B et al (2013) Reward system dysfunction in autism spectrum disorders. Soc Cogn Affect Neurosci 8:565–572. doi:10.1093/scan/nss033

Kohls G, Thönessen H, Bartley GK et al (2014) Differentiating neural reward responsiveness in autism versus ADHD. Dev Cogn Neurosci 10:104–116. doi:10.1016/j.dcn.2014.08.003

Krach S, Paulus FM, Bodden M, Kircher T (2010) The rewarding nature of social interactions. Front Behav Neurosci 4:22. doi:10.3389/fnbeh.2010.00022

Lega B, Kahana M, Jaggi J (2011) Neuronal and oscillatory activity during reward processing in the human ventral striatum 22:795–800. doi:10.1097/WNR.0b013e32834b2975.Neuronal

Liu X, Hairston J, Schrier M, Fan J (2011) NIH Public Access 35:1219–1236. doi:10.1016/j.neubiorev.2010.12.012.Common

Luman M, Tripp G, Scheres A (2010) Identifying the neurobiology of altered reinforcement sensitivity in ADHD: a review and research agenda. Neurosci Biobehav Rev 34:744–754

McGregor IS, Bowen MT (2012) Breaking the loop: oxytocin as a potential treatment for drug addiction. Horm Behav 61:331–339. doi:10.1016/j.yhbeh.2011.12.001

Mier D, Kirsch P (2017) Social-cognitive deficits in schizophrenia. Curr Top Behav Neurosci 30:397–409

Mirenowicz J, Schultz W (1996) Preferential activation of midbrain dopamine neurons by appetitive rather that aversive stimuli. Nature 379:449–451

Moll J, Krueger F, Zahn R et al (2006) Human fronto-mesolimbic networks guide decisions about charitable donation. Proc Natl Acad Sci USA 103:15623–15628. doi:10.1073/pnas.0604475103

Morelli SA, Torre JB, Eisenberger NI (2014) The neural bases of feeling understood and not understood. Soc Cogn Affect Neurosci 1–7. doi:10.1093/scan/nst191

Müller-Pinzler L, Krach S, Krämer UM, Paulus FM (2017) The social neuroscience of interpersonal emotions. Curr Top Behav Neurosci 30:241–256

Paulus FM, Rademacher L, Schäfer TAJ, Müller-Pinzler L, Krach S (2015) Journal impact factor shapes scientists' reward signal in the prospect of publication. PLoS One 10(11):e0142537. doi:10.1371/journal.pone.0142537

Peters J, Bromberg U, Schneider S et al (2011) Lower ventral striatal activation during reward anticipation in adolescent smokers. Am J Psychiatry 168:540–549. doi:10.1176/appi.ajp.2010.10071024

Poustka L, Kamp-Becker I (2017) Current practice and future avenues in autism therapy. Curr Top Behav Neurosci 30:357–378

Preller KH, Herdener M, Schilbach L et al (2014) Functional changes of the reward system underlie blunted response to social gaze in cocaine users. Proc Natl Acad Sci USA 111:2842–2847. doi:10.1073/pnas.1317090111

Rademacher L, Krach S, Kohls G et al (2010) Dissociation of neural networks for anticipation and consumption of monetary and social rewards. Neuroimage 49:3276–3285. doi:10.1016/j.neuroimage.2009.10.089

Rekers Y, Haun DBM, Tomasello M (2011) Children, but not chimpanzees, prefer to collaborate. Curr Biol 21:1756–1758. doi:10.1016/j.cub.2011.08.066

Rilling J, Gutman D, Zeh T et al (2002) A neural basis for social cooperation. Neuron 35:395–405. doi:10.1016/S0896-6273(02)00755-9 [pii]

Robinson TE, Berridge KC (2000) The psychology and neurobiology of addiction: an incentive-sensitization view. Addiction 95(Suppl 2):S91–S117. doi:10.1080/09652140050111681

Sauer C, Montag C, Reuter M, Kirsch P (2013) Imaging oxytocin? dopamine interactions: an epistasis effect of CD38 and COMT gene variants influences the impact of oxytocin on amygdala activation to social stimuli. Front Neurosci 7:1–9. doi:10.3389/fnins.2013.00045

Scheele D, Wille A, Kendrick KM et al (2013) Oxytocin enhances brain reward system responses in men viewing the face of their female partner. Proc Natl Acad Sci USA 110:20308–20313. doi:10.1073/pnas.1314190110

Scheres A, Milham MP, Knutson B, Castellanos FX (2007) Ventral striatal hypo responsiveness during reward anticipation in attention-deficit/hyperactivity disorder. Biol Psychiatry 61:720–724. doi:10.1016/j.biopsych.2006.04.042

Schlagenhauf F, Huys QJM, Deserno L et al (2014) Striatal dysfunction during reversal learning in unmedicated schizophrenia patients. Neuroimage 89:171–180. doi:10.1016/j.neuroimage.2013.11.034

Schultz W (2013) Updating dopamine reward signals. Curr Opin Neurobiol 23:229–238. doi:10.1016/j.conb.2012.11.012

Scott-Van Zeeland AA, Dapretto M, Ghahremani DG et al (2010) Reward processing in autism. Autism Res 3:53–67. doi:10.1002/aur.122

Sescousse G, Li Y, Dreher J-C (2014) A common currency for the computation of motivational values in the human striatum. Soc Cogn Affect Neurosci 1–7. doi:10.1093/scan/nsu074

Skuse DH, Gallagher L (2009) Dopaminergic-neuropeptide interactions in the social brain. Trends Cogn Sci 13:27–35. doi:10.1016/j.tics.2008.09.007

Solomon M, Frank M, Ragland J et al (2015) Feedback-driven trial-by-trial learning in autism spectrum disorders. Am J Psychiatry 172:173–181

Spreckelmeyer KN, Krach S, Kohls G et al (2009) Anticipation of monetary and social reward differently activates mesolimbic brain structures in men and women. Soc Cogn Affect Neurosci 4:158–165. doi:10.1093/scan/nsn051

Stark R, Schienle A, Girod C et al (2005) Erotic and disgust-inducing pictures—differences in the hemodynamic responses of the brain. Biol Psychol 70:19–29. doi:10.1016/j.biopsycho.2004.11.014

Strang S, Park SQ (2017) Human cooperation and its underlying mechanisms. Curr Top Behav Neurosci 30:223–239

Strathearn L (2011) Maternal neglect: oxytocin, dopamine and the neurobiology of attachment. J Neuroendocrinol 23:1054–1065

Strauss GP, Waltz JA, Gold JM (2014) A review of reward processing and motivational impairment in schizophrenia. Schizophr Bull 40:107–116. doi:10.1093/schbul/sbt197

Tripp G, Wickens JR (2008) Research review: dopamine transfer deficit: a neurobiological theory of altered reinforcement mechanisms in ADHD. J Child Psychol Psychiatry Allied Discip 49:691–704

Von Rhein D, Cools R, Zwiers MP et al (2015) Increased neural responses to reward in adolescents and young adults with attention-deficit/hyperactivity disorder and their unaffected siblings. J Am Acad Child Adolesc Psychiatry 54:394–402. doi:10.1016/j.jaac.2015.02.012

Wrase J, Schlagenhauf F, Kienast T et al (2007) Dysfunction of reward processing correlates with alcohol craving in detoxified alcoholics. Neuroimage 35:787–794. doi:10.1016/j.neuroimage.2006.11.043

Zaghloul KA, Blanco JA, Weidemann CT et al (2009) Human substantia nigra neurons encode unexpected financial rewards. Science 323:1496–1499. doi:10.1126/science.1167342

Human Cooperation and Its Underlying Mechanisms

Sabrina Strang and Soyoung Q. Park

Abstract Cooperation is a uniquely human behavior and can be observed across cultures. In order to maintain cooperative behavior in society, people are willing to punish deviant behavior on their own expenses and even without any personal benefits. Cooperation has been object of research in several disciplines. Psychologists, economists, sociologists, biologists, and anthropologists have suggested several motives possibly underlying cooperative behavior. In recent years, there has been substantial progress in understanding neural mechanisms enforcing cooperation. Psychological as well as economic theories were tested for their plausibility using neuroscientific methods. For example, paradigms from behavioral economics were adapted to be tested in the magnetic resonance imaging (MRI) scanner. Also, related brain functions were modulated by using transmagnetic brain stimulation (TMS). While cooperative behavior has often been associated with positive emotions, noncooperative behavior was found to be linked to negative emotions. On a neural level, the temporoparietal junction (TPJ), the striatum, and other reward-related brain areas have been shown to be activated by cooperation, whereas noncooperation has mainly been associated with activity in the insula.

Keywords Cooperation · Second-party punishment · Third-party punishment · Emotions · fMRI · TMS · Striatum · Insula · DLPFC · OFC · TPJ · Dictator game · Ultimatum game · Public goods game · Prisoner's dilemma warm glow

---

S. Strang · S.Q. Park (✉)
Institute of Psychology, University of Lübeck, Lübeck, Germany
e-mail: soyoung.q.park@gmail.com

Curr Topics Behav Neurosci (2016) 30: 223–240
DOI 10.1007/7854_2016_445
Published Online: 28 June 2016

Contents

1 Why Do We Cooperate? ... 224
2 Economic Games as an Empirical Tool to Investigate Human Cooperation ... 226
3 Social Punishment—An Effective Method to Maintain Cooperation ... 230
4 How Does Cooperation Make You Feel?—A Psychological Perspective ... 231
5 Neuroimaging—A Useful Tool to Investigate the Neural Basis of Cooperation ... 232
6 Manipulating Cooperative Behavior by Brain Stimulation ... 235
7 Conclusion ... 237

References ... 237

1 Why Do We Cooperate?

As humans, we share food, goods, power, services, and other resources with our family, friends, and even with unrelated others. Cooperation is a universal human principle and can be found in all cultures (Gächter and Herrmann 2009). Anthropologists, biologists, psychologists, and economists have done extensive research on cooperation and suggested a variety of different definitions of cooperative behavior. For simplicity, we will use the following definition in the course of this chapter; cooperative behavior is any behavior providing benefits to an individual other than the cooperator. The different disciplines provide a multitude of possible motives explaining human cooperation (Axelrod and Hamilton 1981).

One suggested mechanism for cooperation is kin selection. Here, the idea is that humans cooperate with genetic relatives to increase reproduction success. Since the probability of shared genes is higher for siblings than for cousins, we tend to cooperate more with our parents or siblings than with our cousins or more distant relatives (Hamilton 1964). For example, humans report a greater willingness to rescue a genetic relative than a non-relative when they are in danger (Burnstein et al. 1994). However, human cooperation also occurs when relatedness is absent. For example, people cooperate in laboratory settings with other participants they are not related to and not even friends with (see for example; Engel 2011; Guth et al. 1982). Also in everyday life, we do not restrict cooperative behavior to our family members or relatives. We typically cooperate with friends and colleagues on a regular basis. Here, the mechanism of kin selection fails to explain cooperation.

Considering interactions between unrelated individuals, reciprocation theories are predominant in explaining cooperation. Reciprocity is a social rule saying that we should repay kind as well as unkind behavior (Falk and Fischbacher 2006). In other words, we are kind to those persons who were kind to us previously, whereas we are not kind to another unkind person. Reciprocity can be subdivided into direct and indirect reciprocity. Direct reciprocity relies on repeated symmetric interactions between two individuals. In other words, individuals encounter more than once, and both are able to take an action to cooperate. Interactions with friends and colleagues are typical examples of repeated symmetric interactions. In these kinds of interactions, an individual cooperates with another individual because he or she expects the counterpart to cooperate during future encounters. For instance, you do a colleague a favor because you expect help from him or her in the future. The simplest strategy for direct reciprocity is "tit-for-tat"; individuals always start with cooperation and then adapt their behavior according to the behavior of the counterpart. If the counterpart cooperates as well, cooperation continues until the counterpart defects (Axelrod and Hamilton 1981). According to this strategy, people help a colleague as long as the colleague helps them as well. However, when he or she does not show cooperative behavior at an encounter, helping will be aborted. Another strategy of reciprocation is "win-stay, lose-shift." This strategy is similar to "tit-for-tat" but has the advantage that it allows for occasional mistakes (Nowak and Sigmund 1993). In other words, a single noncooperative encounter is excused as a mistake. In this case, cooperative behavior will be continued unless the counterpart shows repeated noncooperative behavior.

What about situations in which humans cooperate with the knowledge that this will remain a single encounter? People also cooperate in asymmetric one-shot interactions (Engel 2011; Guth et al. 1982). In these interactions, individuals only meet once (one-shot) and only one of them is in the position to cooperate. Donations to charity organizations are good examples for this. Here, people help without ever seeing the person they help and without expecting to get help in return. Since direct reciprocity relies on symmetric interactions, it cannot account for these kinds of cooperation. In contrast, indirect reciprocity can explain cooperation in one-shot asymmetric interactions by reputation concerns (Nowak and Sigmund 1998). According to Nowak and Sigmund (1998), people help strangers, because they seek to establish a good reputation for themselves. By establishing a good reputation, they hope to be rewarded by others in turn. In this case, the motives underlying cooperation are rather strategic.

Direct and non-direct reciprocity concerns, however, can explain cooperation only in non-anonymous interactions; they fail to explain why people cooperate in totally anonymous settings, in which the cooperative act remains hidden. In these cases, cooperative behavior can be regarded as purely altruistic or as driven by the good feeling to help someone. Some people might donate money, blood, or other resources anonymously only because they care about helping other persons independent of themselves. On the other hand, others donate because they derive a good feeling from the act of donation rather than caring about the person in need. This "good feeling" is referred to as "warm glow" (Andreoni 1990). In line with this, it has been shown that spending money on others makes people happier (Dunn et al. 2008; Park et al. 2016).

Fehr and Schmidt (1999) proposed an alternative explanation for people's cooperative behavior in non-anonymous as well as anonymous settings, namely inequity aversion. According to this, people dislike unequal outcomes and feel bad when another person is either worse or better off than themselves. Therefore, people make decisions to minimize inequity. Here, the motive for cooperation is not the induction of a good feeling per se, but the avoidance of feeling bad about not cooperating. This builds a contrast to the idea of "warm glow."

In summary, there is a variety of theories about possible motives underlying cooperative behavior, such as kin selection, reputation concerns, "warm glow," or inequity aversion. Although these mechanisms are able to explain cooperation under specific conditions, the quest for a unifying theory explaining all the diverse cases of cooperative behavior is still in progress. One step towards a better understanding of cooperation is to investigate cooperative behavior in laboratory settings. In the following, we will provide an overview of studies investigating human cooperation from different disciplines and perspectives, such as economics, psychology, and neuroscience.

2 Economic Games as an Empirical Tool to Investigate Human Cooperation

Cooperation can be investigated in controlled laboratory settings using economic exchange games derived from game theory such as the prisoner's dilemma, public goods game, dictator game, ultimatum game, or trust game (see Fig. 1). The first two games are used to study cooperation in symmetric interaction, meaning involved persons are able to cooperate to equal extents (Fig. 1a, b). The latter three depict cooperation in asymmetric situations, meaning involved persons are able to cooperate to different extents (Fig. 1c, d, f). All games are typically anonymous, meaning players do not know whom they are playing with. The application of different games enables the investigation of specific aspects of cooperative behavior.

In a public goods game, each participant of a group receives a monetary endowment (*C*), from which they can contribute part, all, or none to a public good (i.e., symmetric interaction). The money contributed (*c*) to the public good by all group members is then multiplied by a given factor (*f*), and the product is distributed equally among all members (see Fig. 1a). Each participant also keeps the rest of the endowment not contributed. The economically rational behavior in this game is no cooperation, i.e., the selfish behavior not to contribute any money to the public good, since this will maximize the individual benefit. Assuming that other group members contribute money to the public good, the selfish person would receive his or her share of the public good financed by the others. Additionally, the selfish player will keep his/her money not contributed and thereby earn more money in sum than those who contributed to the public good. This behavior is referred to as free riding. Prominent examples of free riding in real life are fare or tax evasion; people benefit from a public good but do not contribute to it. Behavioral studies have shown that although some participants are free riders, a substantial fraction of participants cooperates (Andreoni 1988). Cooperative behavior is always observable in the public goods game, but its degree varies with context factors; it is highest in small groups (Carpenter 2007), in repeated interactions (Fehr and Gächter 2000), and in non-anonymous settings (Gächter and Fehr 1999).

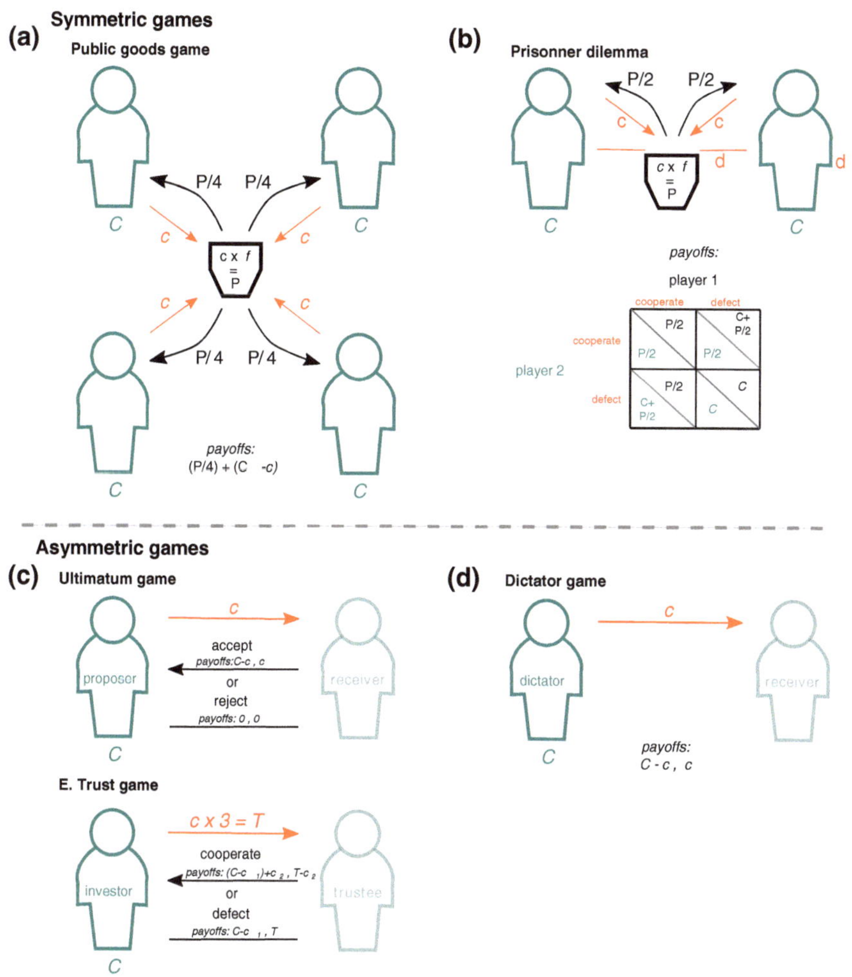

The prisoner's dilemma is a special version of the public goods game with only two players and a limited choice set. Instead of choosing how much to contribute, the participants can only decide between cooperation (contribute everything) and defection (contribute nothing; see Fig. 1b). As in the public goods game, the money contributed (*c*) is multiplied by a given factor (*f*). In this paradigm, the best strategy would be to defect, since this yields a higher payoff independent of what the other player chooses. Let us, for example, assume both player have an endowment (*C*) of 10€ and the multiplying factor *f* is 1.5. In case both players cooperate, they both get a payoff of 15€ (P = (10€ + 10€) × 1.5 = 30€; payoffs = 30€/2 = 15€). However, when only one of them cooperates, the cooperator gets only 7.50€, but the defector gets 17.50€ (P = 10€ x 1.5 = 15€; payoff cooperator = 15€/2 = 7.50€; payoff

◀ **Fig. 1** Schematic illustration of economic games. Symmetric games, **a** public goods game: participants decide how much of their endowment *C* they contribute to a public good *P*. The contributed money *c* is multiplied by a factor *f* (greater than one and less than the number of players), and P is evenly divided among players P/4. Each participant also keeps the rest of *C* not contributed. **b** Prisoners' dilemma: two players with a limited choice set. Participants can only decide between cooperation (c, contribute everything of *C*) or defection (d, contribute nothing). The money contributed (*c*) is multiplied by a factor *f* (greater than one and smaller than two). If both players cooperate, both receive a payoff > *C* (left upper quadrant of matrix), and if both defect, they get *C* (right lower quadrant of matrix). However, if one defects and one cooperates, the one who defects receives a payoff higher as when both cooperate, whereas the one who cooperates receives a payoff between mutual cooperation and defection (right upper and left lower quadrant of matrix). Asymmetric games, **c** ultimatum game: One of two participants (proposer) obtains an endowment *C* and can decide to offer part of this (indicated as *c*) to the other participant (receiver). The receiver can either accept or reject the offer. In case of acceptance, the receiver gets the offered money *c* and the proposer gets the endowment *C* minus the offered *c*. In case of rejection, both do not get any payoff. **d** Dictator game: similar to the ultimatum game, but receiver has no rejection options. The receiver has to take the amount *c* offered by the proposer (here called dictator). **e** Trust game: The money offered by the proposer (here called investor) *c* is multiplied by a factor *f* (usually 3) and transferred to the receiver (here called trustee). The trustee can either cooperate and back transfer some of the money *T* or defect and keep *T*. In case of cooperation, the trustee receives *T* minus the money back transferred c_2 and the investor receives c_2 plus the money not transferred to the trustee (C-c_1). In case of defection, the trustee keeps *T* and the investor only receives the money not transferred to the trustee (C-c_1)

defector = 10€ + (15€/2) = 17.50€). In case of mutual defection, both keep their initial 10€. Although defecting yields higher payoffs for the individual players, many participants choose to cooperate (Neyman 1985). According to a meta-analysis, the mean cooperation rate in the prisoner's dilemma is 47 %; however, in some studies, cooperation rates up to 96 % were observed (Sally 1995).

Since people usually do not know who else is in their group (public goods game) or who is their counter player (prisoner's dilemma; anonymous games), reputation can be ruled out as motivation for cooperation in these two games. While cooperation can be explained by direct reciprocity in repeated versions of these games, it fails to explain cooperation in one-shot versions. In these settings, "warm glow" and inequity aversion are possible motives driving cooperation.

The ultimatum game involves two players: a proposer and a receiver (Fig. 1c). The proposer is endowed with a certain amount of money (*C*) and is asked to divide this between him-/herself and another player, the receiver (i.e., asymmetric interaction, since the receiver is not in the position to act in the first place). In contrast to the public goods game, the transferred amount (*c*) is transmitted unaltered, thus not multiplied by any factor. The receiver can then decide whether to accept the offer or to reject it. In case the receiver accepts the offer from the proposer, both obtain the respective amounts of money. However, in case the receiver rejects the offer, neither proposer nor receiver get anything (see Fig. 1c). This rejection is also referred to as punishment of unfair behavior or social punishment. Rationally, to maximize the individual benefit, both the proposer and the receiver should behave in a selfish manner. A selfish proposer would offer the smallest possible amount to the receiver to maximize his/her benefit. Moreover, a selfish receiver would accept any amount offered, because after all, any amount is better than nothing. However, empirical evidence shows that proposers do offer money (typically around 50 %) to the receivers and receivers tend to reject unfair offers (Camerer 2003a, b; Guth et al. 1982). Usually, allocations of 40–50 % of the proposer's endowment are accepted and allocations below 20 % are rejected (Camerer and Thaler 1995). Rejections in the ultimatum game are considered as social punishment (more details in the next section). The fear of being punished increases cooperation in this setting. However, even without social punishment, cooperation is present, although to a smaller degree.

In the dictator game, another two-person exchange game, the receiver is passive; he or she cannot reject the offer made by the proposer (see Fig. 1d). Although in this game, the proposers do not have to fear any penalty for not cooperating, still many chose to cooperate (Engel 2011). Cooperative behavior in the dictator game cannot be explained by reputation, since the dictator game is typically played as anonymous one-shot version (or with different counter players in each round). However, "warm glow" and inequity aversion are again possible mechanisms driving cooperative behavior.

The trust game is a two player's game as well: an investor and a trustee. Similar to the ultimatum or dictator game, the investor obtains an endowment (C) at the beginning of the experiment and can decide to either transfer nothing, part, or everything to the trustee. Here, the transferred money (c) is multiplied by a given factor. The trustee can then decide either to keep all or most of the money or to return money back to the receiver (see Fig. 1e). Empirical studies show that both investors and trustees show cooperative behavior; most investors transfer money to the trustee, and most trustees return money to the investor (McCabe et al. 1998). According to a meta-analysis, between 40 and 70 % of the endowment are usually transferred to the trustee and the trustee usually back transfers between 30 and 45 % of the received money to the investor (Johnson and Mislin 2011). As the previous games, the trust game is typically played anonymously. Hence, reputation can be ruled out as cooperation motive. However, since the trustee has the option to back transfer money to the investor, direct reciprocity is a possible motive underlying cooperative behavior in this game. According to this, the investor cooperates because he or she expects the trustee to cooperate in the next stage. Moreover, "warm glow" and inequity aversion are possible explanations as well. Taken together, different economic exchange paradigms can be applied to investigate the specific aspects of cooperative behavior in humans. The studies show that on the one hand, a large fraction of people cooperates in all games although they would financially be better off not to do so. On the other hand, there is always a fraction of people who refuses to cooperate. This raises the question of how noncooperation can be minimized in order to maintain cooperation in society.

3 Social Punishment—An Effective Method to Maintain Cooperation

A highly interesting form of human cooperation is social punishment; people even invest their own resources to maintain cooperation. Punishment of noncooperation is a powerful device to enforce cooperation within a society and probably the key determinant for the pervasive cooperative behavior. While on a large scale, cooperation is enforced by law (punishment of tax evasion etc.), on a small scale it is enforced by other members of the society, such as parents, siblings, teachers, friends, or sometimes even strangers. Once a person is directly affected by a noncooperative behavior, he or she can punish the noncooperator. This is called second-party punishment (Egas and Riedl 2008; Fehr and Gächter 2002). As we can observe in reality, sanctioning is a costly matter; people spend time and effort and take the risk of resistance when punishing noncooperators. In order to increase the ecological validity in the laboratory setting, such sanctioning is made to cost for participants (Egas and Riedl 2008; Fehr and Gächter 2002).

One paradigm that enables the investigation of second-party punishment is the ultimatum game. Here, the receiver's decision to reject the offer of the proposer is regarded as costly punishment. The rational receiver would accept any amount offered by the proposer, since this would maximize the receiver's payoff. However, studies have shown that many receivers prefer to reject unfair offers, knowing that the proposer will go home empty-handed as well (Camerer 2003a, b; Guth et al. 1982). Forgoing the offered money in order to zero the proposer's payoff is therefore regarded as costly punishment. Although this punishment behavior seems to be irrational, second-party punishment is thought to foster cooperative behavior in the long run (Fehr and Gächter 2002).

Second-party punishment can also be investigated using an adapted version of the public goods game. Here, participants have the possibility to punish those who contributed less to the public good by decreasing the payoff of the noncooperator. Participants in public good games with punishment option reveal stable contribution rates (Fehr and Gächter 2002). However, when the possibility to punish is eliminated during the experiment, cooperation rates decline (Egas and Riedl 2008). Analogous to the costly sanctioning, participants have to invest part of their own payoff in order to decrease the noncooperator's payoff. In experimental settings, usually a punishment ratio of 1:3 is applied; in other words, one monetary unit invested for punishment decreases the noncooperator's payoff by three monetary units.

An interesting phenomenon of social punishment is the so-called third-party punishment. In other words, people punish others for noncooperation although they themselves are not directly involved in the interaction and are thus not even influenced by the noncooperative behavior of the other person (Fehr and Fischbacher 2004; Hu et al. 2015). A real-life example would be criticizing the ones for not giving a seat to the elderly or disabled people in public transportation. In this case, cooperation is enforced by uninvolved other society members, as confirmed by laboratory experiments (Fehr and Fischbacher 2004; Hu et al. 2015). Although it is costly and there are no direct personal benefits from it, a large number of people punishes noncooperators, probably in order to enforce general cooperative behavior.

It has been shown that cooperative behavior emerges and can be maintained when there is a threat of social punishment (Spitzer et al. 2007). Moreover, cooperation rates are higher in games with punishment compared to the games without punishment option (Egas and Riedl 2008; Fehr and Gächter 2002; Strang et al. 2015). Furthermore, punishment of noncooperators is independent of the number of encounters (Fehr and Gächter 2000). Even if people know that they will never see the noncooperator again (one-shot interactions), implying that they cannot expect any direct benefits from their sanctioning, they are willing to spend own resources to punish this behavior. Taken together, without the threat to be punished, cooperative behavior can deteriorate (Egas and Riedl 2008; Strang et al. 2015). Finally, second- and third-party punishments are universal phenomenon, analogous to cooperation. It has been observed in a variety of societies (Henrich et al. 2006). Populations from very different environments (urban, mangrove forest, prairie, tundra-taiga, etc.) and different countries (the US, Ecuador, and Russia) all demonstrated costly punishment, albeit the magnitude differs (Henrich et al. 2006).

4 How Does Cooperation Make You Feel?—A Psychological Perspective

Besides the altruistic motive to foster cooperation, second- and third-party punishment might additionally be driven by emotions. When we are treated unfairly, we experience negative emotions. Imagine you have to write a group assignment with a fellow student. Since you want to get a good grade, you invest a lot of time and effort in it. In contrast, your fellow students' input is negligible. However, in the end, he or she will benefit from your work, since your work is graded together. How would you feel? As most people, you would probably be upset about the noncooperative behavior of your fellow student. Experimental studies confirm that such negative emotions drive second- and third-party punishment. Physiological measurements associated with negative emotions, such as skin conductance and enhanced activity in the insula, correlate with the extent of second-party punishment (Sanfey et al. 2003; Strang et al. 2016; Van't Wout et al. 2006). Thus, the more people experience negative emotions, the more they are willing to punish other people for their noncooperative behavior. In line with this, regulation of these negative emotions was shown to reduce second-party punishment (Grecucci et al. 2013).

Negative emotions provoked by noncooperative behavior have consequences not only for the noncooperator, but also for uninvolved third persons. Besides punishing the noncooperator directly, people also forward punishment toward uninvolved third persons eliciting a chain of noncooperation and negative emotions

(Gray et al. 2014). A person, who has increased negative emotions due to noncooperation of another person, tends to be noncooperative (punish) to an uninvolved third person, thereby provoking negative emotions in the third person who in turn might forward this noncooperative behavior as well and so forth. However, emotion regulation strategies, such as writing a letter, were shown to reduce the third-party punishment leading to a successful interruption of this chain (Strang et al. 2016).

In contrast to noncooperation, cooperation has been shown to elicit positive emotions. However, cooperation in economic games is usually accompanied by higher monetary payoffs for the receiver. This makes it difficult to disentangle whether people feel good due to the high monetary payoffs they receive or due to the cooperative behavior. Tabibnia et al. (2008) controlled for the potential confound of higher monetary payoffs by varying the proposer's endowment across trials. In this way, the same offer could be a fair/cooperative offer with a small endowment (5€ out of 10€) and an unfair/less cooperative offer with a larger endowment (5€ out of 20€). A difference in emotions between these two types of trials could thus only be attributed to cooperation and not to monetary payoffs. In addition to their decision whether to accept or reject the offers, participants were asked to rate how happy they felt in response to each offer. Their results indicate that happiness correlates positively with cooperation, meaning it is not only the money, but also the cooperation that drives the positive emotion.

Interestingly, not only experiencing cooperative behavior of others but also being the one who cooperates elicits positive emotions. Happiness ratings increase when people cooperate. Donating money to charity as well as cooperation in the ultimatum game, dictator game, trust game, and prisoner's dilemma increases positive emotions or physiological markers of positive emotions in most participants (Dunn et al. 2008; King-Casas et al. 2005; Park et al. 2016; Rilling et al. 2002).

5 Neuroimaging—A Useful Tool to Investigate the Neural Basis of Cooperation

The recent developments of neuroimaging (such as fMRI, EEG and MEG) and brain stimulation techniques opened the possibility to investigate and to alter brain activity involved in cooperative behavior in the human brain. In combination with the economic games described in the previous sections, this allows for a more precise investigation of the underlying neural mechanisms of cooperative behavior. The combination of neuroscientific methods and economic theories is called neuroeconomics. A series of neuroeconomic studies have investigated the underlying brain processes involved in cooperation, using the ultimatum and the dictator game, as well as the trust game and the prisoner's dilemma.

Sanfey and colleagues were the first ones who combined neuroimaging with economic games in order to investigate the neural mechanisms underlying decision making in the ultimatum game (Sanfey et al. 2003). Here, the participants performed the ultimatum game as receivers, receiving offers either from a computer or from a human partner. Half of the offers participants received were fair (50 %), and the other half were unfair (less than 50 %, distributions were 70 %/30 %; 80 %/20 % and 90 %/10 %). When the participants received unfair offers compared to fair offers, several brain regions including the anterior insula (AI), dorsolateral prefrontal cortex (DLPFC), and anterior cingulate cortex (ACC) showed increased activation. Specifically, these areas showed higher activation when the participant played against a human compared to a computer proposer, suggesting that the modulation of brain activation was specific for human cooperation.

As mentioned above, behavioral studies have shown that noncooperative behavior elicits negative emotions, whereas cooperation elicits positive emotions. One brain region frequently associated with negative emotions is the insula; it often shows increased activation during physical and social pain and decreased activation after successful regulation of negative emotions (Goldin et al. 2008; Hein et al. 2010). In the study by Sanfey et al. (2003), noncooperative behavior enhanced the insula activity. Here, the activity in the AI coded the extent of cooperation, showing greater activity for lower offers. Activity in the AI also correlated with rejection rates. A finding that has been replicated in subsequent studies using different versions of the ultimatum game (Hsu et al. 2008; Tabibnia et al. 2008). The authors concluded that the insula might code negative emotions due to noncooperative/unfair behavior (Sanfey et al. 2003; Tabibnia et al. 2008). Their interpretation is in line with the behavioral findings indicating that noncooperation elicits negative emotions. In a recent attempt to regulate activity in the insula by using emotion regulation, Grecucci and colleagues showed that while activity in the AI was only affected by upregulation of negative emotions, the posterior part of the insula was influenced by both up- and downregulation (Grecucci et al. 2013). Activity in the posterior insula was higher when negative emotions were upregulated and lower when negative emotions were downregulated, resembling their behavioral finding; higher acceptance rates of unfair offers when emotions were downregulated and higher rejection rates when emotions were upregulated. Furthermore, activation in the insula was associated with inequity aversion. Using a different paradigm, Hsu et al. (2008) estimated individual inequity parameters based on an inequity aversion model, which correlated with insula activity. Supporting the inequity aversion theory by Fehr and Schmidt (1999), this result suggests that inequity aversion might cause negative emotions due to noncooperation.

In line with the behavioral studies showing that cooperative behavior induces good feelings or positive emotions, cooperative behavior has been shown to activate brain areas that have been previously associated with reward processing (Decety et al. 2004; King-Casas et al. 2005; Park et al. 2016; Rilling et al. 2002; Tabibnia et al. 2008). For example, Tabibnia and colleagues demonstrated (2008) that cooperation in the ultimatum game was accompanied by increased activity in the striatum, amygdala, ventromedial prefrontal, and orbitofrontal cortex (OFC), regions that have been shown to play a key role in reward processing. The OFC and ventromedial prefrontal cortex (VMPFC) were repeatedly shown to be engaged during cooperative behavior in a variety of paradigms (Hare et al. 2010; Sul et al.

2015; Zaki and Mitchell 2011). Voluntary donations to charity organizations (Hare et al. 2010), generous behavior toward another anonymous player (Sul et al. 2015), and equitable interpersonal decisions (Zaki and Mitchell 2011) were all associated with increased activity in these prefrontal areas. Furthermore, when cooperative behavior was directly compared against competitive behavior, an increased activity in the OFC could be observed (Decety et al. 2004).

In an iterated version of the trust game, cooperative behavior of the investor elicits enhanced activity in the striatum of the trustee (King-Casas et al. 2005). Here, the striatal activity level reflected the amount back transferred to the investor. Moreover in the context of mutual cooperation (i.e., the prisoner's dilemma game), cooperation of both players triggered activity in the striatum and other reward-related brain areas (Rilling et al. 2002). Also, the striatal activity diminished when participants experienced the counterplayers' defect, despite of their own cooperative behavior. On the other hand, when the participants were defecting as well, the counterplayers' defect did not induce any change in striatal activity (Rilling et al. 2002). This suggests that not only the behavior of the other player influences our emotional response but also our own behavior and our expectations about the other player's behavior.

Moll and colleagues compared brain activity elicited by making charitable donations with activation elicited by obtaining monetary rewards and could show that the striatum was engaged by donations in the same way as when receiving money (Moll et al. 2006). The authors concluded that monetary donations are rewarding similar to monetary rewards. According to the "warm glow," hypothesis people derive a reward ("a good feeling") only from actively giving money. In contrast, mandatory donations such as taxes are not thought to be rewarding. This hypothesis was tested by Harbaugh et al. (2007), who demonstrated that the striatum shows a greater activation by voluntary donations in comparison with mandatory tax like donations. A recent study investigated the neural mechanism of the causal relationship between cooperative behavior and happiness (Park et al. 2016). This study showed that a commitment to behave generously induces generous behavior and enhances subjective happiness. Importantly, this increase in happiness was reflected in the striatal activity. Thus, studies investigating neural mechanism of cooperation show that cooperation is frequently associated with striatal and other reward-related brain activity, thereby supporting the "warm glow" hypothesis.

Another brain area playing an important role in cooperative behavior is the temporal parietal junction (TPJ; Hutcherson et al. 2015; Morishima et al. 2012; Park et al. 2016; Strombach et al. 2015). The TPJ has been mainly associated with theory of mind, mentalizing and empathy processes (Saxe and Kanwisher 2003; Schurz et al. 2014). These processes are important for cooperative behavior, since in order to cooperate with another person, a representation about the others person's preferences and beliefs is essential. Recent studies have shown that the TPJ plays a distinct role in prosocial cooperative behavior. Functional as well as structural properties of the TPJ have been associated with cooperative behavior (Morishima et al. 2012; Strombach et al. 2015). Strombach and colleagues (2015) suggested that the TPJ modulates activity in reward-related areas, enabling generous behavior.

They showed that the TPJ is more active during generous/cooperative decisions compared to selfish/noncooperative decisions and that connectivity between TPJ and the ventromedial prefrontal cortex is stronger during cooperative compared to noncooperative decisions. According to their interpretation, TPJ overrides selfish impulses during social decisions. Park and colleagues investigating the impact of commitment on generous behavior (Park et al. 2016). In their data, participants who made a commitment to behave more generously showed greater TPJ activation while making cooperative decisions. Importantly, depending on whether they made a generous commitment or not, the TPJ showed different generosity dependent connectivity modulation with OFC and striatum. Specifically, the striatal region that is modulated by TPJ also tracked changes in happiness. The authors therefore suggest that the striatum links happiness and generosity on a neural level.

In line with behavioral results, experiencing the cooperation of others as well as behaving cooperatively elicits activity in brain areas associated with positive emotions, whereas noncooperation elicits activity associated with negative emotions. Most authors conclude from these results that cooperation induces positive emotions, whereas noncooperation induces negative emotions (Decety et al. 2004; King-Casas et al. 2005; Rilling et al. 2002; Sanfey et al. 2003; Tabibnia et al. 2008). Recent studies suggest that the TPJ is involved in cooperative behavior by overriding selfish impulses, while the striatum is associated with an increase in positive emotions due to cooperative behavior (Harbaugh et al. 2007; Hare et al. 2010; Moll et al. 2006; Morishima et al. 2012; Strombach et al. 2015; Sul et al. 2015). Since striatal activity was also shown to be modulated by TPJ, this area is suggested to be the neural correlate linking positive emotions with cooperative behavior (Park et al. 2016). Although these findings improved our understanding of cooperative behavior substantially, there are still a variety of open questions that need to be investigated: How does the interplay between the different brain regions exactly work? How are individual differences in cooperative behavior reflected in the brain? As the described studies show, fMRI is a great tool to measure brain activity correlated with own or others cooperative behavior. However, for some brain areas that show changes in brain activation due to cooperation/noncooperation, the causal involvement remains unclear.

6 Manipulating Cooperative Behavior by Brain Stimulation

Brain stimulation techniques offer the possibility to investigate the causal involvement of brain regions in cooperation. As Sanfey et al. (2003) showed, noncooperative behavior in the ultimatum game is accompanied by activity in the DLPFC. However, the exact role of the DLPFC remained unknown. In order to get a better understanding of the DLPFC role in the decision process as a receiver in the ultimatum game, a transcranial magnetic stimulation (TMS) experiment was conducted, testing two opposing hypotheses (Knoch et al. 2006). One possible hypothesis is that the inhibition of the DLPFC disrupts the control of selfish impulses. Here, the DLPFC disruption would lead to less punishment/lower rejection rates, since rejections are considered to require the control of the selfish impulse, namely maximization of own payoff. Alternatively, the DLPFC might be involved in controlling fairness impulses. According to this, DLPFC inhibition would lead to increases in the punishment of noncooperation/rejection rates. The results supported the first hypothesis; TMS stimulation over the right DLPFC decreased rejection rates in the ultimatum game, implicating that participants attached greater importance on maximizing their payoffs and therefore accepted small unfair offers. Interestingly, participants' reported judgments on fairness were not altered by TMS. Even if the participants were aware of the unfairness of the offers, they still accepted them under the stimulation condition (Knoch et al. 2006).

To understand the specific neural mechanism of DLPFC inhibition in noncooperative behavior, Baumgartner et al. (2011) combined the two methods TMS and fMRI. Here, the authors first stimulated the DLPFC and measured the brain activity using fMRI while participants performed the ultimatum game as receivers. Their results showed that (1) disrupting the right DLPFC led to the absence of enhanced activity by unfair offers and that (2) TMS of the right DLPFC decreased activity in the posterior division of ventromedial prefrontal cortex (pVMPFC). This effect was specific to noncooperative behavior. Further analysis revealed that the connectivity strength between right DLPFC and pVMPFC was modulated by the degree of prosocial punishment. Since pVMPFC has been suggested to be associated with decision values computation (Smith et al. 2010), Baumgartner et al. (2011) suggested that, in the context of the ultimatum game, the pVMPFC might encode the decision value of rejecting an offer. Furthermore, this computation might be regulated through communication with the right DLPFC in case of noncooperative behavior. Similar to this idea, Sul et al. (2015) found that different subparts of the MPFC code for self-regarding and other-regarding values.

Strang et al. (2015) investigated the role of the DLPFC on money allocations as a dictator in the dictator game. Cooperative behavior in the dictator game is also thought to require the control of selfish impulses, a function that is attributed to DLPFC. When inhibiting the right DLPFC, dictator allocations were significantly lower. In conclusion, the right DLPFC seems to be necessary to foster cooperative behavior by allocating fairly (as shown in the dictator game) and by punishing noncooperative behavior (as shown in the ultimatum game). Thus, this area seems to be essential for the establishment of cooperation as well as for the maintenance of cooperative behavior.

7 Conclusion

The motives underlying human cooperation are and continue to be mysterious. However, the recent progress in research on cooperative behavior has helped to solve at least part of the mystery. By combining forces from different disciplines, it could be shown that cooperative behavior increases positive emotions. A network of brain areas, including TPJ, prefrontal areas (OFC, DLPFC and VMFC), and the striatum, has frequently been associated with cooperative behavior. While TPJ activation has been linked to cooperative behavior, the striatum is suggested to be involved in positive emotions. Positive emotions are therefore presumably one major factor that drives cooperative behavior. This phenomenon is called "warm glow." Additional factors underlying cooperative behavior are reciprocity, kin selection, reputations concerns, inequity aversion, and fear of social punishment. Furthermore, the degree of cooperative behavior was shown to be influenced by altering DLPFC activity, suggesting that this region might be involved in overcoming selfish impulses in order to cooperate. Reviewed studies here provide an insight into cooperative human behavior. However, it will need far more studies to disentangle the remaining fragments. Furthermore, although exploring the same question, behavioral, neuroimaging, and brain stimulation studies all have unveiled a different fraction of cooperative behavior. In the end, the mosaic of the combined knowledge of these studies from different disciplines is essential to get the big picture of cooperation and to understand cooperative behavior in detail.

References

Andreoni J (1988) Why free ride? Strategies and learning in public goods experiments. J Public Econ 37:291–304

Andreoni J (1990) Impure altruism and donations to public good: a theory of warm-glow giving. Econ J 100(401):464–477

Axelrod R, Hamilton WD (1981) The evolution of cooperation. Nature 292(5821):291–292

Baumgartner T, Knoch D, Hotz P, Eisenegger C, Fehr E (2011) Dorsolateral and ventromedial prefrontal cortex orchestrate normative choice. Nat Neurosci 14(11):1468–1474

Burnstein E, Crandall C, Kitayama S (1994) Some neo-Darwinian decision rules for altruism: weighing cues for inclusive fitness as a function of the biological importance of the decision. J Pers Soc Psychol 67(5):773–789

Camerer C (2003a) Behavioural studies of strategic thinking in games. Trends Cogn Sci 7(5):225–231

Camerer CF (2003b) Psychology and economics. Strategizing in the brain. Science 300 (5626):1673–1675

Camerer C, Thaler R (1995) Anomalies ultimatums, dictators and manners. J Econ Perspect 9 (2):209–219

Carpenter J (2007) Punishing free-riders: how group size affects mutual monitoring and the provision of public goods. Games Econ Behav 60(1):31–51

Decety J, Jackson PL, Sommerville JA, Chaminade T, Meltzoff AN (2004) The neural bases of cooperation and competition: an fMRI investigation. NeuroImage 23(2):744–751

Dunn EW, Aknin LB, Norton MI (2008) Spending money on others promotes happiness. Science 319(5870):1687–1688

Egas M, Riedl A (2008) The economics of altruistic punishment and the maintenance of cooperation. Proc Biol Sci 275(1637):871–878

Engel C (2011) Dictator games: a meta study. Exp Econ 14(4):583–610

Falk A, Fischbacher U (2006) A theory of reciprocity. Games Econ Behav 54(2):293–315

Fehr E, Fischbacher U (2004) Third-party punishment and social norms. Evol Human Behav 25 (2):63–87

Fehr E, Gächter S (2000) Cooperation and punishment in public goods experiments. Am Econ Assoc 90(4):980–994

Fehr E, Gächter S (2002) Altruistic punishment in humans. Nature 415(6868):137–140

Fehr E, Schmidt K (1999) A theory of fairness, competition and cooperation. Q J Econ 114 (3):817–868

Gächter S, Fehr E (1999) Collective action as a social exchange. J Econ Behav Organ 39(4):341–369

Gächter S, Herrmann B (2009) Reciprocity, culture and human cooperation: previous insights and a new cross-cultural experiment. Philos Trans R Soc 364(1518):791–806

Goldin P, McRae K, Ramel W, Gross J (2008) Suppression of negative emotion. Biol Psychiatry 63(6):577–586

Gray K, Ward AF, Norton MI (2014) Paying it forward: generalized reciprocity and the limits of generosity. J Exp Psychol General 143(1):247–254

Grecucci A, Giorgetta C, Van't Wout M, Bonini N, Sanfey A (2013) Reappraising the ultimatum: an fMRI study of emotion regulation and decision making. Cereb Cortex 23(2):399–410

Gurven M (2004) To give and to give not: the behavioral ecology of human food transfers. Behav Brain Sci 27(04):543–583

Guth W, Schmittberger R, Schwarze B (1982) An experimental analysis of ultimatum bargaining. J Econ Behav Organ 3:367–388

Hamilton WD (1964) The genetical evolution of social behaviour. II. J Theor Biol 7(1):17–52

Harbaugh WT, Mayr U, Burghart DR (2007) Neural responses to taxation and voluntary giving reveal motives for charitable donations. Science 316(5831):1622–1625

Hare TA, Camerer CF, Knoepfle DT, O'Doherty JP, Rangel A (2010) Value computations in ventral medial prefrontal cortex during charitable decision making incorporate input from regions involved in social cognition. J Neurosci 30(2):583–590

Hein G, Silani G, Preuschoff K, Batson C, Singer T (2010) Neural responses to ingroup and outgroup members' suffering predict individual differences in costly helping. Neuron 68 (1):149–160

Henrich J, McElreath R, Barr A, Ensminger J, Barrett C, Bolyanatz A, Ziker J (2006) Costly punishment across human societies. Science 312(5781):1767–1770

Hsu M, Anen C, Quartz S (2008) The right and the good: distributive justice and neural encoding of equity and efficiency. Science 320(5879):1092–1095

Hu Y, Strang S, Weber B (2015) Helping or punishing strangers: neural correlates of altruistic decisions as third-party and of its relation to empathic concern. Front Behav Neurosci 9:1–11

Hutcherson CA, Bushong B, Rangel A (2015) A neurocomputational model of altruistic choice and its implications. Neuron 87(2):451–462

Johnson ND, Mislin AA (2011) Trust games: a meta-analysis. J Econ Psychol 32(5):865–889

King-Casas B, Tomlin D, Anen C, Camerer CF, Quartz SR, Montague PR (2005) Getting to know you: reputation and trust in a two-person economic exchange. Science 308(5718):78–83

Knoch D, Pascual-Leone A, Meyer K, Treyer V, Fehr E (2006) Diminishing reciprocal fairness by disrupting the right prefrontal cortex. Science 314(5800):829–832

McCabe KA, Rassenti SJ, Smith VL (1998) Reciprocity, trust, and payoff privacy in extensive form bargaining. Games Econ Behav 24(1–2):10–24

Moll J, Krueger F, Zahn R, Pardini M, de Oliveira-Souza R, Grafman J (2006) Human fronto-mesolimbic networks guide decisions about charitable donation. Proc Natl Acad Sci 103 (42):15623–15628

Morishima Y, Schunk D, Bruhin A, Ruff CC, Fehr E (2012) Linking brain structure and activation in temporoparietal junction to explain the neurobiology of human altruism. Neuron 75(1):73–79
Neyman A (1985) Bounded complexity justifies cooperation in the finitely repeated prosoner's dilemma. Econ Lett 19:227–229
Nowak M (2006) Five rules for the evolution of cooperation. Science 314(5805):1560–1563
Nowak M, Sigmund K (1993) A strategy of win-stay, lose-shift that outperforms titfortat in the prisoners dilemma game. Nature 364(1):56–58
Nowak M, Sigmund K (1998) Evolution of indirect reciprocity by image scoring. Nature 393 (6685):573–577
Park QS, Kahnt T, Strang S, Seid-Fatemi A, Fehr E, Tobler P (2016) The neural link of generosity and happiness. under review
Rilling J, Gutman D, Zeh T, Pagnoni G, Berns G, Kilts C (2002) A neural basis for social cooperation. Neuron 35(2):395–405
Sally D (1995) Conversation and cooperation in social dilemmas. Rationality Soc 7(1):58–92
Sanfey A, Rilling J, Aronson J, Nystrom L, Cohen J (2003) The neural basis of economic decision-making in the ultimatum game. Science 300(5626):1755–1758
Saxe R, Kanwisher N (2003) People thinking about thinking people—the role of the temporo-parietal junction in "theory of mind". NeuroImage 19(4):1835–1842
Schurz M, Radua J, Aichhorn M, Richlan F, Perner J (2014) Fractionating theory of mind: a meta-analysis of functional brain imaging studies. Neurosci Biobehav Rev 42:9–34
Smith D, Hayden B, Truong T, Song A, Platt M, Huettel S (2010) Distinct value signals in anterior and posterior ventromedial prefrontal cortex. J Neurosci 30(7):2490–2495
Spitzer M, Fischbacher U, Herrnberger B, Grön G, Fehr E (2007) The neural signature of social norm compliance. Neuron 56(1):185–196
Strang S, Gross J, Schuhmann T, Riedl A, Weber B, Sack A (2015) Be nice if you have to—the neurobiological roots of strategic fairness. Soc Cogn Affect Neurosci 10:790–796
Strang S, Grote X, Kuss K, Park SQ, Weber B (2016) Generalized negative reciprocity in the dictator game—how to interrupt the chain of unfairness. Sci Rep 6:1–5
Strombach T, Weber B, Hangebrauk Z, Kenning P, Karipidis II, Tobler PN, Kalenscher T (2015) Social discounting involves modulation of neural value signals by temporoparietal junction. Proc Natl Acad Sci 112(5):1619–1624
Sul S, Tobler PN, Hein G, Leiberg S, Jung D, Fehr E, Kim H (2015) Spatial gradient in value representation along the medial prefrontal cortex reflects individual differences in prosociality. Proc Natl Acad Sci 112(25):7851–7856
Tabibnia G, Satpute A, Lieberman M (2008) The sunny side of fairness: preference for fairness activates reward circuitry (and disregarding unfairness activates self-control circuitry). Psychol Sci 19(4):339–347
Van't Wout M, Kahn R, Sanfey A, Aleman A (2006) Affective state and decision-making in the ultimatum game. Exp Brain Res 169(4):564–568
Zaki J, Mitchell JP (2011) Equitable decision making is associated with neural markers of intrinsic value. Proc Natl Acad Sci 108(49):19761–19766

The Social Neuroscience of Interpersonal Emotions

Laura Müller-Pinzler, Sören Krach, Ulrike M. Krämer and Frieder M. Paulus

Abstract In our daily lives, we constantly engage in reciprocal interactions with other individuals and represent ourselves in the context of our surrounding social world. Within social interactions, humans often experience interpersonal emotions such as embarrassment, shame, guilt, or pride. How interpersonal emotions are processed on the neural systems level is of major interest for social neuroscience research. While the configuration of laboratory settings in general is constraining for emotion research, recent neuroimaging investigations came up with new approaches to implement socially interactive and immersive scenarios for the real-life investigation of interpersonal emotions. These studies could show that among other brain regions the so-called mentalizing network, which is typically involved when we represent and make sense of others' states of mind, is associated with interpersonal emotions. The anterior insula/anterior cingulate cortex network at the same time processes one's own bodily arousal during such interpersonal emotional experiences. Current research aimed to explore how we make sense of others' emotional states during social interactions and investigates the modulating factors of our emotional experiences during social interactions. Understanding how interpersonal emotions are processed on the neural systems level may yield significant implications for neuropsychiatric disorders that affect social behavior such as social anxiety disorders or autism.

L. Müller-Pinzler (✉) · S. Krach · F.M. Paulus
Department of Psychiatry and Psychotherapy, Social Neuroscience Lab, University of Lübeck, Ratzeburger Allee 160, 23538 Lübeck, Germany
e-mail: mueller-pinzler@snl.uni-luebeck.de

S. Krach
e-mail: krach@snl.uni-luebeck.de

F.M. Paulus
e-mail: paulus@snl.uni-luebeck.de

U.M. Krämer
Department of Neurology, University of Lübeck, Ratzeburger Allee 160, 23538 Lübeck, Germany
e-mail: ulrike.kraemer@neuro.uni-luebeck.de

Curr Topics Behav Neurosci (2016) 30: 241–256
DOI 10.1007/7854_2016_437
Published Online: 03 March 2016

Keywords Interpersonal emotions · Embarrassment · Pride · Vicarious emotions · Social interaction · Mentalizing

Contents

1 Social Exclusion 243
2 Embarrassment and Pride 245
3 Guilt 247
4 Inducing Interpersonal Emotions in the Neurosciences 249
5 Vicarious Interpersonal Emotions 250
6 Clinical Applications 252
References 253

Emotions play a central role in regulating and motivating a person's thoughts, feelings, and behavior in almost every aspect of one's life. Examples reach from basic emotions such as fear, which can help us to avoid potentially harmful situations, to rather complex interpersonal emotions such as guilt, motivating restitution behavior for one's mistakes. Interpersonal emotions such as guilt, shame, embarrassment, and pride require the capacity for introspection and self-knowledge and therefore have been considered as unique human emotions. Interpersonal emotions are evoked when we implicitly or explicitly reflect on ourselves and evaluate ourselves in the context of our surrounding social world. This is why they are often also called as "self-conscious emotions" or "social emotions." We, for example, feel ashamed about our own negative attributes or embarrassed by our accidental pratfalls and mishaps. Interpersonal emotions therefore are a motivational force for social behavior providing immediate punishment or reinforcement of real or expected behavioral outcomes and thereby help to retain social structures (Tangney et al. 2007).

The impact of these emotions on interpersonal behavior is well-documented: For example, pride not only increases perseverance on tasks, but also motivates individuals to take responsibility in a group and leads to being perceived as more likeable interaction partners (Williams and DeSteno 2009). Empirical data could show that embarrassment, too, functions as a regulating mechanism for interpersonal behavior. On the one hand, its anticipation and the idea of being embarrassed in front of others drive people to behave in compliance with current social norms. On the other hand, it motivates people to engage in reparative actions to restore their social image once a deviation from a normative standard has been recognized and has triggered embarrassment (Keltner and Buswell 1997). The latter is supported by specific bodily and facial expressions—so-called appeasement gestures—that are clearly linked to embarrassment such as controlled smiles, lowering of the head, and orienting the gaze downwards. Individuals who express their embarrassment through these appeasement gestures in response to an unwanted mishap are later on judged as more likeable and social (Feinberg et al. 2012; Semin and Manstead 1982). Similarly, guilt, which is elicited by a negative evaluation of one's behavior and a sense of responsibility for an unwanted outcome, motivates approach behavior and the attempt to fix the situation by engaging in reparative actions. If the negative appraisal does not concern a specific behavior but one's personality and value as human being as a whole, individuals rather experience shame, resulting in very different behavioral adjustments. In contrast to the experience of guilt, shame triggers distancing behavior with the consequence of social withdrawal (Tangney et al. 1996).

When we experience interpersonal emotions, we need to represent ourselves in the context of our surrounding social world and in relation to others. Therefore, interpersonal emotions are primarily evoked during reciprocal interactions between two or more individuals (Tangney et al. 2007). Despite the constraints of laboratory settings, contemporary emotion research increasingly acknowledges this aspect in the investigation of interpersonal emotions. Instead of making use of hypothetic or script-based scenarios, recent studies rather implemented staged interactions using cover stories and confederates that allow individuals to interact with each other in a controlled environment (e.g., Gilam et al. 2015; Müller-Pinzler et al. 2015a; Williams and DeSteno 2008). The implementation of such interactive paradigms in neuroscience laboratories with EEG or MRI is particularly challenging due to increased spatial restrictions and the specific constraints in the experimental design. Still, in social neuroscience, interactive paradigms and similar concepts have recently gained influence, specifically in context of the "second-person neuroscience" (Krach et al. 2013; Schilbach et al. 2013). In this line, one approach to examine the neural foundations of interpersonal emotions is to immerse participants into a social situation and uphold this state of social immersion for the time of fMRI scanning. In doing so, interaction partners are perceived as active and significant social agents within the situation. Immersive paradigms are thus assumed to increase the participants' emotional engagement and the likelihood to unravel the unique neural pathways of interpersonal emotions. Current neuroimaging investigations increasingly take the social interactive aspect into account and make use of immersive environments as well as more realistic and social paradigms (Gilam et al. 2015; Müller-Pinzler et al. 2015a). In the following, we will introduce social neuroscience approaches to the study of interpersonal emotions and exemplify new avenues in the field.

1 Social Exclusion

We will start this chapter with the introduction of an experimental paradigm that capitalizes on social interaction in order to induce the negative and unpleasant affective experience of being rejected or socially excluded. Humans, similar to many other mammals and living beings, cannot survive without the support of other related individuals. Instead, they depend on their peers in many aspects of their lives. As genuinely social beings, humans have a strong motivation to engage in social relationships and feel social acceptance and appreciation from others. While social connectedness and prosocial behavior have strong reward potential (Krach et al. 2010; see also the chapter of Rademacher et al.), social rejection on the other hand increases distress and negative affect. In many languages, the experience of social exclusion is thereby described with concepts and words closely related to the experience of physical pain (e.g., a "hurt feeling") (Eisenberger et al. 2003). Accordingly, it was suggested that threats of losing one's social connectedness are comparable to the threat of having one's bodily integrity endangered, not only metaphorically speaking but also on the level of brain networks. Social exclusion was therefore often referred to as a form of social pain (Eisenberger et al. 2003; Macdonald and Leary 2005).

Most studies on social exclusion made use of the so-called "cyberball" paradigm (Williams et al. 2000) which is, though being interactive, easily applicable for fMRI and other neuroscience investigations. Cyberball is a virtual ball-tossing game that is used to manipulate the degree of social inclusion or exclusion of a participant. During the game, the participant tosses the ball with two other players, who are either introduced as real-life interaction partners or are simply virtually present like in a multi-player computer game. While the participant thinks that he or she actually throws and catches the ball to or from agents that are remotely controlled from the other players, the frequency of how often the participants receive the ball is actually controlled by an algorithm. In the inclusion condition, participants receive the ball as frequent as other players. However, if the person is excluded, the participant rarely gets the ball and ultimately the fellow players only toss the ball back and forth, thereby excluding the participant from the game and the reciprocal interaction (Hartgerink et al. 2015; Williams et al. 2000). Although this ball-tossing game is simple and easy to implement even in spatially constrained neuroscience laboratories, it gives an idea of the concept of social immersion and how it can be used to induce affective experiences that fundamentally depend on being embedded in a social environment and in direct interaction with other individuals.

Accumulating evidence and recent meta-analytic studies demonstrate that social exclusion, as induced with the cyberball paradigm, is mapped within the same neural circuits processing the affective component of bodily pain, namely the dorsal anterior cingulate cortex (dACC) and anterior insula (AI) (Eisenberger 2012; Cacioppo et al. 2013; Rotge et al. 2014). In addition, more intense cases of acute social rejection, for example when reliving a recent breakup of a romantic relationship, also involve secondary somatosensory cortex areas and the posterior insula (Kross et al. 2011). It was argued that being separated from the social group endangers, our well-being and the "painful" experience of social rejection motivates us to avoid behaviors that might lead to rejection (Eisenberger 2012; Macdonald and Leary 2005). However, there are also investigations that challenge this view on social rejection as being strongly related to physical pain (see, e.g., Iannetti et al. 2013). The overlap between brain regions associated with physical and social pain could also be explained by the fact that both involve processes that are necessary for detecting and paying attention to salient stimuli. Both experiences therefore comprise a component of arousal that is assumed to be mapped within the AI and ACC and these neural activations might be unspecific. In fact, many other affective experiences trigger activity in these brain regions (Lindquist et al. 2012). Another argument states that differential spatial patterns of neural activations for social and physical pain might be found on a more fine-grained level of analysis (Iannetti et al. 2013). The underlying neural activation patterns might therefore be more complex than previously assumed. Further neuroimaging evidence is needed to conclude to which extent social rejection and physical pain are comparable (Iannetti et al. 2013; Rotge et al. 2014) if that is even ever possible by using data obtained by fMRI activation measures due to its spatial resolution.

Irrespective of the conceptual vagueness of labeling the experiences during social exclusion in affective terms, it is obvious that this affective experience can only be examined in the context of social interactions. It is fundamentally necessary to make the individuals believe to interact with other individuals in the neuroscience environment to induce the desired affective reaction. Distinguishing between states of social inclusion and exclusion is impossible in social isolation.

2 Embarrassment and Pride

Embarrassment is an unpleasant feeling that arises when one behaves in a clumsy and unflattering way and fails to uphold one's public image in a social situation (Miller 1996). It functions as a regulating mechanism of interpersonal behavior and is also referred to, as other self-conscious emotions as well, an "emotional moral barometer" that informs us on how we perform according to prevalent norms and moral values (Tangney et al. 2007). Notably, embarrassment is thought to be a genuinely public emotion which requires the presence of an actual or imagined audience that witnesses the norm transgression. An important process that needs to be considered when understanding the neural foundations of embarrassment is that the publicity of the observed behavior motivates individuals to think about others' evaluations of the situation (Miller 1996). We experience embarrassment not only when our actions accidentally deviate from personal standards or we fail to show desirable behavior such as during physical pratfalls or cognitive shortcomings but specifically if we start thinking about how this mishap affects the impression others might have of us.

These two components of the deviation of one's behavior from personal standards and publicity of the behavior during embarrassment map well on the neural systems level. The rather cognitive aspects of embarrassment, comprising thoughts about the expected negative evaluation in-the-eyes-of-others during public failures, essentially require mental-state attributions (Tangney et al. 2007). Studies examining the neural underpinnings of embarrassment could accordingly show that areas of the so-called mentalizing network (Frith and Frith 2003) such as the medial prefrontal cortex (mPFC) and precuneus were activated during the experience of embarrassment (Finger et al. 2006; Müller-Pinzler et al. 2015a; Takahashi et al. 2004). The important role of mentalizing regions for embarrassment is also supported by evidence from clinical and lesion studies: patients with orbitofrontal cortex (OFC) lesions or frontal temporal lobe degeneration show diminished embarrassment expressions, reduced embarrassment recognition, and more often violate social norms while showing less efforts to restore their social image (Beer et al. 2003; Jankowski and Takahashi 2014; Sturm et al. 2006).

Most of the earlier neuroscience studies on embarrassment asked participants to make moral judgments while reading scenarios or asked participants to envision being in an embarrassing situation (Finger et al. 2006; Takahashi et al. 2004). As these paradigms examine participants in social isolation and fundamentally lack any meaningful interaction with other individuals, they were not able to evoke physiological arousal on the neural systems level. More recent investigations have tried to overcome this limitation and implemented more direct and naturalistic social interactions in the MRI. Recently, Müller-Pinzler et al. (2015a) made participants fail in front of a judging audience. In their socially immersive paradigm, participants were engaged with three confederates and asked to estimate properties of objects, i.e., sizes, amounts, or weights. During the scanning session, the confederates stayed outside and constituted the witnessing audience of potential cognitive shortcomings in form of failed property estimations. For each object to be estimated, the participant inside the MRI received manipulated feedback on the accuracy of his or her estimation. An inaccurate and highly deviating estimation induced failure in the sense of a cognitive shortcoming. Importantly, participants were informed whether their performance was also broadcasted to the three confederates outside the scanner or just privately visible to themselves. By doing so, the publicity of the participant's behavior, which is an essential component of the embarrassment experience, was directly manipulated. When failing to show behavior in correspondence with one's own expectations and in compliance with the prevalent social norms, activations of the dorsal AI and ACC were increased, reflecting the affective physiological arousal during embarrassment (Müller-Pinzler et al. 2015a). However, these activations were present not only during failure but also when participants exceeded their expectations and received the feedback of very good estimation performance. Instead of reporting emotion-specific activation of distinct brain regions, this study links embarrassment to increased information transfer between the mentalizing and the arousal processing networks and the ventral aspects of the AI and the amygdala, areas previously associated with emotion processing (Adolphs et al. 1995; Kelly et al. 2012). This increase in functional connectivity was specifically observed during embarrassment and not if participants performed well. This unique pattern of concerted neural activation across these networks might thus represent a neural substrate involved in generating the experience of embarrassment (Müller-Pinzler et al. 2015a, also see Fig. 1).

In a condensed and rather simplistic model, pride could be considered as the positive counterpart of embarrassment. The emotional experience of pride is basically elicited when humans achieve self-relevant goals, instead of experience mishaps and pratfalls. Accordingly, the mPFC and precuneus, areas of the mentalizing network, also show increased activations during pride when participants reflect about their behavior and evaluation in-the-eyes-of-others (Müller-Pinzler et al. 2015a; Takahashi et al. 2008; Zahn et al. 2009). Similar to embarrassment, the arousal component of pride also mapped to greater activation of the AI and ACC (Müller-Pinzler et al. 2015a). Compared to embarrassment, however, the functional

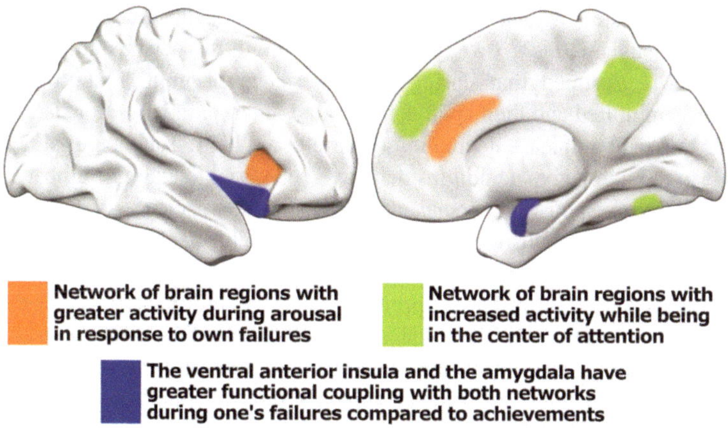

Fig. 1 Brain networks involved in the experience of embarrassment. Dorsal aspects of the anterior insula and the anterior cingulate cortex are associated with the processing of arousal in response to one's failures and show greater activation when participants receive negative feedback. The mPFC and the precuneus are strongly implicated in the process of perspective taking and thinking about the evaluations of others. Together with the fusiform face area in the fusiform gyrus, they are activated while being in the center of attention. Notably, both networks show greater functional coupling with the ventral anterior insula and the amygdala in the (para)limbic systems while performing poorly in a social evaluative context (Müller-Pinzler et al. 2015b). In accordance with a constructivist view on emotions, embarrassment, alike many other interpersonal emotions, might thus result from the orchestrated activity of functionally specialized and spatially distributed brain networks

integration of mentalizing and arousal processing networks with (para-)limbic regions was decreased. At the same time, participants' self-reported experience of pride did not differ in regard to the publicity of their positive feedback. This finding is in line with the notion that publicity of one's behavior and the thoughts about others' evaluations do not influence the experience of pride in the same way as they impact the experience of embarrassment (see Seidner et al. 1988): While the experience of embarrassment strongly depends on the presence of others who witness our failures and pratfalls, the experience of pride does not strictly depend on our image concerns and the positive evaluation of others. Regardless of the publicity of an achievement, pride is typically associated with the activation of striatal areas that are associated with reward processing (Müller-Pinzler et al. 2015a; Zahn et al. 2009). These activations potentially map the rewarding component of pride that might also function as an incentive to uphold socially respected behaviors despite their costs.

3 Guilt

Guilt is an aversive feeling that arises when we believe that we have behaved immorally toward others. The experience of guilt goes hand in hand with a sense of responsibility for the expected or actual negative outcome of one's behavior. It is typically associated with a feeling of regret regarding the transgression and an increased focus on one's own behavior and on those persons that have been harmed. It is assumed that during the experience of guilt, first, transgressive behavior is inhibited and then approach-oriented and reparative actions are initiated in order to fix the unpleasant situation (Fourie et al. 2014; Tangney et al. 2007). In contrast to the interactive paradigms that have been used to examine embarrassment or the affective responses to social exclusion, the neuroscience of guilt currently lacks studies that establish environments where subjects immerse into a social context and interact with other human beings. Most of the previous neuroscience studies on guilt used script-based approaches and mental imagery (see, e.g., Basile et al. 2011; Shin et al. 2000; Takahashi et al. 2004). Given the central role of mental-state attributions during the experience of guilt and the ongoing thoughts about one's own actions and another person's state of mind, most studies found evidence for guilt-related activations of mentalizing areas, such as mPFC, posterior cingulate cortex, and precuneus (Basile et al. 2011; Fourie et al. 2014). In one study, participants were asked to revive past episodes of guilt experiences. Compared to feelings of sadness and shame, participants had greater neural activation in the mPFC and the right OFC (Wagner et al. 2011). These findings support the link between prefrontal functioning and social behavior and might help to explain the behavioral symptoms of patients with OFC damage, who often show altered moral judgments and guilt-related regulation of behavior. Nonetheless, one should keep in mind that these studies examined guilt, a fundamentally interpersonal emotion, in social isolation. In conclusion, it remains to be determined if these studies were able to probe the neural networks as they would be engaged during the full blown guilt experience that would arise in the direct interaction with other human beings.

A recent study aimed to overcome this limitation and induced the experience of guilt in the MRI using faked feedback rather than the earlier script- or memory-based approaches. Fourie et al. (2014) made use of a modified version of the implicit association test (IAT) designed to measure implicit prejudice. In this task, participants were asked to classify positively or negatively valued words together with for example pictures of black- and white-skinned faces. Based on the differences in response times, the degree of implicit racial bias was estimated (for a more in-depth introduction on the IAT see Greenwald et al. 1998). According to the authors, the discrepancy between personal standards (i.e., not being racist) and the actual, however manipulated responses (i.e., revealing putatively implicit racist attitudes) should trigger feelings of guilt in the participants. Among other emotions, such as shame and embarrassment, Fourie et al. (2014) reported successful induction of guilt as the most prominent emotion. Accordingly, the authors found increased activations of the AI and ACC, possibly mapping the arousal component of the guilt experience (Basile et al. 2011; Fourie et al. 2014; Shin et al. 2000). Further, activations of the dorsal ACC, which is assumed to be involved in conflict monitoring, might specifically reflect processes of social response inhibition or reversal in guilt eliciting situations (Fourie et al. 2014). While this elegant study was one of the first to tackle genuine guilt experiences in the MRI, it is questionable if the observed activations are specific to guilt or also relate to the experience of other interpersonal emotions, in particular shame. The feedback of the IAT might have led participants to consider their personality instead of their behavior. Moreover, nobody was directly harmed by the participant's behavior. With respect to the classic conceptualization of guilt, the emotional experience elicited in the Fourie study in its phenomenology rather resembles the definition of the self-conscious emotion of shame (Tangney et al. 1996).

4 Inducing Interpersonal Emotions in the Neurosciences

The research discussed so far highlights a number of challenges associated with studying interpersonal emotions, particularly when using neuroscientific methods. As becomes evident when considering the work of Fourie et al. (2014), the distinction between shame, guilt, and embarrassment is a matter of debate. Researchers have attempted to differentiate shame and guilt based on the types of eliciting situations, based on the publicity or privateness of the situation, or based on the kind of appraisal. The currently predominant view presumes that the focus when feeling ashamed is on the core self, whereas the focus is on one's behavior when experiencing guilt (Tangney et al. 2007). If the failure is considered an accidental mishap on the other hand, it will elicit embarrassment more than shame or guilt. However, which emotion is elicited will depend on the individual's appraisal, which might even change from the first immediate reaction to later emotions. The experimental manipulation in the above-mentioned study of Fourier et al. (2014) may have concerned participants' self-concept (i.e., not being a racist) rather than the evaluation of their specific behavior and thereby elicited shame more than guilt. This illustrates the highly idiosyncratic nature of the emotion shame. Whether failing in a math task, for instance, makes me feel ashamed depends on whether I consider myself a math expert. As shame involves a threat to one's core self, any experimental induction of shame hence has to account for interindividual variability in participants' self-concept.

This is a challenge for neuroscientific studies in particular, as they require highly controlled stimuli which have to be identical across participants and can be repeated several times. This makes imaging studies on social emotions a lot more difficult than behavioral social psychology studies. One approach to face this challenge is exemplified by the already-mentioned study of Wagner et al. (2011). They presented short, but individualized cues being meaningful for the participant only and instructed subjects to relive the associated personal experience. While this method accounts for interindividual variability in shame- and guilt-related situations, it still only assesses "offline" experience of these emotions which will be weaker if not qualitatively different than the emotional experience in the situation. Moreover, such uncontrolled and individualized approaches would be problematic in studies with clinical groups as patients will differ systematically in the quality and severity of the imagined situations. For instance, body-related shame plays a bigger role in certain psychiatric diseases as Borderline Personality disorder or Eating disorder (Tangney et al. 2007), whereas the "typical" student participant might rather remember performance-related shame. Future studies with paradigms that account for the idiosyncratic nature of social emotions and at the same time immerse participants in an active and salient social environment are needed in order to better understand the properties of involved brain networks.

5 Vicarious Interpersonal Emotions

During social interactions, we attempt to make sense of others' sensations, emotions, or thoughts. In this process, we can rely on our ability to share others' feelings, being it embarrassment, social exclusion, or guilt. Mainly, two interacting processes have been described that allow us to empathize in social situations (Keysers and Gazzola 2007; Paulus et al. 2013a; Waytz and Mitchell 2011). First, "mirroring" or "sharing" is assumed to be a direct mapping of others' states as expressed through facial expressions, gestures, sounds, and body postures on one's own neural system resulting in a shared emotional representation between the other person being observed, the social target, and observers themselves. Second, via a mental projection of oneself in the other person's position, mentalizing processes help to reproduce the other's emotions as if they were one's own bodily states. Mentalizing processes thus have been considered to result in comparable internal representations in observers (Hein and Singer 2008; Keysers and Gazzola 2007). Both processes, mirroring and mentalizing, are therefore understood as ongoing simulation processes that enable perceivers to experience another person's mental state (Paulus et al. 2013a; Waytz and Mitchell 2011).

We can thus share others' interpersonal emotions such as embarrassment, shame, pride, and guilt. In this way, similar to the first-hand experience of embarrassment, we are also vicariously embarrassed when we observe others' public failures, pratfalls, or etiquette violations (Miller 1987). We experience vicarious embarrassment even if the observed person is not embarrassed himself or herself, for example, when we observe someone accidentally walking down the street with an open fly (Krach et al. 2011; Müller-Pinzler et al. 2012; Paulus et al. 2013a). While the social target might not be aware of the ongoing mishap, observers indeed are and simulate the threat to the social integrity on their own body. Investigations on vicarious embarrassment could thus show recruitment of the shared circuits of the AI and ACC, higher-order somatosensory cortex areas, as well as regions of the mentalizing network, namely the mPFC and the temporal pole, independently of the emotional state in the observed person (Paulus et al. 2014). The sharing of another's embarrassment, however, involves activation of the pSTS, a brain region that is involved in mirroring bodily and facial expressions as well as multimodal integration. Notably, the functional connectivity of the pSTS with the AI is also increased while sharing another's embarrassment. This suggests that observers might rely more strongly on the social target's gestures and bodily as well as facial expressions and engage in mirroring processes during the presence of embarrassment in the other, compared to those situations where no appeasement gestures can be observed.

Similarly, observing another person who is being socially rejected and suffers from social exclusion can elicit strong feelings of empathic social pain in the observer. As a consequence, we might also feel encouraged to spontaneously help the person in need in order to relief the suffering (Masten et al. 2011). Studies could show that sharing another's experience of social exclusion activated similar regions in the AI and ACC as the first-hand experience of social exclusion (Masten et al. 2011; Meyer et al. 2012). In addition, and similar to the vicarious embarrassment experience, mentalizing areas such as the mPFC, precuneus, and temporal poles were also engaged when vicariously experiencing the stress of another person's social exclusion.

Not only can we share another's social suffering—we can also vicariously experience very positive and rewarding emotions. Mobbs et al. (2009) were wondering, why someone would enjoy watching others win, for example in a game show, without having any personal benefit. They found that similar to the idea of sharing negative affect, the experience of vicarious reward was associated with increased activations of the ventral striatum, a region also activated during the first-hand experience of reward. This response is, however, strongly modulated by the social relation with the person in joy. If we observe gains of unlikeable others, this might in fact elicit envy. Specifically, envy is experienced when the other person is superior in a self-relevant domain. Takahashi et al. (2009) demonstrated that in situations like this, envy is associated with increased activations of the ACC rather than the striatum. In contrast, if an envied person failed or underwent a public mishap, participants experienced the rewarding pleasure of "schadenfreude" with activations in the ventral striatum.

All of these studies demonstrate that the experience of vicarious emotions and their neural foundations are heavily modulated by a multitude of factors. Among others, this modulation depends on characteristics of the social target, e.g., the likeability of the person or the social relationship between observer and social target. This illustrates that not only the first-hand experience of self-conscious emotions but also their vicarious experience need to be examined in socially interactive contexts. Only then can interpersonal modulators be properly considered. For example, Singer et al. (2006) could show how important the valuation of the social target is as a modulatory factor by manipulating the perceived fairness of the social target during prior interaction. They found reduced pain empathy-related activation of the AI in response to another's suffering when the social target previously had behaved in an unfair manner. Other studies could show increased responses of the shared circuits of the AI or ACC when individuals observed friends compared to strangers in physically painful situations (Cheng et al. 2010), when being socially excluded (Meyer et al. 2012), or during embarrassing moments (Müller-Pinzler et al. 2015a and for an in-depth discussion of modulatory factors of vicarious emotions please also see the chapter by Kanske et al.).

6 Clinical Applications

Understanding the neural foundations and modulations of self-conscious emotions has significant implications for neuropsychiatric disorders that affect interpersonal behavior, particularly social anxiety. While almost everybody experiences social anxiety at least occasionally (Leary and Kowalski 1995), excessive and persistent fear of embarrassment and concerns about the evaluations of others are a characteristic of social anxiety disorder. In affected individuals, this can lead to social withdrawal and depression, making social anxiety disorders a major burden for individuals and society (Kessler et al. 2005; Schneier 1992). Psychological models of social anxiety disorders assume that affected individuals have a negatively biased self-image and expect others to have very high and unattainable standards for their behavior. Both factors contribute to the often prevailing and unpleasant experience of failure during social interactions. Thus, if persons with high social anxiety are confronted with an audience, they perceive an inevitable evaluative threat. Attention biases toward information about this social threat further increase this experience and also impact the anticipation of future social situations (Heimberg et al. 2010). Although social anxiety arises during proper social situations, most previous studies on social anxiety neglected this fact and examined social anxiety in social isolation (Blair et al. 2010, 2011). These investigations showed that social anxiety disorder is associated with increased activations of the amygdala, anterior cingulate cortex, and insula in response to fearful facial expressions or written vignettes of embarrassing situations. Still, the main hindrance for individuals suffering from social anxiety disorder lies in the interpersonal nature of everyday life situations.

A few studies attempted to investigate social anxiety with more interpersonal and interactive scenarios. Lorberbaum et al. (2004), for example, studied the neural correlates of social anxiety during the anticipation of giving a public speech. They could find increased activations within the pons, striatum, amygdala, insula, and the temporal poles, regions involved in emotion processing and processing of affective arousal. In contrast, regions within the anterior cingulate and prefrontal cortex showed decreased activations. These results tentatively suggest that enhanced processing of affect and arousal in individuals suffering from social anxiety might decrease their capacity for cognitive control and emotion regulation. In addition, the above-described study on embarrassment provides a more mechanistic understanding how the presence of an audience might impact neural processing in social anxiety (Müller-Pinzler et al. 2015a). During the presence of an audience, participants with higher trait anxiety scores showed longer gaze dwell time on social cues and increased activations of the mentalizing network. Notably, the association of social anxiety with mentalizing activation was mediated by the dwell time on social cues and the attentional bias toward the audience. Thus, attentional shifts toward the audience, the source of the potential evaluative threat, might induce greater mentalizing activation in accordance with more persistent thoughts about the own evaluation in-the-eyes-of-others. These data therefore support the notion that individuals with greater social anxiety pay increased attention to others and hold a negative bias in terms of what the audience might think of them (Heimberg et al. 2010).

In conclusion, these studies nicely demonstrate how more realistic and socially interactive paradigms offer a novel perspective on altered processing of publicity and interpersonal emotions in the pathology of social anxiety and other psychiatric disorders such as autism, with affected individuals showing deficits in their social interaction skills (see also Krach et al. 2015; Paulus et al. 2013b). It will be both challenging and fascinating for future studies to further refine and establish experimental paradigms that allow subjects to immerse into a social context and interact with other human beings. Such interactive settings are crucial to get an unbiased understanding of the neural foundations and behavioral consequences of self-conscious and interpersonal emotions and their alterations in neuro-psychiatric disorders.

References

Adolphs R, Tranel D, Damasio H, Damasio AR (1995) Fear and the human amygdala. J Neurosci 15:5879–5891

Basile B, Mancini F, Macaluso E, Caltagirone C, Frackowiak RSJ, Bozzali M (2011) Deontological and altruistic guilt: evidence for distinct neurobiological substrates. Hum Brain Mapp 32:229–239. doi:10.1002/hbm.21009

Beer JS, Heerey EA, Keltner D, Scabini D, Knight RT (2003) The regulatory function of self-conscious emotion: insights from patients with orbitofrontal damage. J Pers Soc Psychol 85:594–604. doi:10.1037/0022-3514.85.4.594

Blair KS, Geraci M, Hollon N, Otero M, DeVido J, Majestic C, Jacobs M, Blair RJR, Pine DS (2010) Social norm processing in adult social phobia: atypically increased ventromedial frontal cortex responsiveness to unintentional (embarrassing) transgressions. Am J Psychiatry 167:1526–1532. doi:10.1176/appi.ajp.2010.09121797

Blair KS, Geraci M, Korelitz K, Otero M, Towbin K, Ernst M, Leibenluft E, Blair RJR, Pine DS (2011) The pathology of social phobia is independent of developmental changes in face processing. Am J Psychiatry 168:1202–1209. doi:10.1176/appi.ajp.2011.10121740

Cacioppo S, Frum C, Asp E, Weiss RM, Lewis JW, Cacioppo JT (2013) A quantitative meta-analysis of functional imaging studies of social rejection. Sci. Rep. 3:2027. doi:10.1038/srep02027

Cheng Y, Chen C, Lin C, Chou K, Decety J (2010) NeuroImage Love hurts: an fMRI study. Neuroimage 51:923–929. doi:10.1016/j.neuroimage.2010.02.047

Eisenberger NI (2012) Broken hearts and broken bones: a neural perspective on the similarities between social and physical pain. Curr Dir Psychol Sci 21:42–47. doi:10.1177/0963721411429455

Eisenberger NI, Lieberman MD, Williams KD (2003) Does rejection hurt? An FMRI study of social exclusion. Science 302:290–292. doi:10.1126/science.1089134

Feinberg M, Willer R, Keltner D (2012) Flustered and faithful: embarrassment as a signal of prosociality. J Pers Soc Psychol 102:81–97. doi:10.1037/a0025403

Finger EC, Marsh AA, Kamel N, Mitchell DGV, Blair RJR (2006) Caught in the act: the impact of audience on the neural response to morally and socially inappropriate behavior. Neuroimage 33:414–421. doi:10.1016/j.neuroimage.2006.06.011

Fourie MM, Thomas KGF, Amodio DM, Warton CMR, Meintjes EM (2014) Neural correlates of experienced moral emotion: an fMRI investigation of emotion in response to prejudice feedback. Soc Neurosci 9:203–218. doi:10.1080/17470919.2013.878750

Frith U, Frith CD (2003) Development and neurophysiology of mentalizing. Philos Trans R Soc Lond B Biol Sci 358:459–473. doi:10.1098/rstb.2002.1218

Gilam G, Lin T, Raz G, Azrielant S, Fruchter E, Ariely D, Hendler T (2015) Neural substrates underlying the tendency to accept anger-infused ultimatum offers during dynamic social interactions. Neuroimage. doi:10.1016/j.neuroimage.2015.07.003

Greenwald AG, McGhee DE, Schwartz JL (1998) Measuring individual differences in implicit cognition: the implicit association test. J Pers Soc Psychol doi:10.1037/0022-3514.74.6.1464

Hartgerink CHJ, van Beest I, Wicherts JM, Williams KD (2015) The ordinal effects of ostracism: a meta-analysis of 120 cyberball studies. PLoS ONE 10:e0127002. doi:10.1371/journal.pone.0127002

Heimberg RG, Brozovich FA, Rapee RM (2010) A cognitive-behavioral model of social anxiety disorder: update and extension. In: Hofmann SG, DiBartolo PM (eds) Social anxiety: clinical, developmental, and social perspectives. Elsevier, New York, NY, pp 395–422

Hein G, Singer T (2008) I feel how you feel but not always: the empathic brain and its modulation. Curr Opin Neurobiol 18:153–158. doi:10.1016/j.conb.2008.07.012

Iannetti GD, Salomons TV, Moayedi M, Mouraux A, Davis KD (2013) Broken hearts and broken bones: contrasting mechanisms of social and physical pain. Trends Cogn Sci 1–8 doi:10.1016/j.tics.2013.06.002

Jankowski KF, Takahashi H (2014) Cognitive neuroscience of social emotions and implications for psychopathology: examining embarrassment, guilt, envy, and schadenfreude. Psychiatry Clin Neurosci 1–41 doi:10.1111/pcn.12182

Kelly C, Toro R, Di Martino A, Cox CL, Bellec P, Castellanos FX, Milham MP (2012) A convergent functional architecture of the insula emerges across imaging modalities. Neuroimage 61:1129–1142. doi:10.1016/j.neuroimage.2012.03.021

Keltner D, Buswell BN (1997) Embarrassment: its distinct form and appeasement functions. Psychol Bull 122:250–270

Kessler RC, Berglund P, Demler O, Jin R, Merikangas KR, Walters EE (2005) Lifetime prevalence and age-of-onset distributions of DSM-IV disorders in the national comorbidity survey replication. Arch Gen Psychiatry 62:593–602. doi:10.1001/archpsyc.62.6.593

Keysers C, Gazzola V (2007) Integrating simulation and theory of mind: from self to social cognition. Trends Cogn Sci 11:194–196. doi:10.1016/j.tics.2007.02.002

Krach S, Paulus FM, Bodden M, Kircher T (2010) The rewarding nature of social interactions. Front Behav Neurosci 4:3

Krach S, Cohrs JC, de Echeverría Loebell NC, Kircher T, Sommer J, Jansen A, Paulus FM (2011) Your flaws are my pain: linking empathy to vicarious embarrassment. PLoS ONE 6:e18675. doi:10.1371/journal.pone.0018675

Krach S, Müller-Pinzler L, Westermann S, Paulus FM (2013) Advancing the neuroscience of social emotions with social immersion. Behav Brain Sci 36:427–428. doi:10.1017/S0140525X12001951

Krach S, Kamp-Becker I, Einhäuser W, Sommer J, Frässle S, Jansen A, Rademacher L, Müller-Pinzler L, Gazzola V, Paulus FM (2015) Evidence from pupillometry and fMRI indicates reduced neural response during vicarious social pain but not physical pain in autism. Brain Mapp, Hum. doi:10.1002/hbm.22949

Kross E, Berman MG, Mischel W, Smith EE, Wager TD (2011) Social rejection shares somatosensory representations with physical pain. Proc Natl Acad Sci USA 108:6270–6275. doi:10.1073/pnas.1102693108

Leary MR, Kowalski RM (1995) Social anxiety. The Guilford Press, New York

Lindquist KA, Wager TD, Kober H, Bliss-Moreau E, Barrett LF (2012) The brain basis of emotion: a meta-analytic review. Behav Brain Sci 35:121–143. doi:10.1017/S0140525X11000446

Lorberbaum JP, Kose S, Johnson MR, Arana GW, Sullivan LK, Hamner MB, Ballenger JC, Lydiard RB, Brodrick PS, Bohning DE, George MS (2004) Neural correlates of speech anticipatory anxiety in generalized social phobia. NeuroReport 15:2701–2705

Macdonald G, Leary MR (2005) Why does social exclusion hurt? The relationship between social and physical pain. Psychol Bull 131:202–223. doi:10.1037/0033-2909.131.2.202

Masten CL, Morelli SA, Eisenberger NI (2011) An fMRI investigation of empathy for "social pain" and subsequent prosocial behavior. Neuroimage 55:381–388. doi:10.1016/j.neuroimage.2010.11.060

Meyer ML, Masten CL, Ma Y, Wang C, Shi Z, Eisenberger NI, Han S (2012) Empathy for the social suffering of friends and strangers recruits distinct patterns of brain activation. Cogn Affect Neurosci, Soc. doi:10.1093/scan/nss019

Miller RS (1987) Empathic embarrassment: situational and personal determinants of reactions to the embarrassment of another. J Pers Soc Psychol 53:1061–1069

Miller RS (1996) Embarrassment: poise and peril in everyday life. Guilford Press, New York, NY

Mobbs D, Yu R, Meyer M, Passamonti L, Seymour B, Calder AJ, Schweizer S, Frith CD, Dalgleish T (2009) A key role for similarity in vicarious reward. Science 324:900. doi:10.1126/science.1170539

Müller-Pinzler L, Paulus FM, Stemmler G, Krach S (2012) Increased autonomic activation in vicarious embarrassment. Int J Psychophysiol 86:74–82. doi:10.1016/j.ijpsycho.2012.07.183

Müller-Pinzler L, Gazzola V, Keysers C, Sommer J, Jansen A, Frässle S, Einhäuser W, Paulus FM, Krach S (2015a) Neural pathways of embarrassment and their modulation by social anxiety. Neuroimage. doi:10.1016/j.neuroimage.2015.06.036

Müller-Pinzler L, Rademacher L, Paulus FM, Krach S (2015b) When your friends make you cringe: social closeness modulates vicarious embarrassment-related neural activity. Soc Cogn Affect Neurosci. doi:10.1093/scan/nsv130

Paulus FM, Müller-Pinzler L, Westermann S, Krach S (2013a) On the distinction of empathic and vicarious emotions. Front Hum Neurosci 7:196. doi:10.3389/fnhum.2013.00196

Paulus FM, Kamp-Becker I, Krach S (2013b) Demands in reflecting about another's motives and intentions modulate vicarious embarrassment in autism spectrum disorders. Res Dev Disabil 34:1312–1321. doi:10.1016/j.ridd.2013.01.009

Paulus FM, Müller-Pinzler L, Jansen A, Gazzola V, Krach S (2014) Mentalizing and the role of the posterior superior temporal sulcus in sharing others' embarrassment. Cortex, Cereb. doi:10.1093/cercor/bhu011

Rotge J-Y, Lemogne C, Hinfray S, Huguet P, Grynszpan O, Tartour E, George N, Fossati P (2014) A meta-analysis of the anterior cingulate contribution to social pain. Biol Psychiatry 75:198S–198S. doi:10.1093/scan/nsu110

Schilbach L, Timmermans B, Reddy V, Costall A, Bente G, Schlicht T, Vogeley K (2013) Toward a second-person neuroscience. Behav Brain Sci 36:393–414. doi:10.1017/S0140525X12000660

Schneier FR (1992) Social phobia: comorbidity and morbidity in an epidemiologic sample. Arch Gen Psychiatry 49:282. doi:10.1001/archpsyc.1992.01820040034004

Seidner LB, Stipek DJ, Feshbach ND (1988) A developmental analysis of elementary school-aged children's concepts of pride and embarrassment. Child Dev 59:367–377

Semin GR, Manstead ASR (1982) The social implications of embarrassment displays and restitution behaviour. Eur J Soc Psychol 12:367–377. doi:10.1002/ejsp.2420120404

Shin LM, Dougherty DD, Orr SP, Pitman RK, Lasko M, MacKlin ML, Alpert NM, Fischman AJ, Rauch SL (2000) Activation of anterior paralimbic structures during guilt-related script-driven imagery. Biol Psychiatry 48:43–50. doi:10.1016/S0006-3223(00)00251-1

Singer T, Seymour B, O'Doherty JP, Stephan KE, Dolan RJ, Frith CD (2006) Empathic neural responses are modulated by the perceived fairness of others. Nature 439:466–469. doi:10.1038/nature04271

Sturm VE, Rosen HJ, Allison S, Miller BL, Levenson RW (2006) Self-conscious emotion deficits in frontotemporal lobar degeneration. Brain 129:2508–2516. doi:10.1093/brain/awl145

Takahashi H, Matsuura M, Koeda M, Yahata N, Suhara T, Kato M, Okubo Y (2008) Brain activations during judgments of positive self-conscious emotion and positive basic emotion: pride and joy. Cereb Cortex 18:898–903. doi:10.1093/cercor/bhm120

Takahashi H, Kato M, Matsuura M, Mobbs D, Suhara T, Okubo Y (2009) When your gain is my pain and your pain is my gain: neural correlates of envy and schadenfreude. Science 323:937–939. doi:10.1126/science.1165604

Takahashi H, Yahata N, Koeda M, Matsuda T, Asai K, Okubo Y (2004) Brain activation associated with evaluative processes of guilt and embarrassment: an fMRI study. Neuroimage 23:967–974. doi:10.1016/j.neuroimage.2004.07.054

Tangney JP, Miller RS, Flicker L, Barlow DH (1996) Are shame, guilt, and embarrassment distinct emotions? J Pers Soc Psychol 70:1256–1269

Tangney JP, Stuewig J, Mashek DJ (2007) Moral emotions and moral behavior. Annu Rev Psychol 58:345–372. doi:10.1146/annurev.psych.56.091103.070145

Wagner U, N'Diaye K, Ethofer T, Vuilleumier P (2011) Guilt-specific processing in the prefrontal cortex. Cereb Cortex 21:2461–2470. doi:10.1093/cercor/bhr016

Waytz A, Mitchell JP (2011) Two mechanisms for simulating other minds: dissociations between mirroring and self-projection. Curr Dir Psychol Sci 20:197–200. doi:10.1177/0963721411409007

Williams KD, Cheung CK, Choi W (2000) Cyberostracism: effects of being ignored over the Internet. J Pers Soc Psychol doi:10.1037/0022-3514.79.5.748

Williams LA, DeSteno D (2008) Pride and perseverance: the motivational role of pride. J Pers Soc Psychol 94:1007–1017. doi:10.1037/0022-3514.94.6.1007

Williams LA, DeSteno D (2009) Pride: adaptive social emotion or seventh sin? Psychol Sci 20:284–288. doi:10.1111/j.1467-9280.2009.02292.x

Zahn R, Moll J, Paiva M, Garrido G, Krueger F, Huey ED, Grafman J (2009) The neural basis of human social values: evidence from functional MRI. Cereb Cortex 19:276–283. doi:10.1093/cercor/bhn080

Deconstructing Anger in the Human Brain

Gadi Gilam and Talma Hendler

Abstract Anger may be caused by a wide variety of triggers, and though it has negative consequences on health and well-being, it is also crucial in motivating to take action and approach rather than avoid a confrontation. While anger is considered a survival response inherent in all living creatures, humans are endowed with the mental flexibility that enables them to control and regulate their anger, and adapt it to socially accepted norms. Indeed, a profound interpersonal nature is apparent in most events which evoke anger among humans. Since anger consists of physiological, cognitive, subjective, and behavioral components, it is a contextualized multidimensional construct that poses theoretical and operational difficulties in defining it as a single psychobiological phenomenon. Although most neuroimaging studies have neglected the multidimensionality of anger and thus resulted in brain activations dispersed across the entire brain, there seems to be several reoccurring neural circuits subserving the subjective experience of human anger. Nevertheless, to capture the large variety in the forms and fashions in which anger is experienced, expressed, and regulated, and thus to better portray the related underlying neural substrates, neurobehavioral investigations of human anger should aim to further embed realistic social interactions within their anger induction paradigms.

G. Gilam (✉) · T. Hendler (✉)
Functional Brain Center, Wohl Institute for Advanced Imaging,
Tel Aviv Sourasky Medical Center, 6, Weizmann, 64239 Tel Aviv, Israel
e-mail: gadi.gilam@gmail.com

T. Hendler
e-mail: talma@tasmc.health.gov.il

G. Gilam · T. Hendler
School of Psychological Sciences, Tel-Aviv University, P.O. Box 39040,
69978 Tel Aviv, Israel

T. Hendler
Faculty of Medicine, Tel-Aviv University, P.O. Box 39040, 69978 Tel Aviv, Israel

T. Hendler
Sagol School of Neuroscience, Tel-Aviv University, P.O. Box 39040,
69978 Tel Aviv, Israel

Curr Topics Behav Neurosci (2016) 30: 257–274
DOI 10.1007/7854_2015_408
Published Online: 17 December 2015

Keywords Anger · Social interaction · Emotion regulation · Angry brain · Neuroimaging

Contents

1 Introduction ... 258
2 What Is Anger? ... 258
2.1 Causes of Anger ... 259
2.2 Anger Experience ... 260
2.3 Anger Regulation ... 261
2.4 Consequences of Anger ... 262
3 The "Angry Brain" ... 262
3.1 Angry Faces Studies ... 263
3.2 Self-Generated Anger Studies ... 264
3.3 Anger Induction Studies ... 265
4 Concluding Remarks ... 267
References ... 270

1 Introduction

Anybody can become angry, that is easy; but to be angry with the right person, and to the right degree, and at the right time, for the right purpose, and in the right way, that is not within everybody's power and is not easy.
Aristotle

As exemplified in the Iliad, Homer's war epic depicting Achilles's wrath in relation to the events of the Greek-Trojan war, anger is at the core of what it means to be human. Indeed, people report experiencing anger on a daily basis and consider it as one of the most prototypical exemplars of an emotion (Fehr and Baldwin 1996; Scherer and Tannenbaum 1986). Yet, defining anger as a single psychobiological phenomenon has posed considerable theoretical and experimental difficulties. In this chapter, we provide a psychological account of what anger is and review how the subjective experience of anger in healthy humans has been investigated thus far using neuroimaging techniques. We conclude by suggesting the scaffolding for the reconstruction of an "angry brain" which may take into consideration the multi-dimensionality of the anger construct.

2 What Is Anger?

There is much controversy on the theoretical conceptualization of anger, as on defining emotion in general, and while a survey of the emotion literature breaches the scope of this chapter, two renowned theoretical considerations of anger are briefly noted. According to Berkowitz's Cognitive-Neoassociationistic theory (Berkowitz 1990), a primitive form of anger is automatically triggered upon a provocation through an associative network of components including feelings, thoughts, memories, and most emphasized, physiological, and expressive motor reactions. Only with the temporal yet rapid unfolding of the emotional instance, the affected person makes appraisals, interpretations, and causal attributions which enable to construct complex high-order thoughts and feelings related to the actual emotion of anger. Embedded within these later stages is the ability to control and regulate anger and reactions to it. According to Averill's Social-Constructionist theory (Averill 1983), anger is regarded as a social syndrome which cannot be deconstructed into subclasses of physiological, cognitive, or any other element. Averill stresses that social rules govern the organization of the various elements of anger, which is considered in itself as a complete response of the person, and because of the great variety in these various elements, influenced by personal and situational circumstances, it is impossible to define a typical angry experience. Anger can thus be understood only within its specific contextual framework.

Though a clear cognitive–physiological versus social perspectives distinct Berkowitz's and Averill's theories, both agree that there is an intensity element to anger occurrence, from annoyance and irritation to anger and rage. More importantly, Averill and Berkowitz agree (c.f. Berkowitz and Harmon-Jones 2004) in referring to anger as an *emotional syndrome* because of its multidimensional complexity and that a temporal unfolding of the emotion is apparent in its construction. They also both point to the regulatory processes that may intervene along this temporal dynamic. Although theoretical disagreement on the nature and definition of anger remains, the complex and dynamic conceptualization of anger is agreed upon and supported by empirical findings.

2.1 Causes of Anger

Antecedents and instigators of anger may be sorted into three primary categories which support both Berkowitz and Averill's theoretical frameworks. Various exemplars of these categories reappear in anger-inducing paradigms used in the laboratory. These categories are as follows: (1) real or imagined threat such as physical or psychological pain, but also more trivial environmental aversive conditions such as aversive temperature and even polluted air (c.f. Berkowitz and Harmon-Jones 2004), (2) frustration due to goal obstruction (e.g., Carver 2004; Szasz et al. 2011), and (3) perceived personal offense due to unfair treatment, violation of social norms, insults, rejections, criticism, and the likes (e.g., Denson et al. 2009; Memedovic et al. 2010; Porath and Erez 2007; Srivastava et al. 2009).

The first category relating to threat reflects the most basic form of anger, regarded as the instinctive survival response which triggers the fight feature of the fight or flight reaction (Anderson and Bushman 2002; Siever 2008). Reactive aggression triggered by a threat is perhaps the most typical behavioral expression of anger, and thus, anger has been traditionally viewed as interchangeable with aggression. Anger may indeed be pivotal in the generation and propagation of violent acts against the self and others. Nevertheless, aggressive acts may be perpetrated without any trace of anger, and at the same time, anger is an emotional construct in its own right, not necessarily a harbinger of aggression. In contrast, a profound interpersonal foundation is apparent in the third category. In accordance, the expressions of anger have evolved from their primitive forms and adapted to socially accepted norms (Averill 1983; Baumeister et al. 1990; Fehr and Baldwin 1996). For example, people would probably not shout in the middle of a restaurant at a rude waiter, but rather restrain themselves and choose more accepted forms of rebuttal, such as minimizing the tip. This suggests that in order to realistically capture the multifaceted concept of anger, experimental designs should incorporate an interpersonal social interaction and try to dissociate between the experience and the expression of anger.

2.2 *Anger Experience*

During the actual experience of anger, a person is commonly described as having a cluster of physiological, cognitive, and behavioral attributes which are directly related to the temporal dynamics of anger. Physiologically, an angry experience is characterized by an increase in respiration, blood pressure, heart rate, skin and body temperature, and skin conductance (Stemmler 2010) indicating the involvement of both sympathetic and parasympathetic systems of the autonomic nervous system. Other bodily changes include specific facial features and a general muscular tightness (Berkowitz and Harmon-Jones 2004; Scherer and Tannenbaum 1986). Anger is thus generally considered as a very arousing emotional condition.

A negative cognitive appraisal of circumstances characterizes anger. Obsessive and loopy thinking, planning of revenge and retaliation, and judgmental and derogative labeling are just some forms of angry cognitions (Fehr and Baldwin 1996). Such intrusive negative provocation-focused thought patterns during anger are termed together as *rumination* (Rusting and Nolen-Hoeksema 1998) during which people masticate the causes and consequences of the angry event. Rumination also tends to further intensify and prolong the angry experience.

Behaviorally, an angry person is in a general nervous attitude with a proneness to some form of physical or verbal aggression (Deffenbacher et al. 1996). Arguments with yelling and screaming are also very common during anger episodes. However, other expressions may be less confrontational such as using conflict resolution, withdrawing from the situation, or implementing relaxation techniques. Studies have generally found a myriad of behavioral expressions of anger which support Averill's (1983) assertion that "given an adequate provocation, nearly any response, and even no response, can count as a manifestation of anger" (there, p. 1147).

While physiological responses to anger are generally quite short and last up to several minutes, the subjective experience of day-to-day anger typically lasts for about half an hour, during which rumination is common, though duration is correlated with intensity (Potegal 2010). The temporal dynamics of anger experience are also characterized by an escalating property, in which annoyances and irritations accumulate over time, and behavioral responses that begin with mild requests may reach strong angry outbursts (Baumeister et al. 1990). Similar to the folktale of "the straw that broke the camel's back," there seems to be a nonlinear effect in the trajectory of anger in which at the extreme end, a sense of loss of control and irrationality captures the essence of a person's experience, and it is more difficult to be soothed or distracted. While anger rises quickly and declines slowly, it may be terminated by natural decay, quenching, or catharsis, all of which may be considered as forms of anger regulation.

2.3 Anger Regulation

The involvement of processes that control and regulate the experience and expression of anger emerges as a crucial element embedded within this socio-emotional phenomenon. Similar to Aristotle's citation above, Gross (1998) generally defined *emotion regulation* as "the processes by which individuals influence which emotions they have, when they have them, and how they experience and express these emotions" (there, p. 275). Theoretical accounts differ in their view regarding the relation between emotion generation and regulation, as some claim that regulation is an inherent part of the generation process while others separate these two processes, but it is generally accepted that they are both critical in the construction of the emotional episode as it unfolds over time (Gross and Barrett 2011), and this is apparent in our description of anger thus far. Emotion regulation processes or strategies may be automatic or controlled, implicit, or explicit and may modulate the emotion at any stage during the evolvement of its experience and expression. There are many different strategies to regulate emotions, yet *cognitive reappraisal*, in which one changes or reinterprets how she thinks about an emotional situation, has been studied the most.

Examples of laboratory experimentation on anger down-regulation suggest that when facing or recalling an anger provocation, using cognitive reappraisal rather than suppression or rumination (Memedovic et al. 2010; Ray et al. 2008; Szasz et al. 2011) has led to a decrease in reported anger experience and reduced maladaptive cardiovascular response. Other accounts have shown a large variety in actions one may take to cope with anger (Deffenbacher et al. 1996). For example, some actions may be conciliatory in their nature, such as reciprocal communication and talking it over, while other actions may try to create distance and avoidance from the angering stimuli, such as detachment and time-outs; still another set of actions may focus on the physiological aspect, such as relaxation or drug and alcohol consumption. It is also clear that some of these actions are more adaptive and healthy than others.

2.4 *Consequences of Anger*

Anger may have detrimental effects on our lives. It is related to poor quality of life, with people high in trait anger—that is the tendency and frequency of experiencing anger on a daily basis—having impaired psychological and social well-being (Phillips et al. 2006). Anger is implicated in negative health outcomes, most notably in cardiovascular disease (Williams 2010). For example, unrestrained expression or chronic suppression of anger affects essential hypertension and coronary heart disease. Anger irregularity is involved in many psychopathologies, such as psychotic, affective, and personality disorders (Novaco 2010). Even in anxiety disorders such as *Post-Traumatic Stress Disorder* (PTSD), related primarily to abnormal fear, there is a well-documented anger dysregulation which hampers functionality. Anger may also have debilitating effects on cognitive processes, such as in task performance and creativity (Porath and Erez 2007) and judgment and decision making (Lerner and Tiedens 2006).

Surprisingly, although experiencing anger and being the target of another's anger are primarily negative, some episodes of anger are positively evaluated (Averill 1983; Baumeister et al. 1990). Indeed, anger is adaptive and functional and has several positive aspects. It is critical for communicating an offensive event and thus has a role in maintaining status quo. Anger is also an important motivator for taking action and approaching rather than withdrawing away from a possible or actual confrontation (Berkowitz and Harmon-Jones 2004). This may be instrumental in achieving a wide variety of goals. For example, anger has a pivotal role in negotiations, and under certain conditions, expressing anger may lead to beneficial resolutions (Van Dijk et al. 2008). Moreover, anger together with disgust underlies *moral outrage*, the emotional reaction to a perceived moral transgression inflicted by others upon others (Salerno and Peter-Hagene 2013).

3 The "Angry Brain"

Disentangling the causes, consequences, experience, and expression of anger portrays a contextualized multidimensional construct consisting of physiological, cognitive, subjective, and behavioral components. Given the heterogeneous depiction of anger, research on the neural substrates of anger should try to appreciate not only whether and to what extent anger occurs, but even more so what are the forms and fashions in which anger is induced, experienced, expressed, and regulated. For obvious reasons, animal research has been preoccupied with aggression as a behavior rather than the subjective experience of anger. Cannon and Bard's (Bard 1928) classical studies on decorticated cats showed that the hypothalamus is essential for expressing "sham rage" (i.e., aggressive behavior without anger). Seventy-years later, Panksepp (1998), based mostly on studies in rodents, suggested a primitive neural basis for anger shared by all vertebras which in addition to the hypothalamus included the amygdala and periaqueductal gray (PAG). These brain regions seem to be involved in the rapid identification and response to threat in the environment, and thus assumed to have an essential role in the generation of anger and propagation of aggression, which accompany the fight reaction of the fight or flight response. Introduction of noninvasive brain mapping methodologies such as Positron Emission Tomography (PET) and functional Magnetic Resonance Imaging (fMRI) advanced studies to search for the "angry brain" in humans.

3.1 Angry Faces Studies

Neuroimaging studies on the neural substrates of human anger can be generally divided into three types as far as how anger was evoked. The first set of studies used images depicting angry faces. The most robust finding of early studies was that unlike rodent studies, the amygdala did not seem to have a specific involvement in the neural processing of angry faces (e.g., Blair et al. 1999). Recent meta-analyses on hundreds of neuroimaging studies on emotional faces confirm the strong specificity of the amygdala in processing fearful faces, though also associated with both sad and happy faces (Costafreda et al. 2008; Fusar-Poli et al. 2009). On the other hand, angry faces were associated with neural response in regions comprising the middle frontal gyrus (MFG), anterior cingulate cortex (ACC), inferior frontal gyrus (IFG), parahippocampal gyrus (PHG), claustrum, insula, middle temporal gyrus, fusiform gyrus (FFG), and occipital gyrus. It was suggested that while visual regions such as the FFG might be generally relevant for perceptual processing of facial stimuli, paralimbic and insular regions might be involved in processes associated with the generation of anger (though perhaps more relevant in this case is the generation of a general state of arousal), whereas more frontal regions might be involved in processes associated with the conscious experience of the emotion. Nevertheless, these regions were not uniquely associated with angry faces, but seemed to be differentially involved in processing other emotions expressed in human faces. Notwithstanding, more recent studies that were not included in these meta-analyses have shown the involvement of amygdala activation in processing angry faces, especially when considering idiosyncratic personality differences. For example, it was shown that increased amygdala activity (Beaver et al. 2008) and reduced functional connectivity between amygdala and ventromedial prefrontal cortex (vmPFC) (Passamonti et al. 2008) in response to angry faces were correlated with individual differences in behavioral approach, an orientation associated with anger and aggression as mentioned above. Although the use of faces enables highly standardized stimuli across and within subjects, such static stimuli clearly do not capture the complex nature of anger experience, rather enable to investigate the neural mechanisms mediating the perception and recognition of anger in human faces.

3.2 *Self-Generated Anger Studies*

The second set of studies used self-generation of anger by recollecting and imagining personal autobiographic memories or scripted scenarios of angry experiences. PET studies on autobiographical memories of anger (e.g., Damasio et al. 2000; Kimbrell et al. 1999) identified regions in the PFC, especially in the ventral-orbital regions, as well as ACC, temporal poles, regions of the medial temporal lobe (MTL), thalamus and hypothalamus, insula as well as regions in the brainstem and cerebellum. The involvement of the temporal poles in deducing the content of another person's mental state (i.e., mentalizing; Denny et al. 2012) might reflect the engagement in a social interaction during the recollected angry memories. However, the temporal poles as well as the MTL are involved in retrieving declarative memories (Squire et al. 2004) which seems essential in the current paradigm. A recent fMRI study focused on anger regulation via reappraisal and rumination of the autobiographic angry memory (Fabiansson et al. 2012) found activations in orbito-frontal cortex (OFC), IFG, amygdala, thalamus, insula, putamen/caudate shared by regulation strategies. In addition, while reappraisal was more successful in diminishing the subjective experience of anger, there was a specific positive functional coupling between IFG and both amygdala and thalamus during rumination which might portray the failure of such anger-focused thought pattern in attenuating the emotional experience.

While such recollection paradigms enable a more personalized reverberations of anger, these recollections are not entirely standardized across subjects. For example, in one study (Kimbrell et al. 1999), some events involved property loss, others involved being wrongly blamed, and still others generally involved verbal arguments. Such recollections might also be prone to confounds of memory biases and limited introspective insight. A PET study using scripted and more controlled scenarios to elicit the imagined anger experience showed that an unrestrained scenario in which one acts aggressively to express anger was associated with decreased activity in vmPFC and increase in ACC compared to a neural scenario (Pietrini 2000). A more recent fMRI study on social emotions broadly defined similarly found that compared to neutral events, scripted events of social rejection and criticism, which were associated with reports of anger, sadness, and shame, engaged increased activity in vmPFC as well as thalamus, amygdala, precuneus, and posterior cingulate cortex (PCC; Frewen et al. 2011). And yet, these are internally generated paradigms of anger induction and thus still lack the fundamental bluntness of actually being provoked.

3.3 Anger Induction Studies

The third and final set of studies tried to induce anger directly. In the first such study (Denson et al. 2009), participants were requested to solve difficult anagrams and say out loud through a microphone the correct answer or say "no answer" if they did not know the answer. Anger was induced by the experimenter who interrupted participants two times requesting them to speak louder and on a third time stated in a rude and condescending tone of voice "Look, this is the third time I have had to say this! Can't you follow instructions?" The analysis was based on contrasting a baseline period before and after the provocation, during which there was increased activity in the medial and lateral PFC (mPFC and lPFC, respectively), insula, thalamus, hippocampus, ACC, and PCC, of which the dorsal ACC (dACC) positively correlated with self-reported anger and trait aggression and the insula, hippocampus, rostral ACC, and PCC positively correlated with self-reported angry rumination. A very similar pattern of activation was apparent during a condition of angry rumination, during which activity in the mPFC also positively correlated with self-reported angry rumination. A subsequent study asked participants to control their anger in view of such insults and found an increase in self-reported anger compared to baseline, but a smaller effect size compared to the previous study (Denson et al. 2013). A similar pattern of brain activity emerged including the dACC, dorsal mPFC and lPFC (dmPFC and dlPFC, respectively), insula, thalamus, amygdala, and brainstem. Dorsal regions of the PFC and the insula positively correlated with self-reported anger control and negatively correlated with self-reported anger. The brain pattern of anger control was also characterized by a functional coupling between the amygdala and regions of the PFC including dlPFC, dACC, and OFC which may reflect the efforts of PFC regions to exert control over the angering provocation. While these provocation-based anger studies incorporate an interpersonal context, participants remain completely passive while they lay in the MRI scanner; they are subjected to the experimenter's criticism but cannot react. A behavioral measure that may reveal their emotional turmoil is absent.

An additional experimental approach for the interpersonal induction of anger is the classic social decision-making paradigm—the Ultimatum Game (UG) (Güth et al. 1982; Sanfey et al. 2003). In the UG, two players need to agree on how to split a sum of money between themselves in order to actually gain the money. One player makes an offer on how to split the sum, while the second decides whether to accept or reject the offer. Unequal offers of about 25 % and below the total sum are commonly rejected resulting in a monetary loss for both players. Such offers are regarded as unfair offers which violate social norms, elicit anger, and thus result in an aggressive retribution at one's own personal cost. Indeed, it was shown that anger mediated the relationship between the size of offers and rejection rates such that more anger resulted in increased rejections (Srivastava et al. 2009). Congruently, it was shown that unfair UG offers were associated with increased sympathetic arousal as measured by skin conductance response (SCR) (Van't Wout et al. 2006). A recent meta-analysis on the neural structures involved in processing unfair offers compared to fair offers (Feng et al. 2015) found activity in the following regions: dACC, insula, ventrolateral, dorsolateral, and dorsomedial PFC, precuneus, temporal pole, temporalparietal junction (TPJ), and visual regions including the FFG. As detailed above, all these regions have previously been associated with various anger-related contexts.

Though consistent behavioral, psychophysiological, and neural evidence implicated anger with how people cope with unequal offers, it remains true that emotions are not the sole factor in explaining UG behavior (Civai 2013). Factors, such as reward valuation, fairness enforcement norms, and self-involvement, among others, may influence people's behavior in social decision-making paradigms (Rilling and Sanfey 2011). However, for the purpose of inducing anger and as long as such factors are being controlled for, it does not necessarily make a difference if one is angry because of the unfairness of an offer or by self-involvement as both are prerequisites for the subjective experience of anger. In addition to portraying an interpersonal situation, an additional benefit of the UG is that one can experimentally separate between the offer phase, which serves as the anger induction, and the decision-making phase. Thus, the behavior—a decision to accept or reject an offer—may serve as an objective measure of the associated emotional experience.

Indeed, additional evidence for the importance of the emotional response in driving behavior in the UG stems from emotion regulation studies which indicate that regulating anger may be important to the acceptance of unfair offers and that people who are better able to regulate anger associated with such offers are more likely to accept and financially benefit from them (Grecucci and Sanfey 2013). For example, explicitly instructing to use reappraisal to down-regulate emotions associated with unfair offers resulted in increased acceptance rates which were found to correlate with brain activity in an anterior region of the dlPFC (Grecucci et al. 2013). Furthermore, the insula showed effects of emotion modulation as activation decreased when down-regulating and increased when up-regulating. Additional studies on individual differences regarding the tendency to accept or reject unfair offers point at the involvement of ventral regions of the PFC. One such study revealed that activity in the vmPFC/OFC mediated the relationship between pre-UG testosterone levels and rejection rates (Mehta and Beer 2010). Nevertheless, there are several limitations when considering the UG as an anger-inducing paradigm. For one, the induction of anger is strictly focused on the amount of money offered. In addition, especially in the neuroimaging literature, the UG is implemented in a "single-shot" mode in which each offer is from a different, most often a virtual proposer, reducing to almost none the dynamic nature of the interaction. Angering situations, especially in bargaining contexts such as the UG, tend to spiral and escalate due to personal insults and provocations. The Denson and colleagues studies (2009, 2013) similarly lack this basic feature—subjects are provoked, but the naturalistic social-interactive and temporal dynamics of an angry experience is overlooked. A true engagement in social interaction occurs when people can communicate with others in their surroundings, adapting themselves contingently.

To try and tackle some of these limitations, a recent study (Gilam et al. 2015) created a modified version of the UG which incorporated, in a sense, the kind of provocation used by Denson and colleagues (2009, 2013). A repeated version of the UG in which participants decided to accept or reject offers from the same proposer was embedded with online verbal negotiations between the players after each round. Unbeknownst to the participants, the proposer was in fact a professional actor who used scripted and improvised provocations in concert with the sequence of mostly unfair offers to further induce interpersonal anger. All participants reported on anger as the dominant emotional experience, and importantly, anger reports increased as the interaction unfolded. In addition, the idiosyncratic tendencies to accept offers during this anger-infused social interaction, and thus gaining more money along the game, were associated with a balanced emotional profiled including both anger and positive emotions to an equal extent, as well as with increased activity in the anterior vmPFC/OFC and decreased activity in region of the brainstem possibly reflecting the Locus Coeruleus. It was also found that both vmPFC/OFC activity and functional connectivity between the insula and thalamus modulated the emotional experience en route to increased monetary gain. The fact that there were no control conditions, both for the anger induction provocations and for the social interaction, as the lack of clear instructions to regulate emotions, limit the capability to deduce whether these emotionally balanced participants were less angered or actually attenuated their angry response. And yet this paradigm seems to have mimicked realistically the dynamic features of an angry episode within the confined settings of the MRI scanner. A more general criticism, however, to such games as the UG is their excessive emphasis on decision-related processes and material payoffs, which are not a necessary part of real-life emotional experiences and social encounters. These flaws are important to further tweak and improve anger-inducing paradigms in future studies.

4 Concluding Remarks

The uniqueness of anger as an emotion is evident in that it is a negative emotion with a motivationally approach tendency. Furthermore, while anger is an emotion which seems to be apparent also in animals and features a bottom-up arousing component, in humans, anger has evolved into a complex multidimensional emotional construct, highly influenced by sociocultural contexts on the one hand and with profound personal and interpersonal ramifications on the other. Anger is thus inherently subject to and dependant on an individuals' ability to assert control and regulation over it. The wide distribution of brain regions as reviewed above may suggest that brain imaging studies thus far did not adequately dissect the complexity of the anger construct and did not distinguish between different modes of anger manifestation. Notwithstanding, the contextualized multidimensionality of anger may point toward the involvement of several neural circuits in mediating this psychobiological phenomenon. Indeed, there seems to be several findings that are fairly consistent across most sets of neuroimaging studies. Thalamic, limbic, and brainstem regions seem to reflect threat detection network which has a critical role in reactive aggression (Siever 2008). Evidence is most strongly supported by animal models, though it seems a similar role for this network is apparent in humans, in which it is believed to be involved in mediating the experience of anger, especially by generating a state of arousal. Studies on human aggression have shown the involvement of these regions as well as of vmPFC/OFC and ACC. For example, it was recently shown that across participants, activity in the vmPFC while viewing an opponent bearing an angry facial expression compared to a neutral expression during an interactive competitive aggression task was negatively correlated with aggressive behavior (Beyer et al. 2014). In addition, within participants and specifically during the angry opponent trials, activity in the dACC was positively correlated with aggressive behavior. Yet, studying reactive aggression, even in social contexts, does not directly reflect the subjective experience of anger. Similarly, perceiving anger in faces or voices is not necessarily experiencing anger though such stimuli may serve as a social signal of threat. Interestingly though, it does seem that the same brain regions in the PFC are involved in the control and regulation of anger and aggression. Most notably, the vmPFC/OFC and lPFC (including IFG and dlPFC) have been associated with such regulatory functions; the former seems to be associated more specifically with regulation of anger experience and aggressive expressions of anger, while the latter with cognitive control of negative emotions in general (Buhle et al. 2014). The vmPFC/OFC has also been consistently associated with the expected subjective value of many different types of rewards, including monetary payoffs, snacks, and social rewards such as good reputation (Levy and Glimcher 2012). The vmPFC/OFC regulatory role may therefore reflect the expected value of the potential outcome of anger and aggression and thus direct behavior.

The reoccurrence of the insula and the dorsal aspect of the ACC might be related to their joint role in a network dedicated to detect salient sensory events, which has been associated with both physical and social pain (Iannetti and Mouraux 2010), both of which are primary antecedents of anger. A division of labor between these two highly interconnected regions has been suggested in which the insula is associated with the emotional experience, while the dACC is associated with allocation of control and modification of behavioral responses during challenging physical and cognitive situations (Gasquoine 2013). This suggested role of the dACC is congruent with its co-activation with regulatory regions of PFC during both anger and aggression paradigms. Finally, several regions associated with the mentalizing system such as the mPFC, PCC, temporal poles, and the TPJ (Denny et al. 2012) have also reappeared in various anger induction studies, whether self-generated or induced, albeit to a lesser extent. The involvement of this system seems to reflect the interpersonal nature of angering events, but the exact role of mentalizing in the experience, expression, and regulation of anger is still unclear. One important role may be that humans need to attribute the intention to do harm by another person in order to experience anger (Berkowitz and Harmon-Jones 2004). For example, it was shown that unfair UG offers randomly assigned by a computer

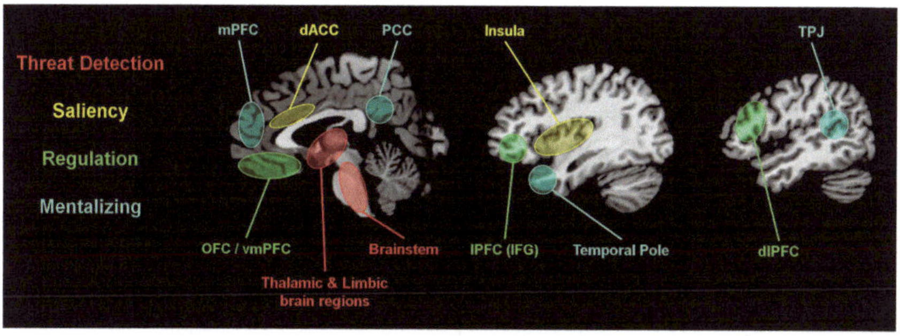

Fig. 1 Schematic scaffolding of the "angry brain." The MRI anatomical scans depict midsagittal (*left* Talairach slice $x = 3$), parasagittal (*middle* Talairach slice $x = 37$), and lateral (*right* Talairach slice $x = 47$) slices of the human brain. Four neural circuits seem to be involved in the subjective experience of anger: (1) threat detection, arousal, and reactive aggression (*red*) include thalamic (e.g., thalamus and hypothalamus), limbic (e.g., amygdala), and brainstem (e.g., PAG, locus coeruleus) regions; (2) saliency and perception of pain (*yellow*) include the insula and the dorsal anterior cingulate cortex (dACC); (3) emotion regulation (*green*) includes orbito-frontal cortex/ventromedial prefrontal cortex (OFC/vmPFC), lateral PFC (lPFC), most notably the inferior frontal gyrus (IFG), and dorsolateral PFC (dlPFC) regions; (4) mentalizing (*cyan*) includes medial PFC (mPFC), posterior cingulate cortex (PCC), temporal poles, and temporoparietal junction (TPJ) regions

were rejected less and also engaged less brain activity in bilateral anterior insula compared to similar offers allegedly made by a human counterpart (Sanfey et al. 2003). In a more specific case of an incidental transgression, mentalizing may have a role in understanding the accidental nature of the event and thus in fact serve as a regulatory mechanism in avoiding or reducing an angry reaction. The simplistic view would contend that mental state attribution is necessary for anger by the mere fact that anger is mostly experienced during social interactions, but this and other questions regarding the interaction between mentalizing and subjective anger deserve further scientific scrutiny.

In this review, we deconstructed human anger revealing its' physiological, cognitive, subjective, and behavioral components, portraying a socially contextualized regulated-prone multidimensional construct. However, most neuroimaging studies to date have focused on limited and specific aspects of the subjective experience of anger and therefore resulted in brain correlates dispersed across the entire brain. And yet, an overview of anger studies in the neuroimaging literature portrays several neural circuits that may provide the scaffolding for the reconstruction of the "angry brain" (Fig. 1). An important limitation to keep in mind regarding this review is that we focused solely on brain mapping techniques in healthy humans and did not integrate knowledge from other experimental modalities, such as electroencephalography or lesion studies, or various patient samples (for a review see Potegal and Stemmler 2010). We emphasize that to capture the large variety in the forms and fashions in which human anger is experienced and expressed and to portray the neurobehavioral substrates of these anger modes and related regulatory processes, studies in both healthy and patient populations must embed realistic interpersonal situations within their paradigms. This reverberates with recent conceptual and empirical advances which emphasize the importance of creating an interactive social context when investigating the neurobiological underpinnings of emotional episodes (Gilam et al. 2015; Müller-Pinzler et al. 2015). An additional aspect of anger which seems to have missed the radar of neuroimaging studies is the necessity to explore the temporal unfolding of the emotional experience and its concomitant neural manifestation (Raz et al. 2012). The prevalence of dysregulated anger in a multitude of psychopathological conditions leads us to hope that future contributions of the neuroscience of anger may be useful not only to better understand this phenomenon but also to promote beneficial products such as improving anger management intervention programs.

Acknowledgments This work was supported by the University of Chicago's Arete Initiative – A New Science of Virtues Program (39174-07); the U.S. Department of Defense award (W81XWH-11-2-0008); EU FP7 Health Cooperation Programme - BrainTrain Project (602186); the I-CORE Program of the Planning and Budgeting Committee (51/11); and the Levy Edersheim Gitter Institute for Neuroimaging and the Adams Super Center for Brain Studies, Tel Aviv University. The authors declare no competing financial interests.

References

Anderson CA, Bushman BJ (2002) Human aggression. Annu Rev Psychol 53(1):27–51. http://doi.org/10.1146/annurev.psych.53.100901.135231

Averill JR (1983) Studies on anger and aggression: implications for theories of emotion. Am Psychol 38(11):1145–1160. http://doi.org/10.1037/0003-066X.38.11.1145

Bard P (1928) A diencephalic mechanism for the expression of rage with special reference to the sympathetic nervous system. Am J Physiol 84:490–515

Baumeister RF, Stillwell A, Wotman SR (1990). Victim and perpetrator accounts of interpersonal conflict: autobiographical narratives about anger. J Pers Soc Psychol 59(5):994–1005. http://doi.org/10.1037/0022-3514.59.5.994

Beaver JD, Lawrence AD, Passamonti L, Calder AJ (2008) Appetitive motivation predicts the neural response to facial signals of aggression. J Neurosci 28(11):2719–2725

Berkowitz L (1990) On the formation and regulation of anger and aggression: a cognitive-neoassociationistic analysis. Am Psychol 45(4):494–503. http://doi.org/10.1037/0003-066X.45.4.494

Berkowitz L, Harmon-Jones E (2004) Toward an understanding of the determinants of anger. Emotion 4(2):107–130. http://doi.org/10.1037/1528-3542.4.2.107

Beyer F, Münte TF, Göttlich M, Krämer UM (2014) Orbitofrontal cortex reactivity to angry facial expression in a social interaction correlates with aggressive behavior. Cereb Cortex, bhu101. http://doi.org/10.1093/cercor/bhu101

Blair RJR, Morris JS, Frith CD, Perrett DI, Dolan RJ (1999). Dissociable neural responses to facial expressions of sadness and anger. Brain 122(5):883–893. http://doi.org/10.1093/brain/122.5.883

Buhle JT, Silvers JA, Wager TD, Lopez R, Onyemekwu C, Kober H, Ochsner KN (2014) Cognitive reappraisal of emotion: a meta-analysis of human neuroimaging studies. Cereb Cortex 24(11):2981–2990. http://doi.org/10.1093/cercor/bht154

Carver CS (2004) Negative affects deriving from the behavioral approach system. Emotion 4(1):3–22. http://doi.org/10.1037/1528-3542.4.1.3

Civai C (2013) Rejecting unfairness: emotion-driven reaction or cognitive heuristic? Front Hum Neurosci 126. http://doi.org/10.3389/fnhum.2013.00126

Costafreda SG, Brammer MJ, David AS, Fu CHY (2008) Predictors of amygdala activation during the processing of emotional stimuli: a meta-analysis of 385 PET and fMRI studies. Brain Res Rev 58(1):57–70. http://doi.org/10.1016/j.brainresrev.2007.10.012

Damasio AR, Grabowski TJ, Bechara A, Damasio H, Ponto LL, Parvizi J, Hichwa RD (2000). Subcortical and cortical brain activity during the feeling of self-generated emotions. Nat Neurosci 3(10):1049–1056. http://doi.org/10.1038/79871

Deffenbacher JL, Oetting ER, Lynch RS, Morris CD (1996) The expression of anger and its consequences. Behav Res Ther 34(7):575–590. http://doi.org/10.1016/0005-7967(96)00018-6

Denny BT, Kober H, Wager TD, Ochsner KN (2012) A meta-analysis of functional neuroimaging studies of self-and other judgments reveals a spatial gradient for mentalizing in medial prefrontal cortex. J Cogn Neurosci 24(8):1742–1752. http://doi.org/10.1162/jocn_a_00233

Denson TF, Pedersen WC, Ronquillo J, Nandy AS (2009) The angry brain: neural correlates of anger, angry rumination, and aggressive personality. J Cogn Neurosci 21(4):734–744. http://doi.org/10.1162/jocn.2009.21051

Denson TF, Ronay R, von Hippel W, Schira MM (2013) Endogenous testosterone and cortisol modulate neural responses during induced anger control. Soc Neurosci 8(2):165–177. http://doi.org/10.1080/17470919.2012.655425

Fabiansson EC, Denson TF, Moulds ML, Grisham JR, Schira MM (2012) Don't look back in anger: neural correlates of reappraisal, analytical rumination, and angry rumination during recall of an anger-inducing autobiographical memory. NeuroImage 59(3):2974–2981. http://doi.org/10.1016/j.neuroimage.2011.09.078

Fehr B, Baldwin MW (1996) Prototype and script analyses of laypeople's knowledge of anger. In: Fletcher GJO, Fitness J (eds) Knowledge structures in close relationships: a social psychological approach. Erlbaum, Hillsdale, pp 219–245

Feng C, Luo YJ, Krueger F (2015) Neural signatures of fairness-related normative decision making in the ultimatum game: a coordinate-based meta-analysis. Hum Brain Mapp 36(2):591–602. http://doi.org/10.1002/hbm.22649

Frewen PA, Dozois DJA, Neufeld, RWJ, Densmore M, Stevens TK, Lanius RA (2011). Neuroimaging social emotional processing in women: fMRI study of script-driven imagery. Soc Cogn Affect Neurosci 6(3):375–392. http://doi.org/10.1093/scan/nsq047

Fusar-Poli P, Placentino A, Carletti F, Landi P, Allen P, Surguladze S, Politi P (2009) Functional atlas of emotional faces processing: a voxel-based meta-analysis of 105 functional magnetic resonance imaging studies. J Psychiatry Neurosci JPN 34(6):418–432

Gasquoine PG (2013) Localization of function in anterior cingulate cortex: from psychosurgery to functional neuroimaging. Neurosci Biobehav Rev 37(3):340–348. http://doi.org/10.1016/j.neubiorev.2013.01.002

Gilam G, Lin T, Raz G, Azrielant S, Fruchter E, Ariely D, Hendler T (2015) Neural substrates underlying the tendency to accept anger-infused ultimatum offers during dynamic social interactions. NeuroImage 120:400–411. http://doi.org/10.1016/j.neuroimage.2015.07.003

Grecucci A, Giorgetta C, van't Wout M, Bonini N, Sanfey AG (2013) Reappraising the ultimatum: an fMRI study of emotion regulation and decision making. Cereb Cortex 23(2):399–410. http://doi.org/10.1093/cercor/bhs028

Grecucci A, Sanfey AG (2013) Emotion regulation and decision-making. Retrieved from http://repository.ubn.ru.nl/handle/2066/134521?

Gross JJ (1998) The emerging field of emotion regulation: an integrative review. Rev Gen Psychol 2(3):271–299. http://doi.org/10.1037/1089-2680.2.3.271

Gross JJ, Barrett LF (2011) Emotion generation and emotion regulation: one or two depends on your point of view. Emot Rev 3(1):8–16. http://doi.org/10.1177/1754073910380974

Güth W, Schmittberger R, Schwarze B (1982). An experimental analysis of ultimatum bargaining. J Econ Behav Org 3(4):367–388. http://doi.org/10.1016/0167-2681(82)90011-7

Iannetti GD, Mouraux A (2010) From the neuromatrix to the pain matrix (and back). Exp Brain Res 205(1):1–12. http://doi.org/10.1007/s00221-010-2340-1

Kimbrell TA, George MS, Parekh PI, Ketter TA, Podell DM, Danielson AL, Post RM (1999) Regional brain activity during transient self-induced anxiety and anger in healthy adults. Biol Psychiatry 46(4):454–465. http://doi.org/10.1016/S0006-3223(99)00103-1

Lerner JS, Tiedens LZ (2006) Portrait of the angry decision maker: how appraisal tendencies shape anger's influence on cognition. J Behav Decis Making 19(2):115–137. http://doi.org/10.1002/bdm.515

Levy DJ, Glimcher PW (2012) The root of all value: a neural common currency for choice. Curr Opin Neurobiol 22(6):1027–1038. http://doi.org/10.1016/j.conb.2012.06.001

Mehta PH, Beer J (2010) Neural mechanisms of the testosterone–aggression relation: the role of orbitofrontal cortex. J Cogn Neurosci 22(10):2357–2368. http://doi.org/10.1162/jocn.2009.21389

Memedovic S, Grisham JR, Denson TF, Moulds ML (2010) The effects of trait reappraisal and suppression on anger and blood pressure in response to provocation. J Res Pers 44(4):540–543. http://doi.org/10.1016/j.jrp.2010.05.002

Müller-Pinzler L, Gazzola V, Keysers C, Sommer J, Jansen A, Frässle S, Krach S (2015) Neural pathways of embarrassment and their modulation by social anxiety. NeuroImage, 119:252–261. http://doi.org/10.1016/j.neuroimage.2015.06.036

Novaco RW (2010) Anger and psychopathology. In: Potegal M, Stemmler G, Spielberger C (eds) International handbook of anger, Springer, New York, pp 465–497. Retrieved from http://www.springerlink.com/content/k0678p82g9276804/abstract/

Panksepp J (1998) Affective neuroscience: the foundations of human and animal emotions, Oxford University Press. Retrieved from http://www.acucentre.com.au/Neuro01/Panksepp/Pank1.pdf

Passamonti L, Rowe JB, Ewbank M, Hampshire A, Keane J, Calder AJ (2008) Connectivity from the ventral anterior cingulate to the amygdala is modulated by appetitive motivation in response to facial signals of aggression. NeuroImage 43(3):562–570. http://doi.org/10.1016/j.neuroimage.2008.07.045

Phillips LH, Henry JD, Hosie JA, Milne AB (2006) Age, anger regulation and well-being. Aging Mental Health 10(3):250–256. http://doi.org/10.1080/13607860500310385

Pietrini P (2000) Neural correlates of imaginal aggressive behavior assessed by positron emission tomography in healthy subjects. Am J Psychiatry 157(11):1772–1781. http://doi.org/10.1176/appi.ajp.157.11.1772

Porath CL, Erez A (2007) Does rudeness really matter? The effects of rudeness on task performance and helpfulness. Acad Manage J 50(5):1181–1197. http://doi.org/10.2307/20159919

Potegal M (2010) The temporal dynamics of anger: phenomena, processes, and perplexities. In: Potegal M, Stemmler G, Spielberger C (eds) International handbook of anger, Springer, New York, pp 385–401. Retrieved from http://www.springerlink.com/content/hp5r2n213gp14176/abstract/

Potegal M, Stemmler G (2010) Constructing a neurology of anger. In: Potegal M, Stemmler G, Spielberger C(eds) International handbook of anger, Springer, New York, pp 39–59. Retrieved from http://link.springer.com/chapter/10.1007/978-0-387-89676-2_4

Ray RD, Wilhelm FH, Gross JJ (2008) All in the mind's eye? Anger rumination and reappraisal. J Pers Soc Psychol 94(1):133–145. http://doi.org/10.1037/0022-3514.94.1.133

Raz G, Winetraub Y, Jacob Y, Kinreich S, Maron-Katz A, Shaham G, Hendler T (2012) Portraying emotions at their unfolding: a multilayered approach for probing dynamics of neural networks. NeuroImage 60(2):1448–1461. http://doi.org/10.1016/j.neuroimage.2011.12.084

Rilling JK, Sanfey AG (2011) The neuroscience of social decision-making. Annu Rev Psychol 62 (1):23–48. http://doi.org/10.1146/annurev.psych.121208.131647

Rusting CL, Nolen-Hoeksema S (1998) Regulating responses to anger: effects of rumination and distraction on angry mood. J Pers Soc Psychol 74(3):790–803. http://doi.org/10.1037/0022-3514.74.3.790

Salerno JM, Peter-Hagene LC (2013) The interactive effect of anger and disgust on moral outrage and judgments. Psychol Sci 24(10):2069–2078. http://doi.org/10.1177/0956797613486988

Sanfey AG, Rilling JK, Aronson JA, Nystrom LE, Cohen JD (2003) The neural basis of economic decision-making in the ultimatum game. Science 300(5626):1755–1758. http://doi.org/10.1126/science.1082976
Scherer KR, Tannenbaum PH (1986) Emotional experiences in everyday life: a survey approach. Motiv Emot 10(4):295–314. http://doi.org/10.1007/BF00992106
Siever LJ (2008) Neurobiology of aggression and violence. Am J Psychiatry 165(4):429–442. http://doi.org/10.1176/appi.ajp.2008.07111774
Squire LR, Stark CEL, Clark RE (2004) The medial temporal lobe*. Annu Rev Neurosci 27 (1):279–306. http://doi.org/10.1146/annurev.neuro.27.070203.144130
Srivastava J, Espinoza F, Fedorikhin A (2009) Coupling and decoupling of unfairness and anger in ultimatum bargaining. J Behav Decis Making 22(5):475–489. http://doi.org/10.1002/bdm.631
Stemmler G (2010) Somatovisceral activation during anger. In: Potegal M, Stemmler, G, Spielberger C (eds) International handbook of anger, Springer, New York, pp 103–121. Retrieved from http://www.springerlink.com/content/t786341m00h646m5/abstract/
Szasz PL, Szentagotai A, Hofmann SG (2011) The effect of emotion regulation strategies on anger. Behav Res Ther 49(2):114–119. http://doi.org/10.1016/j.brat.2010.11.011
Van Dijk E, van Kleef GA, Steinel W, van Beest I (2008) A social functional approach to emotions in bargaining: when communicating anger pays and when it backfires. J Pers Soc Psychol 94 (4):600–614. http://doi.org/10.1037/0022-3514.94.4.600
Van't Wout M, Kahn R, Sanfey A, Aleman A (2006) Affective state and decision-making in the ultimatum game. Exp Brain Res 169(4):564–568. http://doi.org/10.1007/s00221-006-0346-5
Williams JE (2010) Anger/hostility and cardiovascular disease. In: Potegal M, Stemmler, G, Spielberger C (eds) Int handbook of anger, Springer, New York pp 435–447. Retrieved from http://www.springerlink.com/content/v65un18p210w5327/abstract/

On the Control of Social Approach–Avoidance Behavior: Neural and Endocrine Mechanisms

Reinoud Kaldewaij, Saskia B.J. Koch, Inge Volman, Ivan Toni and Karin Roelofs

Abstract The ability to control our automatic action tendencies is crucial for adequate social interactions. Emotional events trigger automatic approach and avoidance tendencies. Although these actions may be generally adaptive, the capacity to override these emotional reactions may be key to flexible behavior during social interaction. The present chapter provides a review of the neuroendocrine mechanisms underlying this ability and their relation to social psychopathologies. Aberrant social behavior, such as observed in social anxiety or psychopathy, is marked by abnormalities in approach–avoidance tendencies and the ability to control them. Key neural regions involved in the regulation of approach–avoidance behavior are the amygdala, widely implicated in automatic emotional processing, and the anterior prefrontal cortex, which exerts control over the amygdala. Hormones, especially testosterone and cortisol, have been shown to affect approach–avoidance behavior and the associated neural mechanisms. The present chapter also discusses ways to directly influence social approach and avoidance behavior and will end with a research agenda to further advance this important research field. Control over approach–avoidance tendencies may serve as an exemplar of emotional action regulation and might have a great value in understanding the underlying mechanisms of the development of affective disorders.

Keywords Approach-avoidance · Emotional action control · Social psychopathology · Anterior prefrontal cortex · Amygdala

R. Kaldewaij (✉) · S.B.J. Koch · I. Toni · K. Roelofs
Donders Institute for Brain, Cognition and Behaviour, Centre for Cognitive Neuroimaging, Radboud University Nijmegen, Nijmegen, The Netherlands
e-mail: r.kaldewaij@donders.ru.nl

I. Volman
Sobell Department of Motor Neuroscience and Movement Disorders, UCL Institute of Neurology, University College London, London, UK

K. Roelofs
Behavioural Science Institute (BSI), Radboud University Nijmegen, Nijmegen, The Netherlands

Curr Topics Behav Neurosci (2016) 30: 275–294
DOI 10.1007/7854_2016_446
Published Online: 28 June 2016

Contents

1 Introduction 276
2 Theoretical Background 277
3 Automatic Approach–Avoidance Behavior 277
3.1 Individual Differences 278
4 Control Over Approach–Avoidance Behavior 279
4.1 Neural Mechanisms 279
5 Manipulating Approach–Avoidance Behavior 282
5.1 Neuroendocrine Effects 282
5.2 Training 285
6 Discussion and Future Horizons 286
6.1 Future Agenda 288
7 Conclusion 289
References 289

1 Introduction

The regulation of social–emotional behavior is crucial for social interactions. Events with a positive or negative valence, such as perceiving a happy or angry facial expression, trigger automatic emotional reactions. It has long been acknowledged that this automatic emotional (or affective) behavior is directed at either reducing or increasing the distance between the self and the object or person that triggered the emotional reaction: For example, people tend to walk away from things that scare them. In other words, positive or negative events or stimuli in the environment trigger automatic approach and avoidance tendencies. These tendencies can be crucial for survival, especially in situations of severe threat. However, in other situations, approach–avoidance tendencies can interfere with goal-directed (instrumental) behavior. In such situations, automatic approach–avoidance behavior needs to be controlled. Control over approach–avoidance tendencies is especially relevant to social interactions, during which overreacting to emotional cues can be at the core of aggression and social avoidance.

Here, we aim to give an overview of research on approach–avoidance behavior, focusing on the underlying neural and endocrine mechanisms, as well as how this behavior is altered in social psychiatric disorders, such as social anxiety and psychopathy. We will briefly discuss the theoretical background of approach–avoidance behavior and discuss how (control over) approach–avoidance tendencies can be experimentally measured and manipulated. Furthermore, we aim to integrate current findings on neural correlates of approach–avoidance behavior in a preliminary model and discuss future perspectives.

2 Theoretical Background

Theories on emotion have stressed the link between affective evaluation and action tendencies (Frijda 1986; Lang et al. 1990). These tendencies prepare for affective behavior (i.e., behavior influenced by emotional processing), which is organized in two motivational systems: approaching or appetitive behavior in response to positive, rewarding stimuli or events, and avoidant or aversive behavior in response to negative, threatening stimuli (Lang et al. 1990). This approach–avoidance distinction of behavior has been used for thousands of years: The ancient Greek philosopher Democritus (460–370 BCE) already described human action as "the pursuit of pleasure and avoidance of pain" (Elliot 2006). The fundamental role of approach–avoidance processes is illustrated by the fact that they are present across species, from single-cell organisms to humans (Elliot 2006). Moreover, they take place at different behavioral levels, from simple reflexes to advanced cognitive actions (Lang et al. 1990).

Gray (1990) described the two motivational systems as a behavioral approach (or activation) system (BAS), responding to stimuli that signal rewards or non-punishments, and a behavioral inhibition system (BIS), responding to signals of punishment and novelty. BAS activation causes an organism to approach a stimulus and has been associated with dopaminergic pathways (Carver et al. 2000). Engagement of the BIS system, in contrast, causes inhibition of ongoing behavior. Gray linked this system to the septo-hippocampal complex, brain stem, and frontal lobe. It has been argued that these systems also relate to the subjective experience of emotion: BAS is associated with positive feelings, such as happiness, whereas BIS is linked to negative feelings, such as fear, anxiety, and sadness (Carver et al. 2000).

3 Automatic Approach–Avoidance Behavior

Approach and avoidance tendencies have been mapped by different paradigms ranging from active approach–avoidance tasks using joysticks to decision paradigms. Using an emotional go-no-go task, Hare et al. (2005) found delayed go-responses for angry faces (versus neutral or happy) to be positively associated with amygdala activation. Berkman and Lieberman (2010) asked participants to imagine being part of a tribe that either liked or disliked certain food types, as appetitive and disgusting cues trigger approach and avoidance, respectively. They found increased left (compared to right) dorsal prefrontal cortex activation during approach (compared to avoidance) action decisions, measured with functional MRI. Experimental paradigms designed to directly tap into intrinsic action tendencies are so-called approach–avoidance tasks (AAT) (Chen and Bargh 1999; Rinck and Becker 2007; Rotteveel and Phaf 2004; Solarz 1960). In this type of task, participants are presented with valenced stimuli, e.g., positive or negative words, emotional faces, and pictures of positive or negative scenes. They are required to respond to these stimuli by flexing or extending of the arm to push a button or move an object, such as a lever or joystick. Flexing the arm is associated with approach behavior, whereas extending of the arm is associated with avoidance. The main outcome measure is reaction (or: initiation) time, i.e., the time that is required to initiate the movement after the presentation of the stimulus. Participants are generally faster at making approaching compared to avoiding movements to positive stimuli such as happy faces, reflecting an approach tendency toward positively evaluated stimuli. Similarly, participants are faster at making avoiding compared to approaching movements toward negative stimuli, such as angry faces, reflecting an avoidance tendency toward negatively evaluated stimuli (Chen and Bargh 1999; Rotteveel and Phaf 2004). These effects are typically found both when emotion evaluation is explicit (pull joystick when face is happy) and to a lesser extent when it is incidental (e.g., when faces are colored and participants push and pull using a color rule)—evidence for this latter category is mixed, and effect seems to depend on the specific task used (Phaf et al. 2014). Similar effects are found in symbolic approach–avoidance tasks where, for example, a manikin on a screen has to be moved toward or away from a negative or positive stimulus using button presses (De Houwer et al. 2001).

3.1 Individual Differences

There are individual differences in approach and avoidance action tendencies. Strong deviations from normal approach–avoidance behavior can be seen in social psychopathologies such as social anxiety, associated with relatively faster avoidance of emotional faces (Heuer et al. 2007; Roelofs et al. 2010), and psychopathy, associated with reduced avoidance of angry faces (Von Borries et al. 2012). Note that angry faces communicate a social challenge, usually eliciting avoidance tendencies in healthy individuals. The psychopathic patients showed no alterations in facial expression recognition, and the lack of avoidance in response to angry faces in psychopaths was associated with self-reported instrumental aggression levels (Von Borries et al. 2012). Together, this may suggest that it is not the face processing itself but the translation from affect to action that is altered in psychopathy (see also Ly et al. 2016). Usually, the altered approach–avoidance tendencies in social psychopathologies are found when angry faces are presented with direct gaze and not with averted gaze, indicating that the anger has to be directed toward the subject to elicit avoidance tendencies and indicating that the elicited avoidance is not simply the result of a stimulus–response compatibility effect (Roelofs et al. 2010; Von Borries et al. 2012). Indeed, using a manikin task and positive and negative words, Krieglmeyer and Deutsch (2010) showed that approach–avoidance tendencies reflect a specific, motivational link between perception of affect and behavior, rather than general stimulus–response compatibility.

4 Control Over Approach–Avoidance Behavior

Although evaluations of emotional situations often happen automatically and implicitly, the associated behavioral action tendency does not necessarily have to be implemented (Gross 2002). If the automatic tendency does not match the behavioral goal, it can be overridden; for example, objects or situations regarded as threatening or disgusting can be approached if one wants to. This ability to arbitrate between automatic affective action tendencies and instrumental goals is crucial for flexible behavior.

An example of affective influences on instrumental behavior is observed in pavlovian–instrumental transfer (PIT) paradigms. Participants first learn to associate specific stimuli (sounds or pictures) with reward (conditioned stimulus, CS+) and non-reward or punishment (CS−), a process which is known as Pavlovian conditioning. When subsequently engaging in a different task, for example, squeezing a handgrip to earn money, they respond more vigorously if the CS+ is presented in the background, even though this stimulus is not related to the monetary outcome at all (Talmi et al. 2008). Affective biasing of instrumental behavior is generally adaptive, and abnormalities in these processes have been associated with psychiatric disorders (Damasio 1997). For example, using a PIT-like paradigm, task-irrelevant happy and angry faces influenced approaching and avoiding whole-body movements to obtain a reward in healthy controls, but not in violent offenders with psychopathic tendencies (Ly et al. 2016, 2014)

The AAT tests the arbitration between automatic affective reactions and instrumental behavior. The task is subdivided into two conditions: an affect-congruent condition, in which participants are required to make approaching movements toward positive stimuli and avoiding movements toward negative stimuli, and an affect-incongruent condition, which requires the exact opposite, i.e., avoiding movements when a positive stimulus is presented and approaching movements when a negative stimulus is presented. The affect-incongruent condition requires control over automatic response tendencies and is therefore more effortful, which is reflected in longer reaction times for the affect-incongruent compared to the affect-congruent condition (Chen and Bargh 1999; Rinck and Becker 2007; Rotteveel and Phaf 2004). This is referred to as the congruency effect (Fig. 1).

4.1 Neural Mechanisms

In terms of brain activity, the anterior prefrontal cortex (aPFC) and the adjacent ventrolateral PFC have consistently been found to be more active during the affect-incongruent compared to the affect-congruent task conditions (Radke et al. 2015; Roelofs et al. 2009a, b; Volman et al. 2011a, b; Volman et al. 2013, 2016). The aPFC has structural connections to the amygdala via the uncinate fasciculus (Von Der Heide et al. 2013) and is highly connected to multiple functional

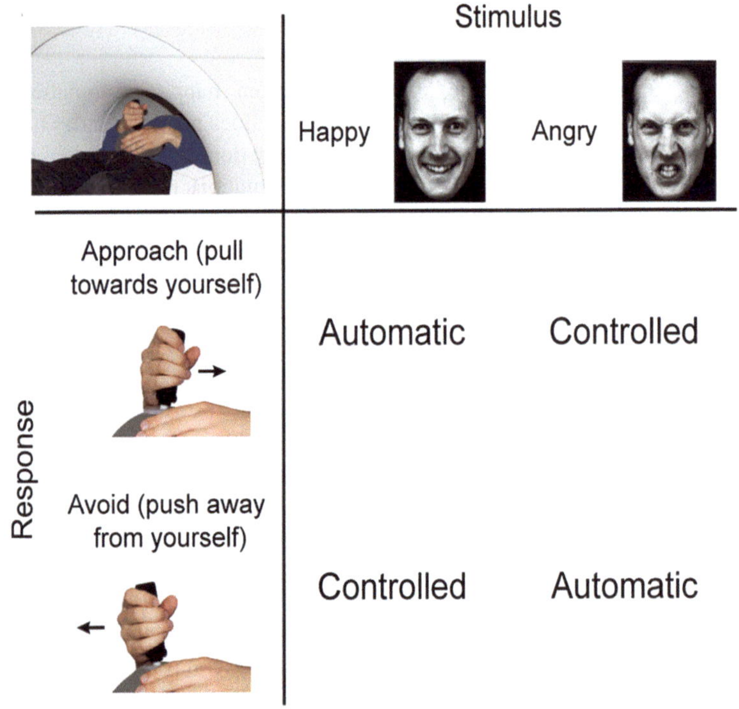

Fig. 1 Schematic overview of the approach–avoidance task (Joystick version). Participants are instructed to pull in response to happy faces or push in response to angry faces (affect-congruent) or vice versa (affect-incongruent). When the task is performed in a scanner, the joystick is placed on the abdomen of the participant. Outside the scanner, the joystick is placed on a table. (Figure from Volman et al. 2013)

networks, including the social–emotional network (see Box 1), cognitive processing network, and default mode network (Liu et al. 2013). It is implicated in the coordination of multiple cognitive processes (Ramnani and Owen 2004) and switching to alternative behaviors (Boorman et al. 2009). In the context of approach–avoidance behavior, the aPFC may coordinate automatic emotional processing with the implementation of a rule that depends on the automatic emotional processing. This coordination is implemented through modulatory effects on regions supporting those two constituent processes, namely the amygdala and the posterior parietal cortex (PPC), respectively (Volman et al. 2011a). Noteworthy, the areas involved in approach–avoidance behavior nicely fit with current models of the neural regions implicated in emotion processing (see Box 1). The dynamics of this network may be altered in individuals suffering from social psychopathologies. For example, psychopaths show less aPFC activation and less functional connectivity between the aPFC and amygdala during incongruent (control-related) trials, compared to healthy controls (Volman et al. 2016). This suggests a link between reduced prefrontal activation and reduced control during situations that provoke emotional reactions.

A study using continuous theta-burst stimulation (cTBS) to inhibit the left aPFC illustrates the central role of this structure in the emotional control over approach–avoidance tendencies. Compared to sham stimulation, active inhibition of this region shortly before the approach–avoidance task resulted in decreased task performance (i.e., more mistakes) during incongruent trials, but not during congruent trials. In addition, cTBS resulted in decreased cerebral blood flow in the bilateral aPFC, as well as increased blood flow in the amygdala. Taken together, these results support the notion that the aPFC is crucial for overriding automatic action tendencies, possibly via inhibition of the amygdala. In contrast, blood flow in the PPC decreased after aPFC inhibition, concordant with its role in rule selection (Volman et al. 2011a).

Box 1 Neural correlates of emotional processing

Over the past decades, there has been extensive research on the neural correlates of automatic emotional processes and control over emotion. The amygdala plays a key role in emotion processing, especially with respect to fear and arousal (LeDoux 2000) and salience processing in general. Salience processing refers to the identification of information relevant to survival (Seeley et al. 2007). The amygdala is connected to areas important for salience detection, such as the anterior insula and dorsal anterior cingulate cortex (Seeley et al. 2007) and responsible for the bodily fear response, such as the hypothalamus and periaqueductal gray (PAG) (Rodrigues et al. 2009). Emotional faces, among other salient stimuli, trigger automatic behavioral tendencies via the amygdala (Adolphs 2002). The prefrontal cortex has widely been implicated in emotion regulation by (inhibitory) control over the amygdala (Etkin et al. 2015), with different subregions serving specific types of regulation. The ventromedial prefrontal cortex (vmPFC) has inhibitory projections to the amygdala and is associated with the process of fear extinction (i.e., the reduction of the fear response toward a conditioned stimulus when the stimulus is no longer threatening) (Rodrigues et al. 2009). Another type of emotion regulation is cognitive reappraisal, which requires the reinterpretation of the meaning of the presented stimulus (Ochsner et al. 2002). Areas related to reappraisal include the dorsolateral prefrontal cortex (dlPFC), inferior frontal gyrus (IFG), and vmPFC (Ochsner et al. 2002). Control over emotional (approach–avoidance) actions implicates down-regulation of the amygdala by more anterior and ventrolateral parts of the prefrontal cortex (Volman et al. 2013).

An illustration of the network involved in control over approach–avoidance behavior is provided by Volman et al. (2013). Dynamic causal modeling (DCM) was used to define the neural circuit and the effective connectivity

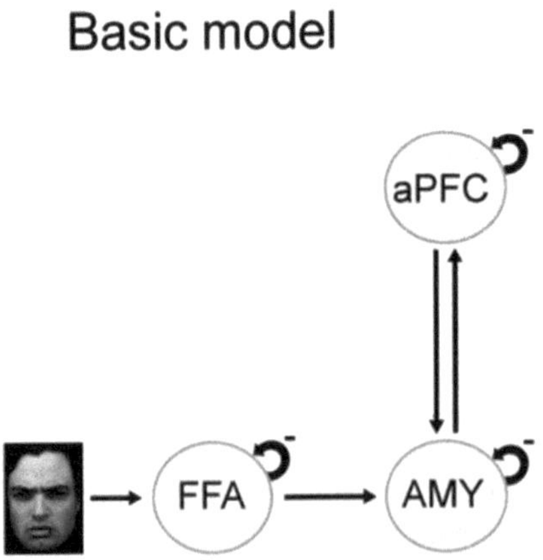

Fig. 2 DCM model as described by Volman et al. (2013). The figure shows a basic neural model of neural activity during the AAT. *aPFC* anterior prefrontal cortex, *FFA* fusiform face area, *AMY* amygdala. (Figure from Volman et al. 2013)

supporting AAT performance. The model focused on three fundamental nodes, namely the fusiform face area (FFA), necessary for processing the facial expressions used in the task (Kanwisher 1997); the amygdala, necessary for emotional processing of the stimulus (Adolphs 2002); and the aPFC, known to interact with the amygdala during AAT (Fig. 2; Volman et al. 2011a, b). Participants of the study performed the approach–avoidance task and showed the normal congruency effect, i.e., longer reaction times in the incongruent compared to congruent conditions, reflecting control over automatic responses during incongruent trials. DCM was used to estimate how the different structures interacted during task performance. Emotional control influenced aPFC activation via the feed-forward input from the amygdala and on the aPFC self-connection. Modulation of the amygdala occurred via the aPFC (Fig. 2). Interestingly, the inhibitory connection from the aPFC to the amygdala was weaker in carriers of a short allele (S-carriers) of the 5-HTTLPR gene compared to non-carriers. S-carriers are associated with reduced serotonin transporter availability and serotonin reuptake and are more likely to develop social psychopathologies after stressful events (Canli and Lesch 2007). Moreover, the amygdala response to threatening stimuli is increased in S-carriers (Hariri et al. 2002). The reduced inhibitory control of the aPFC over the amygdala in this group provides an explanation for the increased amygdala response and may be implicated in emotional vulnerability in general.

5 Manipulating Approach–Avoidance Behavior

5.1 Neuroendocrine Effects

Testosterone and cortisol are known to modulate emotional reactions. Neuroendocrine influences on approach–avoidance behavior have been investigated by determining effects of endogenous hormone levels and by drug administration studies. Noteworthy, effects of hormones have been shown to be especially relevant to approach–avoidance behavior in psychopathology related to social–emotional behavior.

5.1.1 Testosterone

Testosterone is released from the gonads as the end product of the hypothalamic–pituitary–gonadal (HPG) axis (Campbell et al. 2009) and is strongly associated with aggressive and approach-related behavior (Archer 2006; Bos et al. 2012). The hormone and its metabolites can influence neural activity via various neurochemical mechanisms. Baseline endogenous testosterone levels have shown to be positively correlated to activation of the amygdala and the orbitofrontal cortex, which is part of the prefrontal cortex (Mehta and Beer 2010; Van Wingen et al. 2011). Moreover, both the administration of testosterone and baseline testosterone levels are associated with decreased connectivity between the amygdala and prefrontal cortex (Peper et al. 2011)

Endogenous testosterone in male participants modulates the strength of the AAT congruency effect in the aPFC: Lower baseline (pretask) testosterone levels in saliva were associated with higher aPFC activity during incongruent (compared to congruent) trials. Moreover, functional connectivity between the aPFC and amygdala was modulated by endogenous baseline testosterone levels in this task. Lower baseline testosterone levels were associated with more negative functional connectivity between the aPFC and amygdala, suggesting a negative correlation between endogenous testosterone and prefrontal inhibitory control (Volman et al. 2011b). Psychopaths show a similar negative correlation between baseline testosterone levels and local activity in the aPFC as well as aPFC–amygdala connectivity (Volman et al. 2016). This underlines the impact of testosterone levels on control over emotional behavior. These findings are complemented by testosterone administration studies. Compared to placebo, sublingual testosterone administration in healthy female participants leads to reduced avoidance tendencies in response to angry faces, but not in response to happy faces (Enter et al. 2014). Neural effects of testosterone administration were specific to angry faces as well: In incongruent trials (approach-angry), testosterone administration resulted in increased amygdala activation (Radke et al. 2015). These testosterone administration studies indicate that testosterone biases behavior toward the approach of social threat, which may underlie mechanisms of social dominance and aggression. The approach-related effects of testosterone may be beneficial for individuals suffering from social anxiety disorder (SAD): A recent study by Enter et al. (2016) showed that SAD patients show increased approach tendencies toward angry faces after sublingual testosterone administration compared to placebo, indicating that testosterone reduces the social-avoidant tendencies in these patients.

5.1.2 Cortisol

The glucocorticoid cortisol is the end product of the hypothalamic–pituitary–adrenal (HPA) axis. The HPA axis plays an important role in responding to stressful events (De Kloet et al. 2005). Perception of a stressor leads to the activation of a cascade of hormones starting at the hypothalamus, ultimately leading to the release of cortisol by the adrenal glands. Cortisol increases activity of the sympathetic nervous system, which is important for fight and flight behavior. In the brain, cortisol binds to areas that contain glucocorticoid or mineralocorticoid receptors, such as the amygdala, hippocampus, and frontal areas (Lupien et al. 2007). As cortisol plays especially an important role in the context of responding to stress, studies have focused on cortisol levels in response to stress induction and cortisol administration rather than baseline cortisol levels. For example, cortisol administration has been shown to reduce amygdala activity in response to emotional expressions (Henckens et al. 2010).

Cortisol levels can be increased by stress induction, for example, by the Trier Social Stress Test (TSST), during which participants have to give a speech and do a mental arithmetic task in front of an audience (Kirschbaum et al. 1993). Stress-induced cortisol in male and female patients with SAD was associated with stronger avoidance in response to angry faces, compared to healthy controls (Roelofs et al. 2009). Also, cortisol administration enhances the behavioral congruency effect specifically for angry faces but only in highly avoidant healthy subjects: Male participants with a high score on the behavioral inhibition scale (BIS) were slower to approach an angry face after cortisol administration, compared to placebo (van Peer et al. 2007). This behavioral effect was accompanied by increased event-related potentials (ERPs), P150 and P3, in response to angry faces when an avoidant movement has to be made, suggesting increased processing of threat stimuli. Similar results were obtained in a cortisol administration study with male and female patients with SAD, in which an increased avoidance tendency was found in response to angry faces only. Additionally in SAD patients with high symptom severity, there was also an effect of cortisol on the P150 amplitude (i.e., increased visual processing) during avoidance, especially in response to angry faces (van Peer et al. 2009). Taken together, the effects of cortisol on threat avoidance seem to be specific for highly anxious individuals and might be an underlying factor in the maintenance of social avoidance in SAD.

5.1.3 Oxytocin

The neuropeptide oxytocin is synthesized in the hypothalamus and acts on various brain structures, such as the amygdala, hippocampus, and brainstem (Meyer-Lindenberg et al. 2011). Oxytocin has prosocial and anxiolytic properties (Koch et al. 2014; Meyer-Lindenberg et al. 2011) and has been implicated in social approach behavior (Heinrichs et al. 2009). After intranasal oxytocin administration, low socially anxious males showed increased approaching tendencies in response to specifically angry faces, while they showed the normal tendency to avoid angry faces in the placebo condition (Radke et al. 2013). This effect was absent in high socially anxious males. In addition, no effects of oxytocin on approach tendencies toward happy faces were found, which might have been expected given the prosocial effect of oxytocin. Therefore, the beneficial effects on approach behavior seem mainly to be due to the anxiolytic—and not the prosocial—effects of oxytocin (Radke et al. 2013).

5.2 Training

Approach–avoidance tendencies can be changed with behavioral training, in which an adapted version of the approach–avoidance task is used. The aim of the training could be to reduce an avoidance bias or to increase an approach bias. For instance, in the case of anxiety, participants are instructed to pull the stimulus that they would normally avoid (e.g., a spider in spider phobia) and to push away a neutral or positive control stimulus. The aim of the training could also be reducing an approach bias. For instance, in the case of addiction, exactly opposite instructions are used: Participants have to push away (abuse-related) stimuli that they would normally approach (e.g., a picture of an alcoholic drink in alcohol abuse).

Promising results have been obtained, for example, in the training of highly socially anxious individuals, who are inclined to avoid happy and angry faces. When highly socially anxious males and females are trained to approach happy male and female faces and avoid checkerboards, they approach female happy faces faster, compared to individuals who received the opposite training (i.e., avoid happy faces and approach checkerboards). Importantly, after training, they show less self-rated emotional vulnerability and anxiety in response to a social stress task (Rinck et al. 2013). Similar trainings have been shown to be beneficial for treatment outcome in alcohol-dependent inpatients. Alcohol-dependent/alcoholic patients who received a training using the AAT in which they have to avoid (i.e., push away) pictures of alcoholic drinks, showed an alcohol avoidance bias, instead of their pretraining alcohol approach bias. Moreover, after regular treatment, they showed less relapse after one-year follow-up. No effect on alcohol avoidance bias was found in the control groups of alcoholic patients receiving sham training or no training at all (Wiers et al. 2011). Approach–avoidance training has also been successfully applied in improving racial attitudes. Non–African American students were either trained to approach photographs of African American persons and avoid Caucasian persons, or approach Caucasian and avoid African American, or move the joystick sideways. Compared to the other conditions, participants in the approach-African American condition showed reduced implicit racial prejudice during an implicit association task and increased immediacy during interaction with a African American confederate (i.e., decrease in distance and increase in body orientation directed toward the confederate) (Kawakami et al. 2007).

6 Discussion and Future Horizons

Approach–avoidance tendencies result from the evaluation of positive and negative stimuli or events. Positive evaluations are associated with approaching behavior, whereas negative evaluations are related to avoidance. Behavior opposite to these automatic tendencies requires emotional control, reflected in slower action initiation. In the brain, the regulation of approach–avoidance behavior is marked by inhibition of the amygdala by the aPFC (see Fig. 3). The amygdala plays a central role in the network underlying emotional action processing, which also contains visual processing areas such as the FFA. The aPFC coordinates the implementation of task instructions, associated with PPC activation, and emotional action tendencies supported by the amygdala. Decoupling of functional connectivity between the aPFC and amygdala is associated with reduced control over emotional actions. The fact that control of approach–avoidance behavior requires involvement of these anterior regions of the PFC suggests that control over social approach and avoidance may involve coordination of multiple hierarchically organized processes (Koechlin and Summerfield 2007). Future studies should investigate the content and the format of the neural representations that are coordinated during emotional control. It is also important to understand whether those control circuits are used only when social approach–avoidance behavior needs to be controlled. The interplay of aPFC, PPC, and amygdala eventually leads to behavior via the motor system. The PMC is a likely candidate for this integration (Boorman et al. 2009). It is important to understand how motor synergies for approach/avoidance, computed in the PMC (Graziano 2006), are accessed by the aPFC during emotional

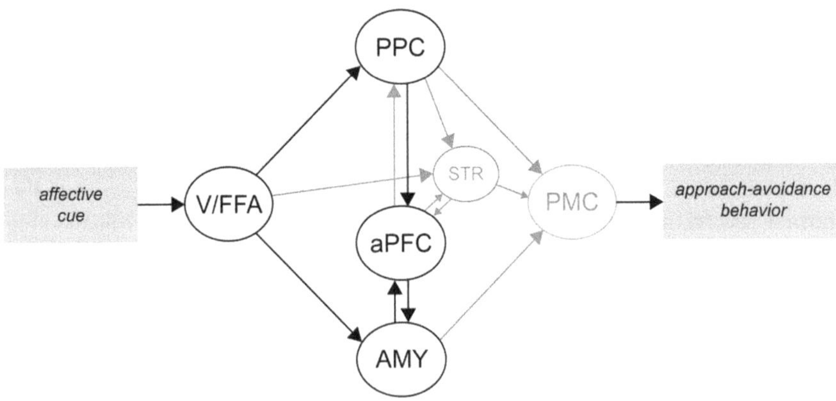

Fig. 3 Schematic model of brain regions involved in (control over) approach–avoidance behavior. Affective cues trigger emotional action tendencies via the amygdala. The aPFC arbitrates between these automatic tendencies and current behavioral goals, associated with posterior parietal cortex activation. Areas and connections in gray are hypothesized output regions. *PPC* posterior parietal cortex, *aPFC* anterior prefrontal cortex, *V/FFA* visual processing areas/fusiform face area, *AMY* amygdala, *PMC* premotor cortex, *STR* striatum

regulation. A likely route through which the aPFC could influence the PMC is provided by fronto-striatal loops (Haber and Knutson 2009). The striatum plays a central role in value-based action decisions (Rangel et al. 2008) and actions to obtain reward or avoid punishment in monetary and social incentive delay tasks (Spreckelmeyer et al. 2009). In AAT studies, it has also been found to be more active during affect-incongruent trials, in which the behavioral goal conflicts with the automatic tendency (Volman et al. 2016).

Social approach–avoidance behavior and the dynamics of the underlying neural network have been found to be altered in several (social–emotional) psychiatric disorders and to be influenced by (exogenous administration of) hormones such as testosterone and cortisol. Increased approach-related behavior toward angry faces has been found in patients with psychopathy, as well as in healthy female participants after testosterone administration. Similarly, reduced functional aPFC–amygdala connectivity and reduced aPFC activation during situations that require control (i.e., incongruent trials) were associated with both psychopathy and increased endogenous testosterone levels. In sum, testosterone is anxiolytic and promotes social approach behavior. Increased levels of cortisol, caused by stress induction or cortisol administration, enhance the tendency to avoid angry faces in anxious males and females. Moreover, cortisol led to higher amplitudes of ERPs related to visual processing, during the avoidance of angry faces. Taken together, cortisol seems to enhance threat processing and already existent avoidant behavior. Similar to testosterone, oxytocin has been shown to have anxiolytic effects, resulting in more approach to social threat, but only in low anxious individuals.

In this chapter, we did not go into detail about findings on lateralization of approach–avoidance tendencies. It has been proposed that the different motivational systems correlate with different sites of the prefrontal cortex: Approach motivation relates to left-sided activation of this brain area, whereas avoidance/withdrawal motivation is associated with right-sided activation (Davidson 2004). This notion of lateralized motivational systems is supported by electroencephalography (EEG) studies of baseline prefrontal activation asymmetry. For example, depressed patients show less left than right frontal activity compared to healthy controls (Davidson 1998). Moreover, behavioral approach sensitivity, measured on the BAS scale, relates to greater left than right frontal cortical activity (Harmon-Jones and Allen 1997). From functional MRI research, however, there is little evidence for lateralization of these motivational systems or affect in general (Wager et al. 2003), although one study found increased left (compared to right) dorsal prefrontal cortex activation during approach (compared to avoidance) action decisions (Berkman and Lieberman 2010). However, the paradigm used in that study is not likely to elicit automatic approach–avoidance action tendencies, because participants were required to imagine themselves being part of a tribe that made different approach or avoidance decisions than they would normally do.

6.1 Future Agenda

Research on the regulation of emotion has mainly focused on cognitive strategies related to the altering the affective evaluation of salient stimuli or events, such as emotion suppression, emotion reappraisal, and redirection of attention (Gross 2002). The control over emotional actions that follow such affective evaluations has remained a largely understudied aspect of emotion regulation. This is surprising, given the importance of these action tendencies in our understanding of emotional processing and social–emotional behavior. The control over automatic approach–avoidance tendencies, as operationalized in so-called approach–avoidance tasks, may serve as an exemplar of this type of emotional control. More research on the underlying neural correlates of controlling approach–avoidance and its relationship with other aspects of emotion regulation such as reappraisal and attention redirection will further our understanding on how individuals deal with emotional events they encounter. First hints that these processes may be related come from a study showing that neural medial and lateral PFC effects during appraisal are partly explained by regulating (avoidant) eye movements (van Reekum et al. 2007).

The studies discussed here indicate that endocrine influences on approach–avoidance behavior may play an important role in the development and maintenance of problems in social–emotional behavior. However, hormones can also have a beneficial effect on exaggerated approach–avoidance tendencies, as has been shown in the study by Enter and colleagues (2014, 2016): The approach-related effects of testosterone counteracted the avoidant tendencies toward angry faces in healthy subjects (Enter et al. 2014) and in SAD patients (Enter et al. 2016). This result shows the potential of hormones in the enhancement of existing therapies, such as cognitive–behavioral therapy. For example, it is worth investigating if testosterone might help patients during exposure to situations that are experienced as socially threatening.

Another way of improving maladaptive approach or avoidance tendencies is by behavioral training, which has been shown to be effective for alcohol addiction and social anxiety disorder (Rinck et al. 2013; Wiers et al. 2011). To date, little is known about the effects of such training on the underlying neural mechanisms. It would be insightful, for example, in case of social anxiety, to know whether reductions in avoiding social threat after training originate from increased top-down inhibitory control from the aPFC over the amygdala, or from reduced salience of the social threatening stimulus or event. The first explanation would indicate that approach–avoidance training increases emotional control over specific automatic tendencies, whereas the latter explanation would suggest that the affective evaluation of the stimulus itself has changed. A recent study on the effects of approach–avoidance training in alcoholism showed a reduction in amygdala activation in response to alcohol cues, suggesting devaluation of the stimulus (Wiers et al. 2014). More understanding about the underlying mechanism of approach–avoidance training could also be beneficial for the development of specific combination treatments. For example, administration of testosterone in SAD patients before approach–avoidance training of emotional faces might make the training even more effective.

Although it is clear that psychiatric disorders characterized by deficits in social–emotional functioning are associated with altered approach–avoidance behavior, little is known about the causal role of approach–avoidance tendencies and the (in) ability to control them. It would be valuable to know whether a lack of control over these tendencies can actually predict emotional vulnerability and/or the development of social psychopathology. Some evidence hints in that direction, for example, the study by Volman et al. (2013), showing reduced top-down control over approach–avoidance tendencies in carriers of a polymorphism associated with increased risk of social psychopathology, such as anxiety, depression, and aggression-related disorders, after stressful events. Longitudinal investigations could shed light on whether abnormal approach–avoidance tendencies relate pre-existing vulnerability for social psychopathology or that they are acquired abnormalities. Such a study could be performed in groups at risk of social psychopathologies or individuals who are likely to encounter stressful events, such as soldiers or policemen. More knowledge about the underlying mechanisms of the development and maintenance of social psychopathology is crucial for prevention of these disorders and the development of novel treatment approaches.

7 Conclusion

The study of the regulation of approach–avoidance tendencies has been shown to be instrumental for understanding affective behavior. Individual differences in (control over) approach–avoidance behavior are associated with levels of hormones such as testosterone and cortisol, and abnormal social interactions, as observed in social anxiety and psychopathy. The amygdala, triggering action tendencies in response to emotionally salient stimuli, and anterior PFC, inhibiting the amygdala when appropriate, play an important role in the neural network underlying the regulation of approach–avoidance tendencies. We have only just begun to explore the potential of the approach–avoidance paradigm in understanding the role of emotional action regulation in normal and affective behavior.

References

Adolphs R (2002) Recognizing emotion from facial expressions: Psychological and neurological mechanisms. Behav Cogn Neurosci Rev 1:21–62

Archer J (2006) Testosterone and human aggression: an evaluation of the challenge hypothesis. Neurosci Biobehav Rev 30(3):319–345. doi:10.1016/j.neubiorev.2004.12.007

Berkman ET, Lieberman MD (2010) Approaching the bad and avoiding the good: lateral prefrontal cortical asymmetry distinguishes between action and valence. J Cogn Neurosci 22 (9):1970–1979. doi:10.1162/jocn.2009.21317

Boorman ED, Behrens TEJ, Woolrich MW, Rushworth MFS (2009) How green is the grass on the other side? Frontopolar cortex and the evidence in favor of alternative courses of action. Neuron 62(5):733–743. doi:10.1016/j.neuron.2009.05.014
Bos PA, Hermans EJ, Ramsey NF, van Honk J (2012) The neural mechanisms by which testosterone acts on interpersonal trust. NeuroImage 61(3):730–737. doi:10.1016/j.neuroimage.2012.04.002
Campbell DW, Sareen J, Stein MB, Kravetsky LB, Paulus MP, Hassard ST, Reiss JP (2009) Happy but not so approachable: the social judgments of individuals with generalized social phobia. Depress Anxiety 26(5):419–424. doi:10.1002/da.20474
Canli T, Lesch K-P (2007) Long story short: the serotonin transporter in emotion regulation and social cognition. Nat Neurosci 10(9):1103–1109. doi:10.1038/nn1964
Carver CS, Sutton SK, Scheier MF (2000) Action, emotion, and personality. Pers Soc Psychol Bull. doi:10.1177/0146167200268008
Chen M, Bargh JA (1999) Consequences of automatic evaluation: immediate behavioral predispositions to approach or avoid the stimulus. Pers Soc Psychol Bull 25(2):215–224. doi:10.1177/0146167299025002007
Damasio A (1997) Neuropsychology. towards a neuropathology of emotion and mood. Nature 386 (6627):769–770. doi:10.1038/386769a0
Davidson RJ (1998) Affective style and affective disorders: perspectives from affective neuroscience. Cogn Emot 12(3):307–330. doi:10.1080/026999398379628
Davidson RJ (2004) What does the prefrontal cortex "do" in affect: perspectives on frontal EEG asymmetry research. Biol Psychol 67(1–2):219–234. doi:10.1016/j.biopsycho.2004.03.008
De Houwer J, Crombez G, Baeyens F, Hermans D (2001) On the generality of the affective Simon effect. Cogn Emotion 15:189–206. doi:10.1080/02699930125883
De Kloet ER, Joëls M, Holsboer F (2005) Stress and the brain: from adaptation to disease. Nat Rev Neurosci 6(6):463–475. doi:10.1038/nrn1683
Elliot AJ (2006) The hierarchical model of approach-avoidance motivation. Motiv Emotion 30 (2):111–116. doi:10.1007/s11031-006-9028-7
Enter D, Spinhoven P, Roelofs K (2014) Alleviating social avoidance: effects of single dose testosterone administration on approach-avoidance action. Horm Behav 65(4):351–354. doi:10.1016/j.yhbeh.2014.02.001
Enter D, Spinhoven P, Roelofs K (2016) Dare to approach single dose testosterone administration promotes threat approach in patients with social anxiety disorder. Clin Psychol Sci. doi:10.1177/2167702616631499
Etkin A, Büchel C, Gross JJ (2015) The neural bases of emotion regulation. Nat Rev Neurosci 16:693–700. doi:10.1038/nrn4044
Frijda NH (1986) The emotions. Cambridge University Press, London
Gray JA (1990) Brain systems that mediate both emotion and cognition. Cogn Emot 4(3):269–288. doi:10.1080/02699939008410799
Graziano M (2006) The organization of behavioral repertoire in motor cortex. Annu Rev Neurosci 29(1):105–134. doi:10.1146/annurev.neuro.29.051605.112924
Gross JJ (2002) Emotion regulation: affective, cognitive, and social consequences. Psychophysiology 39(3):281–291. doi:10.1017/S0048577201393198
Haber SN, Knutson B (2009) The reward circuit: linking primate anatomy and human imaging. Neuropsychopharmacol Official Publ Am Coll Neuropsychopharmacol 35(1):1–23. doi:10.1038/npp.2009.129
Hare TA, Tottenham N, Davidson MC, Glover GH, Casey BJ (2005) Contributions of amygdala and striatal activity in emotion regulation. Biol Psychiatry 57(6):624–632. doi:10.1016/j.biopsych.2004.12.038
Hariri AR, Mattay VS, Tessitore A, Kolachana B, Fera F, Goldman D, Weinberger DR (2002) Serotonin transporter genetic variation and the response of the human amygdala. Sci (New York, N.Y.) 297(5580):400–403. doi:10.1126/science.1071829

Harmon-Jones E, Allen JJ (1997) Behavioral activation sensitivity and resting frontal EEG asymmetry: covariation of putative indicators related to risk for mood disorders. J Abnorm Psychol 106(1):159–163. doi:10.1037/0021-843X.106.1.159
Heinrichs M, von Dawans B, Domes G (2009) Oxytocin, vasopressin, and human social behavior. Front Neuroendocrinol 30(4):548–557. doi:10.1016/j.yfrne.2009.05.005
Henckens MJAG, van Wingen GA, Joëls M, Fernández G (2010) Time-dependent effects of corticosteroids on human amygdala processing. J Neurosci 30(38):12725–12732. doi:10.1523/JNEUROSCI.3112-10.2010
Heuer K, Rinck M, Becker ES (2007) Avoidance of emotional facial expressions in social anxiety: the approach-avoidance task. Behav Res Ther 45(12):2990–3001. doi:10.1016/j.brat.2007.08.010
Kanwisher N (1997) The fusiform face area: a module in human extrastriate cortex specialized for face perception. J Neurosci 17(11):4302–4311
Kawakami K, Phills CE, Steele JR, Dovidio JF (2007) (Close) distance makes the heart grow fonder: improving implicit racial attitudes and interracial interactions through approach behaviors. J Pers Soc Psychol 92(6):957–971. doi:10.1037/0022-3514.92.6.957
Kirschbaum C, Pirke KM, Hellhammer DH (1993) The "trier social stress test"–a tool for investigating psychobiological stress responses in a laboratory setting. Neuropsychobiology. doi:10.1159/000119004
Koch SBJ, van Zuiden M, Nawijn L, Frijling JL, Veltman DJ, Olff M (2014) Intranasal oxytocin as strategy for medication-enhanced psychotherapy of PTSD: salience processing and fear inhibition processes. Psychoneuroendocrinology 40:242–256. doi:10.1016/j.psyneuen.2013.11.018
Koechlin E, Summerfield C (2007) An information theoretical approach to prefrontal executive function. Trends Cogn Sci 11(6):229–235. doi:10.1016/j.tics.2007.04.005
Krieglmeyer R, Deutsch R (2010) Comparing measures of approach–avoidance behaviour: the manikin task vs. two versions of the joystick task. Cogn Emot 24(5):810–828. doi:10.1080/02699930903047298
Lang PJ, Bradley MM, Cuthbert BN (1990) Emotion, attention, and the startle reflex. Psychol Rev 97(3):377–395. doi:10.1037/0033-295X.97.3.377
LeDoux J (2000) Emotion circuits in the brain. Ann Rev Neurosci, 155–184
Liu H, Qin W, Li W, Fan L, Wang J, Jiang T, Yu C (2013) Connectivity-based parcellation of the human frontal pole with diffusion tensor imaging. J Neurosci Official J Soc Neurosci 33(16):6782–6790. doi:10.1523/JNEUROSCI.4882-12.2013
Lupien SJ, Maheu F, Tu M, Fiocco A, Schramek TE (2007) The effects of stress and stress hormones on human cognition: implications for the field of brain and cognition. Brain Cogn 65(3):209–237. doi:10.1016/j.bandc.2007.02.007
Ly V, Huys QJM, Stins JF, Roelofs K, Cools R (2014) Individual differences in bodily freezing predict emotional biases in decision making. Front Behav Neurosci 8(July):1–9. doi:10.3389/fnbeh.2014.00237
Ly V, Von Borries AKL, Brazil IA, Bulten HB, Cools R, Roelofs K (2016) Reduced transfer of affective value toinstrumental behavior in violent offenders. Submitted
Mehta PH, Beer J (2010) Neural mechanisms of the testosterone-aggression relation: the role of orbitofrontal cortex. J Cogn Neurosci 22(10):2357–2368. doi:10.1162/jocn.2009.21389
Meyer-Lindenberg A, Domes G, Kirsch P, Heinrichs M (2011) Oxytocin and vasopressin in the human brain: social neuropeptides for translational medicine. Nat Rev Neurosci 12(9):524–538. doi:10.1038/nrn3044
Ochsner KN, Bunge SA, Gross JJ, Gabrieli JDE (2002) Rethinking feelings: an FMRI study of the cognitive regulation of emotion. J Cog Neurosci 14(8):1215–29
Peper JS, van den Heuvel MP, Mandl RCW, Hulshoff Pol HE, van Honk J (2011) Sex steroids and connectivity in the human brain: a review of neuroimaging studies. Psychoneuroendocrinology 36(8):1101–1113. doi:10.1016/j.psyneuen.2011.05.004

Phaf RH, Mohr SE, Rotteveel M, Wicherts JM (2014) Approach, avoidance, and affect: a meta-analysis of approach-avoidance tendencies in manual reaction time tasks. Front Psychol 5 (MAY):1–16. doi:10.3389/fpsyg.2014.00378

Radke S, Roelofs K, de Bruijn ERA (2013) Acting on anger: social anxiety modulates approach-avoidance tendencies after oxytocin administration. Psychol Sci 24(8):1573–1578. doi:10.1177/0956797612472682

Radke S, Volman I, Mehta P, van Son V, Enter D, Sanfey A, Toni I, de Bruijn ERA, Roelofs K (2015) Testosterone biases the amygdala toward social threat approach. Sci Adv 1(5): e1400074

Ramnani N, Owen AM (2004) Anterior prefrontal cortex: insights into function from anatomy and neuroimaging. Nat Rev Neurosci 5(3):184–194. doi:10.1038/nrn1343

Rangel A, Camerer C, Montague PR (2008) A framework for studying the neurobiology of value-based decision making. Nat Rev Neurosci 9(7):545–556. doi:10.1038/nrn2357

Rinck M, Becker ES (2007) Approach and avoidance in fear of spiders. J Behav Ther Exp Psychiatry 38(2):105–120. doi:10.1016/j.jbtep.2006.10.001

Rinck M, Telli S, Kampmann IL, Woud ML, Kerstholt M, Te Velthuis S, Becker ES (2013) Training approach-avoidance of smiling faces affects emotional vulnerability in socially anxious individuals. Front Human Neurosci 7:481. doi:10.3389/fnhum.2013.00481

Rodrigues SM, LeDoux JE, Sapolsky RM (2009) The influence of stress hormones on fear circuitry. Annu Rev Neurosci 32:289–313. doi:10.1146/annurev.neuro.051508.135620

Roelofs K, Minelli A, Mars RB, van Peer J, Toni I (2009a) On the neural control of social emotional behavior. Soc Cog Affect Neurosci 4(1):50–58. doi:10.1093/scan/nsn036

Roelofs K, van Peer J, Berretty E, Jong P De, Spinhoven P, Elzinga BM (2009b) Hypothalamus-pituitary-adrenal axis hyperresponsiveness is associated with increased social avoidance behavior in social Phobia. Biol Psychiatry 65(4):336–343. doi:10.1016/j.biopsych.2008.08.022

Roelofs K, Putman P, Schouten S, Lange WG, Volman I, Rinck M (2010) Gaze direction differentially affects avoidance tendencies to happy and angry faces in socially anxious individuals. Behav Res Ther 48(4):290–294. doi:10.1016/j.brat.2009.11.008

Rotteveel M, Phaf RH (2004) Automatic affective evaluation does not automatically predispose for arm flexion and extension. Emotion (Washington, D.C.) 4(2):156–172. doi:10.1037/1528-3542.4.2.156

Seeley WW, Menon V, Schatzberg AF, Keller J, Glover GH, Kenna H, Greicius MD (2007) Dissociable intrinsic connectivity networks for salience processing and executive control. J Neurosci 27(9):2349–2356. doi:10.1523/JNEUROSCI.5587-06.2007

Solarz AK (1960) Latency of instrumental responses as a function of compatibility with the meaning of eliciting verbal signs. J Exp Psychol 59(4):239–245. doi:10.1037/h0047274

Spreckelmeyer KN, Krach S, Kohls G, Rademacher L, Irmak A, Konrad K, Grunder G (2009) Anticipation of monetary and social reward differently activates mesolimbic brain structures in men and women. Soc Cog Affect Neurosci 4(2):158–165. doi:10.1093/scan/nsn051

Talmi D, Seymour B, Dayan P, Dolan RJ (2008) Human Pavlovian instrumental transfer. J Neurosci 28(2):360–368. doi:10.1523/JNEUROSCI.4028-07.2008

van Peer JM, Roelofs K, Rotteveel M, van Dijk JG, Spinhoven P, Ridderinkhof KR (2007) The effects of cortisol administration on approach-avoidance behavior: an event-related potential study. Biol Psychol 76(3):135–146. doi:10.1016/j.biopsycho.2007.07.003

van Peer JM, Spinhoven P, van Dijk JG, Roelofs K (2009) Cortisol-induced enhancement of emotional face processing in social phobia depends on symptom severity and motivational context. Biol Psychol 81(2):123–130. doi:10.1016/j.biopsycho.2009.03.006

van Reekum CM, Johnstone T, Urry HL, Thurow ME, Schaefer HS, Alexander AL, Davidson RJ (2007) Gaze fixations predict brain activation during the voluntary regulation of picture-induced negative affect. NeuroImage 36(3):1041–1055. doi:10.1016/j.neuroimage.2007.03.052

Van Wingen GA, Ossewaarde L, Bäckström T, Hermans EJ, Fernández G (2011) Gonadal hormone regulation of the emotion circuitry in humans. Neuroscience 191:38–45. doi:10.1016/j.neuroscience.2011.04.042

Volman I, Roelofs K, Koch S, Verhagen L, Toni I (2011a) Anterior prefrontal cortex inhibition impairs control over social emotional actions. Curr Biol CB 21(20):1766–1770. doi:10.1016/j.cub.2011.08.050

Volman I, Toni I, Verhagen L, Roelofs K (2011b) Endogenous testosterone modulates prefrontal-amygdala connectivity during social emotional behavior. Cereb Cortex (New York, N.Y., 1991) 21(10):2282–90. doi:10.1093/cercor/bhr001

Volman I, Verhagen L, den Ouden HEM, Fernández G, Rijpkema M, Franke B, Roelofs K (2013) Reduced serotonin transporter availability decreases prefrontal control of the amygdala. J Neurosci Official J Soc Neurosci 33(21):8974–9. doi:10.1523/JNEUROSCI.5518-12.2013

Volman I, Louise K, von Borries A, Hendrik Bulten B, Jan Verkes R, Toni I, Roelofs K (2016) Testosterone modulates altered prefrontal control of emotional actions in psychopathic offenders. Eneuro. doi:10.1523/ENEURO.0107-15.2016

Von Borries AKL, Volman I, de Bruijn ERA, Bulten BH, Verkes RJ, Roelofs K (2012) Psychopaths lack the automatic avoidance of social threat: Relation to instrumental aggression. Psychiatry Res 200(2–3):761–766. doi:10.1016/j.psychres.2012.06.026

Von Der Heide RJ, Skipper LM, Klobusicky E, Olson IR (2013) Dissecting the uncinate fasciculus: disorders, controversies and a hypothesis. Brain 136(6):1692–1707. doi:10.1093/brain/awt094

Wager TD, Phan KL, Liberzon I, Taylor SF (2003) Valence, gender, and lateralization of functional brain anatomy in emotion: a meta-analysis of findings from neuroimaging. NeuroImage 19(3):513–531. doi:10.1016/S1053-8119(03)00078-8

Wiers RW, Eberl C, Rinck M, Becker ES, Lindenmeyer J (2011) Retraining automatic action tendencies changes alcoholic patients' approach bias for alcohol and improves treatment outcome. Psychol Sci 22(4):490–497. doi:10.1177/0956797611400615

Wiers CE, Stelzel C, Gladwin TE, Park SQ, Pawelczack S, Gawron CK, Bermpohl F (2014) Effects of cognitive bias modification training on neural alcohol cue reactivity in alcohol dependence. Am J Psychiatry. doi:10.1176/appi.ajp.2014.13111495

Mapping Social Interactions: The Science of Proxemics

Cade McCall

Abstract Interpersonal distance and gaze provide a wealth of information during face-to-face social interactions. These "proxemic" behaviors offer a window into everyday social cognition by revealing interactants' affective states (e.g., interpersonal attitudes) and cognitive responses (e.g., social attention). Here we provide a brief overview of the social psychological literature in this domain. We focus on new techniques for experimentally manipulating and measuring proxemics, including the use of immersive virtual environments and digital motion capture. We also discuss ways in which these approaches can be integrated with psychophysiological and neuroimaging techniques. Throughout, we argue that contemporary proxemics research provides psychology and neuroscience with a means to study social cognition and behavior as they naturally emerge and unfold in vivo.

Keywords Proxemics · Social psychology · Social neuroscience · Social interaction

Contents

1 Proxemics: Individual and Group Processes ... 297
2 Social Interaction, Proxemics, and Challenge of Experimental Control ... 298
3 Measuring Social Space ... 301
4 Future Directions ... 305
References ... 306

Whether supportive, strategic, combative, or romantic, social interactions are at the core of everyday experience. They provide the contexts in which our social capabilities evolved and develop. Nevertheless, research in psychology and neuroscience seldom examines social interaction directly. Instead, we tend to focus on the indi-

C. McCall (✉)
Max Planck Institute for Human Cognitive and Brain Sciences, Stephanstraße 1A, 04103 Leipzig, Germany
e-mail: mccall.cade@gmail.com

Curr Topics Behav Neurosci (2016) 30: 295–308
DOI 10.1007/7854_2015_431
Published Online: 04 January 2016

vidual's responses to abstracted social stimuli such as photos, films, or vignettes, without examining the cognitive, affective, and behavioral processes that unfold during an actual interaction. The reason for this shortcoming is clear: Social interactions are hard targets. They are difficult to control, multidimensional, and have a give-and-take quality that makes them inherently nonlinear. Given these facts, identifying meaningful measures that account for both individual and collective responses during an interaction is an important challenge for empirical research.

The study of proxemics addresses that challenge by looking at the ways in which people use physical space during face-to-face encounters (Hall 1969, 1973). Proxemic behaviors include the degree to which interactants approach or avoid each other, orient toward or away from each other, and engage in mutual, joint, or averted gaze. These behaviors provide continuously observable targets during an interaction. Unlike subjective measures, proxemic measures do not necessarily rely upon the conscious reflection of subjects; they are relatively automatic and reflect a variety of underlying social and affective processes (e.g., Dotsch and Wigboldus 2008; Dovidio et al. 1997). Unlike many physiological measures, proxemics can be measured in vivo, during actual social interactions, and without hooking participants up to monitoring equipment or placing them in a confined space (i.e., an MRI scanner). In this sense, proxemic research allows us to take a more ethological approach to study interactions (Kingstone et al. 2008) and to conduct research with a great deal of face validity.

Heider and Simmel's classic research on anthropomorphism (Heider and Simmel 1944) nicely illustrates the meaning inherent in simple proxemic behaviors. The films in those studies showed little more than geometrical shapes approaching and avoiding each other over a brief period of time. But as anyone who has watched the films knows, one automatically infers the thoughts and feelings of those shapes. When we see a triangle approach a circle and then watch the two shapes move across the screen together, we see a relationship forming between two living agents. Their simple proxemic behaviors are sufficient to convey a relatively rich social narrative. This intuitive mapping of social space drove early proxemics research in which participants were asked to depict social relationships with cartoons or by arranging felt figures on a board (Kuethe 1962; Pedersen and Shears 1973). While these methods are rather crude and do not always predict real-world proxemic behavior (Hayduk 1978), they demonstrate that people spontaneously map complex constructs such as intimacy, social schemata, and interpersonal attitudes onto distances and orientations. Proximity provides an intuitive metaphor and a physical analog to psychological representations of relationships.

This emphasis on proximity in social representations is not surprising given the central role of approach and avoidance in affect (e.g., Elliot 2008). While approach and avoidance motivations are often discussed in relatively abstract terms (Carver and Harmon-Jones 2009; Elliot 2006), they take very literal form in physical behavior. For example, positive and negative attitudes (Chen and Bargh 1999; McCall et al. 2012; Solarz 1960) and emotional displays (e.g., sad vs. angry facial expressions) (e.g., Marsh et al. 2005) can automatically facilitate approach or avoidant motor behaviors. During a social interaction, these basic processes emerge in proxemic behavior via the actions we automatically use to increase or decrease interpersonal distance and engagement with others.

Gaze, the other central ingredient of proxemic behavior, is also rich in social meaning. We are acutely attuned to the direction of others' gaze direction and head orientation (Loomis et al. 2008; Schilbach 2015). Even when only observing another person, we use the direction of their gaze to infer their attentional focus and, in turn, to direct our own attention to relevant information in the environment (Frischen et al. 2007). This engagement in joint gaze is one of our most basic social processes and is core to the evolution of social cognition (Tomasello 1995). Mutual gaze, directly gazing into another's eyes, is another powerful feature of nonverbal communication. Even infants prefer looking at faces with direct as compared to averted gaze (Farroni et al. 2002) and direct gaze facilitates a wide variety of cognitive processes including memory encoding and facial expression recognition (Schilbach 2015; Senju and Johnson 2009).

Together these brief snapshots of the literature demonstrate that gaze and interpersonal proximity provide important social cues to which we readily attend, from which we draw a variety of inferences about the social context, and through which we reveal cognitive and affective responses to others.

1 Proxemics: Individual and Group Processes

So how do we study proxemics in the context of actual social interactions? We can address this question at two levels. First, we can study the proxemics of the individual. How and why does an individual express specific traits and cognitive or affective states through her proxemic behavior? And do those patterns in proxemic behavior predict more explicit or substantive behaviors (e.g., social decision-making)? Second, we can study proxemics at the level of the dyad or group. What are the emergent dynamics of proxemic behavior within an interaction? How do behaviors of individuals within the group affect each other and how does this give-and-take influence the perceptions and outcomes of the interaction? The history of proxemics research provides examples at both of these levels (Harrigan 2005; Hayduk 1983) with researchers studying both individual differences (e.g., age, gender, personality, and clinical condition), the influence of internal states of individuals (e.g., emotions), and the influence of the nature of the relationship itself (e.g., power and status).

Research on prejudice toward stigmatized others provides examples of how proxemics can examine both the underlying processes within the individual and the emergent dynamics of multiple interactants. At an individual level, prejudiced individuals tend to avoid gaze and to distance themselves from stigmatized others. For example, in one older study (Worthington 1974), a confederate stood in an airport lobby and waved participants over to ask for directions. For half of the participants, the confederates bore a physical stigma (a disability) and for the other half they did not. Participants approached the stigmatized individuals more slowly and kept a greater distance than when the confederates did not bear the stigma. More recent work has extended these types of findings by linking patterns of avoidance to implicit measures of prejudice. For example, participants high in anti-fat prejudice attitudes chose to sit further from fat women during an ostensibly unrelated task (Bessenoff and Sherman 2000). This relationship between implicit prejudice and avoidance has also been found in gaze patterns. In one study (Dovidio et al. 1997), White and Black confederates interviewed White participants. Participants also completed both explicit and implicit (responses latency) measures of racial prejudice. The participants who scored high on those implicit (but not explicit) measures of prejudice engaged in less eye contact and blinked more when speaking to the Black versus White interviewers (Dovidio et al. 1997). Notably, these types of prejudiced responses emerge even when participants explicitly deny such prejudices (Bessenoff and Sherman 2000; Dovidio et al. 1997), suggesting that attitudes might automatically influence proxemic behavior during an interaction.

Research on prejudice also illustrates how proxemic patterns can function at a group or dyadic level. In a classic study on self-fulfilling prophecies in interracial interactions, Word and colleagues (Word et al. 1974) first ran an experiment in which they staged a job interview between White interviewers (the participants) and either Black or White applicants (confederates). Coding of the nonverbal behavior in those interactions revealed that the interviewers kept significantly further from Black (as compared to the White) applicants. A second experiment staged another set of job interviews, but this time White confederates were the interviewers and the participants (all White) were the applicants. The critical manipulation in this experiment was of the interviewers' behavior. For half of the participants, the interviewer behaved in an avoidant fashion, like the participants interviewing Black applicants in the first experiment. For the other half of participants, the interviewer behaved like the participants who had interviewed the White applicants. When those White participants were treated like the Black applicants in the first experiment, they responded with avoidant behaviors of their own (i.e., placing their chairs further from the interviewer). The interviewers' behaviors apparently set into motion a dynamic and the applicants responded in turn, producing a simple but powerful demonstration of how proxemic behaviors within a group or dyad can create an emergent pattern with important social implications (i.e., in the perpetuation of racial prejudices and stereotypes).

2 Social Interaction, Proxemics, and Challenge of Experimental Control

While these studies on prejudice and proxemics were obviously fruitful, they possess methodological shortcomings that are common across the field. By their nature, actual, free-flowing social interactions are difficult to control. This point becomes clear when we consider the role of the confederate in laboratory research. No matter how dedicated and virtuous the confederate, interactions are a minefield of social meaning with manifold potential for feedback loops between interactants. Problems emerge even with a simple manipulation of confederate identity. On the one hand, it might be critical to manipulate the identity (i.e., race and gender) of a confederate in order to make causal inferences about the effect of that identity on participant behavior. On the other hand, it is very difficult to manipulate a confederate's identity while maintaining the double-blind nature of a study. A confederate will most likely know if she possesses a given identity or stigma (but for a clever approach, see Blascovich et al. 2001) and this knowledge may, in turn, implicitly affect her behavior. For example, when a White confederate enters a social interaction with a White participant, she brings with her a history of social interactions with White people that may be different than that of a confederate with a different set of experiences. This background may, in turn, bias her performance in such a way that the manipulation of identity becomes a manipulation of behavior.

Deliberately manipulating confederate behavior is equally, if not more, problematic. It is difficult to do consciously control one's behavior during a task as cognitively demanding as a social interaction and it is impossible to be sure that one is only manipulating the critical features of behavior for the given study. While a confederate may try to simply lean toward a participant in a given condition, she may inadvertently invoke a broader behavioral schema associated with approach motivation more generally (Price et al. 2012). Nonverbal and verbal behaviors have many facets to which social perceivers are exquisitely sensitive. Any facet of behavior that bleeds through a confederate's performance could thus bias results. For example, in the case of the self-fulfilling prophecy experiments (Word et al. 1974), the White versus Black "applicants" may have behaved differently from each other in some unconscious fashion, just as the "interviewers" in the second experiment may have gone beyond their scripts in their attempt to emulate the biased behavior of participants in the first experiment. Any number of small changes along these lines could put into motion a dynamic that critically affected participant responses. Indeed, the ability of an individual's nonverbal behavior to alter that of the dyad or group is exactly what makes the self-fulfilling prophecy experiment an important study.

In recent years, immersive virtual environments have helped address these problems of control in research on social interactions. Immersive virtual environments use head-mounted displays or "cave" environments to place participants in digital worlds whose visual and auditory features are designed and animated by the developer. In the social domain, this technology grants experimenters the ability to manipulate confederate behavior and identity with a radical level of control (Blascovich et al. 2002; McCall and Blascovich 2009). Experimenters can select the race, gender, or other social identifiers of virtual humans while independently manipulating critical features of nonverbal behavior via their animations. Virtual humans can be controlled by tracking and animating an actual human's "real-world" nonverbal and verbal behavior, by animating it entirely by algorithm, or by selectively mixing the two methods of control (Bailenson et al. 2004). Given these abilities, experimenters can manipulate the social identity of the confederate while maintaining a double-blind, controlled design and can further manipulate specific features of confederate nonverbal behavior in a discrete and selective fashion. In addition to social manipulations, experimenters can use virtual environments to stage experiments in any context that can be animated, i.e., at a bus stop or on a busy street (Dotsch and Wigboldus 2008; Gillath et al. 2008), allowing researchers to study social interaction far outside of the confines of the classic psychology laboratory.

The use of virtual environments and virtual humans in social interaction research raises an obvious question: how "real" are these interactions? Do participants treat agents and avatars as they treat people in the physical world? Along these lines, early work using virtual environments demonstrated that participants keep greater distance from the front of agents, walk further from agents who engage them in mutual gaze, and recoil when virtual humans invade their personal space (Bailenson et al. 2003). Moreover, research on prejudice replicates and extends the above-mentioned research on prejudice conducted using more traditional paradigms.

In one study (Dotsch and Wigboldus 2008), experimenters placed (White) Dutch participants at a bus stop with virtual humans who possessed White versus Moroccan physical traits but otherwise behaved in an identical fashion. This simple manipulation of identity revealed prejudice-based responses that parallel earlier work on stigma (Bessenoff and Sherman 2000; Worthington 1974) with participants keeping a greater distance from the Moroccan agents. Moreover, this relationship was predicted by higher skin conductance responses as well as implicit prejudices against Moroccans as measured by an implicit association test (Bluemke and Friese 2008). Further research demonstrated that these subtle differences in proxemic behavior predict more obviously substantives expressions of bias. In this study (McCall et al. 2009), participants encountered Black or White virtual humans in a brief social interaction before engaging in a violent shooting game with those agents. Participants who kept a greater distance from the Black agents as well as participants who averted their gaze from directly facing Black agents were more aggressive toward those Black agents during the shooting task. Notably, both these studies (Dotsch and Wigboldus 2008; McCall et al. 2009) place participants in situations that would be extremely difficult to create in a traditional laboratory environment.

Another recent study (Kane et al. 2012) demonstrates the utility of manipulating nonverbal behavior in virtual environments in order to study dynamics in the relationship between two interactants. Manipulating behavior during a face-to-face interaction between two members of a romantic couple is impossible to do with any degree of control. (Imagine instructing a man to ignore his wife during a given task.) Virtual environments, however, make this manipulation possible. In this study, participants came to the laboratory with their close relationship partners. They were then separated into two different physical rooms, but placed in the same virtual world where both people were represented by an avatar. Within the world, the participant was led to believe that the behavior of her partner's avatar was controlled by the partner actual movements via motion capture animation. The participant thus believed that the orientation and approach of their partner's avatar reflected their actual physical behavior. In fact, the experiments controlled all of the partner's animations. While the participant completed a stressful task (walking precariously on the edge of a cliff), the partner's behavior was manipulated such that it was either nonverbally supportive (i.e., facing the participant and nodding) or inattentive (i.e., facing away from the participant to look out over the rest of the virtual world).

This manipulation set into motion a nonverbal dynamic between the partners in two ways. First, participants with the inattentive partner continued to look toward their partner for the duration of the stressful task despite the fact the partner continued to ignore them, suggesting that participants with the unsupportive partners continued to seek nonverbal support. Second, participants with the inattentive partner kept further away from those partners in a subsequent interaction in the virtual world. Most importantly, this dynamic had consequences for participants' affective state; participants with the attentive partner reported feeling more comforted and supported by their partners. Together, these findings illustrate not only the utility of virtual environments for studying proxemic dynamics, but also the fact that subtle differences in nonverbal and proxemic behaviors can set into motion a nonverbal dynamic with greater implications for interpersonal relationships.

3 Measuring Social Space

Measurement is another critical challenge for proxemics research. Early research relied upon projective techniques such as drawing or placing figures on a board (e.g., Kuethe 1962) to evaluate participants' ideas about personal space and expectancies regarding interpersonal distance. Not surprisingly, observing actual interactions has proven more reliable (Harrigan 2005; Hayduk 1983) and is certainly more ecologically valid. Nevertheless, measuring behavior during something as complex and multidimensional as a social interaction is a difficult business. One approach is to use human observers to code or evaluate recordings of social interactions. Coding schemes can focus on specific, physical features of the interaction (such as the amount of time mutual gaze occurred) or more general impressions of the interaction as a whole (e.g., Dovidio et al. 1997; Harrigan 2005). Coding or evaluation approaches allow for a nuanced description of behavior, but that nuance comes at a cost in terms of the time and personal resources necessary to code entire interactions. For better or for worse, human coding is also dependent to some extent on the observers' subjective responses.

This is not to say, however, that objective measurement of proxemic behavior is either straightforward or superior to observer accounts. The quality and richness of objectively gathered proxemic data vary greatly. Early work opted for measures of interpersonal distance such as seat choice in a lobby or chair adjustment during an interview (e.g., Bessenoff and Sherman 2000). While these types of approaches are unobtrusive and produce clear quantitative values, they are not particularly precise.

Precision is nevertheless important to objective measurement of proxemic behaviors given that the range is quite narrow of a variable such as interpersonal distance during a conversation (e.g., Hall 1969; Kennedy et al. 2009). Furthermore, measures such as chair choice are "one-shot" measures. They provide a relatively limited portrait of an interaction by relying upon a single moment instead of examining the length of the interaction. Social interactions are of course not single events, but a series of events that unfold over time. Given that fact, one-shot measures necessarily omit a great deal of potentially critical information.

Advances in digital motion tracking address these concerns. Motion tracking can provide precise and continuous measurement of interactants' orientation, gaze, and position over the course of an entire interaction, offering a boon for the objective measurement of proxemic behavior. The benefits of motion tracking are evident in the recent research using immersive virtual environments, which necessarily relies upon this technology to track and render the user's viewpoint. Many of the abovementioned studies (Bailenson et al. 2003; Dotsch and Wigboldus 2008; Gillath et al. 2008; Kane et al. 2012; McCall et al. 2009) capitalized on motion tracking to gather precise data on interpersonal distance and gaze.

Nevertheless, the advent of motion tracking also creates the ironic problem of having too much data. After decades of manual coding and measuring seat distance, we now have a stream of data from multiple bodies over the entire length of the interaction. This fact creates the challenge of reducing data to a manageable and interpretable form. Most research using this technology has boiled tracking data down into a limited set of aggregated numbers. These include, for example, the mean or minimum interpersonal distance between a participant and a target or the average looking angle toward the person. These approaches are less than ideal for several reasons. First and most obviously, the entire interaction is reduced to one or two numbers that may also be one-shot measurements (e.g., minimum distance). That reduction may wash out meaningful results, makes data vulnerable to noise, and robs us of much nuance.

Second, these individual measures separate interpersonal distance and orienting angle despite the fact that these two proxemic variables are inherently dependent. For example, it is well established that the comfortable range of interpersonal distance varies depending upon whether or not we are facing or are turned away from the other individual (Bailenson et al. 2003; Hayduk 1983). This fact is intuitively obvious if you imagine a crowded elevator. You may be comfortable standing within inches of the back of a stranger, but if he were to turn around and face you, things would become exceedingly awkward. Nevertheless, proxemics research using minimum distance and gaze tends to ignore this fact or to look exclusively at distances in the fronts of the interactants.

Finally, while tracking data might provide objective measures of an individual's orientation or gaze toward another interactant, this provides only a fraction of what's important about gaze. An interactant's gaze is most interesting in terms of how it relates to the other interactant's gaze at that moment. In other words, although I might gaze at you during half of an interaction, the meaning of that gaze varies dramatically if I did so when you were looking away versus gazing back at me. Were we engaged in mutual, joint, or averted gaze? Accordingly, proxemic measures should account for the emergent relationships between all interactants gaze and orienting patterns.

Proxemic imaging, a new method for proxemic analysis (McCall and Singer 2015), attempts to address these challenges. Proxemic images use motion tracking data to create frequency maps depicting both interpersonal distance and orientation during a dyadic interaction. Motion tracking data provide the location and orientation of the head in allocentric space. Proxemic imaging takes these data and transforms them into "dyadic space." Dyadic space is defined by three dimensions: (1) the interpersonal distance between Person A and Person B, (2) the angular distance of Person A from directly facing Person B, and (3) the angular distance of Person B from directly facing Person A. Given a time series of tracking data, each sample is binned into a specific location in these three dimensions. Collapsing across a given dimension of this structure produces a different type of proxemic image. Collapsing across either of the gaze dimensions creates egocentric maps of Person A or Person B's personal space. Collapsing across the interpersonal distance dimension creates a "gaze map" which depicts the degree to which the interactants engaged in mutual, joint, or averted gaze. Figure 1 illustrates dyadic space and provides two examples of where a given moment in an interaction would emerge on the different proxemics maps.

The primary advantage of proxemic imaging is that it provides a relatively nuanced look at a dyadic interaction that accounts for intra- and interpersonal dependencies between gaze and movement. A recent study on nonverbal responses to fairness violations demonstrates this point. Fairness violations create an interesting context for proxemic behavior, because unfair behavior elicits both approach-related responses such as anger and retaliation (Fehr and Gachter 2002; Pillutla and Murnighan 1996), and avoidance-related responses such as disgust and negative interpersonal attitudes (Chapman et al. 2009; Singer et al. 2006). In this study, participants first played an economic game with two individuals, one of whom behaved fairly and the other of whom did not. After the game, participants encountered those other players in a virtual art museum. In a final, surprise task, participants were given the opportunity to financially punish either of the players by paying to subtract money from their earnings. As one would expect from the literature on proxemics and attitudes (e.g., Bessenoff and Sherman 2000; Dotsch and Wigboldus 2008; McCall et al. 2009), participants generally avoided the unfair as compared to the fair players. The proxemic images of the participants' egocentric space revealed that participants kept the fair player close to their side as compared to the unfair player. Proxemic imaging also revealed a distinct pattern among participants who chose to punish the unfair player (i.e., the more retaliative participants). Those high punishers exhibited more approach behavior, spending significantly more time than low punishers in front unfair players. The gaze maps further revealed that high punishers were more likely to engage in mutual gaze with both players and, additionally, to turn their backs to the unfair players' gaze. This rich pattern of behaviors was uniquely apparent in the proxemic images of the interaction, and not in more traditional measures.

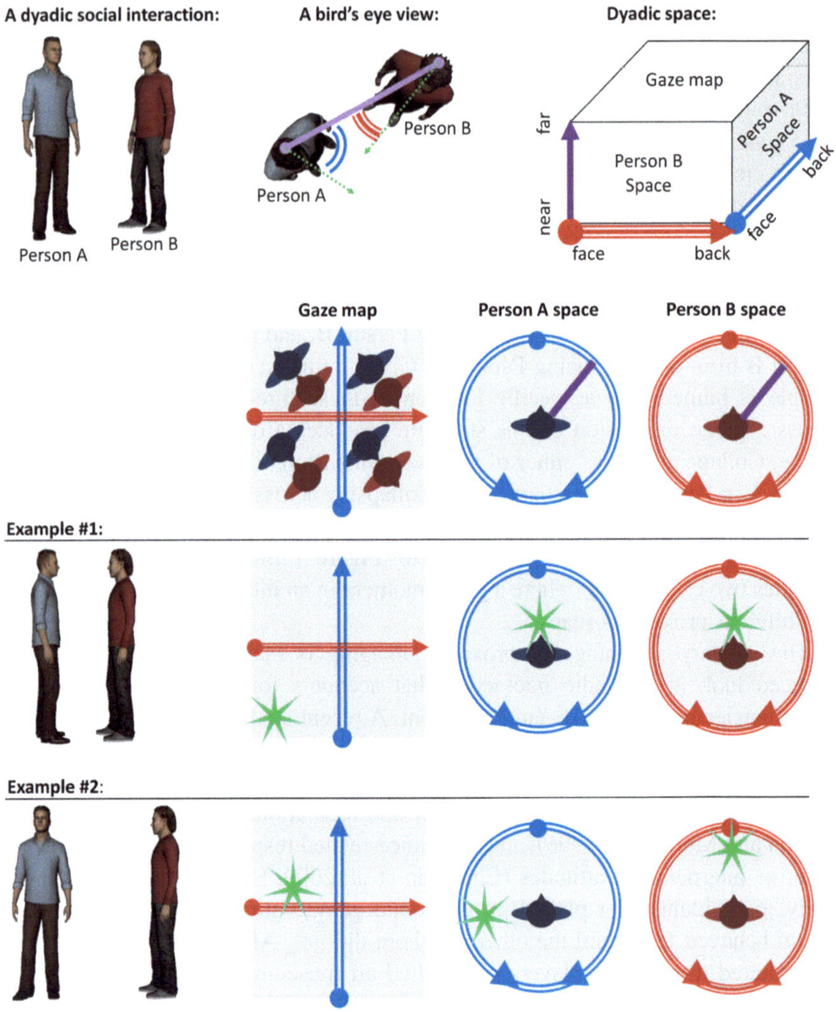

Fig. 1 Proxemic imaging. Proxemic images are frequency maps of motion capture data from a dyadic social interaction. Data are binned into "dyadic space," which is defined by three dimensions: the distance between the interactants (the *solid purple line*) and, for each interactant, the angular distance from directly facing the other (the *blue* and *red striped arrows*). Each sample of tracking data is binned into dyadic space. Collapsing across one dimension produces a gaze map, or maps of either individual's egocentric space. The examples illustrate where the given moment would emerge in the proxemics maps (the *green stars*)

These data suggest that proxemic imaging may provide a nuanced method for analyzing proxemic behavior that accounts for the inherent dependency of gaze and distance as well as the critical interrelationship between two individuals' gaze behavior. Further work using this method will need to test its utility in examining real-world interactions and interactions with more lively interactive content.

4 Future Directions

Researchers across a range of disciplines have been studying proxemics for decades (Harrigan 2005; Hayduk 1983). Here we have presented only a sliver of those findings, with a particular focus on approaches and methods that lend themselves to social psychology and neuroscience. While the content of this chapter is by no means a review of the literature, it points to several avenues for future research.

For one, proxemics research should continue to implement the technologies discussed here (i.e., immersive virtual environments, motion tracking, and proxemic imaging) to address basic questions about proxemic behavior's underlying roots in cognition, affect, and physiology. One critical domain that merits more work is the proxemic correlates of clinical phenomena. Chief among these is Autism Spectrum Disorder, which is associated with anomalies in both interpersonal distance (e.g., Perry et al. 2015) and the tendency to engage in mutual gaze (e.g., Spezio et al. 2007). More controlled and nuanced proxemic analyses might help explain variance within and between clinical populations such as these. Furthermore, the same technologies used to manipulate and measure proxemics might be used, in turn, to design training programs for helping clinical populations develop appropriate communication skills (e.g., Irish 2013; Schwartz et al. 2010).

More generally, future research will also need to develop better approaches to capturing the dynamic and interactive nature of proxemic behavior. As others have discussed (e.g., De Jaegher et al. 2010; Froese 2013; Schilbach et al. 2013), psychology and neuroscience need new ways of understanding the give and take of social interaction. It is clear from the work cited here (Kane et al. 2012; Word et al. 1974) that individual proxemic behaviors can set into motion feedback loops whereby the actions of one interactant play off of the others. These group level processes necessarily emerge over time. Accordingly, the field must start using methods that examine dynamics as they unfold. Time series analyses, perhaps of proxemic images, might help explain how interactions of different types (e.g., cooperative, competitive, and successfully communicative) evolve and how that evolution shapes their outcomes. Dynamics in proxemic behavior could further be linked to peripheral physiology and subjective experience. Over the years, psychologists have used a variety of different methods for eliciting continuous first person accounts of social interactions that, in turn, can be integrated with physiological time series (e.g., Ickes et al. 1990; Levenson and Ruef 1992). The next wave of research on proxemics should expand on this methodological tradition to better understand how experience, physiology, and behavior converge over the course of actual, naturalistic, face-to-face interactions.

Finally, the time is ripe for the neuroscience of proxemics. Plenty of research has explored the neural correlates of gaze perception, biological motion perception, and other core features of nonverbal communication, but rarely in the context of actual physical social interactions. As a consequence, we know relatively little about the neural underpinnings of proxemic behaviors or proxemic dynamics as they emerge in everyday life. Studies of personal space violations among patients with amygdala damage (Kennedy et al. 2009) and studies of reciprocal eye gaze and interpersonal distance gathered within MR scanners (Kennedy et al. 2009; Schilbach et al. 2010) already suggest neural correlates to proxemic responses. Future work can build on this by taking advantage of recent technological innovations. By combining motion capture or immersive virtual environments with mobile neuroimaging techniques such as functional near-infrared spectroscopy (Ferrari and Quaresima 2012; Piper et al. 2014) or portable EEG technologies (Debener et al. 2012; Gevins et al. 2012), we could measure brain activity while participants freely and naturalistically engage in actual face-to-face interactions. Such an approach would not only improve our understanding of proxemics, but would also liberate social neuroscience from the relatively impoverished environments in which it is traditionally studied, allowing us to examine social interactions within the types of environments in which they naturally occur.

References

Bailenson JN, Blascovich J, Beall AC, Loomis JM (2003) Interpersonal distance in immersive virtual environments. Perso Soc Psychol Bull 29:819–834

Bailenson JN, Beall AC, Loomis J, Blascovich J, Turk M (2004) Transformed social interaction: decoupling representation from behavior and form in collaborative virtual environments. Presence: Teleoperators Virtual Environ 13(4):428–441

Bessenoff GR, Sherman JW (2000) Automatic and controlled components of prejudice toward fat people: evaluation versus stereotype activation. Soc Cognition 18(4):329–353

Blascovich J, Mendes WB, Hunter SB, Lickel B, Kowai-Bell N (2001) Perceiver threat in social interactions with stigmatized others. J Pers Soc Psychol 80(2):253

Blascovich J, Loomis J, Beall AC, Swinth KR, Hoyt CL, Bailenson JN (2002) Immersive virtual environment technology as a methodological tool for social psychology. Psychol Inq 13 (2):103–124

Bluemke M, Friese M (2008) Reliability and validity of the Single-Target IAT (ST-IAT): assessing automatic affect towards multiple attitude objects. Eur J Soc Psychol 38(6):977–997

Carver CS, Harmon-Jones E (2009) Anger is an approach-related affect: evidence and implications. Psychol Bull 135(2):183

Chapman HA, Kim DA, Susskind JM, Anderson AK (2009) In bad taste: evidence for the oral origins of moral disgust. Science 323(5918):1222–1226

Chen M, Bargh JA (1999) Consequences of automatic evaluation: immediate behavioral predispositions to approach or avoid the stimulus. Pers Soc Psychol Bull 25(2):215–224

De Jaegher H, Di Paolo E, Gallagher S (2010) Can social interaction constitute social cognition? Trends Cogn Sci 14(10):441–447

Debener S, Minow F, Emkes R, Gandras K, Vos M (2012) How about taking a low-cost, small, and wireless EEG for a walk? Psychophysiology 49(11):1617–1621

Dotsch R, Wigboldus DH (2008) Virtual prejudice. J Exp Soc Psychol 44(4):1194–1198

Dovidio JF, Kawakami K, Johnson C, Johnson B, Howard A (1997) On the nature of prejudice: automatic and controlled processes. J Exp Soc Psychol 33(5):510–540

Elliot AJ (2006) The hierarchical model of approach-avoidance motivation. Motiv Emot 30 (2):111–116

Elliot AJ (2008) Handbook of approach and avoidance motivation. Taylor & Francis, New York

Farroni T, Csibra G, Simion F, Johnson MH (2002) Eye contact detection in humans from birth. Proc Natl Acad Sci 99(14):9602–9605

Fehr E, Gachter S (2002) Altruistic punishment in humans. Nature 415(6868):137–140
Ferrari M, Quaresima V (2012) A brief review on the history of human functional near-infrared spectroscopy (fNIRS) development and fields of application. Neuroimage 63(2):921–935
Frischen A, Bayliss AP, Tipper SP (2007) Gaze cueing of attention: visual attention, social cognition, and individual differences. Psychol Bull 133(4):694
Froese T (2013) Interactively guided introspection is getting science closer to an effective consciousness meter. Conscious Cogn 22(2):672–676
Gevins A, Chan CS, Sam-Vargas L (2012) Towards measuring brain function on groups of people in the real world. PLoS ONE 7(9):e44676
Gillath O, McCall C, Shaver PR, Blascovich J (2008) What can virtual reality teach us about prosocial tendencies in real and virtual environments? Media Psychol 11(2):259–282
Hall ET (1969) The hidden dimension, vol 1990. Anchor Books, New York
Hall ET (1973) The silent language. Anchor, New York
Harrigan JA (2005) Proxemics, kinesics, and gaze. The new handbook of methods in nonverbal behavior research, pp 137–198
Hayduk LA (1978) Personal space: an evaluative and orienting overview. Psychol Bull 85(1):117
Hayduk LA (1983) Personal space: where we now stand. Psychol Bull 94(2):293
Heider F, Simmel M (1944) An experimental study of apparent behavior. Am J Psychol 243–259
Ickes W, Stinson L, Bissonnette V, Garcia S (1990) Naturalistic social cognition: empathic accuracy in mixed-sex dyads. J Pers Soc Psychol 59(4):730
Irish JE (2013) Can I sit here? A review of the literature supporting the use of single-user virtual environments to help adolescents with autism learn appropriate social communication skills. Comput Human Behav 29(5):A17–A24
Kane HS, McCall C, Collins NL, Blascovich J (2012) Mere presence is not enough: responsive support in a virtual world. J Exp Soc Psychol 48(1):37–44
Kennedy DP, Gläscher J, Tyszka JM, Adolphs R (2009) Personal space regulation by the human amygdala. Nat Neurosci 12(10):1226–1227
Kingstone A, Smilek D, Eastwood JD (2008) Cognitive ethology: a new approach for studying human cognition. Br J Psychol 99(3):317–340
Kuethe JL (1962) Social schemas. J Abnorm Soc Psychol 64(1):31
Levenson R, Ruef A (1992) Empathy: a physiological substrate. J Pers Soc Psychol 63:234–246
Loomis JM, Kelly JW, Pusch M, Bailenson JN, Beall AC (2008) Psychophysics of perceiving eye-gaze and head direction with peripheral vision: Implications for the dynamics of eye-gaze behavior. Perception 37:1443–1457
Marsh AA, Ambady N, Kleck RE (2005) The effects of fear and anger facial expressions on approach-and avoidance-related behaviors. Emotion 5(1):119
McCall C, Blascovich J (2009) How, when, and why to use digital experimental virtual environments to study social behavior. Soc Influence 4:138–154
McCall C, Singer T (2015) Facing off with unfair others: introducing proxemic imaging as an implicit measure of approach and avoidance during social interaction. PLoS ONE
McCall C, Blascovich J, Young A, Persky S (2009) Proxemic behaviors as predictors of aggression towards Black (but not White) males in an immersive virtual environment. Soc Influence 4(2):138–154
McCall C, Tipper CM, Blascovich J, Grafton ST (2012) Attitudes trigger motor behavior through conditioned associations: neural and behavioral evidence. Soc Cogn Affect Neurosci 7 (7):841–849
Pedersen DM, Shears LM (1973) A review of personal space research in the framework of general system theory. Psychol Bull 80(5):367
Perry A, Levy-Gigi E, Richter-Levin G, Shamay-Tsoory SG (2015) Interpersonal distance and social anxiety in autistic spectrum disorders: a behavioral and ERP study. Soc Neurosci 10 (4):354–365
Pillutla MM, Murnighan JK (1996) Unfairness, anger, and spite: emotional rejections of ultimatum offers. Organ Behav Hum Decis Process 68(3):208–224

Piper SK, Krueger A, Koch SP, Mehnert J, Habermehl C, Steinbrink J, Schmitz CH (2014) A wearable multi-channel fNIRS system for brain imaging in freely moving subjects. Neuroimage 85:64–71

Price TF, Peterson CK, Harmon-Jones E (2012) The emotive neuroscience of embodiment. Motiv Emot 36(1):27–37

Schilbach L (2015) Eye to eye, face to face and brain to brain: novel approaches to study the behavioral dynamics and neural mechanisms of social interactions. Curr Opin Behav Sci 3:130–135

Schilbach L, Wilms M, Eickhoff SB, Romanzetti S, Tepest R, Bente G, Vogeley K (2010) Minds made for sharing: initiating joint attention recruits reward-related neurocircuitry. J Cogn Neurosci 22(12):2702–2715

Schilbach L, Timmermans B, Reddy V, Costall A, Bente G, Schlicht T, Vogeley K (2013) Toward a second-person neuroscience. Behav Brain Sci 36(04):393–414

Schwartz C, Bente G, Gawronski A, Schilbach L, Vogeley K (2010) Responses to nonverbal behaviour of dynamic virtual characters in high-functioning autism. J Autism Dev Disord 40 (1):100–111

Senju A, Johnson MH (2009) The eye contact effect: mechanisms and development. Trends Cogn Sci 13(3):127–134

Singer T, Seymour B, O'Doherty JP, Stephan KE, Dolan RJ, Frith CD (2006) Empathic neural responses are modulated by the perceived fairness of others [Research Support, Non-U.S. Gov't]. Nature 439(7075):466–469. doi:10.1038/nature04271

Solarz AK (1960) Latency of instrumental responses as a function of compatibility with the meaning of eliciting verbal signs. J Exp Psychol 59(4):239

Spezio ML, Adolphs R, Hurley RS, Piven J (2007) Analysis of face gaze in autism using "Bubbles". Neuropsychologia 45(1):144–151

Tomasello M (1995) Joint attention as social cognition. Joint attention: its origins and role in development, pp 103–130

Word CO, Zanna MP, Cooper J (1974) The nonverbal mediation of self-fulfilling prophecies in interracial interaction. J Exp Soc Psychol 10(2):109–120

Worthington ME (1974) Personal space as a function of the stigma effect. Environ Behav 6:289–294

Part III
Clinical Implications

Genetic Animal Models for Autism Spectrum Disorder

Jan C. Schroeder, Dominik Reim, Tobias M. Boeckers and Michael J. Schmeisser

Abstract Autism spectrum disorder (ASD) affects approximately 1 % of the human population and has a strong genetic component. Hence, the recent discovery of major "ASD genes" has subsequently resulted in the generation of several genetic animal models of ASD. Careful analysis of behavioral phenotypes and characterization of the underlying neurobiological mechanisms in these models should further help us to identify novel therapeutic targets and develop more effective strategies in the future to ameliorate or even reverse core symptoms and comorbidities of ASD. In this review, we will focus on the mutant mouse as animal model and outline how to characterize both behavioral and neurobiological phenotypes in this organism. We will further discuss a selection of major ASD mutant mouse lines. Our conclusions will finally address the current goals and perspectives in the field to obtain a more comprehensive and possibly also converging picture of ASD pathogenesis, which could be most useful for the desired bench-to-bedside strategy of translational medicine for this complex disorder.

Keywords Autism spectrum disorder · Fragile X · Tuberous sclerosis · Rett syndrome · Neuroligin3 · Shank3

Contents

1 Introduction 312
  1.1 Autism Spectrum Disorder 312
  1.2 Genetic Basis 312
  1.3 Animal Models for ASD 313
2 How to Characterize an Animal Model for ASD 314
  2.1 Behavioral Characterization 314

J.C. Schroeder · D. Reim · T.M. Boeckers · M.J. Schmeisser (✉)
Institute for Anatomy and Cell Biology, Ulm University, Ulm, Germany
e-mail: michael.schmeisser@uni-ulm.de

M.J. Schmeisser
Department of Neurology, Ulm University, Ulm, Germany

Curr Topics Behav Neurosci (2016) 30: 311–324
DOI 10.1007/7854_2015_407
Published Online: 20 November 2015

2.2 Neurobiological Characterization ... 316
3 Major Genetic Mouse Models for ASD ... 317
3.1 Fmr1 Mutant Mice ... 317
3.2 Tsc1 and Tsc2 Mutant Mice ... 317
3.3 Mecp2 Mutant Mice ... 319
3.4 Cntnap2 Mutant Mice ... 320
3.5 Neuroligin3 and Neuroligin4 Mutant Mice ... 320
3.6 Shank2 and Shank3 Mutant Mice ... 321
4 Conclusions ... 321
References ... 322

1 Introduction

1.1 Autism Spectrum Disorder

Autism spectrum disorder (ASD) comprises several disease entities all sharing two core features as outlined in DSM-5: (1) persistent deficits in social communication and social interaction across multiple contexts and (2) stereotyped, repetitive behavior and restricted interests. With respect to (1), autistic individuals show a lack in social reciprocity during verbal and nonverbal communication and struggle with social approach and relationships, for example. With respect to (2), they often display motor stereotypies, perseverative interests for particular entities, and inflexibility with regard to their daily routine. Furthermore, there is a wide array of comorbidities, especially intellectual disability (ID), seizures, hyperactivity, impaired motor coordination, and altered perception of sensory information. Current data point toward an ASD prevalence of 1 in 100 individuals with an overrepresentation of males. The term ASD includes syndromic autism, i.e., autism is part of a complex syndromic neurodevelopmental disorder such as tuberous sclerosis, fragile X syndrome or Phelan–McDermid syndrome, and nonsyndromic autism, i.e., autism is the only phenotypical feature of an affected individual. Importantly, recent advances implicate that ASD has a strong genetic component, which opens up the possibility of targeted treatment based on genetically defined biological mechanisms.

1.2 Genetic Basis

In order to understand a disorder and to find effective treatments, a major task is to elucidate its etiology and the underlying pathomechanisms. Genetic studies have provided powerful insights into the association of genes and the development of ASD: Up to now, 25 % of cases are caused by genetic alterations such as coding sequence mutations, copy number variants, or chromosomal rearrangements (Bourgeron 2015; Chen et al. 2015; Huguet et al. 2013). However, not one or several major, but a vast and heterogenous number of genes and genetic loci independently related to ASD have been identified, thereby contributing to the complexity of the disorder. Despite this complexity, the identification of ASD pathomechanisms and the development of targeted therapies are big goals in the field to eventually help an enormous amount of affected individuals and their families to significantly increase their quality of life.

One of the best examples in this context is fragile X syndrome (FXS), a syndromic genetic disorder and the most common inherited cause of ASD and ID. After the identification of *FMR1* gene silencing as the underlying genetic cause, *Fmr1*-knockout mice were subsequently generated and characterized with respect to behavioral and neurobiological phenotypes. Indeed, it has been possible to delineate a pathomechanism including a molecular target in these mutants, which has been repeatedly validated in preclinical studies and is now used for developing promising pharmacological agents to ameliorate or even reverse FXS symptoms in FXS patients (Bhakar et al. 2012).

1.3 Animal Models for ASD

The gold standard for the identification of pathomechanisms caused by genetic alterations and resulting in behavioral phenotypes is the generation of a valid animal model. Genetic design can be based on either a knockout approach, i.e., the disruption of the gene of interest, or a knock-in approach, i.e., the introduction of a particular mutation into the gene of interest. Subsequent analysis of each animal model will further reveal deficits on distinct levels, from behavior to anatomy to cellular physiology and molecular signaling. Furthermore, animal models are essential tools to identify potential target sites, to screen for compounds, and to finally develop novel treatment strategies (Kas et al. 2014; Kleijer et al. 2014). In many cases, the most widely used organism is the mouse, Mus musculus, due to rather simple means of genetic manipulation on the one hand and due to the advantages in housing, breeding, and handling on the other hand. It is important to mention at this point that further organisms including rat, prairie vole, zebra fish, songbirds, invertebrates, or even nonhuman primates are also used in ASD research. In this review though, we will basically focus on the mouse as the primary ASD model system and discuss the principles of ASD mouse mutant characterization.

How do we experimentally refine and condense the enormous number of known ASD genes in a way that allows both profound investigation and possibly also converging conclusions? One possibility is to produce valid ASD mouse mutants, based on the most frequent genetic mutations known from human studies, and to characterize them in parallel in the search for an overlap or even convergence in disease pathomechanisms.

Several mutant lines have been generated up to this day and have been validated as powerful tools for the investigation of ASD pathophysiology. A common approach is to generate mice based on syndromic ASDs, as it is estimated that approximately 15 % of ASDs can be attributed to monogenic heritable disorders such as FXS. Moreover, statistical methods integrating the prevalence of given ASD-associated mutations with algorithms predicting their functional implication and likelihood to be detrimental can be exploited. For example, it has recently been demonstrated that ASD-causing mutations were enriched in key genes highly conserved throughout evolution and that these genes control three crucial processes during neural development: synaptogenesis, transcriptional regulation, and chromatin remodeling (De Rubeis et al. 2014). By analyzing mutants for these key genes and comparing their phenotypes among each other, we might be able to identify a pathomechanistic pattern of convergence, which could in the end help to understand not only certain ASD variants, but possibly also ASD in general.

2 How to Characterize an Animal Model for ASD

Animal models are of most value if the observed phenotypes reliably reflect the disorder in human individuals. The most important indicators of a promising model are construct validity (the same biological origin of the disorder), face validity (parallel phenotypes), and predictive validity (similar reaction to treatment). Construct validity is achieved in ASD mouse models by targeting the same genes that are mutated in human individuals. To determine face validity, behavioral characterization of the mutants is performed. In this context, at least the two core features of ASD symptomatology according to DSM-5 are to be replicated: (1) deficits in social interaction and (2) stereotyped, repetitive behavior. Finally, due to a lack of effective treatment options for ASD in humans, it is still difficult to evaluate predictive validity in ASD mouse models thus far. Therefore, translational ASD research currently focuses on the identification of behavioral phenotypes, corresponding to the clinical manifestation of the disorder and the subsequent neurobiological characterization with respect to neuroanatomy, physiology, and molecular mechanisms with the major goal to identify neurobiological targets for drug development (for a schematic overview of "How to characterize an animal model for ASD," see Fig. 1).

2.1 Behavioral Characterization

Despite the fact that the mouse brain is by far less complex than the human brain, mice are social animals and display a remarkable set of social behaviors, as outlined in the first part of this book. The assays presented here demonstrate how neuroscientists selected behaviors in mice that currently serve as prototypes for the diagnostic categories of ASD in humans and that are accessible to standardized tests (Silverman et al. 2010; Wohr and Scattoni 2013).

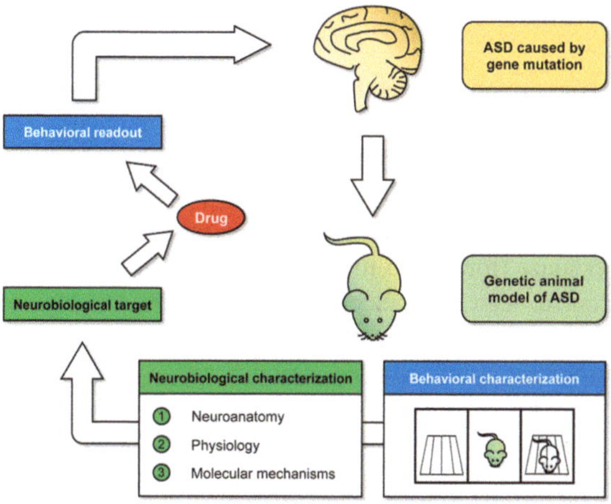

Fig. 1 How to characterize an animal model for ASD. In a significant amount of cases, ASD is caused by a gene mutation, which enables researchers to generate a genetic animal model of ASD, e.g., a mutant mouse (construct validity). Mutants are then characterized with respect to autistic-like behavior (face validity) and further analyzed with respect to neurobiological phenotypes. This might result in the identification of a promising neurobiological target for later drug development and the effective treatment of behavioral phenotypes both in mutant mice and in human individuals (predictive validity)

2.1.1 Core Features

A widely used assay to characterize social interaction deficits in mouse models of ASD is the three-chamber assay developed by Jacqueline Crawley's laboratory. In this assay, sociability is tested as a subject mouse can choose between interaction with an unknown mouse or an empty chamber. The assay can be expanded to test social novelty recognition as the subject mouse can further choose between interaction with a familiar and an unfamiliar mouse. An ASD mutant typically shows impairments in both sociability and social novelty recognition. Of course, this is just one example of a social interaction paradigm, and there are several more to evaluate impaired social behavior. Importantly, also ultrasonic vocalizations and olfactory scent marking—two different ways how mice socially communicate with each other—can be recorded during such paradigms and are frequently altered in ASD mutants.

Motor activities such as spontaneous self-grooming or burying objects are common in mice. It is frequently observed that ASD mutants spend more time with these activities, which is then interpreted as stereotyped, repetitive behavior. Appropriate paradigms include quantifying the time a mouse spends self-grooming in an empty cage or counting the amount of marbles left unburied by a mouse previously placed in a cage with a given number of marbles, for example.

2.1.2 Comorbidities

In addition to the aforementioned core features of ASD, there are frequent comorbidities that can also be evaluated in ASD mutants, i.e., ID, seizures, hyperactivity, anxiety, impaired motor coordination, altered perception of sensory information, and a disruption of circadian rhythms. ID, for example, can be addressed by testing cognitive functions in mice in well-known paradigms such as the Morris water maze (for spatial memory) or the Y-maze (for working memory). Seizures, of course, can be detected by electroencephalography (EEG) and hyperactivity by counting the distance and velocity of a mouse freely moving in an open field arena within a defined time period.

2.2 Neurobiological Characterization

After successful generation and behavioral characterization, neurobiological analysis starts in an ASD mutant line. This should be done on the macroscopical, microscopical, and molecular level in a brain region or neuronal subpopulation-specific manner to finally get a comprehensive picture of the corresponding phenotypes.

Macroscopic and microscopic neuroanatomy of the ASD mutant brain can be evaluated by state-of-the-art brain imaging including cutting-edge molecular imaging, by the application of classical histological methods including Nissl and Golgi staining, and by electron microscopy-based ultrastructural analysis. Important parameters that might be subjected to changes in an ASD mutant are, among others, neuron number and morphology including dendritic tree complexity, spine density per dendrite length, spine morphology, spine motility, synapse turnover, and the morphology of subcellular structures including dendritic and axonal organelles.

Physiological analysis of well-investigated neural circuitry, including the trisynaptic pathway of the hippocampus, frontostriatal connectivity, or cerebellar in- and output, can further reveal neuronal and/or synaptic phenotypes.

Finally, analysis of RNA and protein from subcellular compartments of different brain regions or neuronal subpopulations often leads to the identification of molecular changes pointing toward defined alterations of signaling pathways that could further be used as neurobiological target for therapy development.

Some elegant works providing intriguing neurobiological insight into the ASD mutant brain are studies involving *Fmr1* (Dolen et al. 2007; Koekkoek et al. 2005), *Tsc* (Tang et al. 2014; Tsai et al. 2012), *Mecp2* (Chang et al. 2006; Chao et al. 2010), *Cntnap2* (Penagarikano et al. 2011), *Neuroligin* (Baudouin et al. 2012; Rothwell et al. 2014; Tabuchi et al. 2007), and *Shank* (Peca et al. 2011; Schmeisser et al. 2012; Won et al. 2012) mutant mice.

3 Major Genetic Mouse Models for ASD

Within this section, we will give a brief overview of major genetic mouse models for ASD (for short summary, see Table 1).

3.1 *Fmr1 Mutant Mice*

FXS, a major genetic cause for ASD and ID, is caused by the expansion of a triplet repeat (CGG) and subsequent hypermethylation of the promoter region of the X-chromosomal *FMR1* gene. This results in the transcriptional silencing and absence of the gene product fragile X mental retardation protein (FMRP), which acts as an mRNA binding protein involved in translational regulation in neurons. Deletion of *Fmr1* in mice leads to behavioral phenotypes analogous to the clinical symptoms of FXS patients, including social impairment, repetitive behaviors, cognitive deficits, hyperactivity, and a largely decreased seizure threshold. On the neurobiological level, the main phenotypes of *Fmr1* mutants are increased number of immature dendritic spines and abnormally elevated local protein synthesis at the synapse. Importantly, both behavioral and neurobiological phenotypes can be reversed by inhibiting metabotropic glutamate receptor 5 (mGluR5), either by pharmacological antagonism or by the deletion of the *Grm5* gene in *Fmr1* mutants (Bhakar et al. 2012). As several recent studies point toward the fact that a dysregulation of translational control in neurons could be a general pathomechanistic concept in ASD (Santini and Klann 2014), it is of high interest that mGluR5 antagonism corrects local protein synthesis at synapses of *Fmr1* mutants. Moreover, several other pharmacological agents have been successfully used to ameliorate or reverse *Fmr1* mutant phenotypes thus far (Wang et al. 2015).

3.2 *Tsc1 and Tsc2 Mutant Mice*

Tuberous sclerosis complex (TSC) is another frequent genetic cause for ASD and ID. It is induced by the heterozygous loss of either *TSC1* or *TSC2*, whose gene products normally form a complex, which inhibits mammalian target of rapamycin (mTOR). Reduced inhibition leads to mTOR hyperactivity, dysregulation of protein synthesis, and enhanced proliferation. TSC patients therefore develop benign tumors called hamartomas in various tissues. In addition, they exhibit autism, intellectual disability, and epilepsy (Ehninger 2013). *Tsc1* or *Tsc2* mutant mice do also replicate clinical features of TSC such as impaired social interaction, repetitive behaviors, cognitive deficits, and a high frequency of seizures. Neurobiologically, phenotypes include abnormal neuronal migration and lamination, increased cell size, hypomyelination, reduced number of dendritic spines, and abnormally reduced

Table 1 Behavioral phenotypes and neurobiological targets in major ASD mutant mouse lines

	Fmr1	*Tsc1*	*Tsc2*	*Mecp2*	*Cntnap2*	*Nlgn3*	*Nlgn4*	*Shank2*	*Shank3*
Behavior: social impairment	+	+	+	+	+	+	+	+	+
Behavior: repetitive behavior	+	+	+	+	+	+	+	+	+
Behavior: cognitive deficits	+	+	+	+		+		+	+
Behavior: epileptic seizures	+	+	+	+	+	–			
Behavior: hyperactivity	+	–	–	–	+	+		+	–
Neurobiology: molecular target	mGluR5	mTOR	mTOR			mGluR1		NMDAR	NMDAR

A box is filled with a "+" if a behavioral phenotype has been observed at least once in a study involving the indicated mutant line and with a "–" if it has never been observed; the molecular targets are clearly outlined. If a behavioral phenotype or molecular target has not been evaluated/identified yet, the box is empty local protein synthesis at the synapse. Both behavioral and neurobiological phenotypes can be reversed in *Tsc* mutants by correcting local protein synthesis with either rapamycin, which blocks mTOR hyperactivity, or positive allosteric modulators of mGluR5 (Davis et al. 2015). Interestingly and importantly, this again points toward the fact that ASD might at least in part be caused by dysregulated translational control in neurons, which is strongly supported by a study showing that the offspring of *Fmr1* (synaptic protein synthesis high) and *Tsc2* (synaptic protein synthesis low) mutant crossbreeding no more exhibits ASD-related phenotypes (Auerbach et al. 2011).

3.3 *Mecp2 Mutant Mice*

Mutations in the X-chromosomal *MECP2* gene encoding methyl-CpG-binding protein 2, which is involved in DNA methylation and chromatin remodeling, cause Rett syndrome, a syndromic neurodevelopmental disorder affecting only females. Clinical manifestations include autistic-like features, severe intellectual disability, and seizures. Another characteristic feature is an early regression of acquired skills throughout development (Lyst and Bird 2015; Pohodich and Zoghbi 2015). It should be noted that Rett syndrome is not classified as ASD according to DSM-5, but listed as a separate disorder. However, since *Mecp2* mutant mice are of great value for the understanding of ASD-like phenotypes, they have also been included in this review. *Mecp2* mutant mice recapitulate many of the aforementioned behavioral phenotypes and are used to investigate the underlying pathomechanisms. On the neurobiological level, key features in *Mecp2* mutants are reduced dendritic branching, aberrant formation of glutamatergic synapses, imbalances between excitatory and inhibitory transmission, and reduced expression of the brain-derived neurotrophic factor (BDNF). Although the underlying molecular mechanisms largely remain unclear, they most probably include impaired epigenetic transcriptional regulation and altered chromatin remodeling due to the involvement of Mecp2 in both biological processes. Two intriguing studies further underlined that Rett syndrome phenotypes might be much more dynamical in nature as previously thought: It could be shown that re-expression of Mecp2 in adult *Mecp2* mutant mice restores and Mecp2 deletion in adult mice induces both behavioral and neurobiological phenotypes (Guy et al. 2007; McGraw et al. 2011). Interestingly, enhancing BDNF expression or direct application of insulin-like growth factor 1 (IGF-1) are two promising options for Rett syndrome treatment as they have already been effectively used to antagonize *Mecp2* mutant phenotypes (Wang et al. 2015).

3.4 *Cntnap2* Mutant Mice

CNTNAP2, the largest gene in the human genome, encodes an intriguing molecule: Contactin-associated protein-like 2 (CNTNAP2) belongs to the neurexin superfamily of presynaptic cell adhesion molecules and was initially identified as a regulator of ion channel composition at the nodes of Ranvier. Studies linking *CNTNAP2* mutations to ASD implicate an additional role for CNTNAP2 in brain development. In line, *Cntnap2* mutant mice do show autistic-like phenotypes including social interaction deficits and repetitive behaviors as well as comorbidities such as hyperactivity and epilepsy. Neurobiologically, neuronal migration defects, a reduction in the number of GABAergic interneurons, and a highly asynchronous cortical neuronal activity are hallmarks of *Cntnap2* mutants (Penagarikano and Geschwind 2012). In addition, the number of oxytocin-expressing neurons is reduced in the paraventricular nucleus of the hypothalamus (Penagarikano et al. 2015). Most interestingly, oxytocin has been successfully used to antagonize the social interaction deficits and risperidone to antagonize repetitive behaviors and hyperactivity in *Cntnap2* mutants (Penagarikano and Geschwind 2012; Penagarikano et al. 2015). These pharmacological studies highly support the hypothesis that the disruption of different neurobiological pathways is involved in the development of the different behavioral phenotypes of ASD.

3.5 *Neuroligin3* and *Neuroligin4* Mutant Mice

The neuroligins (NLGN1-4) are postsynaptic cell adhesion molecules forming transsynaptic signaling complexes with the neurexins, their presynaptic counterparts. This interaction is essential for both the formation and maintenance of functional synaptic connections (Sudhof 2008). Since mutations in the X-chromosomal *NLGN3* and *NLGN4* genes were linked to ASD, several mouse models have been generated to evaluate the deletion of either *Nlgn3* or *Nlgn4* or the R451C substitution based on a mutation in *NLGN3*. Overall, autistic-like behavior including core symptoms such as social interaction deficits and repetitive behaviors could be observed in most of the studies conducted (El-Kordi et al. 2013; Radyushkin et al. 2009; Tabuchi et al. 2007). Nevertheless, in some cases, it was not possible to replicate the behavioral phenotype of the same *Nlgn* mutants when analyzed in other independent laboratories (Chadman et al. 2008; Ey et al. 2012). A combination of methodological and environmental differences possibly underlies this discrepancy and is pointing toward high phenotypical variability even within the same ASD mutant line. To identify robust phenotypes more effectively, it might therefore be essential in the future to assess mutant behavior in several independent cohorts of the same strain at different sites. Main neurobiological phenotypes in *Nlgn3* mutant mice include region- and circuit-specific alterations on the synaptic level (Etherton et al. 2011; Rothwell et al. 2014), while *Nlgn4* mutants have not been investigated in this context yet. *Nlgn3* knockouts further exhibit a similar synaptic phenotype at the PF-PC synapse in the cerebellum as *Fmr1* mutants (increased mGluR1 and occlusion of mGluR1 LTD), which is reversible after the completion of neurodevelopment (Baudouin et al. 2012). This strongly supports both converging synaptic pathomechanisms among different ASD mutants and the dynamical nature of "neurodevelopmental" phenotypes.

3.6 Shank2 and Shank3 Mutant Mice

Shank2 and *Shank3* encode proline-rich scaffold proteins located at the postsynaptic density (PSD) of excitatory synapses, building large molecular interaction platforms for various other proteins including glutamate receptors and members of the actin cytoskeleton (Verpelli et al. 2012). As mutations in all three *SHANK* genes show a clear association with ASD (Leblond et al. 2014), several *Shank* mouse mutants have been generated for characterization of both autistic-like behaviors and neurobiological phenotypes (Jiang and Ehlers 2013; Schmeisser 2015). Although behavioral phenotypes reminiscent of ASD can be found in all mutants, phenotypical variability is very high in *Shank2* and *Shank3* mutant lines, especially with respect to neurobiology. This most likely reflects the individual contribution of each *SHANK* gene and its corresponding isoforms to synaptic function in a circuit-specific manner, which is differentially disrupted in each line. Albeit this complexity, the NMDA receptor is a promising molecular target as most neurobiological phenotypes among *Shank2* and *Shank3* mutant lines are related to alterations in NMDA receptor signaling. In addition, based on the reversal of both synaptic and behavioral phenotypes in a *Shank3* mutant line, IGF-1 might be beneficial not only for Rett syndrome, but also for *SHANK3*-associated ASD including Phelan–McDermid syndrome (Bozdagi et al. 2013).

4 Conclusions

Based on genetic studies and modern technology, several robust genetic animal models for ASD have been generated, advancing our understanding of ASD pathomechanisms and enabling us to discover novel targets for promising therapies.

While in this review, we gave a general introduction to ASD animal model characterization and outlined some of the major mutant mouse lines, the main challenges for the future are the following:

(1) Can we efficiently map the neural circuits responsible for the observed phenotypes in each mutant line?

(2) Do the defective pathways among different mutant lines converge on common molecular mechanisms?
(3) How do these mechanisms evolve throughout development and can we identify vulnerable time frames when treatment is most promising?

By using technologies including the Cre-lox and TetR systems, which convey spatial and temporal control of gene expression and by careful and detailed analysis of molecular pathways, we need to stratify our knowledge of behavioral and neurobiological phenotypes in ASD animal models with respect to neuroanatomical origins, converging molecular mechanisms, vulnerable time frames, and reversibility. This might finally enable us to develop not only individual, but possibly also even pleiotropic therapeutic strategies for ASD.

References

Auerbach BD, Osterweil EK, Bear MF (2011) Mutations causing syndromic autism define an axis of synaptic pathophysiology. Nature 480(7375):63–68

Baudouin SJ, Gaudias J, Gerharz S, Hatstatt L, Zhou K, Punnakkal P, Tanaka KF, Spooren W, Hen R, De Zeeuw CI, Vogt K, Scheiffele P (2012) Shared synaptic pathophysiology in syndromic and nonsyndromic rodent models of autism. Science 338(6103):128–132

Bhakar AL, Dolen G, Bear MF (2012) The pathophysiology of fragile X (and what it teaches us about synapses). Annu Rev Neurosci 35:417–443

Bourgeron T (2015) From the genetic architecture to synaptic plasticity in autism spectrum disorder. Nat Rev Neurosci 16(9):551–563

Bozdagi O, Tavassoli T, Buxbaum JD (2013) Insulin-like growth factor-1 rescues synaptic and motor deficits in a mouse model of autism and developmental delay. Mol Autism 4(1):9

Chadman KK, Gong S, Scattoni ML, Boltuck SE, Gandhy SU, Heintz N, Crawley JN (2008) Minimal aberrant behavioral phenotypes of neuroligin-3 R451C knockin mice. Autism Res 1 (3):147–158

Chang Q, Khare G, Dani V, Nelson S, Jaenisch R (2006) The disease progression of Mecp2 mutant mice is affected by the level of BDNF expression. Neuron 49(3):341–348

Chao HT, Chen H, Samaco RC, Xue M, Chahrour M, Yoo J, Neul JL, Gong S, Lu HC, Heintz N, Ekker M, Rubenstein JL, Noebels JL, Rosenmund C, Zoghbi HY (2010) Dysfunction in GABA signalling mediates autism-like stereotypies and Rett syndrome phenotypes. Nature 468 (7321):263–269

Chen JA, Penagarikano O, Belgard TG, Swarup V, Geschwind DH (2015) The emerging picture of autism spectrum disorder: genetics and pathology. Annu Rev Pathol 10:111–144

Davis PE, Peters JM, Krueger DA, Sahin M (2015) Tuberous sclerosis: a new frontier in targeted treatment of autism. Neurotherapeutics 12(3):572–583

De Rubeis S, He X, Goldberg AP, Poultney CS, Samocha K, Cicek AE, Kou Y, Liu L, Fromer M, Walker S, Singh T, Klei L, Kosmicki J, Shih-Chen F, Aleksic B, Biscaldi M, Bolton PF, Brownfeld JM, Cai J, Campbell NG, Carracedo A, Chahrour MH, Chiocchetti AG, Coon H, Crawford EL, Curran SR, Dawson G, Duketis E, Fernandez BA, Gallagher L, Geller E, Guter SJ, Hill RS, Ionita-Laza J, Jimenz Gonzalez P, Kilpinen H, Klauck SM, Kolevzon A, Lee I, Lei I, Lei J, Lehtimaki T, Lin CF, Ma'ayan A, Marshall CR, McInnes AL, Neale B, Owen MJ, Ozaki N, Parellada M, Parr JR, Purcell S, Puura K, Rajagopalan D, Rehnstrom K, Reichenberg A, Sabo A, Sachse M, Sanders SJ, Schafer C, Schulte-Ruther M, Skuse D, Stevens C, Szatmari P, Tammimies K, Valladares O, Voran A, Li-San W, Weiss LA, Willsey AJ, Yu TW, Yuen RK, Study DDD, Consortium UK, Cook EH, Freitag CM, Gill M,

Hultman CM, Lehner T, Palotie A, Schellenberg GD, Sklar P, State MW, Sutcliffe JS, Walsh CA, Scherer SW, Zwick ME, Barett JC, Cutler DJ, Roeder K, Devlin B, Daly MJ, Buxbaum JD (2014) Synaptic, transcriptional and chromatin genes disrupted in autism. Nature 515(7526):209–215

Dolen G, Osterweil E, Rao BS, Smith GB, Auerbach BD, Chattarji S, Bear MF (2007) Correction of fragile X syndrome in mice. Neuron 56(6):955–962

Ehninger D (2013) From genes to cognition in tuberous sclerosis: implications for mTOR inhibitor-based treatment approaches. Neuropharmacology 68:97–105

El-Kordi A, Winkler D, Hammerschmidt K, Kastner A, Krueger D, Ronnenberg A, Ritter C, Jatho J, Radyushkin K, Bourgeron T, Fischer J, Brose N, Ehrenreich H (2013) Development of an autism severity score for mice using Nlgn4 null mutants as a construct-valid model of heritable monogenic autism. Behav Brain Res 251:41–49

Etherton M, Foldy C, Sharma M, Tabuchi K, Liu X, Shamloo M, Malenka RC, Sudhof TC (2011) Autism-linked neuroligin-3 R451C mutation differentially alters hippocampal and cortical synaptic function. Proc Natl Acad Sci USA 108(33):13764–13769

Ey E, Yang M, Katz AM, Woldeyohannes L, Silverman JL, Leblond CS, Faure P, Torquet N, Le Sourd AM, Bourgeron T, Crawley JN (2012) Absence of deficits in social behaviors and ultrasonic vocalizations in later generations of mice lacking neuroligin4. Genes Brain Behav 11 (8):928–941

Guy J, Gan J, Selfridge J, Cobb S, Bird A (2007) Reversal of neurological defects in a mouse model of Rett syndrome. Science 315(5815):1143–1147

Huguet G, Ey E, Bourgeron T (2013) The genetic landscapes of autism spectrum disorders. Annu Rev Genomics Hum Genet 14:191–213

Jiang YH, Ehlers MD (2013) Modeling autism by SHANK gene mutations in mice. Neuron 78 (1):8–27

Kas MJ, Modi ME, Saxc MD, Smith DG (2014) Advancing the discovery of medications for autism spectrum disorder using new technologies to reveal social brain circuitry in rodents. Psychopharmacology 231(6):1147–1165

Kleijer KT, Schmeisser MJ, Krueger DD, Boeckers TM, Scheiffele P, Bourgeron T, Brose N, Burbach JP (2014) Neurobiology of autism gene products: towards pathogenesis and drug targets. Psychopharmacology 231(6):1037–1062

Koekkoek SK, Yamaguchi K, Milojkovic BA, Dortland BR, Ruigrok TJ, Maex R, De Graaf W, Smit AE, VanderWerf F, Bakker CE, Willemsen R, Ikeda T, Kakizawa S, Onodera K, Nelson DL, Mientjes E, Joosten M, De Schutter E, Oostra BA, Ito M, De Zeeuw CI (2005) Deletion of FMR1 in Purkinje cells enhances parallel fiber LTD, enlarges spines, and attenuates cerebellar eyelid conditioning in Fragile X syndrome. Neuron 47(3):339–352

Leblond CS, Nava C, Polge A, Gauthier J, Huguet G, Lumbroso S, Giuliano F, Stordeur C, Depienne C, Mouzat K, Pinto D, Howe J, Lemiere N, Durand CM, Guibert J, Ey E, Toro R, Peyre H, Mathieu A, Amsellem F, Rastam M, Gillberg IC, Rappold GA, Holt R, Monaco AP, Maestrini E, Galan P, Heron D, Jacquette A, Afenjar A, Rastetter A, Brice A, Devillard F, Assouline B, Laffargue F, Lespinasse J, Chiesa J, Rivier F, Bonneau D, Regnault B, Zelenika D, Delepine M, Lathrop M, Sanlaville D, Schluth-Bolard C, Edery P, Perrin L, Tabet AC, Schmeisser MJ, Boeckers TM, Coleman M, Sato D, Szatmari P, Scherer SW, Rouleau GA, Betancur C, Leboyer M, Gillberg C, Delorme R, Bourgeron T (2014) Meta-analysis of SHANK mutations in autism spectrum disorders: a gradient of severity in cognitive impairments. PLoS Genet 10(9):e1004580

Lyst MJ, Bird A (2015) Rett syndrome: a complex disorder with simple roots. Nat Rev Genet 16 (5):261–275

McGraw CM, Samaco RC, Zoghbi HY (2011) Adult neural function requires MeCP2. Science 333 (6039):186

Peca J, Feliciano C, Ting JT, Wang W, Wells MF, Venkatraman TN, Lascola CD, Fu Z, Feng G (2011) Shank3 mutant mice display autistic-like behaviours and striatal dysfunction. Nature 472(7344):437–442

Penagarikano O, Abrahams BS, Herman EI, Winden KD, Gdalyahu A, Dong H, Sonnenblick LI, Gruver R, Almajano J, Bragin A, Golshani P, Trachtenberg JT, Peles E, Geschwind DH (2011) Absence of CNTNAP2 leads to epilepsy, neuronal migration abnormalities, and core autism-related deficits. Cell 147(1):235–246

Penagarikano O, Geschwind DH (2012) What does CNTNAP2 reveal about autism spectrum disorder? Trends Mol Med 18(3):156–163

Penagarikano O, Lazaro MT, Lu XH, Gordon A, Dong H, Lam HA, Peles E, Maidment NT, Murphy NP, Yang XW, Golshani P, Geschwind DH (2015) Exogenous and evoked oxytocin restores social behavior in the Cntnap2 mouse model of autism. Sci Transl Med 7(271):271ra8

Pohodich AE, Zoghbi HY (2015) Rett syndrome: disruption of epigenetic control of postnatal neurological functions. Hum Mol Genet

Radyushkin K, Hammerschmidt K, Boretius S, Varoqueaux F, El-Kordi A, Ronnenberg A, Winter D, Frahm J, Fischer J, Brose N, Ehrenreich H (2009) Neuroligin-3-deficient mice: model of a monogenic heritable form of autism with an olfactory deficit. Genes Brain Behav 8 (4):416–425

Rothwell PE, Fuccillo MV, Maxeiner S, Hayton SJ, Gokce O, Lim BK, Fowler SC, Malenka RC, Sudhof TC (2014) Autism-associated neuroligin-3 mutations commonly impair striatal circuits to boost repetitive behaviors. Cell 158(1):198–212

Santini E, Klann E (2014) Reciprocal signaling between translational control pathways and synaptic proteins in autism spectrum disorders. Sci Signal 7(349):re10

Schmeisser MJ (2015) Translational neurobiology in Shank mutant mice–model systems for neuropsychiatric disorders. Ann Anat 200:115–117

Schmeisser MJ, Ey E, Wegener S, Bockmann J, Stempel AV, Kuebler A, Janssen AL, Udvardi PT, Shiban E, Spilker C, Balschun D, Skryabin BV, Dieck S, Smalla KH, Montag D, Leblond CS, Faure P, Torquet N, Le Sourd AM, Toro R, Grabrucker AM, Shoichet SA, Schmitz D, Kreutz MR, Bourgeron T, Gundelfinger ED, Boeckers TM (2012) Autistic-like behaviours and hyperactivity in mice lacking ProSAP1/Shank2. Nature 486(7402):256–260

Silverman JL, Yang M, Lord C, Crawley JN (2010) Behavioural phenotyping assays for mouse models of autism. Nat Rev Neurosci 11(7):490–502

Sudhof TC (2008) Neuroligins and neurexins link synaptic function to cognitive disease. Nature 455(7215):903–911

Tabuchi K, Blundell J, Etherton MR, Hammer RE, Liu X, Powell CM, Sudhof TC (2007) A neuroligin-3 mutation implicated in autism increases inhibitory synaptic transmission in mice. Science 318(5847):71–76

Tang G, Gudsnuk K, Kuo SH, Cotrina ML, Rosoklija G, Sosunov A, Sonders MS, Kanter E, Castagna C, Yamamoto A, Yue Z, Arancio O, Peterson BS, Champagne F, Dwork AJ, Goldman J, Sulzer D (2014) Loss of mTOR-dependent macroautophagy causes autistic-like synaptic pruning deficits. Neuron 83(5):1131–1143

Tsai PT, Hull C, Chu Y, Greene-Colozzi E, Sadowski AR, Leech JM, Steinberg J, Crawley JN, Regehr WG, Sahin M (2012) Autistic-like behaviour and cerebellar dysfunction in Purkinje cell Tsc1 mutant mice. Nature 488(7413):647–651

Verpelli C, Schmeisser MJ, Sala C, Boeckers TM (2012) Scaffold proteins at the postsynaptic density. Adv Exp Med Biol 970:29–61

Wang H, Pati S, Pozzo-Miller L, Doering LC (2015) Targeted pharmacological treatment of autism spectrum disorders: fragile X and Rett syndromes. Front Cell Neurosci 9:55

Wohr M, Scattoni ML (2013) Behavioural methods used in rodent models of autism spectrum disorders: current standards and new developments. Behav Brain Res 251:5–17

Won H, Lee HR, Gee HY, Mah W, Kim JI, Lee J, Ha S, Chung C, Jung ES, Cho YS, Park SG, Lee JS, Lee K, Kim D, Bae YC, Kaang BK, Lee MG, Kim E (2012) Autistic-like social behaviour in Shank2-mutant mice improved by restoring NMDA receptor function. Nature 486 (7402):261–265

Treatment Approaches in Rodent Models for Autism Spectrum Disorder

Susanna Pietropaolo, Wim E. Crusio and Francesca R. D'amato

Abstract Recent years have seen an impressive amount of research devoted to the developing of therapies to treat autism spectrum disorder (ASD). This work has been largely based on rodent models, employing a multitude of genetic and environmental manipulations. Unfortunately, the task of identifying suitable treatments for ASD is extremely challenging, due to a variety of problems specific to the research in this field. Here, we first discuss these problems, including (I) the presence of a large variety of rodent models (often without universal consensus on their validity), (II) the difficulties in choosing the most appropriate behavioural markers to assess the efficacy of possible treatments, (III) the limited knowledge we still have of the neurobiological bases of ASD pathology and of its aetiology, and (IV) the complexity of ASD itself, including a highly heterogeneous group of disorders sometimes with markedly different symptoms (therefore unlikely to be treated with the same approaches). Second, we give a critical overview of the most relevant advances in designing treatments for ASD, focusing on the most commonly used animal model, the laboratory mouse. We include pharmacological and non-pharmacological approaches, underlining their specific advantages, but also their current limitations especially in relation to the problems discussed before. Finally, we highlight the theoretical (e.g. the combination of multiple rather than single treatments) and methodological (e.g. use of single-gene mouse models)

S. Pietropaolo (✉) · W.E. Crusio
INCIA, University of Bordeaux, Bat B2, Allée Geoffroy St. Hilaire, CS 50023,
33615 Pessac Cedex, France
e-mail: susanna.pietropaolo@u-bordeaux.fr

S. Pietropaolo · W.E. Crusio
INCIA, UMR 5287, CNRS, Bat B2, Allée Geoffroy St. Hilaire, CS 50023,
33615 Pessac Cedex, France

F.R. D'amato
CNR, Cell Biology and Neurobiology Institute, IRCCS, Santa Lucia Foundation,
Via del Fosso di Fiorano 64, 00143 Rome, Italy

F.R. D'amato
Department of Psychiatry and Neurosciences, Laval University, Québec City, Canada

Curr Topics Behav Neurosci (2016) 30: 325–340
DOI 10.1007/7854_2015_433
Published Online: 29 January 2016

approaches that seem more promising to us, suggesting various strategies that can be adopted to simplify the complex field of research on treatments for ASD.

Keywords Autism · Social interactions · Mouse behaviour · Gene–environment interactions · Neuronal hyper-excitability · Behavioural therapies · Developmental disorders

Contents

1 Introduction 326
2 Problems of Research on Treatments for ASD 328
2.1 Constructing Valid Mouse Models 328
2.2 Demonstrating the Efficacy of a Treatment 329
2.3 The Unknown Aetiology of ASD 330
2.4 The Complexity of ASD 331
3 Major Therapeutic Approaches to ASD 332
3.1 Pharmacological Approaches 332
3.2 Non-pharmacological Therapies 333
4 Possible Ways Forward 335
4.1 Multiple Combined Rather Than Single Treatments 335
4.2 Trying to Simplify the Task: Reducing the Levels of Complexity 336
5 Conclusions and Perspectives 337
References 337

1 Introduction

Writing about the therapies for autism spectrum disorder (ASD) is at first sight a paradoxical task, since no accepted treatment is currently available for ASD. Yet, in the last few years an impressive amount of research has focused on trying to identify possible treatments to alleviate the dramatic symptoms of this pathology. The difficulty of this research stems from the fact that the aetiology of ASD is extremely complex and still largely unknown; furthermore, no biomarkers are available, so that the efficacy of a treatment can only be judged based on the remission of symptoms, i.e. the behavioural alterations affecting ASD patients.

Although finding treatments for mental disorders is never an easy task, the case of ASD represents a unique example of research challenges that involves the combined efforts of physicians and behavioural and molecular scientists in a variety of fields. Pre-clinical research is especially relevant, with a major role for investigations on rodent models for ASD, because clinical research on ASD is complicated by the developmental nature of the pathology, requiring the involvement of very young patients and their families, with all the obvious related difficulties. Furthermore, the complexity of ASD aetiopathology involves multiple genetic and environmental factors that can be more easily controlled for by using rodent models, not to mention the direct possibility of studying the brain and molecular correlates of the pathology offered exclusively by animal research.

However, pre-clinical research faces its own difficulties, as psychiatric disorders are notoriously difficult to model in animals (Crusio 2015). Because the exact aetiology of ASD is unknown, creating an animal model with construct validity is near impossible. However, several single-gene disorders are known that, among other symptoms, include ASD or ASD-like problems. We have argued recently that a good approach to the modelling problem would be the investigation of mouse models of these single-gene disorders (Oddi et al. 2013). Whereas most animal models of ASD only have face validity at best, these single-gene models have the much more important construct validity that is necessary to have a reasonable chance of the model having any predictive validity. Of course, this construct validity cannot necessarily be extended beyond the specific monogenic disorder being modelled, but this approach nevertheless seems to be a useful one.

This chapter will therefore critically illustrate the main problems of identifying therapeutic approaches to ASD, with a special emphasis on animal research, especially the most widely used model species, the laboratory mouse. This rodent species has quickly become the unchallenged protagonist of ASD research, due to its unique suitability for genetic manipulations and easiness of rapid laboratory breeding and husbandry. Paradoxically, social interactions, one of the major ASD-relevant behaviours, have traditionally been studied mainly in rats, with mice for a long time being considered a species with a restricted social repertoire. Although some specific social behaviours, such as context-dependent ultrasonic vocalizations and juvenile play activity, are indeed less complex in mice than in rats (Scattoni et al. 2009; Servadio et al. 2015), there is growing evidence that laboratory mice are suitable to study social behaviours, including complex ones, such as empathic responses (Gonzalez-Liencres et al. 2014; Langford et al. 2006; Sanders et al. 2013; Watanabe 2011). Nonetheless, experimental protocols to study social behaviour need to be adapted properly from rats to mice, a process that still needs refinement and is the object of increasing research efforts (Wohr and Scattoni 2013).

It is not our goal to provide the field of ASD research with an exhaustive and detailed description of the variety of treatments that have been recently proposed from mouse models, since several articles comprehensively reviewing the literature on this topic already exist (see e.g. Ruhela et al. 2015; Servadio et al. 2015; Silverman and Crawley 2014). We will instead give an overview of the major therapeutic treatments, including both pharmacological and behavioural/environmental approaches, highlighting how they can be affected by the problems of ASD research described before. Despite the limitations signalled above, we will conclude the chapter on an optimistic note by suggesting possible methodological and theoretical strategies to improve ASD-related mouse research and highlighting the currently most promising approaches to develop effective treatments.

2 Problems of Research on Treatments for ASD

2.1 Constructing Valid Mouse Models

The first problem that ASD pre-clinical research has to face lies in the choice of the mouse model used to validate the treatment of interest. As reviewed in several recent articles (Banerjee et al. 2014; Bey and Jiang 2014; Ellegood and Crawley 2015; Moy et al. 2006), a huge variety of mouse models have been proposed in the last decades, based on very different approaches.

One strategy is based on identifying mouse strains that spontaneously present some major ASD-like symptoms, such as low sociability, poor communication abilities, and high levels of repetitive/inflexible behaviours. The best-known example of this approach is the BTBR strain (Mcfarlane et al. 2008; Meyza et al. 2013), widely employed in ASD research and used to identify the novel genes involved in the control of social behaviour. The main disadvantages of this model are as follows: (i) the problem of identifying an appropriate control strain and (ii) the possibility that strain-specific phenotypes of emotionality or territoriality may explain the supposed ASD-like behavioural profiles (Oddi et al. 2013).

Another approach consists in manipulating the mouse genome to import in the animal model some of the genetic alterations associated with ASD, e.g. gene deletion or silencing, or chromosomal abnormalities. The validity of this experimental strategy also has some problems, including (i) the confounding impact of the genetic background of the engineered strain that can affect the animals' phenotype (Crusio et al. 2009; Gerlai 1996; Pietropaolo et al. 2011), and (ii) the actual complexity and heterogeneity of the genetic alterations known to be somehow involved in ASD, including almost all genes affecting neuronal function and development, the single-specific contribution of which often not being clear (Abrahams and Geschwind 2008). Other mouse models rely on the exposure (mostly prenatally) to adverse environmental factors (Dufour-Rainfray et al. 2011; Patterson 2011), such as teratogens and viral infections, suspected to contribute to the aetiology of ASD on the basis of epidemiological data. Similar to the previous one, this approach oversimplifies the complexity of the, in this case environmental, risk factors involved in ASD aetiology and is further complicated by interference in mother–infant interactions that is highly critical in the mouse species, due to the sensorial immaturity of newborn pups (Oddi et al. 2013).

Finally, other mouse models have been created by focusing on other developmental pathologies having symptoms in common with ASD (e.g. fragile X or Rett syndrome), but with well-established mouse models and known monogenic causal factors (Oddi et al. 2013). A criticism to this research strategy is the fact that the genetic mutations induced in these models (e.g. *Fmr1* deletion for fragile X syndrome) are not invariably present in ASD patients and may therefore not be universally valid in modelling the pathology, but only a specific subset of patients (Budimirovic and Kaufmann 2011).

Whatever the approach chosen, the major problem is that there is no general consensus on the validity of the existing mouse models, so that none is universally recognized as a "good" model and none can be definitively discarded as a "bad" one. In this context, it is evident that demonstrating the efficacy of a treatment in only one model cannot be considered sufficient, so that multiple models should be concomitantly used to support positive conclusions on the value of a proposed therapy. The general criterion suggested to consider a mouse model as valid is that it presents alterations in at least the three core domains affected in ASD, i.e. social interaction, communication, and behavioural flexibility/repetitive behaviours (Crawley 2004, 2007; Wohr and Scattoni 2013). The presence of additional behavioural abnormalities, such as hyperactivity, memory deficits, or enhanced anxiety, may increase the validity of the model, but is not required (Moy et al. 2006). Nonetheless, this criterion is not rigid, and it is not uncommon to consider a mouse line with clear and marked deficits in social interaction or social interest of relevance for ASD research, even in the absence of the deficits in the other behavioural domains. The logical consequence of this chaotic situation is that potential treatments for ASD are tested on whatever mouse line presents some behavioural deficits that resemble those observed in ASD, often assessed with incredibly heterogeneous methods.

2.2 *Demonstrating the Efficacy of a Treatment*

The most striking characteristic of mouse research on ASD treatments is probably the lack of consensus on the behavioural alterations that a proposed therapy should cure. It should be obvious that an ideal treatment should significantly improve all core ASD symptoms, as well as some additional ones. However, most studies assessing the effects of treatments focus only on social interactions, occasionally adding other behaviours from the core autistic domains and/or the additional ones. This makes it almost impossible to compare the efficacy of different proposed treatments and sometimes difficult to judge the real therapeutic impact of a specific one. Should we then discard all treatments that do not fully rescue all the main ASD-like behavioural alterations? It is unlikely that this research strategy would be successful, also because it is becoming increasingly clear that ASD cannot be treated with a single approach, but only with a combination of treatments. Hence, interventions that rescue only the social symptoms should not be overlooked because they can potentially be combined with other treatments improving, for example, repetitive/inflexible behaviours (Gross et al. 2012).

This lack of consensus exists not only for which behavioural domain should be affected by a proposed treatment, but also for the specific methods used to test these behaviours. Indeed, even a rapid overview of the literature shows that nearly all possible tests for core ASD-like symptoms have been used in mouse therapeutic studies (for a review of the behavioural methods used in ASD research see Crawley 2004, 2007; Silverman et al. 2010). Rescue of social deficits, for example, has been demonstrated with tests of direct social interactions requiring manual scoring and using a variety of methods (e.g. male–male or male–female interactions, resident–intruder or neutral cage tests, heterogenous ASD versus control mouse or homogeneous dyads), or automatized tests such as the 3-compartment test or the tube-dominance test. Similarly, the presence of repetitive/inflexible behaviours is not uniformly assessed, with testing methods ranging from the simple analysis of self-grooming to complex protocols of working memory and habit formation. The analysis of communication deficits has received growing attention in ASD research, mostly by investigating ultrasonic vocalizations (but scent marking can be also used), although this is often neglected in therapeutic studies. This is especially surprising because ultrasonic communication is a behaviour that can be easily studied in developing mice and could therefore be of high utility in assessing therapeutic strategies for a developmental disorder as ASD. The developmental phase is, in fact, insufficiently considered in ASD studies, which mostly use adult mice. This is a major limitation of current therapeutic studies that we should try to overcome in the future.

Further confounding factors, in addition to age, are the genetic background on which a mutation in a given ASD mouse model is studied, as already mentioned above, and the sex of the mice is tested. Most ASD mouse studies use male subjects, first because a huge part of research on behavioural phenotyping tends to use the male sex to avoid controlling for oestrous cycle differences in female mice, and second because of the higher incidence of ASD in boys compared to girls. However, even when testing social interactions is limited to male laboratory mice, age can be a confounding factor since the expression of these behaviours follows a precise developmental-stage-related pattern (Terranova et al. 1998), with affiliative motivation being high in young animals and decreasing with sexual maturity, and aggression becoming more and more evident with increasing age. Furthermore, the use of male mice for ASD studies limits their translational power, since even though in lower proportions, girls are also affected by ASD and it is not granted that therapeutic interventions will have the same effect in patients of both sexes.

In conclusion, future studies should use multiple tests for a single behavioural domain, making a maximal effort to standardize the procedures and controlling as much as possible for the confounding experimental variables described here and try to include both sexes as well as developmental analyses.

2.3 *The Unknown Aetiology of ASD*

The elucidation of the causal molecular mechanisms responsible for an observed pathological phenotype is usually necessary to design potential therapeutic interventions. This is not possible in the case of ASD, because this disorder is due to a combination of genetic and environmental factors, manifesting through a variety of complex mechanisms, including imbalances of excitation and inhibition as well as hypo-/hyper-connectivity (Pardo and Eberhart 2007). In addition, the genetic causes of ASDs show a high degree of heterogeneity, with hundreds of ASD-associated genes now identified (Abrahams and Geschwind 2008). The list of possible pathogenic mechanisms has dramatically increased, including multiple genetic mutations, combinations thereof, and gene–environment interactions (Mcomish et al. 2014). It is hoped, and recent studies indeed suggest, that all this may functionally converge into a relatively smaller subset of cellular and biochemical pathways affecting distinct neuronal functions (Zoghbi and Bear 2012). The functional convergence on particular signalling pathways and the shared synaptopathology of ASDs even have raised the hope that similar therapeutic strategies may be effective for different forms of ASDs which are related, but genetically distinct (Zoghbi and Bear 2012).

The role of environmental insults (e.g. prenatal exposure to teratogens/contaminants, viral/bacterial infections, heavy stressors, and traumas) is also far from being elucidated, with currently an overall tendency of animal research on ASD to preferentially focus on genetic models, as a result of the general emphasis on the genetic causes of mental disorders. Environmental factors however play an obviously central role in the aetiopathology of ASD, and their study is of special interest in the context of understanding the great inter-individual variability observed in the severity and quality of the symptoms displayed by ASD patients. In this respect, the investigation of gene–environment interactions, i.e. of the combination of genetic risk factors and environmental adversity, is receiving increasing attention in ASD research (Happe et al. 2006), and it is considered a key issue to understand the individual differences in resilience observed in patients affected by this disorder (Mcomish et al. 2014; Nithianantharajah and Hannan 2006).

2.4 The Complexity of ASD

ASD is now defined as a single disorder that includes pathologies that were previously considered separate, that is autism, Asperger's syndrome, childhood disintegrative disorder, and pervasive developmental disorder not otherwise specified. The term "spectrum" refers to the wide range of symptoms and severity characterizing ASD patients, and an impressive variability affects the expression of both core (i.e. the "triad" defining the diagnostic criterion of ASD) and additional symptoms (Rapin 1991). Language deficits and delays are, for example, completely absent in patients with Asperger's syndrome, as is mental retardation. Also, ASD patients may either underreact or overreact to sensory stimuli, and they may show highly variable intelligence levels, from deficits to unevenly developed cognitive skills.

This high heterogeneity of ASD itself is obviously related to the fact that it is a disorder that has been artificially defined as a single disorder, but in reality is a combination of divergent disorders that all have very variable symptoms. Hence, a more promising approach could be to discard any unitary explanation of ASD and instead try to treat specific symptoms, rather than the whole disorder (Happe et al. 2006).

3 Major Therapeutic Approaches to ASD

Independent of the choice of the specific treatment, two major therapeutic approaches to ASD can be identified, one focused on rescuing the pathologic phenotype once it is already manifest, and the other aimed to prevent the appearance of the symptoms in individuals at risk. Recent research has in fact challenged the notion that ASD is an irreversible neurodevelopmental disorder: a growing number of mouse studies have shown that certain neuronal defects can be reversed in the mature brain, either by pharmacological or environmental interventions (Delorme et al. 2013). The efficacy of the treatments obviously depends on their timing: clinicians encourage beginning intervention in children with ASD as soon as the first signs are manifest, knowing that a reliable diagnosis of the disease is possible as early as 18–24 months of age.

3.1 *Pharmacological Approaches*

Pharmacological therapies currently used in the clinic are mainly limited to the treatment of some relatively minor symptoms, including excitation, irritability, and aggressiveness. Many patients with ASD are treated with psychoactive drugs or anticonvulsants, the most common drug classes being antidepressants, stimulants, and antipsychotics. These therapies obviously do not act on any causal factor of the disorder and have severe side effects, especially in children. It is therefore evident that pre-clinical research will have a pivotal role in testing treatments with more specific molecular targets and therefore with high therapeutic potential.

The main and also most recent approach of animal research on pharmacological therapies for ASD is to focus on the genes known to be altered in this disorder and try to identify a common mechanism mediated by the large variety of single genes involved. Most studies following this approach agree that dysregulation of activity-dependent protein synthesis is a central factor in the pathogenesis of this disorder. Most ASD genes have in fact effects on this major process either by affecting protein synthesis intrinsically or by impairing synaptic function in general. These processes can be targeted by specific compounds from various directions, e.g. through inhibition of the mTOR pathway using rapamycin, or metabotropic glutamate receptor-related modulation of synaptic activity using mGluR1 and mGluR5 antagonists, or stimulating specific ionic channels, just to mention some of the best-known examples. A detailed overview of these compounds and their mechanisms of action can be found elsewhere (e.g. Delorme et al. 2013; Kleijer et al. 2014).

One critical problem with this "genetic" approach is that it is still largely unknown in which phases of development the mutations found in ASD produce a detectable phenotype. In rodents, most genes involved in synaptogenesis are strongly expressed after birth, reaching a plateau in adulthood, while these genes are switched on very early in utero in humans, and reach a plateau at 6–9 months of foetal life. This may also be the case for most ASD susceptibility genes. Thus, although mouse models remain the most valuable tool to screen the efficacy of possible therapeutic agents for ASD, species-specific differences might limit direct extrapolations to humans. Results from clinical trials are a necessity to obtain reliable information about a specific potential treatment.

Clinical trials are, on the one hand, facilitated by the fact that most of the compounds tested in ASD mouse models are already clinically available, although for other pathologies. For example, rapamycin was developed to prevent transplant rejection, and BMS, an agonist of a potassium channel that, as we have recently shown, improves behavioural deficits in a mouse model of fragile X syndrome (Hebert et al. 2014; Zhang et al. 2014), was used to treat stroke patients. On the other hand, possible side effects of the proposed treatments must be carefully evaluated because of the developmental nature of the disorder and the likely life-long duration of the pharmacological administration. In this respect, animal studies often test a treatment only for its acute effects; this is surprising, since this approach is sufficient only to suggest a mechanistic role of the molecule of interest, but does not resemble the requirements of the clinic. Animal studies on chronic administration should therefore always be performed to inform clinical trial protocols, as they are critical to highlight the potential adverse effects of prolonged and/or early treatment.

3.2 Non-pharmacological Therapies

Given the developmental age of the patients when the first symptoms become manifest and the limitations and side effects of the available pharmacological treatments, there is a high interest in alternative interventions based on non-pharmacological treatments (Lofthouse et al. 2012). Among these, behavioural therapies, such as structured teaching, speech, and language therapy, or pet and occupational therapies, are those that deserve the highest attention, since they are overall recognized as the currently most effective treatments for ASD (Myers et al. 2007). These forms of behavioural stimulation, especially when started early in life, can help ASD children acquire self-care, improve functioning, and decrease maladaptive behaviours, i.e. reducing the major impact of the pathology on the quality of life of patients and their caretakers.

Mouse studies have tried to study the brain mechanisms underlying behavioural therapies, modelling these interventions through protocols of environmental stimulation applied in animal genetic models for ASD. The exposure to an enriched environment, such as housing conditions with toys, tunnels, and stimulating objects, has been applied to different mouse models of ASD, such as the BTBR strain and mouse models of Rett and fragile X syndromes, where they demonstrated beneficial effects on some behavioural abnormalities as well as on electrophysiological disturbances and neurotrophin imbalances in specific brain areas (e.g. Lonetti et al. 2010; Restivo et al. 2005; Reynolds et al. 2013). Environmental stimulation, in the form of increased maternal care, has been applied also in the very early post-natal phase, thus allowing investigation of the effects of this intervention during development, and not only later in adulthood. Interestingly, this early enrichment rescued at long term not only the social, motor, and cognitive alterations in behaviour (Fig. 1), but also the abnormalities in dendritic spine morphology in the hippocampus and amygdala of a mouse model of fragile X syndrome (Oddi et al. 2015).

These data on enrichment effects in mouse models of autism have interesting implications for the debate on theories trying to give a unitary explanation of the neurobiology of ASD. Among these, the Intense World Theory (Markram and Markram 2010) proposed that the autistic brain is excessively reactive to the environment, hyper-responsive to stimuli, and hyper-plastic. As a consequence, memory formation is exaggerated in autistic patients, especially when related to emotional experiences, and brain maturation is accelerated, making the surrounding environment painfully intense. Consequently, in a sensory enriched and changing environment, the prognosis is then expected to become worse. It is evident that the enrichment data obtained so far from both ASD mouse models and patients are in clear contradiction with this theory (Woo et al. 2015; Woo and Leon 2013).

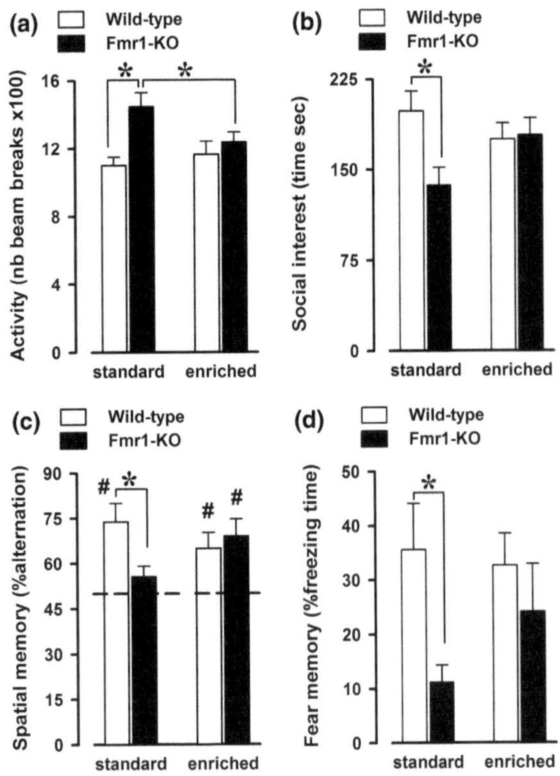

Fig. 1 Long-term behavioural effects of early enrichment in the Fmr1-KO mouse model for fragile X syndrome and autism. Early enrichment rescued the behavioural abnormalities displayed by adult Fmr1-KO mice, reared under standard conditions including their hyperactivity (**a**), reduced levels of social interactions (**b**), deficits in spatial memory in the T-maze (**c**), and in fear conditioning (**d**). # = significantly different from chance level, represented by the *dotted line*. Modified from (Oddi et al. 2015)

However, two important caveats need to be made. First, the enriched environment used in mouse research not only consists of a more stimulating situation than the standard one, but also provides the animal with a better control of its environment, e.g. providing shelters to hide from the experimenters and for resting. The enriched environments commonly used in mice also provide more opportunities for physical exercise (e.g. through running wheels and tunnels), known to reduce emotional reactivity to sensory stimuli. It is therefore possible that environmental enrichment provides a selective enhancement in specific types of sensory stimulation that rescues the excitatory imbalance of an autistic brain. Second, the described theory has been formulated mainly based on the results obtained with the valproic acid (VPA) rat model, which is indeed characterized by the enhanced fear conditioning, whereas the ASD mouse models used so far in enrichment studies present memory deficits, including this cognitive domain (see e.g. Fig. 1d), and have morphological brain abnormalities, suggesting a delayed, rather than an accelerated maturation of the circuits. Therefore, these models themselves already do not present the phenotype hypothesized by the intense world theory.

Another interesting non-pharmacological therapy is represented by dietary interventions. Among these, dietary supplementation with Omega-3 polyunsaturated fatty acid (n-3 PUFAs) is one of the most studied, because of its ability to stimulate synaptic function and plasticity and reduce inflammation in a number of pathologies. We recently reported that n-3 PUFA dietary supplementation, started at weaning, partially rescued the behavioural abnormalities and neuroinflammatory imbalances displayed by a mouse model of fragile X syndrome (Pietropaolo et al. 2014). Unfortunately, to our knowledge, this therapy has been tested neither in other mouse models for ASD, nor at other starting ages, so further studies are needed to confirm the efficacy of this promising approach.

4 Possible Ways Forward

4.1 Multiple Combined Rather Than Single Treatments

Our discussion until now clearly shows that it probably is utopian to orient ASD research towards a unitary therapeutic approach. Pharmacological and non-pharmacological approaches present advantages and limitations at the same time, as schematized in Fig. 2.

We are increasingly convinced that a combination of multiple treatments, probably of different nature (pharmacological and other), is needed to meaningfully alter the expression of this complex pathology. This approach may possibly require "breaking" the disease into its specific symptoms and focus on treating them separately, rather than ASD as a whole.

Some animal and human studies have, for example, focused on sensory–motor disturbances, considering them a key alteration somehow inducing many other

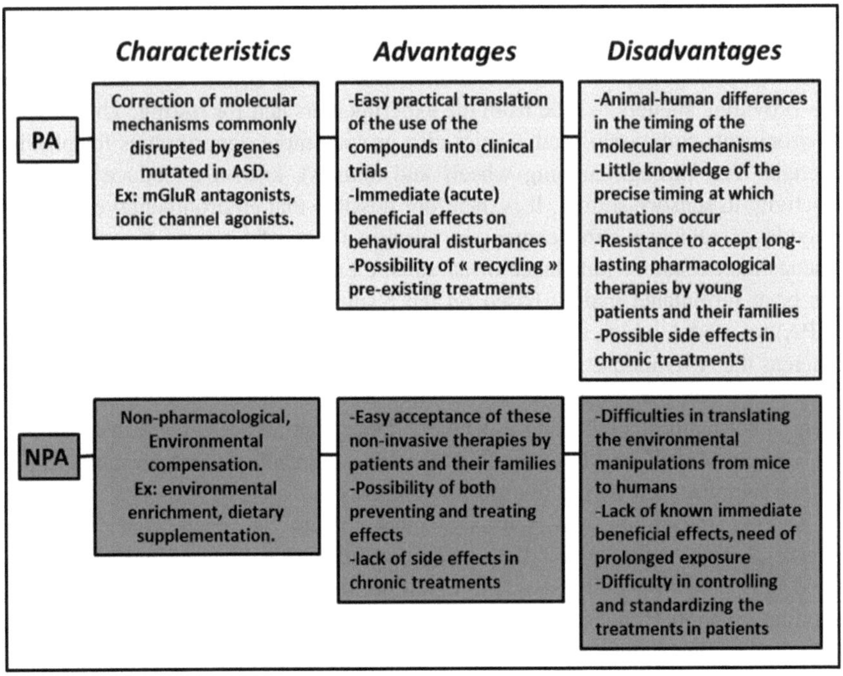

Fig. 2 Major advantages and limitations of different therapeutic strategies to ASD using animal models. Summary of the main characteristics, qualities, and disadvantages of both pharmacological (PA) and non-pharmacological approaches (NPA) to treat ASD

ASD symptoms through a disrupted perception of the environment. Indeed, a protocol of sensory–motor enrichment based on animal studies has been shown to induce markedly positive effects in ASD patients (Woo et al. 2015). Other studies have highlighted the importance of treating the social deficits, as the major and most invalidating ASD symptoms, as permitted, for example, by oxytocin inhalation (Modi and Young 2012), while others have pointed to the disruption of biological/circadian rhythms, easily treated with melatonin (Tordjman et al. 2015).

4.2 *Trying to Simplify the Task: Reducing the Levels of Complexity*

As already mentioned, it is possible that a "reductionist" approach to ASD may help to reduce the complexity of the problem and hence to identify novel therapies. As mentioned in our brief overview of the available treatments, several studies have benefited from the use of mouse models of other pathologies, such as fragile X and

Rett syndromes, having some symptoms in common with ASD, but characterized by lower levels of complexity and a clearer aetiology.

The main advantage of this approach consists in the fact that these pathologies have known monogenic causes and well-established mouse models, thus allowing to overcome the two major problems of ASD research. Indeed, this approach has helped to identify a variety of treatments, including pharmacological interventions and environmental stimulation, and some of these fundamental studies have led to the development of drugs that are currently in clinical trials (Dolen et al. 2007; Krueger and Bear 2011).

5 Conclusions and Perspectives

As we have argued above, more attention should be paid to fine-grained analysis of the time-dependent expression of both the symptoms and the effects of the proposed treatments. Developmental analyses should be applied to animal models, aiming to identify early behavioural markers and therefore the most interesting time windows for therapeutic interventions. In parallel, both pharmacological and non-pharmacological therapies should be tested in multiple mouse models at different critical time windows in order to increase the efficacy of the interventions and to investigate both their short- and long-term effects. As for possible pharmacological treatments, up till now most efforts have involved acute treatments, but patients would have to take medications chronically and our knowledge of long-term effects of the molecules tested is woefully inadequate. Chronic treatments may have as yet undetected side effects or, more positively, could be effective using lower doses than acute treatments. At the same time, the different tests used to evaluate autistic-like behaviours should be more thoroughly validated and cross-validated. This should lead to a better standardization of tests and the behavioural endpoints used to assess possible therapeutic effects. Finally, more attention should be paid to genotype–environment interactions, in attempts to increase our understanding of the aetiopathology of the disorder, the large phenotypical variability of patients, the complexity of symptoms as well as in studies of possible therapeutic effects of proposed treatments.

References

Abrahams BS, Geschwind DH (2008) Advances in autism genetics: on the threshold of a new neurobiology. Nat Rev Genet 9:341–355

Banerjee S, Riordan M, Bhat MA (2014) Genetic aspects of autism spectrum disorders: insights from animal models. Front Cell Neurosci 8:58

Bey AL, Jiang YH (2014) Overview of mouse models of autism spectrum disorders. Curr Protoc Pharmacol/Editorial Board, S.J. Enna, 66:5.66.1–5.66.26

Budimirovic DB, Kaufmann WE (2011) What can we learn about autism from studying fragile X syndrome? Dev Neurosci 33:379–394
Crawley JN (2004) Designing mouse behavioral tasks relevant to autistic-like behaviors. Ment Retard Dev Disabil Res Rev 10:248–258
Crawley JN (2007) Mouse behavioral assays relevant to the symptoms of autism. Brain Pathol 17:448–459
Crusio WE (2015) Key issues in contemporary behavioral genetics. Curr Opin Behav Sci 2:89–95
Crusio WE, Goldowitz D, Holmes A, Wolfer D (2009) Standards for the publication of mouse mutant studies. Genes Brain Behav 8:1–4
Delorme R, Ey E, Toro R, Leboyer M, Gillberg C, Bourgeron T (2013) Progress toward treatments for synaptic defects in autism. Nat Med 19:685–694
Dolen G, Osterweil E, Rao BS, Smith GB, Auerbach BD, Chattarji S, Bear MF (2007) Correction of fragile X syndrome in mice. Neuron 56:955–962
Dufour-Rainfray D, Vourc'h P, Tourlet S, Guilloteau D, Chalon S, Andres CR (2011) Fetal exposure to teratogens: evidence of genes involved in autism. Neurosci Biobehav Rev 35:1254–1265
Ellegood J, Crawley JN (2015) Behavioral and neuroanatomical phenotypes in mouse models of autism. Neurother J Am Soc Exp Neurother 12:521–533
Gerlai R (1996) Gene-targeting studies of mammalian behavior: is it the mutation or the background genotype? Trends Neurosci 19:177–181
Gonzalez-Liencres C, Juckel G, Tas C, Friebe A, Brune M (2014) Emotional contagion in mice: the role of familiarity. Behav Brain Res 263:16–21
Gross C, Berry-Kravis EM, Bassell GJ (2012) Therapeutic strategies in fragile X syndrome: dysregulated mGluR signaling and beyond. Neuropsychopharmacol Official Publ Am Coll Neuropsychopharmacol 37:178–195
Happe F, Ronald A, Plomin R (2006) Time to give up on a single explanation for autism. Nat Neurosci 9:1218–1220
Hebert B, Pietropaolo S, Meme S, Laudier B, Laugeray A, Doisne N, Quartier A, Lefeuvre S, Got L, Cahard D, Laumonnier F, Crusio WE, Pichon J, Menuet A, Perche O, Briault S (2014) Rescue of fragile X syndrome phenotypes in Fmr1 KO mice by a BKCa channel opener molecule. Orphanet J Rare Dis 9:124
Kleijer KTE, Schmeisser M, Krueger DD, Boeckers TM, Scheiffele P, Bourgeron T, Brose N, Burbach JPH (2014) Neurobiology of autism gene products: towards pathogenesis and drug targets. Psychopharmacology 231:1037–1062
Krueger DD, Bear MF (2011) Toward fulfilling the promise of molecular medicine in fragile X syndrome. Annu Rev Med 62:411–429
Langford DJ, Crager SE, Shehzad Z, Smith SB, Sotocinal SG, Levenstadt JS, Chanda ML, Levitin DJ, Mogil JS (2006) Social modulation of pain as evidence for empathy in mice. Science 312:1967–1970
Lofthouse N, Hendren R, Hurt E, Arnold LE, Butter E (2012) A review of complementary and alternative treatments for autism spectrum disorders. Autism Res Treat 2012:870391
Lonetti G, Angelucci A, Morando L, Boggio EM, Giustetto M, Pizzorusso T (2010) Early environmental enrichment moderates the behavioral and synaptic phenotype of MeCP2 null mice. Biol Psychiatry 67:657–665
Markram K, Markram H (2010) The intense world theory—a unifying theory of the neurobiology of autism. Frontiers Hum Neurosci 4:224
McFarlane HG, Kusek GK, Yang M, Phoenix JL, Bolivar VJ, Crawley JN (2008) Autism-like behavioral phenotypes in BTBR T+ tf/J mice. Genes Brain Behav 7:152–163
McOmish CE, Burrows EL, Hannan AJ (2014) Identifying novel interventional strategies for psychiatric disorders: integrating genomics, 'enviromics' and gene-environment interactions in valid preclinical models. Br J Pharmacol 171:4719–4728
Meyza KZ, Defensor EB, Jensen AL, Corley MJ, Pearson BL, Pobbe RL, Bolivar VJ, Blanchard DC, Blanchard RJ (2013) The BTBR T+ tf/J mouse model for autism spectrum disorders-in search of biomarkers. Behav Brain Res 251:25–34

Modi ME, Young LJ (2012) The oxytocin system in drug discovery for autism: animal models and novel therapeutic strategies. Horm Behav 61:340–350

Moy SS, Nadler JJ, Magnuson TR, Crawley JN (2006) Mouse models of autism spectrum disorders: the challenge for behavioral genetics. Am J Med Genet C Semin Med Genet 142C:40–51

Myers SM, Johnson CP (2007) Management of children with autism spectrum disorders. Pediatrics 120:1162–1182 (American Academy of Pediatrics Council on Children)

Nithianantharajah J, Hannan AJ (2006) Enriched environments, experience-dependent plasticity and disorders of the nervous system. Nat Rev Neurosci 7:697–709

Oddi D, Crusio WE, D'Amato FR, Pietropaolo S (2013) Monogenic mouse models of social dysfunction: implications for autism. Behav Brain Res 251:75–84

Oddi D, Subashi E, Middei S, Bellocchio L, Lemaire-Mayo V, Guzman M, Crusio WE, D'Amato FR, Pietropaolo S (2015) Early social enrichment rescues adult behavioral and brain abnormalities in a mouse model of fragile X syndrome. Neuropsychopharmacol Official Publ Am Coll Neuropsychopharmacol 40:1113–1122

Pardo CA, Eberhart CG (2007) The neurobiology of autism. Brain pathol 17:434–447

Patterson PH (2011) Maternal infection and immune involvement in autism. Trends Mol Med 17:389–394

Pietropaolo S, Goubran MG, Joffre C, Aubert A, Lemaire-Mayo V, Crusio WE, Laye S (2014) Dietary supplementation of omega-3 fatty acids rescues fragile X phenotypes in Fmr1-Ko mice. Psychoneuroendocrinology 49:119–129

Pietropaolo S, Guilleminot A, Martin B, D'Amato FR, Crusio WE (2011) Genetic-background modulation of core and variable autistic-like symptoms in Fmr1 knock-out mice. PLoS ONE 6: e17073

Rapin I (1991) Autistic children: diagnosis and clinical features. Pediatrics 87:751–760

Restivo L, Ferrari F, Passino E, Sgobio C, Bock J, Oostra BA, Bagni C, Ammassari Teule M (2005) Enriched environment promotes behavioral and morphological recovery in a mouse model for the fragile X syndrome. Proc Natl Acad Sci USA 102:11557–11562

Reynolds S, Urruela M, Devine DP (2013) Effects of environmental enrichment on repetitive behaviors in the BTBR T+ tf/J mouse model of autism. Autism Res Official J Int Soc Autism Res 6:337–343

Ruhela RK, Prakash A, Medhi B (2015) An urgent need for experimental animal model of autism in drug development. Ann Neurosci 22:44–49

Sanders J, Mayford M, Jeste D (2013) Empathic fear responses in mice are triggered by recognition of a shared experience. PLoS ONE 8:e74609

Scattoni ML, Crawley J, Ricceri L (2009) Ultrasonic vocalizations: a tool for behavioural phenotyping of mouse models of neurodevelopmental disorders. Neurosci Biobehav Rev 33:508–515

Servadio M, Vanderschuren LJ, Trezza V (2015) Modeling autism-relevant behavioral phenotypes in rats and mice: do 'autistic' rodents exist? Behav Pharmacol 26:522–540

Silverman JL, Crawley JN (2014) The promising trajectory of autism therapeutics discovery. Drug Discovery Today 19:838–844

Silverman JL, Yang M, Lord C, Crawley JN (2010) Behavioural phenotyping assays for mouse models of autism. Nat Rev Neurosci 11:490–502

Terranova ML, Laviola G, de Acetis L, Alleva E (1998) A description of the ontogeny of mouse agonistic behavior. J Comp Psychol 112:3–12

Tordjman S, Davlantis KS, Georgieff N, Geoffray MM, Speranza M, Anderson GM, Xavier J, Botbol M, Oriol C, Bellissant E, Vernay-Leconte J, Fougerou C, Hespel A, Tavenard A, Cohen D, Kermarrec S, Coulon N, Bonnot O, Dawson G (2015) Autism as a disorder of biological and behavioral rhythms: toward new therapeutic perspectives. Frontiers Pediatr 3:1

Watanabe S (2011) Empathy and reversed empathy of stress in mice. PLoS ONE 6:e23357

Wohr M, Scattoni ML (2013) Behavioural methods used in rodent models of autism spectrum disorders: current standards and new developments. Behav Brain Res 251:5–17

Woo CC, Donnelly JH, Steinberg-Epstein R, Leon M (2015) Environmental enrichment as a therapy for autism: a clinical trial replication and extension. Behav Neurosci 129:412–422

Woo CC, Leon M (2013) Environmental enrichment as an effective treatment for autism: a randomized controlled trial. Behav Neurosci 127:487–497

Zhang Y, Bonnan A, Bony G, Ferezou I, Pietropaolo S, Ginger M, Sans N, Rossier J, Oostra B, LeMasson G, Frick A (2014) Dendritic channelopathies contribute to neocortical and sensory hyperexcitability in Fmr1(-/y) mice. Nat Neurosci 17:1701–1709

Zoghbi HY, Bear MF (2012) Synaptic dysfunction in neurodevelopmental disorders associated with autism and intellectual disabilities. Cold Spring Harb Perspect biol 4(3):a009886

Neuroimaging-Based Phenotyping of the Autism Spectrum

Boris C. Bernhardt, Adriana Di Martino, Sofie L. Valk and Gregory L. Wallace

Abstract Recent advances in neuroimaging have offered a rich array of structural and functional markers to probe the organization of regional and large-scale brain networks. The current chapter provides a brief introduction into these techniques and overviews their contribution to the understanding of autism spectrum disorder (ASD), a neurodevelopmental condition associated with atypical social cognition, language function, and repetitive behaviors/interests. While it is generally recognized that ASD relates to structural and functional network anomalies, the extent and overall pattern of reported findings have been rather heterogeneous. Indeed, while several attempts have been made to label the main neuroimaging phenotype of ASD (e.g., 'early brain overgrowth hypothesis', 'amygdala theory', 'disconnectivity hypothesis'), none of these frameworks has been without controversy. Methodological sources of inconsistent results may include differences in subject inclusion criteria, variability in image processing, and analysis methodology. However, inconsistencies may also relate to high heterogeneity across the autism spectrum itself. It, therefore, remains to be investigated whether a consistent imaging phenotype that adequately describes the entire autism spectrum can, in fact, be established. On the other hand, as previous

B.C. Bernhardt (✉) · S.L. Valk
Department of Social Neuroscience, Max Planck Institute for Human Cognitive and Brain Sciences, Leipzig, Germany
e-mail: bernhardt@cbs.mpg.de

B.C. Bernhardt
McConnell Brain Imaging Centre, Montreal Neurological Institute and Hospital, McGill University, Montreal, Quebec, Canada

A. Di Martino
Autism Research Program at the NYU Child Study Center, New York University Langone Medical Center, New York, NY, USA

G.L. Wallace
Department of Speech and Hearing Sciences, George Washington University, Washington, DC, USA

Curr Topics Behav Neurosci (2016) 30: 341–356
DOI 10.1007/7854_2016_438
Published Online: 24 February 2016

findings clearly emphasize the value of neuroimaging in identifying atypical brain morphology, function, and connectivity, they ultimately support its high potential to identify biologically and clinically relevant endophenotypes.

Keywords Autism · ASD · MRI · Connectome · Biomarkers

Contents

1 Introduction 342
2 Probing Regional Morphology 344
3 Disturbances in Inter-regional Networks 349
4 Summary and Outlook 351
References 352

1 Introduction

Autism spectrum disorder (ASD) encompasses a group of early-onset, lifelong, neurodevelopmental conditions currently diagnosed in >1 % of children (CDC 2014). Core behavioral impairments include atypical social cognition, language function, and sensory abnormalities, often together with narrow and repetitive behaviors and interests.

ASD is highly heterogeneous, with significant phenotypic variations across affected individuals. Individuals with ASD show an elevated risk for psychiatric comorbidities and neurological disorders. For example, around 30 % of children with ASD also meet criteria for attention-deficit/hyperactivity disorder (ADHD) (Simonoff et al. 2008); similarly, prevalent subgroups of individuals with ASD present with epileptic seizures (Bolton et al. 2011) or anxiety (van Steensel et al. 2011). Although comorbid conditions often co-occur (Simonoff et al. 2008), the extent to which they cluster in specific patterns is to date unknown. The etiology of ASD is complex and likely not uniform. To that end, an increasing catalog of genetic and developmental risk factors has been identified (De Rubeis and Buxbaum 2015). This remarkable neurobiological variability has challenged the development of novel diagnostic procedures and targeted therapies (Ecker et al. 2015).

Quantitative neuroimaging, particularly magnetic resonance imaging (MRI), offers a rich array of markers that describe the structural and functional organization of the human brain with increasing spatial resolution and validity (Fig. 1). These metrics promise to capture important anatomical aspects in both the typical and atypical brain, and have the potential to be translated into powerful tools that are able to stratify the affected individuals into biological subtypes. This chapter aims to overview the current state of neuroimaging data in ASD. Generally, findings have shown diverse brain anomalies in ASD, ranging from atypical brain growth

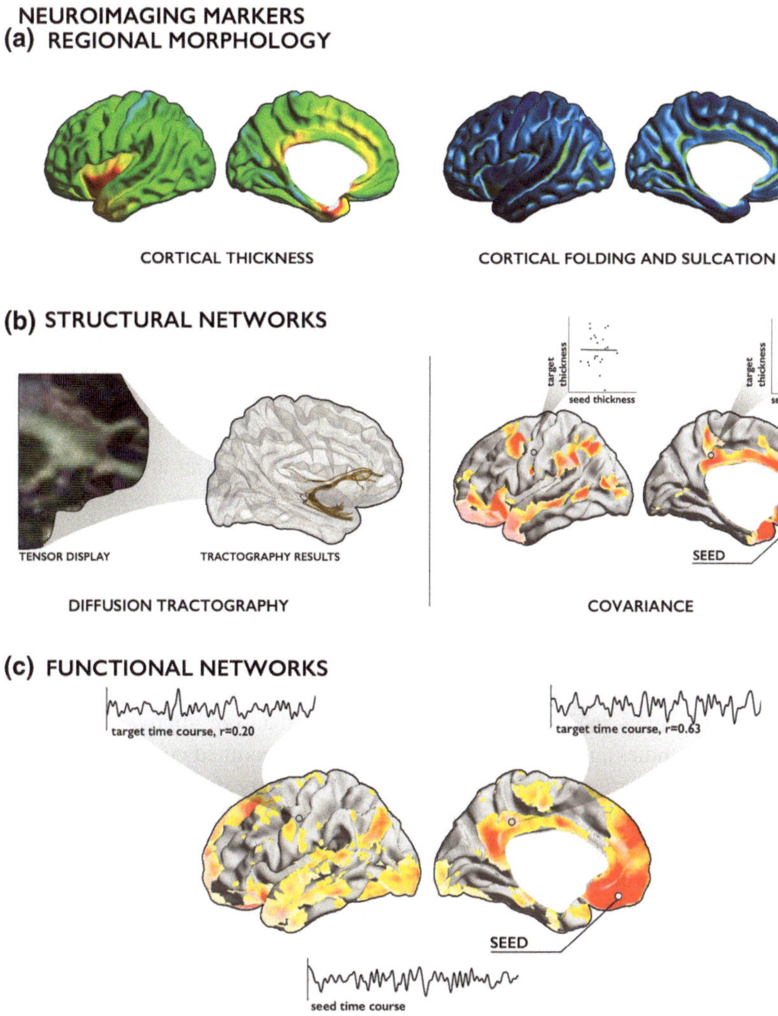

Fig. 1 Neuroimaging approaches previously applied to study ASD. **a** Markers of regional morphology: surface-based cortical thickness and cortical folding indices; **b** Neuroimaging-based structural networks have been derived mainly from diffusion MRI tractography and structural covariance analysis. Diffusion MRI tractography approximates plausible white matter pathways by following the directionality of water diffusion; covariance analysis infers inter-regional networks through the analysis of across-subjects correlations in morphological quantities. **c** Functional networks can be derived from inter-regional time series correlations

patterns, morphological anomalies in both cortical and subcortical regions, and inter-regional structural as well as functional network alterations. However, a consistent, universally accepted imaging phenotype that adequately describes all facets of the disorder (and reliably characterizes all individuals with ASD) is lacking. While outlining major findings in ASD, we will specifically attempt to emphasize methodological sources contributing to the somewhat low reproducibility of findings in ASD.

2 Probing Regional Morphology

Among the earliest and most extensively examined structural findings in ASD are those suggesting *brain overgrowth*. Assessing global brain volume based on MRI, Piven, and colleagues noted increased brain volumes in males with ASD relative to controls. These findings were complemented by cross-sectional assessments of head circumference (a proxy for overall brain growth) by Courchesne et al. (2001), who reported increased head size in up to 90 % of 2–4-year-old boys. As circumference was relatively normal in separate cohorts assessed shortly after birth and during adolescence, the authors suggested that ASD might relate to an atypical developmental trajectory characterized by early overgrowth, followed by normalization in later childhood and adolescence. In a follow-up study, the same group provided more direct evidence for this claim by using longitudinal analyses and pinpointed increases in head size to an interval between 1 and 14 months after birth (Courchesne et al. 2003). Notably, findings indicative of brain overgrowth were subsequently confirmed by several groups (Palmen et al. 2004). Nevertheless, a recent systematic review highlighted potential biases, when comparing head circumference measures in ASD relative to normative data (an approach chosen by some previous studies), in contrast to assessing locally recruited controls (Raznahan et al. 2013). Noteworthy, a recent longitudinal MRI study in newborns with ASD has demonstrated that increases in brain volume may co-occur with excessive cerebrospinal fluid (CSF), particularly in the frontal lobes, from 6 months onward (Shen et al. 2013). On the one hand, these findings suggest that CSF concentrations may confound findings based on global measures of head size. On the other hand, they point to the potential role of CSF circulation in the pathogenesis of ASD and may impact neurogenesis and neuronal migration patterns, eventually contributing to atypical cortical morphology in the disorder. Despite the support for brain overgrowth in ASD, remaining inconsistencies emphasize the need for future studies to prospectively enroll both individuals with ASD and controls, ideally from fetal stages onward to clarify trajectories of early brain development. To identify possible mechanisms that relate to overgrowth in subgroups with ASD, these studies would provide not only head circumference metrics, but also regionally specific measures of brain morphology and connectivity (see below), and are complemented by genetic assessments and monitoring of environmental factors.

Several quantitative neuroimaging analysis approaches have been brought forward to provide a more precise window into the morphology of individual brain regions. Common techniques include volumetry (the manual or automated tracing of a structure of interest to estimate its volume), voxel-based morphometry (*VBM*, an automated technique that statistically compares spatially normalized estimates of gray matter across subjects), together with surface-based measures of cortical thickness and surface area (the two components of gray matter volume), and gyrification as well as subcortical shape analysis. Using such regionally specific approaches, several studies have suggested gray matter increases in ASD compared to controls, particularly in frontal and temporal lobes in both children (Raznahan et al. 2010) and adults (Ecker et al. 2012). Given the putative functional roles of these areas, fronto-temporal anomalies might be compatible with marked impairments in social cognition and with atypical language often seen in autism (Rojas et al. 2006). Several assessments have indeed suggested a potential link between brain structure and atypical behavior in ASD. For example, fronto-temporal structural changes have been shown to correlate with abnormal sociocognitive and language function (Lai et al. 2014). Moreover, structural imaging studies in healthy populations have reported associations between the structure of similar regions known to play an important role in social cognition, such as the temporoparietal junction (TPJ) and superior temporal sulcus, and autistic traits both cross-sectionally (von dem Hagen et al. 2011) and longitudinally (Wallace et al. 2012)

Despite the frequency of fronto-temporal anomalies being reported in ASD at the group level, they may not be present in all affected individuals. A VBM study in adults and adolescents, for example, has suggested rather diffuse ASD-specific gray matter increases in all but the frontal lobes (Piven et al. 1996). Similarly, a study in children has reported temporal and parietal cortical thickening, while frontal and occipital regions appeared morphologically unaffected (Hardan et al. 2006). The difficulty in synthesizing a consistent pattern of findings is further increased by several studies reporting rather less and not more gray matter in ASD, both in children (Hardan et al. 2006) and in adult samples (Wallace et al. 2010). This high degree of divergence was also revealed in previous meta-analyses, suggesting a rather complex pattern of ASD-related regional anomalies with only modest convergence across studies (Duerden et al. 2012).

It has frequently been suggested that inconsistent findings might have arisen as a function of age of the included cohorts. In other words, individuals with ASD may show a different pattern of changes relative to typical controls when tested at different neurodevelopmental stages and throughout aging. This hypothesis would be compatible with findings suggestive of altered developmental and age-related trajectories in ASD. Indeed, several studies have suggested divergent age effects on brain morphology changes in ASD relative to controls (Raznahan et al. 2010; Wallace et al. 2010; Duerden et al. 2012). Studying a cross-sectional sample of adolescents and young adults with ASD and controls, for example, Wallace and colleagues observed more rapid age-related cortical thinning in temporal and parietal regions compared to controls (Wallace et al. 2010). In two separate longitudinal studies, one including individuals aged 3–39 years (Zielinski et al. 2014) and the other including individuals aged 14–24 years (Wallace et al. 2015), researchers have also demonstrated accelerated thinning in ASD in adolescence. Overall, these data indicate that a comprehensive understanding of ASD may likely require accounting for its impact on dynamic brain changes from childhood into adulthood, in which complex morphological anomalies may vary over time throughout the life span.

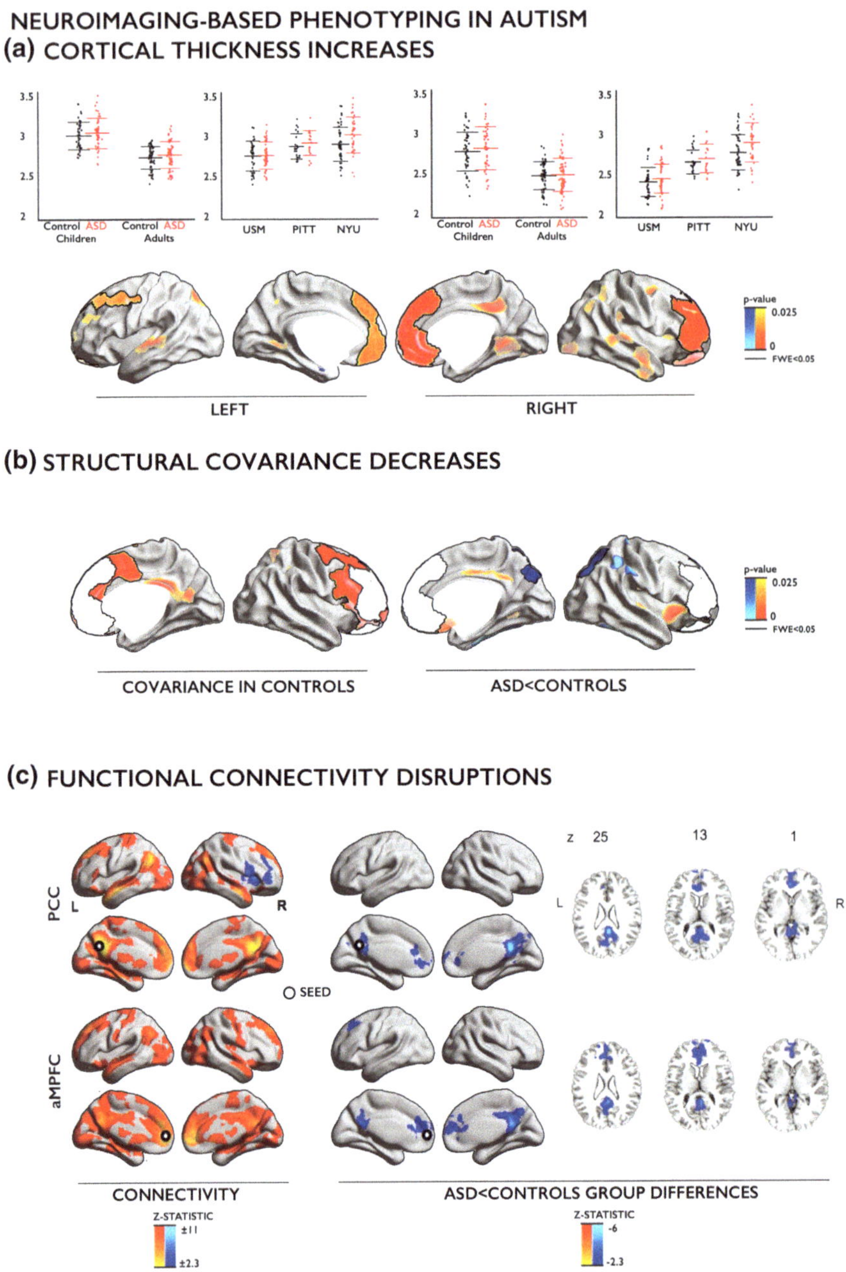
NEUROIMAGING-BASED PHENOTYPING IN AUTISM
(a) CORTICAL THICKNESS INCREASES
3.5
3
2.5
2
Control ASD
Children
Control ASD
Adults
USM
PITT
NYU
p-value
0.025
0
FWE<0.05
LEFT
RIGHT
(b) STRUCTURAL COVARIANCE DECREASES
p-value
0.025
0
FWE<0.05
COVARIANCE IN CONTROLS
ASD<CONTROLS
(c) FUNCTIONAL CONNECTIVITY DISRUPTIONS
PCC
L
R
z 25
13
1
SEED
aMPFC
CONNECTIVITY
Z-STATISTIC
±11
±2.3
ASD<CONTROLS GROUP DIFFERENCES
Z-STATISTIC
-6
-2.3

◀ **Fig. 2** Recent findings in ASD based on the autism brain imaging data exchange (ABIDE) repository. **a** Cortical thickness increases in 107 ASD relative to 113 controls, based on the three sites that included children and adult data in both groups. Findings were consistent across all sites studied and measurable in children as well as adults, yet of variable effect size. **b** Structural covariance alterations in ASD, showing largely decreased covariance in ASD between medial/lateral prefrontal seed regions and posterior midline targets corresponding to the larger precuneus area. Similar to the thickness findings (*panel A*), findings were consistent across the three sites but of variable effect size. **c** Functional connectivity alterations showing primarily decreased connectivity between medial frontal and posterior midline regions in ASD relative to controls. Adapted from Valk et al. (2015) and Di Martino et al. (2014) with permission

A further example of heterogeneity of findings in morphological investigations of ASD comes from the study of the amygdala, a structure with a key role in emotional processing. A careful review of the initial inconsistent volumetric findings of amygdala suggested that age of the sample examined may affect the results (Haar et al. 2014). This notion was supported by the reports of increased amygdala volume in toddlers and preschoolers which were related to later measures of joint attention (Mosconi et al. 2009). Beyond age, other potential sources of variability on volumetric findings of the amygdala, such as the role of comorbidity with anxiety disorders or alexithymia (Bird et al. 2010), have been suggested but have not been fully clarified.

A generally limiting factor of reproducibility of findings relates to low statistical power resulting from relatively small samples studied, a condition often imposed by high costs and challenges associated with recruitment and scanning of individuals with ASD. Recently, the Autism Brain Imaging Data Exchange (ABIDE) initiative made a multicenter repository containing neuroimaging data from 539 individuals with ASD and 573 controls, together with standardized behavioral phenotypic information, freely available (Di Martino et al. 2014). Based on data from three independent ABIDE sites, Valk and colleagues performed a surface-based cortical thickness analysis and assessed overall patterns of findings as well as across-site reproducibility (Valk et al. 2015). Assessing data from more than 200 male participants that passed quality control and surface correction procedures, the authors observed cortical thickness increases in ASD relative to controls, particularly in medial and lateral prefrontal cortices, when pooling data across all imaging sites (Fig. 2a). While the authors reported a generally consistent direction of effects across each of the analysis sites, they emphasized variability in effects sizes across the sites (that may be due to variability in scanning parameters and inclusion criteria), together with increased effects in children compared to adults (Valk et al. 2015).

In addition to methodological factors, diverse findings in human studies may be driven by intrinsic variability across the autism spectrum, which should be more directly addressed for a better understanding of the disorder. At the level of histopathology, several studies have pointed toward a rather complex substrate with multiple etiologies. A *postmortem* study by Bailey and colleagues, for instance, reported increased cortical thickness and/or increased neuronal numbers in the frontal cortex in 3/6 of cases (Bailey et al. 1998). Another study observed laminar rearrangement together with a poorly defined gray and white matter interface in some specimens (Avino and Hutsler 2010). Such cortical interface *blurring* is generally considered a common sign of atypical migration and aberrant organization, emerging in prenatal developmental stages. These findings may suggest an increased frequency of processes resembling known malformations of cortical development (MCD, generally considered to be a common cause of epilepsy) in ASD. Notably, several MCDs have indeed been shown to frequently co-occur with ASD, ranging from disruptions of cell proliferation, such as tuberous sclerosis complex, to those occurring during migration and cortical organization, such as polymicrogyria and schizencephaly. Their co-occurrence with ASD underlines that specific, possibly genetically mediated, developmental disruptions may result in variable neuropsychiatric and neurological phenotypes, emphasizing etiological commonalities across different clinicopathological entities.

Variable histopathological findings motivate the use of novel imaging biomarkers in ASD. Surface-based approaches most frequently applied to measure cortical thickness also provide a range of meaningful indices relating to cortical surface geometry (Hong et al. 2015), such as the folding complexity, sulcal depth, surface area, and geodesic distance mapping. Several recent studies have reported alterations in these features (Nordahl et al. 2007) in ASD, enriching the possibilities to perform an integrative neuroimaging phenotyping of the disorder. In a series of studies, Ecker and colleagues compared the spatial distribution of alterations in cortical thickness, volume, and surface area when comparing ASD to controls and reported relatively non-overlapping morphological anomalies across these feature dimensions (Ecker et al. 2013). Moreover, they have shown that patients can be accurately discriminated from controls based on supervised pattern learners that took advantage of these surface-based markers (Ecker et al. 2010). However, replicating these findings and demonstrating their discriminability from a plethora of other neurodevelopmental disorders (akin to differential diagnoses made in the clinic setting) is required before their potential clinical utility can be truly evaluated. It is conceivable that different morphological indices probe complementary facets of cortical architecture, myelination, connectivity, and development. Cortical thickness, for example, is assumed to reflect combinations of neuronal density, dendritic arborization, and myelin (Huttenlocher et al. 1982). Conversely, folding anomalies may indicate disrupted mechanical properties of the cortex, possibly secondary to disruptions in underlying white matter connectivity (Nordahl et al. 2007). Lastly, surface area changes may be driven specifically by radial unit progenitor cells that divide at the ventricular surface, thereby generating more proliferation units that ultimately lead to increased numbers of cortical columns (Ecker et al. 2015). Notably, a previous postmortem study of Casanova and colleagues suggested a smaller width but increased number of so-called *mini-columns*, neuronal assemblies centered on radially oriented pyramidal neurons, in ASD (Casanova et al. 2002).

3 Disturbances in Inter-regional Networks

Complex patterns of structural alterations in ASD may be reflective of reconfigurations of large-scale brain networks, a key determinant of functional, behavioral, and clinical outcomes. Recent advances in imaging acquisition and analysis techniques offer several opportunities to assess inter-regional structural and functional brain networks.

To study brain structural networks in the living human brain, the most well-established approaches are diffusion MRI tractography and structural MRI covariance analysis. Diffusion MRI data allows for the estimation of directionality and magnitude of water diffusion at each imaging voxel. As diffusivity is influenced by membrane permeability, myelination, and fiber packing (Concha et al. 2010), diffusion markers can serve as proxies of fiber microstructure and architecture. The most widely used parameters are fractional anisotropy (FA), estimating the deviation of water diffusion from random displacement, and mean diffusion (MD), a marker of bulk diffusion. Via tractographic techniques (Mori et al. 1999), it is possible to reconstruct plausible fiber tracts running through the white matter. While challenged close to gray mater regions, such as the neocortex and subcortical nuclei, where multiple fiber populations may intersect or merge (Jones et al. 2012), tractography within deep white matter achieves good correspondence with anatomically plausible tracts. Tractography of several deep bundles has furthermore been directly cross-validated using several different paradigms, for example, with functional connectivity analysis in humans (Johansen-Berg et al. 2005) and using manganese tracing in pigs (Knosche et al. 2015).

Structural networks can also be derived from covariance of MRI-derived morphology across subjects, such as cortical thickness or gray matter density (Lerch et al. 2006). In contrast to diffusion tractography, this framework is not tailored to an approximation of the course of anatomical connections between regions and only shows partial correspondence to diffusion-derived structural connectivity. On the other hand, high correlations in structural markers have been shown to relate to manifestations of persistent functional-trophic cross-talk, maturational interchange, together with common developmental and pathological influences (Alexander-Bloch et al. 2013). Compared to the diffusion anomalies, covariance analysis offers a direct seeding from cortical gray matter in a high-resolution space that only suffers from limited geometric distortions. It, thus, represents a meaningful complementary statistical approach to network mapping. Moreover, given the ubiquity of T1-weighted images in almost every clinical and research protocol, covariance analysis represents a pragmatic and cost-effective approach for network mapping in large populations.

In ASD, both approaches have generally suggested large-scale structural network breakdowns, contributing to the *underconnectivity theory* of autism. Several diffusion MRI studies, for example, have reported alterations in multiple fiber bundles in ASC (Fletcher et al. 2010; Barnea-Goraly et al. 2004; Sundaram et al. 2008). A recent analysis observed convergent decreases in FA, gray matter volume, and functional integration of the TPJ area in adults with ASD, and suggested an association between diffusion alterations and decreased emotionality (Mueller et al. 2013). Diffusion alterations have recently been confirmed by a systematic meta-analysis of 14 studies; in that report, consistent abnormalities were highlighted especially in pathways mediating frontal connectivity, such as in superior longitudinal and uncinate fasciculi as well as the corpus callosum (Aoki et al. 2013). Interestingly, a recent study showed a similar structural compromise in children with ASD and their unaffected siblings, suggesting a possible genetic basis for white matter anomalies in these conditions (Barnea-Goraly et al. 2010). Complementing diffusion MRI work, covariance analyses in ASD were also suggestive of compromised inter-regional structural integration (Bernhardt et al. 2014). A majority of covariance studies reported reduced network-level embedding of regions primarily involved in social cognition and affective processes, including the TPJ area (Bernhardt et al. 2014) and medial and lateral prefrontal cortices (Valk et al. 2015). In the latter study, the authors observed consistently decreased covariance between medial prefrontal and midline parietal regions across three independent imaging sites, indicating that structural segregation of these networks may be a reproducible finding in ASD (Fig. 2b) (Valk et al. 2015).

In the functional domain, several groups have studied inter-regional functional MRI time series, particularly during task-free conditions. Advantages of such *resting-state* over task-related paradigms include the possibility to examine multiple cortical areas in one relatively short session and relatively little demands on individuals with a reduced ability to perform tasks. Resting-state networks have been shown to be highly reproducible across subjects and are thought to correspond closely to systems engaging in specific tasks (Smith et al. 2009). Comparative (Mantini et al. 2011) and modeling (Honey et al. 2007) studies have furthermore suggested that anatomical pathways, in part, determine functional connections. In ASD, atypicalities in functional coupling of several large-scale networks have been reported, with the majority of findings supporting underconnectivity (Geschwind and Levitt 2007; Gotts et al. 2012), particularly of long-range inter-regional functional associations. Yet, several studies have emphasized that differential head motion may contribute to observing connectivity differences (Power et al. 2012), a finding important to consider given that many studies include children and the elevated comorbidity of ADHD with ASD. Moreover, systematic analyses and multi-method simulations have suggested a sizable impact of analytical choices on group differences reported in functional imaging studies, supporting efforts to establish consistent analytical routines, transparent reporting practices, work on openly accessible data, and more rigorous attempts to cross-validate functional connectivity measures (Muller et al. 2011). Based on ABIDE data, for example, the multi-site analysis by Di Martino and colleagues reconciled seemingly inconsistent findings of under- and overconnectivity in ASD. Evidence of both mechanisms were present but varies as a function of the circuits involved. Underconnectivity encompassed cortico-cortical connections and was predominant; yet, overconnectivity was also present and largely characteristic of subcortico-cortical circuits (Fig. 2c) (Di Martino et al. 2014).

Task-free functional studies suggesting connectivity differences have also been complemented by results based on alternative analytical paradigms and neuroimaging modalities. Task-based studies, for example, have suggested underconnectivity during social cognition tasks in ASD (Kana et al. 2009). On the other hand, a recent magnetoencephalography study suggested that connectivity alterations in ASD relative to controls may vary as a function of the frequency band. Moreover, findings may be different depending on the inclusion of frontal nodes in the analysis (Kitzbichler et al. 2015).

Methodological options for evaluating large-scale networks in ASD have recently been extended by the introduction of graph theory to neuroimaging (Bullmore and Sporns 2009). Systematic connectivity mapping among all pairs of regions (usually derived from parcellations) can be used to build a connectivity matrix, the substrate for graph-theoretical analyses. Brain graphs representing these matrices can be visualized, which facilitate an intuitive understanding of topological properties. Fundamental topological parameters include *clustering coefficient*, a measure of local network integration that also relates to network stability, and *characteristic path length*, a proxy for global network efficiency. Graph theory can also be used to calculate the *centrality* metrics that quantify the embedding of regions within the network; high centrality scores are a feature of hub regions relevant for the overall network architecture and communication.

Graph-theoretical analyses of healthy human brain networks have consistently demonstrated a *small-world* topology that is characterized by separate groups of densely clustered regions (modules) interconnected by short paths passing through core hub regions (such as the midline parietal cortex). Several neurological and neuropsychiatric conditions, including schizophrenia (Bassett et al. 2008), ADHD (Cao et al. 2013), and epilepsy (Bernhardt et al. 2011, 2015a), have recently been associated with topological disruptions based on this framework. While only relatively few graph-theoretical assessments have been published in the ASD literature to date, initial findings have shown decreased clustering, suggesting that overall connectomic alterations may be indicative of network reorganization toward a more random arrangement (Rudie et al. 2012). These findings have recently been confirmed (Itahashi et al. 2014); moreover, they have been complemented by data indicating shifts in the overall hub organization in ASD (Di Martino et al. 2013).

4 Summary and Outlook

Neuroimaging has greatly advanced our understanding of brain anomalies associated with ASD. As detailed above, the previous literature has revealed a rather complex pattern of structural and functional alterations in cohorts of individuals with ASD. Among structural MRI studies, a frequently reported finding has been gray matter increases in frontal and temporal regions that might be developmentally constrained and related to aberrant growth patterns in early age and possibly to malformation processes. Investigations of functional and structural network alterations have primarily suggested global underconnectivity, together with islands of focal connectivity increases. Initial graph-theoretical reports have suggested an association between ASD and topological randomization, together with shifts in the spatial distribution of network hubs. Nevertheless, findings at all levels have not been free from controversy, and ongoing as well as future efforts to increase homogeneity and transparency of subject inclusion criteria as well as analysis routines are highly recommended.

ASD is likely best understood as a developmentally dynamic disorder, motivating longitudinal studies that carefully assess brain development and aging-related change in large cohorts across the life span. The increasing use of high and ultra-high field MRI has the potential to achieve an adequate resolution to address layer-specific structural as well functional alterations and connectivity arrangements in vivo. This scale of analysis is likely needed to detect possibly an elusive pathological substate in specific ASD subgroups (e.g., those with subtle cortical malformations). Based on increasingly complex and high-dimensional neuroimaging datasets, machine-learning techniques offer an appropriate analytic framework to integrate data and to discover and evaluate imaging biomarkers of ASD. In other brain disorders, these techniques are able to discover latent subtypes within seemingly homogeneous cohorts (Bernhardt et al. 2015b). The high variability documented across prior studies of ASD is indicative of claims that this population is very likely composed of different biological subtypes. Variability across the spectrum, thus, needs to be explicitly addressed in research rather than ignored. In this light, cross-site collaborative efforts to aggregate and share large datasets, such as the ABIDE initiative, represent useful and urgent avenues to pursue in our efforts to better understand the neural underpinnings of the complex autism spectrum.

References

Alexander-Bloch A, Giedd JN, Bullmore E (2013) Imaging structural co-variance between human brain regions. Nat Rev Neurosci 14:322–336

Aoki Y, Abe O, Nippashi Y, Yamasue H (2013) Comparison of white matter integrity between autism spectrum disorder subjects and typically developing individuals: a meta-analysis of diffusion tensor imaging tractography studies. Mol Autism 4:25

Avino TA, Hutsler JJ (2010) Abnormal cell patterning at the cortical gray-white matter boundary in autism spectrum disorders. Brain Res 1360:138–146

Bailey A, Luthert P, Dean A et al (1998) A clinicopathological study of autism. Brain 121(Pt 5):889–905

Barnea-Goraly N, Kwon H, Menon V, Eliez S, Lotspeich L, Reiss AL (2004) White matter structure in autism: preliminary evidence from diffusion tensor imaging. Biol Psychiatry 55:323–326

Barnea-Goraly N, Lotspeich LJ, Reiss AL (2010) Similar white matter aberrations in children with autism and their unaffected siblings: a diffusion tensor imaging study using tract-based spatial statistics. Arch Gen Psychiatry 67:1052–1060

Bassett DS, Bullmore E, Verchinski BA, Mattay VS, Weinberger DR, Meyer-Lindenberg A (2008) Hierarchical organization of human cortical networks in health and schizophrenia. J Neurosci 28:9239–9248

Bernhardt BC, Chen Z, He Y, Evans AC, Bernasconi N (2011) Graph-theoretical analysis reveals disrupted small-world organization of cortical networks in temporal lobe epilepsy. Cerebral Cortex 21:2147–57

Bernhardt BC, Valk SL, Silani G, Bird G, Frith U, Singer T (2014) Selective disruption of sociocognitive structural brain networks in autism and alexithymia. Cereb Cortex 24:3258–3267

Bernhardt BC, Bernasconi N, Hong S, Dery S, Bernasconi A (2015a) Subregional mesiotemporal network topology is altered in temporal lobe epilepsy. Cereb Cortex (in press)

Bernhardt BC, Hong SJ, Bernasconi A, Bernasconi N (2015b) Magnetic resonance imaging pattern learning in temporal lobe epilepsy: classification and prognostics. Ann Neurol 77(3):436–46

Bird G, Silani G, Brindley R, White S, Frith U, Singer T (2010) Empathic brain responses in insula are modulated by levels of alexithymia but not autism. Brain J Neurol 133:1515–1525

Bolton PF, Carcani-Rathwell I, Hutton J, Goode S, Howlin P, Rutter M (2011) Epilepsy in autism: features and correlates. Br J Psychiatry 198:289–294

Bullmore E, Sporns O (2009) Complex brain networks: graph theoretical analysis of structural and functional systems. Nat Rev Neurosci 10:186–198

Cao Q, Shu N, An L et al (2013) Probabilistic diffusion tractography and graph theory analysis reveal abnormal white matter structural connectivity networks in drug-naive boys with attention deficit/hyperactivity disorder. J Neurosci 33:10676–10687

Casanova MF, Buxhoeveden DP, Switala AE, Roy E (2002) Minicolumnar pathology in autism. Neurology 58:428–432

CDC (2014) Prevalence of autism spectrum disorder among children aged 8 years—autism and developmental disabilities monitoring network, 11 sites, United States, 2010. Morbidity and mortality weekly report Surveillance summaries 63:1–21

Concha L, Livy DJ, Beaulieu C, Wheatley BM, Gross DW (2010) In vivo diffusion tensor imaging and histopathology of the fimbria-fornix in temporal lobe epilepsy. J Neurosci 30:996–1002

Courchesne E, Karns CM, Davis HR et al (2001) Unusual brain growth patterns in early life in patients with autistic disorder: an MRI study. Neurology 57:245–254

Courchesne E, Carper R, Akshoomoff N (2003) Evidence of brain overgrowth in the first year of life in autism. JAMA 290:337–344

De Rubeis S, Buxbaum JD (2015) Genetics and genomics of autism spectrum disorder: embracing complexity. Human molecular genetics 24(R1):R24–R31

Di Martino A, Zuo XN, Kelly C et al (2013) Shared and distinct intrinsic functional network centrality in autism and attention-deficit/hyperactivity disorder. Biological psychiatry

Di Martino A, Yan CG, Li Q et al (2014) The autism brain imaging data exchange: towards a large-scale evaluation of the intrinsic brain architecture in autism. Mol Psychiatry 19:659–667

Duerden EG, Mak-Fan KM, Taylor MJ, Roberts SW (2012) Regional differences in grey and white matter in children and adults with autism spectrum disorders: an activation likelihood estimate (ALE) meta-analysis. Autism Res 5:49–66

Ecker C, Marquand A, Mourao-Miranda J et al (2010) Describing the brain in autism in five dimensions–magnetic resonance imaging-assisted diagnosis of autism spectrum disorder using a multiparameter classification approach. J Neurosci 30:10612–10623

Ecker C, Suckling J, Deoni SC et al (2012) Brain anatomy and its relationship to behavior in adults with autism spectrum disorder: a multicenter magnetic resonance imaging study. Arch Gen Psychiatry 69:195–209

Ecker C, Ginestet C, Feng Y et al (2013) Brain surface anatomy in adults with autism: the relationship between surface area, cortical thickness, and autistic symptoms. JAMA Psychiatry 70:59–70

Ecker C, Bookheimer SY, Murphy DG (2015) Neuroimaging in autism spectrum disorder: brain structure and function across the lifespan. Lancet neurology 14(11):1121–34

Fletcher PT, Whitaker RT, Tao R et al (2010) Microstructural connectivity of the arcuate fasciculus in adolescents with high-functioning autism. NeuroImage 51:1117–1125
Geschwind DH, Levitt P (2007) Autism spectrum disorders: developmental disconnection syndromes. Curr Opin Neurobiol 17:103–111
Gotts SJ, Simmons WK, Milbury LA, Wallace GL, Cox RW, Martin A (2012) Fractionation of social brain circuits in autism spectrum disorders. Brain 135:2711–2725
Haar S, Berman S, Behrmann M, Dinstein I (2014) Anatomical abnormalities in autism? Cereb Cortex
Hardan AY, Muddasani S, Vemulapalli M, Keshavan MS, Minshew NJ (2006a) An MRI study of increased cortical thickness in autism. Am J Psychiatry 163:1290–1292
Hardan AY, Girgis RR, Lacerda AL et al (2006b) Magnetic resonance imaging study of the orbitofrontal cortex in autism. J Child Neurol 21:866–871
Honey CJ, Kotter R, Breakspear M, Sporns O (2007) Network structure of cerebral cortex shapes functional connectivity on multiple time scales. Proc Natl Acad Sci USA 104:10240–10245
Hong S-J, Bernhardt B, Schrader D, Caldairou B, Bernasconi N, Bernasconi A (2015) MRI-based lesion profiling of epileptogenic cortical malformations. In: Navab N, Hornegger J, Wells W, Frangi A (eds) Medical image computing and computer-assisted intervention—MICCAI 2015, Springer, Berlin, pp 501–509
Huttenlocher PR, de Courten C, Garey LJ, Van der Loos H (1982) Synaptogenesis in human visual cortex–evidence for synapse elimination during normal development. Neurosci Lett 33:247–252
Itahashi T, Yamada T, Watanabe H et al (2014) Altered network topologies and hub organization in adults with autism: a resting-state fMRI study. PLoS ONE 9:e94115
Johansen-Berg H, Behrens TE, Sillery E et al (2005) Functional-anatomical validation and individual variation of diffusion tractography-based segmentation of the human thalamus. Cereb Cortex 15:31–39
Jones DK, Knosche TR, Turner R (2012) White matter integrity, fiber count, and other fallacies: the do's and don'ts of diffusion MRI. NeuroImage 73:239–254
Kana RK, Keller TA, Cherkassky VL, Minshew NJ, Just MA (2009) Atypical frontal-posterior synchronization of theory of mind regions in autism during mental state attribution. Soc Neurosci 4:135–152
Kitzbichler MG, Khan S, Ganesan S et al (2015) Altered development and multifaceted band-specific abnormalities of resting state networks in autism. Biol Psychiatry 77:794–804
Knosche TR, Anwander A, Liptrot M, Dyrby TB (2015) Validation of tractography: comparison with manganese tracing. Hum Brain Mapp 36:4116–4134
Lai MC, Lombardo MV, Ecker C et al (2014) Neuroanatomy of individual differences in language in adult males with autism. Cereb Cortex
Lerch JP, Worsley K, Shaw WP et al (2006) Mapping anatomical correlations across cerebral cortex (MACACC) using cortical thickness from MRI. NeuroImage 31:993–1003
Mantini D, Gerits A, Nelissen K et al (2011) Default mode of brain function in monkeys. J Neurosci Official J Soci Neurosci 31:12954–12962
Mori S, Crain BJ, Chacko VP, van Zijl PC (1999) Three-dimensional tracking of axonal projections in the brain by magnetic resonance imaging. Ann Neurol 45:265–269
Mosconi MW, Cody-Hazlett H, Poe MD, Gerig G, Gimpel-Smith R, Piven J (2009) Longitudinal study of amygdala volume and joint attention in 2- to 4-year-old children with autism. Arch Gen Psychiatry 66:509–516
Mueller S, Keeser D, Samson AC et al (2013) Convergent findings of altered functional and structural brain connectivity in individuals with high functioning autism: a multimodal MRI study. PLoS ONE 8:e67329
Muller RA, Shih P, Keehn B, Deyoe JR, Leyden KM, Shukla DK (2011) Underconnected, but how? A survey of functional connectivity MRI studies in autism spectrum disorders. Cereb Cortex 21:2233–2243
Nordahl CW, Dierker D, Mostafavi I et al (2007) Cortical folding abnormalities in autism revealed by surface-based morphometry. J Neurosci 27:11725–11735

Palmen SJ, Hulshoff Pol HE, Kemner C et al (2004) Larger brains in medication naive high-functioning subjects with pervasive developmental disorder. J Autism Dev Disord 34:603–613
Piven J, Arndt S, Bailey J, Andreasen N (1996) Regional brain enlargement in autism: a magnetic resonance imaging study. J Am Acad Child Adolesc Psychiatry 35:530–536
Power JD, Barnes KA, Snyder AZ, Schlaggar BL, Petersen SE (2012) Spurious but systematic correlations in functional connectivity MRI networks arise from subject motion. NeuroImage 59:2142–2152
Raznahan A, Toro R, Daly E et al (2010) Cortical anatomy in autism spectrum disorder: an in vivo MRI study on the effect of age. Cereb Cortex 20:1332–1340
Raznahan A, Wallace GL, Antezana L et al (2013) Compared to what? Early brain overgrowth in autism and the perils of population norms. Biol Psychiatry 74:563–575
Rojas DC, Peterson E, Winterrowd E, Reite ML, Rogers SJ, Tregellas JR (2006) Regional gray matter volumetric changes in autism associated with social and repetitive behavior symptoms. BMC Psychiatry 6:56
Rudie JD, Shehzad Z, Hernandez LM et al (2012) Reduced functional integration and segregation of distributed neural systems underlying social and emotional information processing in autism spectrum disorders. Cereb Cortex 22:1025–1037
Shen MD, Nordahl CW, Young GS et al (2013) Early brain enlargement and elevated extra-axial fluid in infants who develop autism spectrum disorder. Brain 136:2825–2835
Simonoff E, Pickles A, Charman T, Chandler S, Loucas T, Baird G (2008) Psychiatric disorders in children with autism spectrum disorders: prevalence, comorbidity, and associated factors in a population-derived sample. J Am Acad Child Adolesc Psychiatry 47:921–929
Smith SM, Fox PT, Miller KL et al (2009) Correspondence of the brain's functional architecture during activation and rest. Proc Natl Acad Sci USA 106:13040–13045
Sundaram SK, Kumar A, Makki MI, Behen ME, Chugani HT, Chugani DC (2008) Diffusion tensor imaging of frontal lobe in autism spectrum disorder. Cereb Cortex 18:2659–2665
Valk SL, Di Martino A, Milham MP, Bernhardt BC (2015) Multicenter mapping of structural network alterations in autism. Hum Brain Mapp 36:2364–2373
van Steensel FJ, Bogels SM, Perrin S (2011) Anxiety disorders in children and adolescents with autistic spectrum disorders: a meta-analysis. Clin Child Fam Psychol Rev 14:302–317
von dem Hagen EA, Nummenmaa L, Yu R, Engell AD, Ewbank MP, Calder AJ (2011) Autism spectrum traits in the typical population predict structure and function in the posterior superior temporal sulcus. Cereb Cortex 21:493–500
Wallace GL, Dankner N, Kenworthy L, Giedd JN, Martin A (2010) Age-related temporal and parietal cortical thinning in autism spectrum disorders. Brain 133:3745–3754
Wallace GL, Shaw P, Lee NR et al (2012) Distinct cortical correlates of autistic versus antisocial traits in a longitudinal sample of typically developing youth. J Neurosci 32:4856–4860
Wallace GL, Eisenberg IW, Robustelli B et al (2015) Longitudinal cortical development during adolescence and young adulthood in autism spectrum disorder: increased cortical thinning but comparable surface area changes. J Am Acad Child Adolesc Psychiatry 54:464–469
Zielinski BA, Prigge MB, Nielsen JA et al (2014) Longitudinal changes in cortical thickness in autism and typical development. Brain 137:1799–1812

Current Practice and Future Avenues in Autism Therapy

L. Poustka and I. Kamp-Becker

Abstract Autism spectrum disorders (ASDs) are neurodevelopmental disorders with early onset, characterized by deficits in social communication and repetitive and restricted interests and activities. A growing number of studies over the last 10 years support the efficacy of behaviorally based interventions in ASD for the improvement of social communication and behavioral functioning. In contrast, research on neurobiological based therapies for ASD is still at its beginnings. In this article, we will provide a selective overview of both well-established evidence-based treatments and novel interventions and drug treatments based on neurobiological principles aiming at improving core symptoms in ASD. Directions and options for future research on treatment in ASD are discussed.

Keywords ASD · Behavioral intervention · Neurobiological based intervention · Neurodevelopmental disorders

Contents

1 Introduction 358
2 Current Practice in Therapy 360
  2.1 Comprehensive Interventions 360
  2.2 Skill-Based Interventions 362
  2.3 Interventions for Accompanying Symptoms 363
3 Future Avenues in Therapy 364
  3.1 Neurofeedback Training of Deficient μ-Suppression 364
  3.2 Variations: Real-Time fMRI and NIRS Neurofeedback 366

L. Poustka (✉) · I. Kamp-Becker
Clinic for Child and Adolescent Psychiatry, Medical University of Vienna, Währinger Gürtel 18-20, 1090 Vienna, Austria
e-mail: luise.poustka@meduniwien.ac.at
URL: http://www.meduniwien.ac.at/kjp

Curr Topics Behav Neurosci (2016) 30: 357–378
DOI 10.1007/7854_2016_442
Published Online: 19 April 2016

3.3 Computer-Supported Cognitive Training of Basic Affect Recognition 367
3.4 Early Intervention and Adjuvant Pharmacotherapy 368
3.5 Outlook 370
References 372

1 Introduction

In this chapter, we will, after a brief introduction on autism spectrum disorder (ASD) symptomatology and etiology, provide a selective overview of established and evidence-based interventions in ASD. Then, we will discuss several novel treatments and trainings based on neurobiological principles as well as drug therapies which are currently developed aiming at improving core symptoms in ASD.

ASD is a severe, lifelong, and highly cost-intensive neurodevelopmental disorder characterized by impairments in social interaction (e.g., deficits in appropriate eye contact, facial expression, emotion perception, gesture, and social and emotional reciprocity) and communication (e.g., echolalia, stereotyped language, reduced reciprocal conversation), in combination with restricted and repetitive behaviors (e.g., rigid preferences for routines, repetitive motor mannerisms) (Lai et al. 2014). In the early years of life, impairments of social attention and reciprocity, such as reduced preference and attention to persons and other social stimuli, are observed (Jones and Klin 2013). Although ASD is often considered a childhood disorder, it persists throughout the life span (Howlin 2013). There is considerable **heterogeneity** in the expression and severity of the core and associated symptoms of ASD (Charman et al. 2011). For example, variability in the social interaction domain ranges from a near absence of interest in interacting with others to more subtle difficulties in managing complex social interactions that require an understanding of others' goals and intentions or other cues of social context. Heterogeneity is further affected by variability in other factors that are heterogeneous themselves, such as developmental trajectories, gender, level of language, cognitive functioning, adaptive behaviors, and comorbidity of ASD. The phenotypic variability is also paralleled by great genetic heterogeneity in ASD (for review see: Betancur 2011; Jeste and Geschwind 2014). Moreover, ASD has been associated with a variety of genetic diseases, and findings from numerous genetic studies point toward 200–1000 genes that are involved in ASD susceptibility, both accounting for high heterogeneity in ASD (Berg and Geschwind 2012). Recent estimates suggest that nearly one half of the individuals with ASD may have an intelligence quotient (IQ) under average, but nearly one-third has an average IQ and about three percent an IQ above average (Charman et al. 2011). Furthermore, patients with ASD show a high rate (up to 70–85 %) of accompanying comorbid disorders. Comorbid disorders that often occur together with ASD are attention deficit/hyperactivity disorder (ADHD), tic disorders, obsessive–compulsive disorders as well as mood and anxiety disorders (Abdallah et al. 2011; Gjevik et al. 2011; Simonoff et al. 2008). Moreover, individuals with ASD often show motor impairments and suffer from self-injury as well as sleep problems (Matson et al. 2011). This additional psychopathology does not only hamper the course of development, but it also impedes the effect of therapeutic interventions (Antshel et al. 2011). The long-term development of ASD is substantially mediated by the level of language and intelligence, with an IQ < 70 predicting a significantly worse outcome (Anderson et al. 2014; Magiati et al. 2014). Additionally, the presence of stereotyped and repetitive behaviors and comorbidities influences the development (Hofvander et al. 2009; Kamp-Becker et al. 2009). As adults, many people with ASD, including those with an average IQ, are significantly disadvantaged regarding employment, social relationships, physical and mental health, and quality of life (Howlin et al. 2013, 2014).

The prevalence of ASD has steadily increased over the last decade particularly in individuals without intellectual disability, the current median worldwide prevalence being 0.62–0.70 % (Elsabbagh et al. 2012). The increase may partly be a result of changes in diagnostic concepts and criteria (Fisch 2013). However, the rise is probably also due to improved awareness and recognition, changes in diagnosis, and younger age of diagnosis (Lai et al. 2014).

It is now assumed that ASD is an equally strong genetically and environmentally determined disorder (Hallmeyer et al. 2011). Certain environmental factors in the etiology are increasingly discussed, with regard to rising paternal and maternal age (Parner et al. 2012) or medication during pregnancy (Gardener et al. 2009). Genes contribute to behavior and cognition in ASD via their effects on brain structure and development. Various ASD susceptibility genes and linked biological coherent functional pathways have been identified in the past decade. Findings form genetic studies point toward a "many genes, common pathways" hypothesis (Geschwind 2008), referring to a primary deficit early in neural development and in the development of synaptic functions, resulting in deviant cortical development and later abnormal cortical connectivity (Parikshak et al. 2013). Continuing comparisons to other neuropsychiatric diseases, including ADHD, schizophrenia, and ID, will also be important to identify features specific to ASD.

Several atypical neurocognitive profiles/models and neurophysiological alterations that have been obtained from neuroimaging, eye tracking, and electrophysiological studies have been reported (e.g., Dichter 2012; Neuhaus et al. 2010): deficits in *theory of mind, empathy* (e.g., Pelphrey et al. 2011; Krach et al. 2015), and *social motivation* (e.g., Dichter et al. 2012; Kohls et al. 2012); *weak central coherence* (Happe and Frith 2006); and *executive function* (Pellicano 2012). The heterogeneity of ASD may not only underlie the insufficiency of single-cause neurocognitive models in explaining the ASD phenotype, but may also underlie the considerable variability in treatment response and outcome (Fava and Strauss 2014). There is no doubt that a complex disorder such as ASD requires a multifaceted treatment approach. Therefore, treatment individualization is crucial to understand which child will benefit most from which intervention and what intensity and length of treatment is necessary.

2 Current Practice in Therapy

Central aims of interventions in ASD are to strengthen an individual's functional independence and to optimize quality of life through improvement of social skills, communication and language, and reduction of co-occurring behavioral problems and disorders. Additionally, individuals with ASD should be supported to fulfill their specific strengths (Lai et al. 2014).

The multiple developmental and behavioral problems associated with ASD require multidisciplinary care, coordination of services, and advocacy for concerned individuals with ASD and their families. Although ASD is a neurobiological disorder, psychological and educational interventions are currently the primary treatments for addressing the core deficits in individuals with ASD. The pervasive and severe deficits of concerned people with ASD are associated with decreased parenting efficacy, increased parenting stress, and an increase in mental and physical health problems compared to parents of both typically developing children and children with other developmental disorders (Karst and Van Hecke 2012). For these reasons, parents are important participants in the intervention, parent-training methods typically being a genuine part of therapist-delivered treatment programs (Dawson and Bernier 2013). Although ASD is considered not curable (with some exceptions) with the help of suitable interventions, it is certainly possible to achieve considerable improvements in quality of life and the psychosocial level of functioning of people with ASD. Interventions are usually administered in different contexts or settings (at home, at school, in an institution, at work), supported by different persons (therapist, teacher/educator, peers), in individual, and group contexts. Different behavioral methods are used, and the interventions vary in terms of start point, duration, and intensity. Furthermore, they differ in whether they target the core ASD symptoms comprehensively, specific skills (e.g., social skills), or accompanying symptoms.

2.1 Comprehensive Interventions

While there are a high number of treatments available in the field of ASD, only a small proportion has been tested scientifically in terms of efficacy. Research suggests that the best empirical evidence could be shown for behavioral programs that are implemented very early in life and are highly intensive and individualized (named early intensive behavioral intervention, EIBI, Dawson et al. 2012). EIBI is characterized by the active engagement of the child for many hours per week (usually 20 up to 40 h) in a planned educational intervention with specific goals derived from assessment results. The intervention is undertaken in direct one by one adult-to-child instruction, initially teaching simple skills and progressing to more complex skills. Progress and outcome are continuously measured. Within this framework, there are different programs which vary according to the specific curriculum and teaching procedures, but which all involve general principles of learning theory and behavior therapy named "applied behavior analysis" (ABA) (Volkmar et al. 2014). Main elements of EIBI programs are as follows: curriculum content (focusing attention to social stimuli, imitation, receptive and expressive language and communication, appropriate play, social interaction, etc.); highly supportive teaching environments and generalization strategies; predictability and routine; functional approach to problem behaviors; and family involvement (parents as cotherapists). These interventions can be effective in improving cognitive, adaptive, and social–communicative outcomes in young children with ASD (Dawson and Bernier 2013; Virues-Ortega 2010; Vismara and Rogers 2010; Warren et al. 2011). There is some evidence that EIBI is more likely to produce substantial improvements in young children with ASD than common eclectic interventions, even when these are intensive (Howard et al. 2014). However, analysis of treatment response at the individual level indicates that while some children show obvious improvements, some show moderate gains, and others only show minimal or no treatment gains (Howlin et al. 2009; Vivanti et al. 2014). Correlates of gains that are reported most often are pre-treatment cognitive abilities, level of adaptive behaviors, and language abilities (Virues-Ortega 2010; Vivanti et al. 2014). In particular, children with higher cognitive levels (IQ > 70) seem to be able to transfer their acquired social communication skills into daily functioning (Ben-Itzchak et al. 2014). Data suggest that subgroups of children display more prominent gains across studies, but participant characteristics associated with greater gains are not well understood (Warren et al. 2011). There is an urgent need to index and predict treatment response in order to assign treatment specifically to different subgroups.

Some recent studies demonstrate the utility of behavioral based interventions in altering not only the course of behavioral development but additionally the course of brain development. After following two years of EIBI intervention, children showed normalized spectral power in the alpha and theta ranges in an EEG paradigm while viewing faces vs. objects; brain activity was comparable to typically developing children, while this was not the case for children with ASD, who did not receive EIBI training (Dawson et al. 2012). It is hypothesized that (a) when intervention is provided early and intensively for at least two years, a normalization of brain activity related to social processing is possible, (b) learning strategies have to address core deficits in social motivation through an emphasis on positive social engagement and arousal modulation, (c) promotion of complex neural networks and connectivity are possible through thematic, multisensory, and multidomain teaching approaches (Sullivan et al. 2014).

Several intervention approaches focus on **teaching parents** interventions that can be used in home and community settings. Parent-delivered interventions can enhance generalization of skills, efficiency of delivery, and self-efficacy for parents. There is some evidence for the benefit of brief, targeted, parent-mediated interventions on child outcomes (McConachie and Diggle 2007; Oono et al. 2013). In particular, parent trainings lead to improved communicative behavior in children, increased maternal knowledge of ASD, enhanced maternal communication style and parent–child interaction, and maybe reduced maternal depression. Accompanying symptoms in children with ASD (problem behaviors and disruptive/antisocial behaviors, anxiety) can be reduced, and improved skills in behavior management in parents are achieved (Pillay et al. 2011). The well-established parenting program "Stepping Stones/Triple P" seems to be a promising intervention for parents of children with ASD (Tellegen and Sanders 2014; Whittingham et al. 2009).

The intervention program "Treatment and Education of Autistic and Related Communication Handicapped Children" (TEACCH) makes use of structured teaching. Three factors are reported to be essential: (a) organization of the physical environment in a way that is consistent with the needs of the concerned individual (e.g., minimizing possible distractions), (b) arrangement of activities in a predictable manner (e.g., use of visual schedules of daily routines), and (c) structured organization of the materials and tasks to promote independence from adult directions/prompts (e.g., use visual materials if the student is more able to benefit from them). A meta-analysis of 13 studies demonstrated that TEACCH could reach moderate-to-large improvements in social behavior and decrease of maladaptive behavior, but only small positive effects on adaptive behavioral repertoires including communication, activities of daily living, and motor functioning. Effects on perceptual, motor, verbal, and cognitive skills were small (Virues-Ortega et al. 2013).

2.2 *Skill-Based Interventions*

The Picture Exchange Communication System (PECS) is a communication-training program for young nonverbal children with ASD using a picture/icon aided augmentative communication system. PECS aims to teach spontaneous social communication skills by means of symbols, pictures, or icons. Teaching relies on behavioral principles, particularly reinforcement techniques. The few existing randomized controlled trials found small-to-moderate gains in communication, while progress in speech was small to negative (Flippin et al. 2010). However, again, treatment response is variable; effects of the training are potentially moderated by baseline factors (Carr and Felce 2007; Gordon et al. 2011).

People with ASD have profound difficulties in understanding the intention emotions, feelings, beliefs, and thoughts of other people and themselves. This is assumed to be an explanation for social communication deficits (theory of mind (ToM) model, see above). Thus, successful interventions to **teach ToM functions** might improve these deficits. However, a systematic review demonstrates that there is only little evidence of maintenance of ToM functions, generalization of training effects to other settings, or developmental effects on related skills. As the study quality was rated as low, further longitudinal designs and larger samples are needed (Fletcher-Watson et al. 2014).

A lack of social skills and deficits in social reciprocity contributes to the overall and stable impairments of individuals with ASD, especially for those with average or above average cognitive skills (Magiati et al. 2014). Children and adolescents with ASD usually fail to acquire appropriate social skills and lack opportunities for positive peer interactions. Social skills training (SST) involves teaching specific skills such as maintaining eye contact and initiating a conversation through behavioral and social learning. SST has been reported to be an effective component of treatment regimens for many childhood disorders, and group-based SST, which aims to improve communication skills and social interaction abilities, is a highly recommended intervention for children and adolescents with ASD (Reichow et al. 2012; Wang et al. 2013). Within group-based SSTs, children and adolescents have the opportunity to practice newly learned skills in a naturalistic context that promotes interaction with peers. The effectiveness of SST in ASD has been proven in several randomized, controlled trials RCTs. In particular, evidence shows that SST improves social competence and friendship quality (DeRosier et al. 2011) as well as nonverbal communication, empathic responding, and social relations (Soorya et al. 2015). Nevertheless, other impairments in the social domain, like emotion recognition ability or social communication skills, show less improvement after SST, and some patients benefit more than others from SST intervention, depending on gender, level, and quality of comorbidity, sensory characteristics, and other factors, which are still not well understood (McMahon et al. 2013; Reichow et al. 2012).

2.3 *Interventions for Accompanying Symptoms*

Children and adolescents with high-functioning ASD are at high risk for developing accompanying symptoms/disorders, such as anxiety, depression, or behavior problems. **Cognitive behavior therapies** have resulted in significant reductions in anxiety (Reaven et al. 2012), improvements in anger management (Sofronoff et al. 2007), reduction of the frequency of self-isolation, and improvements in time spent with peers as well as in the frequency of positive or appropriate interactions with peers (Wood et al. 2014).

Although the core symptoms of ASD can barely be influenced by medication and therefore, currently, there is no pharmacological agent approved for the treatment of the core symptoms of ASD, drug treatment can be used as a valuable adjunct therapy (McPheeters et al. 2011; Poustka et al. 2011). Frequent targets for pharmacologic intervention include associated comorbid conditions (e.g., anxiety, depression, Attention deficit/hyperactivity disorder) and behavior problems (aggression, self-injurious behavior, lack of impulse control, tantrums, compulsive-like behaviors, repetitive or stereotyped behaviors, sleep disturbances). There is a growing body of controlled evidence for pharmacological intervention (Volkmar et al. 2014).

3 Future Avenues in Therapy

As ASD is a predominantly neurobiological determined disorder, neurobiological based approaches should, alongside behavior-based methods, be considered in the treatment of ASD. In the following, some innovative neurobiological based approaches for interventions, including pharmacotherapy targeting core symptoms in ASD, are discussed.

3.1 Neurofeedback Training of Deficient μ-Suppression

Neurofeedback allows patients to learn self-regulation over several training sessions by increasing control over critical target aspects of their brain activity, thus presenting a possible alternative for behavioral therapy for ASD in order to achieve normalization of social behavior. Brain activity is not otherwise directly accessible to conscious attention and control, but can be measured and fed back in near real time using EEG parameters. Oscillations in the beta or theta frequency range and slow cortical potentials (SCP) have all been related to activation or deactivation and are thus typical targets. The training is based on operant conditioning and cortical plasticity: Producing "desired" EEG activity while suppressing "undesired" activity is facilitated through the presentation of pleasant, easily perceivable, and comprehensible animations or other feedback signals.

As imitation and face processing are impaired in ASD, it has been speculated that a deficient mirror neuron system (MNS) could contribute to typical impairments in this condition. The μ-rhythm of the EEG could reflect such a MNS dysfunction in ASD (Oberman et al. 2005). This rhythm over the central motor region is suppressed (μ-suppression) when one performs voluntary movements like closing a hand. Such μ-suppression typically also appears when observing (hand) movements of other people. In individuals with ASD, this μ-suppression has been reported to be decreased when observing the movements of other persons. The degree of μ-suppression correlates with the ability to imitate movements and gestures (Bernier et al. 2007) and increases with the perceived familiarity of the observed person (Oberman et al. 2008). This relationship between ASD, mirror neurons, and the μ-rhythm provides a potential rationale for a disorder-specific form of neurofeedback in ASD, although findings concerning mirror neurons are conflicting (Fan et al. 2010; Dinstein et al. 2010). If the function of the mirror neurons in ASD is not completely impaired, the suppression of μ-waves might be trained by means of neurofeedback, with the goal of reactivating the mirror neurons in such a way that the imitation ability improves.

Neurofeedback requires an EEG system with computer-based online signal analysis for feedback of the desired and undesired brain activities in the form of animations embedded in a game or task presented to the patients. While these gamelike animations of brain activity are presented on the patient's monitor, the original EEG along with some control data is visualized simultaneously on a second monitor for the therapist. Prior to the neurofeedback training, a baseline EEG is recorded in order to determine individual thresholds for the target activities. These thresholds are usually adjusted in the course of the training. The change in EEG during the training dominates the first phase of neurofeedback. Following this, the goal is to generalize the achieved changes to everyday situations. To this aim, practice rounds without feedback can be performed (transfer), and the application of the self-regulation ability is trained in the school environment. All three steps (self-regulation with feedback, transfer, and experience of self-efficacy in everyday life) should be contained in the therapy plan.

The empirical basis for neurofeedback treatment of the socio-communicative core symptoms of ASD is still limited. While first pilot studies showed promising results, but also considerable methodological problems (Jarusiewicz 2002; Coben and Padolsky 2007), more systematic work was reported by Pineda et al. (2008). Their neurofeedback studies aimed to achieve changes in autistic symptoms through an improved suppression of the μ-rhythm, in order improve imitation abilities. High μ-power is suggested to be an indicator of a relaxed and yet focused state and is thus assumed to be a prerequisite for successful μ-suppression, which is suggestive of more activation of the MNS (Oberman et al. 2008). This in turn should lead to better imitation behavior. Following 15 h of neurofeedback to strengthen 8- to 13-Hz rhythms in the EEG over the right central region, a reactivation of the previously attenuated μ-suppression emerged; i.e., in the observation of the movements of unfamiliar persons, a decrease of the μ-rhythm was shown, which could not previously be detected. The imitation ability enhanced, although this effect was also discernible in the placebo group. In a follow-up study with a sample of 19 participants with high-functioning autism using a randomized, double-blind design, the effects on imitation ability could not be replicated, despite improved μ-suppression in the experimental group. However, the studies by Pineda et al. (2008) demonstrated the effects of neurofeedback on ADHD symptoms in ASD in a stable manner. In a next step, the same research group used a bidirectional EEG training for μ-suppression based on a game encouraging social interaction, involving feedback based on imitation and emotional responsiveness. Results demonstrated increased μ-suppression, but also improved ASD-related symptoms and increased social responsiveness and parent-rated improvement of adaptive behavior in a group of 13 children with ASD (Friedrich et al. 2015). Altogether, recent findings suggest considerable improvement in social interactions and communication skills through EEG—based neurofeedback; still, randomized, controlled, large-scale trials are needed in order to verify the effectiveness of neurofeedback training in ASD.

3.2 Variations: Real-Time fMRI and NIRS Neurofeedback

Still, relatively new variants of neurofeedback based on *real-time-fMRI or fNIRS (functional near infrared spectroscopy)* (Holper et al. 2012) are of particular interest with regard to autistic disorders. While conventional neurofeedback is based on neuroelectrical brain activity recorded as EEG, the (equally noninvasive) real-time fMRI and fNIRS use the localized hemodynamic BOLD signals measured in a scanner (fMRI) or through optodes mounted on the scalp (fNIRS) for feedback. The participant is thus trained through feedback to volitionally control the brain activity or connectivity patterns in specific areas or networks (Sitaram et al. 2007). This imaging has the advantage of high spatial resolution and precision compared to EEG derivations and may thus be used to target the dysfunctions of the brain structures and networks implicated in ASD more precisely. While fMRI-based approaches have comparable spatial resolution for cortical and subcortical brain structures, the scalp recorded fNIRS activity is only sensitive to more superficial cortical activity up to a depth of about 2.5 cm. Originally developed for use in motor disorders and brain–computer interfaces, these techniques are increasingly being tested in psychiatric disorders. Experimental studies show a trainability of, e.g., the anterior insula (AI) (Caria et al. 2006, 2007), the anterior cingulate (Weiskopf et al. 2003), right inferior frontal gyrus (Rota et al. 2009), and individually localized emotional networks (Johnston et al. 2011). Rota et al. (2011) reported on improvements in the ability to identify emotional intonations after training healthy subjects to deliberately increase activation in their right inferior frontal gyrus (rIFG) using *real-time fMRI.* Moreover, the training obviously enhanced and lateralized connectivity of the rIFG to the right hemisphere. The authors interpreted this as a possibility of cortical reorganization in a functionally specific manner (Rota et al. 2011). Given that ASD is widely acknowledged as a disorder of synaptic connectivity resulting in altered functioning of distributed neural networks (Mueller 2007), *real-time fMRI* training might also be a promising intervention strategy for some aspects of ASD-related behavior, in order to enhance the development of "normal connectivity" and lateralization. In a very recent article, Caria and de Falco (2015) suggested the AI in particular to be a promising target region for *real-time fMRI* training in ASD. This derives from the specific role of the AI for the processing of emotional and social information by supporting the neural representation of the own physiological state. Thus, the volitional control of AI via *real-time fMRI* training may lead to changes in emotional behavior, such as self-evaluation of emotionally salient stimuli, which has so far been tested in studies with healthy participants (Caria et al. 2010) as well as patients with schizophrenia (Ruiz et al. 2013). Emotion regulation (ER), the ability to modify one's own emotional state, which promotes adaptive and goal-directed behavior, is increasingly investigated in ASD (Mazefsky et al. 2013). Impairments in ER might be an underlying factor in many maladaptive behaviors observed in individuals with ASD, such as disruptive behavior, anger, and aggression. While intervention studies addressing improvement of ER in ASD mainly focused on CBT (e.g., Scarpa and Reyes 2011), above-mentioned r*eal-time fMRI* training of the AI might be a promising strategy for the enhancement of regulation abilities in individuals with ASD.

3.3 Computer-Supported Cognitive Training of Basic Affect Recognition

Due to the lasting problems that people with ASD have in terms of affect recognition, and due to the associated hypoactivation of the fusiform gyrus and (at least partially) the amygdala, a series of training programs have been developed to train such skills. Many of these programs are now delivered via computer-based instructions, which offer some advantages especially for individuals with ASD. Computers are motivating for most autistic individuals and are a preferred medium of learning and communication. They offer the possibility to present information and data in a certain way (amount, speed, appearance). Computers are free from social signals, demands, and compulsions, they generate consistent and predictable reactions, and they allow the simulation of real situations in the form of virtual reality. Moreover, level of complexity can be flexibly adapted to the cognitive needs of the user.

These characteristics are in line with the environment preferred by people with ASD, which should be formal and rational in terms of information exchange (no irony, no sarcasm, no ambiguous information). Therefore, computer-based instructions and similar technologies offer suitable preconditions and stimulating environments to foster communicative, social, playful, and imaginative skills in ASD in a "protected space." As in other clinical groups, doubts and concerns have been expressed regarding the increased use of computers for ASD patients. Dangers are particularly seen in the fact that computers might increase social isolation and obsessions in ASD. It has also been indicated that computers might be used as a substitute for real encounters with people rather than as a supplement. Such concerns are legitimate and necessary, but none of the empirical studies so far have produced evidence for such negative evaluations of the use of computer-based trainings in ASD. On the contrary, studies show that by using computers-based training programs for individuals with ASD, a higher motivation and attention level is achieved, more social interactions come about, instructions and directions are more easily understood, and better training results are achieved (Bernard-Opitz et al. 2001; Williams et al. 2002). It is nevertheless possible that some individuals with ASD require special support to use computer-based trainings, and where appropriate, the selected software training should carefully be chosen with caution to the executive deficits in people with ASD (Grynszpan et al. 2007).

There are various computerized programs or games for the training of emotion recognition in ASD, e.g., *The Cambridge Mindreading (CAM) Face-Voice Battery* (Golan and Baron-Cohen 2006), *The Transporters* (Golan et al. 2010), *Let's face it* (Tanaka et al. 2010), or *The Social Cognition Training Tool (SCOTT)* (Dziobek et al.

2011). Ramdoss et al. (2012) performed a first systematic analysis of 11 studies involving computer-based interventions in ASD including a variety of different programs and a wide range of age and intellectual abilities in trained individuals with ASD. The authors reported rather mild effects on social and emotional skills in participants with ASD, which may be at least partly due to problems of generalization of newly acquired skills in real-life situations (Ramdoss et al. 2012).

Bölte et al. (2006, 2015) examined the effectiveness of a computer-based program for the training of basic affect recognition in ASD (Frankfurt Test and Training of Facial Affect Recognition, FEFA, Bölte and Poustka 2003) on the behavioral and neurobiological level. The researchers addressed the question of whether positive behavioral effects, paralleled by increased activation in the fusiform gyrus, could be achieved through intensive affect recognition training. The FG is a structure in the occipito-temporal cortex which is particularly involved in recognition of facial affect. The FEFA training module (Bölte and Poustka 2003) comprises approximately 500 photographs of faces and 500 of the eye region, classified according to the concept of the 6 (+ 1) basic emotions: joy, sadness, anger, disgust, surprise, and fear (+neutral). The task of the FEFA consists in assessing the respective basic emotion in faces or eye areas including visual and acoustic feedback. In the first pilot study in 2006, clear improvements in affect recognition in the FEFA face test to the amount of 1–2 standard deviations (normative level) were found in trained individuals, together with activation changes in the superior parietal lobe and the medial occipital gyrus. Contrary to expectation, no increased activation of the fusiform gyrus was found. In a large-scale replication study, with an improved design on 32 high-functioning individuals with ASD and 25 controls, the replicated behavioral effects were also linked to a significantly increased post-training activation of the fusiform gyrus and the amygdala (Bölte et al. 2015). This suggests brain plasticity of key regions for recognition of facial affect in ASD, including the tempting possibility of normalizing brain activation related to social processing via intensive training of facial affect recognition. Notably, participants in this study were 19.3 years old on average (range 14–33); while comprehensive intervention programs are usually recommended to start in early childhood to be effective, findings of this study suggest that training effects including associated changes of brain activation can also be achieved in adolescents and young adults.

3.4 Early Intervention and Adjuvant Pharmacotherapy

In clinical practice, successful psychopharmacological treatment frequently represents the foundation for successful psychotherapeutic and educational interventions. The core symptoms of ASD—possibly with the exception of repetitive sand stereotyped behavior—cannot be sufficiently treated with medication. However, various pharmacological agents have been proven to be effective in the treatment of associated disorders in ASD. Moreover, medication can support the psychotherapeutic efforts to achieve improvements in core deficits of ASD such as social interaction and communication abilities.

In relation to neurochemical approaches, an intensified controversial discussion is currently taking place regarding the possibility of early pharmacological treatment in the sensitive phases of an increased plasticity of the brain (Canitano and Bozzi 2015; Bethea and Sikich 2007). The aim is to modify central developmental processes with key functions for the brain development of young children with ASD. Various neurochemical processes are being discussed in terms of the underlying pathophysiology of autism. These include, for instance, a disorder of the glutamate/GABA metabolism. In particular, an early cortical imbalance of excitatory versus inhibitory (GABA) neurotransmission during critical periods of brain development may compromise "normal" brain plasticity and the differentiation of primary sensory functioning and in consequence cause impairment of higher-order processing (LeBlanc and Fagiolini 2011). Currently, amantadine and memantine, as non-competitive antagonists of the NMDA receptor, are being tested; first investigations regarding memantine, which has up to now been used in the treatment of dementia, show positive effects on attention and withdrawal behavior (Erickson et al. 2007) as well as on memory functions (Owley et al. 2006). D-cycloserine, a partial NMDA agonist with differing affinity to NMDA receptor subtypes (traditionally used for the treatment of tuberculosis), was also tested for the treatment of negative symptoms in patients with schizophrenia and should have a positive effect on some aspects of social impairment in autistic disorders (Urbano et al. 2015). A similar line of research concentrates on GABAergic signaling in relation to excitatory versus inhibitory imbalance in ASD (Coghlan et al. 2013). In this respect, arbaclofen, a GABA receptor agonist which is suggested to increase inhibitory neurotransmission, has been tested in patients with FraX syndrome (Berry-Krevis et al. 2012) as well as in an exploratory open-label trial with 32 children and adolescents with non-syndromic ASD (Erickson et al. 2013). Results suggested improvements in both irritability and global functioning as well as some core domains of ASD, but large-scale studies are needed to confirm these findings.

Other attempts concern the two neuropeptides oxytocin (OXT) and vasopressin. Both peptides are synthesized in the hypothalamus, secreted by the hypophysis and play a central role, among other things, in human attachment behavior and social memory. In particular, OXT has over the past years been increasingly examined in relation to social anxieties, to social cognition, and to the core symptoms of autistic disorder (current overviews by Baribeau and Anagnostou 2015; Guastella and Hickie 2015). OXT is of particular relevance in ASD, as it has been identified as a powerful enhancer of neural activity related to social cognition, the formation of social bonds, and socially reinforced learning (Groppe et al. 2013; Kirsch et al. 2005; Meyer-Lindenberg et al. 2011). Further, recent findings from genetic, animal and single-dose intervention studies suggest OXT to be of therapeutic potential for the improvement of social deficits in ASD. In the last decade, a rapidly growing number of interdisciplinary (pharmacokinetics, (epi)genetics, neuroimaging, imaging genetics, clinical) studies have consistently indicated that OXT plays an important role in modulating human social behaviors with translational relevance for understanding ASD. Pioneering but strong evidence in patients with ASD suggests that OXT has the potential to enhance motivation and attention to social cues (Yamasue et al. 2012), thereby potentially impacting processing of affiliative emotions, social reward, and higher cognitive functions such as empathy and theory of mind (ToM) in the long run. A recently published meta-analysis (Bakermans-Kranenburg and van IJzendoorn 2013a) summarized recent studies on pharmacotherapeutic applications of OXT treatment to explore its potentials and limitations, concluding that studies on ASD showed significant effect sizes. Additionally, it has been demonstrated that (epi)genetic factors affect the response of OXT, either by directly acting on OXT genes or via regulating genes in pathways related to OXT (Kumsta and Heinrichs 2013; Macdonald 2012). As results of genotype effects are not entirely consistent (Bakermans-Kranenburg and Van IJzendoorn 2013b), epigenetic factors (e.g., methylation differences) have to be taken into account to identify which (epi)genetic factors contribute to the acute and long-term effects of intranasal OXT. However, despite the growing number of studies on OXT in populations with ASD (for very recent findings see, e.g., Guastella et al. 2015; Yatawara et al. 2015; Watanabe et al. 2015), only few studies have tested the efficacy of OXT in combination with psychotherapeutic interventions (e.g., Dadds et al. 2014). Watanabe et al. (2015) conducted a study with once-weekly administered intranasal OTX to 20 adults with ASD over a period of 6 weeks. They found an OTX-induced significant reduction in ASD-specific symptoms, as well as enhancement of functional connectivity between anterior cingulate cortex and dorso-medial prefrontal cortex. By contrast, Guastella et al. (2015) reported no effects on social responsiveness or social cognition of twice daily administered OTX over a period for 8 weeks in a placebo-controlled trial on 52 adolescents with ASD. Notably, parents reports about improvement were significantly influenced by expectations whether their child had received OTX or placebo. The only study to date examining whether OXT administration enhances the effects of behavioral interventions was performed by Dadds et al. (2014). They tested daily administration of OXT in a sample of boys with ASD during parent–child interaction training over a period of four days (N = 38, age 7–16 with a total of four doses of OXT per person). Compared to placebo, there was no significant effect of OXT on social interaction skills or emotion recognition capability. Although some weaknesses of the study by Dadds must be considered, benefits from single-dose studies have apparently not translated to benefits when examining the literature of extended OTX treatment, and further studies on the combination of OTX plus psychotherapeutic interventions are needed.

3.5 Outlook

While a growing number of studies support the efficacy of behavior-based interventions in ASD, research on neurobiological based interventions or on the combination of these psychotherapeutic interventions with pharmacological treatments is still sparse. Only few studies so far have shown that administration of additional medication may have the potential to enhance the effects of psychotherapy/behavioral therapies. Moreover, there is very limited information regarding the extent to which training effects are modulated by variables such as intelligence, developmental state, medication, psychiatric comorbidity, symptom profile, and severity or genetic liability. As treatment response even for well-validated intervention programs is highly variable within treatment groups (Magiati et al. 2007), the possibility of using biomarkers such as changes in brain activity or genetic risk variants in order to index (see Dawson et al. 2012; Bölte et al. 2015) and predict treatment response is a crucial factor in intervention research. Thus, the prediction of treatment response in different subtypes of ASD should be included in well-designed intervention studies in order to correctly allocate individuals to treatment settings.

While basic research is progressing at a rapid pace, neurobiologically oriented psychotherapy or "neuropsychotherapy" can still be seen as a young discipline of applied psychology and psychiatry, with a limited number of available studies (Linden 2006). In child and adolescent psychiatry, empirical studies on morphological or functional neuronal correlates of psychotherapy are indeed still rare, with some exceptions such as studies on neurofeedback (Holtmann et al. 2014), working memory training (Klingberg et al. 2005), or behavioral interventions (Siniatchkin et al. 2012) in ADHD. Recently, trainings for pre-readers and in individuals with dyslexia were also examined in terms of neurobiological correlates (Brem et al. 2010; Spironelli et al. 2010). Despite the long way from functional to structural integrity changes, structural gray matter changes in language regions reflect second language learning within one year (Stein et al. 2012), and Keller and Just (2009) even demonstrated altered functional connectivity in poor readers after 10 weeks of reading intervention. It is thus tempting to speculate about possibilities of enhancing the development of "normal connectivity" through early training of regions involved in social cognition. To date, there are no studies examining training effects of common early intervention programs on structural or functional connectivity in very young children with ASD. Many methodological challenges and open questions remain to be resolved before neuropsychotherapeutic approaches can be recommended as sufficiently evaluated interventions for ASD. The preliminary evidence on the use of neurofeedback and cognitive training in ASD still has to be viewed as limited due to a lack of larger randomized controlled studies. Secondly, in future studies on so-called neuropsychotherapy in ASD, it should be examined to what extent neurobiological effects are modulated by variables such as intelligence, developmental state, medication, psychiatric comorbidity, symptom profile, and severity. In any case, research on neuropsychotherapy in ASD should also assess the neural effects of those techniques with the best documented efficiency on the behavioral level. Psychotherapeutic interventions may also lead to neurobiological changes reflecting (partial) normalization or compensation (see studies by Dawson et al. 2012, or Bölte et al. 2015). The potential to distinguish between alternative mechanisms of effective treatment at the neurobiological level additionally shows that psychotherapeutic and neurobiological research complements one another.

References

Abdallah MW, Greaves-Lord K et al (2011) Psychiatric comorbidities in autism spectrum disorders: findings from a Danish Historic Birth Cohort. Eur Child Adolesc Psychiatry 20(11–12):599–601

Anderson DK, Liang JW et al (2014) Predicting young adult outcome among more and less cognitively able individuals with autism spectrum disorders. J Child Psychol Psychiatry 55 (5):485–494

Antshel KM, Polacek C et al (2011) Comorbid ADHD and anxiety affect social skills group intervention treatment efficacy in children with autism spectrum disorders. J Dev Behav Pediatr 32(6):439–446

Bakermans-Kranenburg MJ, van IJzendoorn MH (2013a) Sniffing around oxytocin: review and meta-analyses of trials in healthy and clinical groups with implications for pharmacotherapy. Transl Psychiatry 3:e258

Bakermans-Kranenburg MJ, Van IJzendoorn MH (2013b) A sociability gene? Meta-analysis of oxytocin receptor genotype effects in humans. Psychiatr Genet. doi:10.1097/YPG0b013e3283643684

Baribeau DA, Anagnostou E (2015) Oxytocin and vasopressin: linking pituitary neuropeptides and their receptors to social neurocircuits. Front Neurosci 9:335

Ben-Itzchak E, Watson LR et al (2014) Cognitive ability is associated with different outcome trajectories in autism spectrum disorders. J Autism Dev Disord 44(9):2221–2229

Berg J, Geschwind D (2012) Autism genetics: searching for specificity and convergence. Genome Biol 13:1–16

Bernard-Opitz V, Sriram N, Nakhoda-Sapuan S (2001) Enhancing social problem solving in children with autism and normal children through computer-assisted instruction. J Autism Dev Disord 31(4):377–384

Bernier R, Dawson G, Webb S, Murias M (2007) EEG mu rhythm and imitation impairments in individuals with autism spectrum disorder. Brain Cogn 64(3):228–237

Berry-Kravis EM, Hessl D, Rathmell B, Zarevics P, Cherubini M, Walton-Bowen K et al (2012) Effects of STX209 (arbaclofen) on neurobehavioral function in children and adults with fragile X syndrome: a randomized, controlled, phase 2 trial. Sci Transl Med 4(152):ra127

Betancur C (2011) Etiological heterogeneity in autism spectrum disorders: more than 100 genetic and genomic disorders and still counting. Brain Res 1380:42–77

Bethea TC, Sikich L (2007) Early pharmacological treatment of autism: a rationale for developmental treatment. Biol Psychiatry 61(4):521–537

Bölte S, Poustka F (2003) The recognition of facial affect in autistic and schizophrenic subjects and their first-degree relatives. Psychol Med 33(5):907–915

Bölte S, Hubl D, Feineis-Matthews S, Prvulovic D, Dierks T, Poustka F (2006) Facial affect recognition training in autism: can we animate the fusiform gyrus? Behav Neurosci 120 (1):211–216

Bölte S, Ciaramidaro A, Schlitt S, Hainz D, Kliemann D, Beyer A, Poustka F, Freitag C, Walter H (2015) Training-induced plasticity of the social brain in autism spectrum disorder. Br J Psychiatry 2:149–157

Brem S, Bach S, Kucian K, Guttorm TK, Martin E, Lyytinen H, Brandeis D, Richardson U (2010) Brain sensitivity to print emerges when children learn letter-speech sound correspondences. Proc Natl Acad Sci USA 107(17):7939–7944

Canitano R, Bozzi Y (2015) New treatment perspectives in autism spectrum disorders. Front Pediatr 3:22

Caria A, de Falco S (2015) Anterior insular cortex regulation in autism spectrum disorders. Front Behav Neurosci 6(9):38

Caria A, Veit R, Gaber T, Sitaram R, Kuebler A Birbaumer N (2006) Can we learn to increase our emotional involvement? Real-time fMRI of anterior cingulate cortex during emotional processing. Human Brain Mapping, Florence

Caria A, Veit R, Sitaram R, Lotze M, Weiskopf N, Grodd W, Birbaumer N (2007) Regulation of anterior insular cortex activity using real-time fMRI. Neuroimage 35(3):1238–1246

Caria A, Sitaram R, Veit R, Begliomini C, Birbaumer N (2010) Volitional control of anterior insula activity modulates the response to aversive stimuli. A real-time functional magnetic resonance imaging study. Biol Psychiatry 68:425–432

Carr D, Felce J (2007) Brief report: increase in production of spoken words in some children with autism after PECS teaching to Phase III. J Autism Dev Disord 37(4):780–787

Charman T, Jones CRG et al (2011) Defining the cognitive phenotype of autism. Brain Res 1380:10–21

Coben R, Padolsky I (2007) Assessment-guided neurofeedback for autistic spectrum disorder. J Neurotherapy 11:5–23

Coghlan S, Horder J, Inkster B, Mendez MA, Murphy DG, Nutt DJ (2013) GABA system dysfunction in autism and related disorders: from synapse to symptoms. Neurosci Biobehav Rev 36(9):2044–2055

Dadds MR, MacDonald E, Cauchi A, Williams K, Levy F, Brennan J (2014) Nasal oxytocin for social deficits in childhood autism: a randomized controlled trial. J Autism Dev Disord 44 (3):521–531

Dawson G, Bernier R (2013) A quarter century of progress on the early detection and treatment of autism spectrum disorder. Dev Psychopathol 25(4pt2):1455–1472

Dawson G, Jones EJH et al (2012) Early behavioral intervention is associated with normalized brain activity in young children with autism. J Am Acad Child Adolesc Psychiatry 51 (11):1150–1159

DeRosier ME, Swick DC et al (2011) The efficacy of a social skills group intervention for improving social behaviors in children with high functioning autism spectrum disorders. J Autism Dev Disord 41(8):1033–1043

Dichter GS (2012) Functional magnetic resonance imaging of autism spectrum disorders. Dialogues Clin Neurosci 14(3):319–351

Dichter GS, Felder JN et al (2012) Reward circuitry function in autism spectrum disorders. Soc Cogn Affect Neur 7(2):160–172

Dinstein I, Thomas C, Humphreys K, Minshew N, Behrmann M, Heeger DJ (2010) Normal movement selectivity in autism. Neuron 66(3):461–469

Dziobek I, Rosenblau G, Kliemann D, Kappelhoff H, Heekeren H (2011) Ein neues Softwaretraining für komplexe Emotionserkennung: social cognition training tool (SCOTT). DGPPN Kongress 2011

Elsabbagh M, Divan G et al (2012) Global prevalence of autism and other pervasive developmental disorders. Autism Res: Official J Int Soc Autism Res 5(3):160–179

Erickson CA, Posey DJ, Stigler KA, Mullett J, Katschke AR, McDougle CJ (2007) A retrospective study of memantine in children and adolescents with pervasive developmental disorders. Psychopharmacology 191:141–147

Erickson CA, Veenstra-Vanderweele JM, Melmed RD, McCracken JT, Ginsberg LD, Sikich L et al (2013) STX209 (arbaclofen) for autism spectrum disorders: an 8-week-open label study. J Autism Dev Disord 44:958–964

Fan Y-T, Decety J, Yang C-Y, Liu J-L, Cheng Y (2010) Unbroken mirror neurons in autism spectrum disorders. J Child Psychol Psychiatry 51(9):981–988

Fava L, Strauss K (2014) Response to early intensive behavioral intervention for autism–an umbrella approach to issues critical to treatment individualization. Int J Dev Neurosci 39:49–58

Fisch GS (2013) Nosology and epidemiology in autism: classification counts. Am J Med Genet C Semin Med Genet 161(9):2399

Fletcher-Watson S, McConnell F et al (2014) Interventions based on the theory of mind cognitive model for autism spectrum disorder (ASD). Cochrane Database Syst Rev 3 CD008785

Flippin M, Reszka S et al (2010) Effectiveness of the picture exchange communication system (PECS) on communication and speech for children with autism spectrum disorders: a meta-analysis. Am J Speech Lang Pathol 19(2):178–195

Friedrich EV, Sivanathan A, Lim T, Suttie N, Louchart S, Pillen S, Pineda JA (2015) An effective neurofeedback intervention to improve social interactions in children with autism spectrum disorder. J Autism Dev Disord 45:4084–4100

Gardener H, Spiegelman D, Buka S (2009) Prenatal risk factors for autism: comprehensive meta-analysis. Br J Psychiatry 195(1):7–14

Geschwind DH (2008) Autism: many genes, common pathways? Cell 135:391–395

Gjevik E, Eldevik S et al (2011) Kiddie-SADS reveals high rates of DSM-IV disorders in children and adolescents with autism spectrum disorders. J Autism Dev Disord 41(6):761–769

Golan O, Baron-Cohen S (2006) Systemizing empathy: teaching adults with Asperger syndrome or high functioning autism to recognize complex emotions using interactive multimedia. Dev Psychopathol 18(2):591–617

Golan O, Ashwin E, Granader Y, McClintock S, Day K, Leggett V, Baron-Cohen S (2010) Enhancing emotion recognition in children with autism spectrum conditions: an intervention using animated vehicles with real emotional faces. J Autism Dev Disord 40(3):269–279

Gordon K, Pasco G et al (2011) A communication-based intervention for nonverbal children with autism: what changes? Who benefits? J Consult Clin Psychol 79(4):447–457

Groppe SE, Gossen A et al (2013) Oxytocin influences processing of socially relevant cues in the ventral tegmental area of the human brain. Biol Psychiatry 74:172–179

Grynszpan O, Martin JC, Nadel J (2007) Exploring the influence of task assignment and output modalities on computerized training for autism. Interact Stud 8:241–266

Guastella AJ, Hickie IB (2015) Oxytocin treatment, circuitry and autism: a critical review of the literature placing oxytocin into the autism context. Biol Psychiatry 79(3):234–242

Guastella AJ, Gray KM, Rinehart NJ, Alvares GA, Tonge BJ, Hickie IB, Keating CM, Cacciotti-Saija C, Einfeld SL (2015) The effects of a course of intranasal oxytocin on social behaviors in youth diagnosed with autism spectrum disorders: a randomized controlled trial. J Child Psychol Psychiatry 56(4):444–452

Hallmayer J, Cleveland S et al (2011) Genetic heritability and shared environmental factors among twin pairs with autism. Arch Gen Psychiatry 68(11):1095–1102

Happe F, Frith U (2006) The weak coherence account: detail-focused cognitive style in autism spectrum disorders. J Autism Dev Disord 36(1):5–25

Hofvander B, Delorme R et al (2009) Psychiatric and psychosocial problems in adults with normal-intelligence autism spectrum disorders. BMC Psychiatry 9:35

Holper L, Scholkmann F, Wolf M (2012) Between-brain connectivity during imitation measured by fNIRS. Neuroimage 63(1):212–222

Holtmann M, Sonuga-Barke E, Cortese S, Brandeis D (2014) Neurofeedback for ADHD: a review of current evidence. Child Adolesc Psychiatr Clin N Am 23(4):789–806

Howard JS, Stanislaw H et al (2014) Comparison of behavior analytic and eclectic early interventions for young children with autism after three years. Res Dev Disabil 35 (12):3326–3344

Howlin P (2013) Social disadvantage and exclusion: adults with autism lag far behind in employment prospects. J Am Acad Child Adolesc Psychiatry 52(9):897–899

Howlin P, Magiati I et al (2009) Systematic review of early intensive behavioral interventions for children with autism. Am J Intellect Dev Disabil 114(1):23–41

Howlin P et al (2013) Social outcomes in mid- to later adulthood among individuals diagnosed with autism and average nonverbal IQ as children. J Am Acad Child Adolesc Psychiatry 52 (6):572–581

Howlin P, Savage S et al (2014) Cognitive and language skills in adults with autism: a 40-year follow-up. J Child Psychol Psychiatry 55(1):49–58
Jarusiewicz B (2002) Efficacy of neurofeedback for children in the autistic spectrum: a pilot study. J Neurotherapy 6:39–49
Jeste SS, Geschwind DH (2014) Disentangling the heterogeneity of autism spectrum disorder through genetic findings. Nat Rev Neurol 10(2):74–81
Johnston S, Linden DE, Healy D, Goebel R, Habes I, Boehm SG (2011) Upregulation of emotion areas through neurofeedback with a focus on positive mood. Cogn Affect Behav Neurosci 11 (1):44–51
Jones W, Klin A (2013) Attention to eyes is present but in decline in 2–6-month-old infants later diagnosed with autism. Nature 504(7480):427–431. doi:10.1038/nature12715
Kamp-Becker I, Ghahreman M et al (2009) Dimensional structure of the autism phenotype: relations between early development and current presentation. J Autism Dev Disord 39 (4):557–571
Karst JS, Van Hecke AV (2012) Parent and family impact of autism spectrum disorders: a review and proposed model for intervention evaluation. Clin Child Fam Psychol Rev 15(3):247–277
Keller TA, Just MA (2009) Altering cortical connectivity: remediation-induced changes in the white matter of poor readers. Neuron 64(5):624–631
Kirsch P, Esslinger C et al (2005) Oxytocin modulates neural circuitry for social cognition and fear in humans. J Neurosci 25(49):11489–11493
Klingberg T, Fernell E, Olesen PJ, Johnson M, Gustafsson P, Dahlstrom K, Gillberg CG, Forssberg H, Westerberg H (2005) Computerized training of working memory in children with ADHD–a randomized, controlled trial. J Am Acad Child Adolesc Psychiatry 44(2):177–186
Kohls G, Schulte-Rüther M et al (2012) Reward system dysfunction in autism spectrum disorders. Soc Cogn Affect Neurosci. doi:10.1093/scan/nss033
Krach S, Kamp-Becker I, Einhäuser W, Sommer J, Frässle S, Jansen A, Rademacher L, Müller-Pinzler L, Gazzola V, Paulus FM (2015) Evidence from pupillometry and fMRI indicates reduced neural response during vicarious social pain but not physical in autism. Hum Brain Mapp 36(11):4730–4744
Kumsta R, Heinrichs M (2013) Oxytocin, stress and social behavior: neurogenetics of the human oxytocin system. Curr Opin Neurobiol 23(1):11–16
Lai M-C, Lombardo MV et al (2014) Autism. Lancet 383(9920):896–910
LeBlanc JJ, Fagiolini M (2011) Autism: a “critical period” disorder? Neural Plast 2011, 921680
Linden DE (2006) How psychotherapy changes the brain—the contribution of functional neuroimaging. Mol Psychiatry 11(6):528–538
Macdonald KS (2012) Sex, receptors, and attachment: a review of individual factors influencing response to oxytocin. Front Neurosci 6:194
Magiati I, Charman T et al (2007) A two-year prospective follow-up study of community-based early intensive behavioural intervention and specialist nursery provision for children with autism spectrum disorders. J Child Psychol Psy 48(8):803–812
Magiati I, Tay XW et al (2014) Cognitive, language, social and behavioural outcomes in adults with autism spectrum disorders: a systematic review of longitudinal follow-up studies in adulthood. Clin Psychol Rev 34(1):73–86
Matson ML, Matson JL et al (2011) Comorbidity of physical and motor problems in children with autism. Res Dev Disabil 32(6):2304–2308
Mazefsky CA, Herrington J, Siegel M, Scarpa A, Maddox BB, Scahill L, White SW (2013) The role of emotion regulation in autism spectrum disorder. J Am Acad Child Adolesc Psychiatry 52(7):679–688
McConachie H, Diggle T (2007) Parent implemented early intervention for young children with autism spectrum disorder: a systematic review. J Eval Clin Pract 13(1):120–129
McMahon CM, Lerner MD, Britton N (2013) Group-based social skills interventions for adolescents with higher-functioning autism spectrum disorder: a review and looking to the future. Adolesc Health Med Ther 4:23–28

McPheeters ML, Warren Z et al (2011) A systematic review of medical treatments for children with autism spectrum disorders. Pediatrics 127(5):e1312–e1321

Meyer-Lindenberg A, Domes G et al (2011) Oxytocin and vasopressin in the human brain: social neuropeptides for translational medicine. Nat Rev Neurosci 12(9):524–538

Müller RA (2007) The study of autism as a distributed disorder. Ment Retard Dev Disabil Res Rev 13(1):85–95

Neuhaus E, Beauchaine TP et al (2010) Neurobiological correlates of social functioning in autism. Clin Psychol Rev 30(6):733–748

Oberman LM, Hubbard EM, McCleery JP, Altschuler EL, Ramachandran VS, Pineda JA (2005) EEG evidence for mirror neuron dysfunction in autism spectrum disorders. Cogn Brain Res 24 (2):190–198

Oberman LM, Ramachandran VS, Pineda JA (2008) Modulation of mu suppression in children with autism spectrum disorders in response to familiar or unfamiliar stimuli: the mirror neuron hypothesis. Neuropsychologia 46(5):1558–1565

Oono IP, Honey EJ et al (2013) Parent-mediated early intervention for young children with autism spectrum disorders (ASD). Cochrane Database Syst Rev 4:CD009774

Owley T, Salt J, Guter S, Grieve A, Walton L, Ayuyao N, Leventhal BL, Cook EH Jr (2006) A prospective, open-label trial of memantine in the treatment of cognitive, behavioral, and memory dysfunction in pervasive developmental disorders. J Child Adolesc Psychopharmacol 16(5):517–524

Parikshak NN, Luo R, Zhang A, Won H, Lowe JK et al (2013) Integrative functional genomic analyses implicate specific molecular pathways and circuits in autism. Cell 155:1008–1021

Parner ET, Baron-Cohen S et al (2012) Parental age and autism spectrum disorders. Ann Epidemiol 22(3):143–150

Pellicano E (2012) The development of executive function in autism. Autism Res Treat Print 2012, 146132

Pelphrey KA, Shultz S et al (2011) Research review: constraining heterogeneity: the social brain and its development in autism spectrum disorder. J Child Psychol Psychiatry 52(6):631–644

Pillay M, Alderson-Day B et al (2011) Autism spectrum conditions—enhancing nurture and development (ASCEND): an evaluation of intervention support groups for parents. Clin Child Psychol Psychiatry 16(1):5–20

Pineda JA, Brang D, Hecht E, Edwards L, Carey S, Bacon M, Futagaki C, Suk S, Tom J, Birnbaum C, Rork A (2008) Positive behavioral and electrophysiological changes following neurofeedback training in children with autism. Res Autism Spectr Disord 2:557–581

Poustka L, Banaschewski T et al (2011) Psychopharmacology of autism spectrum disorders. Nervenarzt 82(5):582–589

Ramdoss S, Machalicek W, Rispoli M, Mulloy A, Lang R, O'Reilly M (2012) Computer-based interventions to improve social and emotional skills in individuals with autism spectrum disorders: a systematic review. Dev Neurorehabil 15(2):119–135

Reaven J, Blakeley-Smith A et al (2012) Facing your fears in adolescence: cognitive-behavioral therapy for high-functioning autism spectrum disorders and anxiety. Autism Res Treat Print 2012, 423905

Reichow B, Steiner AM et al (2012) Social skills groups for people aged 6 to 21 with autism spectrum disorders (ASD). Cochrane Database Syst Rev 7:CD008511

Rota G, Sitaram R, Veit R, Erb M, Weiskopf N, Dogil G, Birbaumer N (2009) Self-regulation of regional cortical activity using real-time fMRI: the right inferior frontal gyrus and linguistic processing. Hum Brain Mapp 30(5):1605–1614

Rota G, Handjaras G, Sitaram R, Birbaumer N, Dogil G (2011) Reorganization of functional and effective connectivity during real-time fMRI-BCI modulation of prosody processing. Brain Lang 117(3):123–132

Ruiz S, Lee S, Soekadar SR, Caria A, Veit R, Kircher T et al (2013) Acquired self-control of insula cortex modulates emotion recognition and brain network connectivity in schizophrenia. Hum Brain Mapp 34:200–212

Scarpa A, Reyes NM (2011) Improving emotion regulation with CBT in young children with high functioning autism spectrum disorders: a pilot study. Behav Cogn Psychother 39(4):495–500

Simonoff E, Pickles A et al (2008) Psychiatric disorders in children with autism spectrum disorders: prevalence, comorbidity, and associated factors in a population-derived sample. J Am Acad Child Psy 47(8):921–929

Siniatchkin M, Glatthaar N, von Muller GG, Prehn-Kristensen A, Wolff S, Knochel S, Steinmann E, Sotnikova A, Stephani U, Petermann F, Gerber WD (2012) Behavioural treatment increases activity in the cognitive neuronal networks in children with attention deficit/hyperactivity disorder. Brain Topogr. doi:10.1007/s10548-012-0221-6

Sitaram R, Caria A, Veit R, Gaber T, Rota G, Kuebler A, Birbaumer N (2007) FMRI brain-computer interface: a tool for neuroscientific research and treatment. Comput Intell Neurosci 25487

Sofronoff K, Attwood T et al (2007) A randomized controlled trial of a cognitive behavioural intervention for anger management in children diagnosed with Asperger syndrome. J Autism Dev Disord 37(7):1203–1214

Soorya LV, Siper PM et al (2015) Randomized comparative trial of a social cognitive skills group for children with autism spectrum disorder. J Am Acad Child Adolesc Psychiatry 54 (3):208–216

Spironelli C, Penolazzi B, Vio C, Angrilli A (2010) Cortical reorganization in dyslexic children after phonological training: evidence from early evoked potentials. Brain 133(11):3385–3395

Stein M, Federspiel A, Koenig T, Wirth M, Strik W, Wiest R, Brandeis D, Dierks T (2012) Structural plasticity in the language system related to increased second language proficiency. Cortex 48(4):458–465

Sullivan K, Stone WL et al (2014) Potential neural mechanisms underlying the effectiveness of early intervention for children with autism spectrum disorder. Res Dev Disabil 35 (11):2921–2932

Tanaka JW, Wolf JM, Klaiman C, Koenig K, Cockburn J, Herlihy L, Brown C, Stahl S, Kaiser MD, Schultz RT (2010) Using computerized games to teach face recognition skills to children with autism spectrum disorder: the Let's Face It! program. J Child Psychol Psychiatry 51(8):944–952

Tellegen CL, Sanders MR (2014) A randomized controlled trial evaluating a brief parenting program with children with autism spectrum disorders. J Consult Clin Psychol 82(6):1193–1200

Urbano M, Okwara L, Manser P, Hartmann K, Deutsch SI (2015) A trial of d-cycloserine to treat the social deficit in older adolescents and young adults with autism spectrum disorders. J Neuropsychiatry Clin Neurosci 27(2):133–138

Virues-Ortega J (2010) Applied behavior analytic intervention for autism in early childhood: meta-analysis, meta-regression and dose-response meta-analysis of multiple outcomes. Clin Psychol Rev 30(4):387–399

Virues-Ortega J, Julio FM et al (2013) The TEACCH program for children and adults with autism: a meta-analysis of intervention studies. Clin Psychol Rev 33(8):940–953

Vismara LA, Rogers SJ (2010) Behavioral treatments in autism spectrum disorder: what do we know? Annu Rev Clin Psycho 6:447–468

Vivanti G, Prior M et al (2014) Predictors of outcomes in autism early intervention: why don't we know more? Front Pediatr 2:58

Volkmar F, Siegel M et al (2014) Practice parameter for the assessment and treatment of children and adolescents with autism spectrum disorder. J Am Acad Child Adolesc Psychiatry 53 (2):237–257

Wang S-Y, Parrila R et al (2013) Meta-analysis of social skills interventions of single-case research for individuals with autism spectrum disorders: results from three-level HLM. J Autism Dev Disord 43(7):1701–1716

Warren Z, McPheeters ML et al (2011) A systematic review of early intensive intervention for autism spectrum disorders. Pediatrics 127(5):e1303–e1311

Watanabe T, Kuroda M, Kuwabara H, Aoki Y, Iwashiro N, Tatsunobu N, Takao H, Nippashi Y, Kawakubo Y, Kunimatsu A, Kasai K, Yamasue H (2015) Clinical and neural effects of six-week administration of oxytocin on core symptoms of autism. Brain 138(Pt 11):3400–3412

Weiskopf N, Veit R, Erb M, Mathiak K, Grodd W, Goebel R, Birbaumer N (2003) Physiological self-regulation of regional brain activity using real-time functional magnetic resonance imaging (fMRI): methodology and exemplary data. Neuroimage 19(3):577–586

Whittingham K, Sofronoff K et al (2009) Stepping stones triple P: an RCT of a parenting program with parents of a child diagnosed with an autism spectrum disorder. J Abnorm Child Psychol 37(4):469–480

Williams C, Wright B, Callaghan G, Coughlan B (2002) Do children with autism learn to read more readily by computer assisted instruction or traditional book methods? A pilot study. Autism 6(1):71–91

Wood JJ, Fujii C et al (2014) Impact of cognitive behavioral therapy on observed autism symptom severity during school recess: a preliminary randomized, controlled trial. J Autism Dev Disord 44(9):2264–2276

Yamasue H, Yee JR et al (2012) Integrative approaches utilizing oxytocin to enhance prosocial behavior: from animal and human social behavior to autistic social dysfunction. J Neurosci 32 (41):14109–14117

Yatawara CJ, Einfeld SL, Hickie IB, Davenport TA, Guastella AJ (2015) The effect of oxytocin nasal spray on social interaction deficits observed in young children with autism: a randomized clinical crossover trial. Mol Psychiatry, Oct 27 [Epub ahead of print]

The Social Context Network Model in Psychiatric and Neurological Diseases

Sandra Baez, Adolfo M. García and Agustín Ibanez

Abstract The role of contextual modulations has been extensively studied in basic sensory and cognitive processes. However, little is known about their impact on social cognition, let alone their disruption in disorders compromising such a domain. In this chapter, we flesh out the social context network model (SCNM), a

---

S. Baez (✉) · A.M. García
Laboratory of Experimental Psychology and Neuroscience (LPEN), Institute of Translational and Cognitive Neuroscience (INCyT), INECO Foundation, Favaloro University, Pacheco de Melo 1860, 1126 Buenos Aires, Argentina
e-mail: sbaez@ineco.org.ar

A.M. García
e-mail: adolfomartingarcia@gmail.com

S. Baez · A.M. García · A. Ibanez
National Scientific and Technical Research Council (CONICET), Buenos Aires, Argentina

S. Baez · A.M. García
UDP-INECO Foundation Core on Neuroscience (UIFCoN), Diego Portales University, Santiago, Chile

A.M. García
Faculty of Elementary and Special Education (FEEyE), National University of Cuyo (UNCuyo), Mendoza, Argentina

A. Ibanez (✉)
Institute of Translational and Cognitive Neuroscience (INCyT), INECO Foundation, Favaloro University, Buenos Aires, Argentina
e-mail: aibanez@ineco.org.ar

A. Ibanez
Universidad Autónoma Del Caribe, Barranquilla, Colombia

A. Ibanez
Laboratory of Neuroscience, Adolfo Ibáñez University, Santiago, Chile

A. Ibanez
Australian Research Council (ARC) Centre of Excellence in Cognition and Its Disorders, Sydney, Australia

S. Baez
Grupo de Investigación Cerebro Y Cognición Social, Bogotá, Colombia

Curr Topics Behav Neurosci (2016) 30: 379–396
DOI 10.1007/7854_2016_443
Published Online: 25 April 2016

neuroscientific proposal devised to address the issue. In SCNM terms, social context effects rely on a fronto-temporo-insular network in charge of (a) updating context cues to make predictions, (b) consolidating context–target associative learning, and (c) coordinating internal and external milieus. First, we characterize various social cognition domains as context-dependent phenomena. Then, we review behavioral and neural evidence of social context impairments in behavioral variant frontotemporal dementia (bvFTD) and autism spectrum disorder (ASD), highlighting their relation with key SCNM hubs. Next, we show that other psychiatric and neurological conditions involve context-processing impairments following damage to the brain regions included in the model. Finally, we call for an ecological approach to social cognition assessment, moving beyond widespread abstract and decontextualized methods.

Keywords Social cognition · The social context network model · Psychatric disorders · Neurological disorders · Context processing

Contents

1 Introduction 380
2 Contextual Social Cognition 381
3 The Social Context Network Model 382
3.1 The Role of Frontal Lobes in Contextual Integration 382
3.2 Target–Context Associations in the Temporal Lobes 384
3.3 The Insular Cortex as a Convergence Area 385
4 The Social Context Network Model in Psychiatric and Neurological Diseases 385
4.1 Behavioral Variant Frontotemporal Dementia 386
4.2 Autism Spectrum Disorder (ASD) 387
4.3 Other Neuropsychiatric Disorders 390
5 Toward an Ecological Assessment of Social Context 391
6 Conclusions 392
References 393

1 Introduction

Social interactions naturally occur in context-rich settings (Ibanez and Manes 2012) which modulate all aspects of cognition, from basic perception to interpersonal domains. Perception of specific objects and colors always occurs in the context of other objects and colors. Listening and speaking processes are driven by the anticipation of upcoming discourse choices. Facial emotions are perceived and interpreted under the influence of verbal, visual, and gestural information. Likewise, adequate social behavior requires the integration of explicit and implicit contextual cues to properly deploy politeness, humor, irony, agreement, disagreement, or even silence.

In neuroscientific research, the role of context has been widely explored in low-level domains (e.g., visual perception), with relatively exhaustive anatomical and explanatory models (Bar 2004). Strikingly, although some authors (e.g., Adolphs 2009; Todorov et al. 2006) have addressed the relevance of contextual modulations in social cognition, no model has attempted to explain how specific mechanisms and brain regions contribute to context processing. In this chapter, we describe the social context network model (SCNM), which proposes that contextual modulations of social cognitive processes depend on a fronto-temporo-insular network (Ibanez and Manes 2012). Moreover, we propose that psychiatric and neurological disorders can be described as featuring marked contextual social cognition impairments. Finally, we present evidence that highlights the importance of assessing social cognition by means of ecological instruments.

2 Contextual Social Cognition

Social cognition is deployed in specific scenarios which mold meanings during interaction (Ibanez and Manes 2012; Kennedy and Adolphs 2012). The notion of social cognition involves several domains, such as emotion recognition, theory of mind (ToM), empathy, moral judgment, and decision making. Relevant cognitive operations involve perceiving, inferencing, and interpreting situational elements to derive an integrated construal of context which biases an action's meaning (Adolphs 2009; Ibanez and Manes 2012). For instance, structural processing of faces and subsequent recognition of their emotions occur against a backdrop of gestural, lexical, prosodical, and otherwise situational clues, all of which influence emotional assessment (Barrett et al. 2011). As a further example, consider empathy for pain. This highly contextual phenomenon (Melloni et al. 2014) depends on the recognition and vicarious experiencing of another person's physical (Lamm et al. 2007) and psychological (Masten et al. 2011) suffering. If we witness an assault in which the victim gets beaten, our initial empathy may be modulated by situational variables, to the extent that it can be accompanied or superseded by feelings of risk or the urge to escape or attack. Moreover, empathy is modulated by perceived fairness of others (Singer et al. 2006). Thus, empathic neural responses are reduced when observing an unfair person receiving pain. In addition, neural activity related to affective sharing and empathy for social pain is sensitive to the degree of closeness with the other person (Muller-Pinzler et al. 2015).

No less important is the impact of context on ToM, namely the capacity to attribute cognitive and affective states to oneself and others. Indeed, beliefs, intentions, and other mental states are more reliably inferred when embedded in contextual frames. The same is true for social decision making. In some cases, this process is guided by ambiguous clues in the absence of predictions about the outcome; in other situations, however, decisions cannot be made without full awareness of the risks and potential consequences. More generally, our gregarious behavior at large is driven by context-sensitive social norms. The joyful demeanor we adopt at a birthday party would be less than decorous at a funeral. Context also shapes our emotional responses while witnessing threats to another's social integrity. For instance, different behavioral reactions and neural circuits are involved in embarrassment with and embarrassment for another person's mishaps (Paulus et al. 2015). Furthermore, the feelings we express about other "racial," ethnic, or social groups other than our own are partially determined by situational constraints, in general, and the presence of members of these groups, in particular. In this vein, Johns et al. (2005) found that social emotions such as guilt or shame for the negative actions of in-group members are modulated by identity with the group and the perceived negativity of the event.

In sum, social cognition processes seem to be embedded in specific contextual circumstances that help to build intrinsic social meanings. To account for this crucial observation and based on multiple sources of evidence (detailed below), the SCNM proposes that the contextual influence on social cognitive processing is mediated by a fronto-temporo-insular network which (a) updates context cues to make predictions, (b) supports context–target associative learning, and (c) coordinates internal and external milieus.

3 The Social Context Network Model

During everyday interactions, common sense and implicit experience-based associative learning come together so that contextual frames can be updated to predict the meaning of probable socially relevant events. In terms of the SCNM, such contextual associations are mediated by a cortical network (Fig. 1) engaging frontal, temporal, and insular regions (Ibanez and Manes 2012). The updating of ongoing contextual information and its association with episodic memory supports target–context relations driven by activity in frontal regions (e.g., orbitofrontal cortex, lateral prefrontal cortex, superior orbital sulcus). The value of target–context associations is indexed in temporal circuits distributed throughout the amygdala, the hippocampus, and the perirhinal and parahippocampal cortices. Finally, internal and external milieus are coordinated by the insula to trigger internal motivational states.

3.1 The Role of Frontal Lobes in Contextual Integration

Access to episodic information is updated via context-driven predictions mediated by orbitofrontal, lateral prefrontal, and superior orbital regions (Watanabe and Sakagami 2007). In nonhuman animals, prefrontal neurons rapidly adapt to circumstantial meanings in short-term context paradigms, and they seem to update the same targets in different contexts (Sigala et al. 2008). Neurons in the orbitofrontal cortex are attuned to information about the organism's motivational context (Watanabe and Sakagami 2007). In primates, lateral prefrontal neurons fire in a

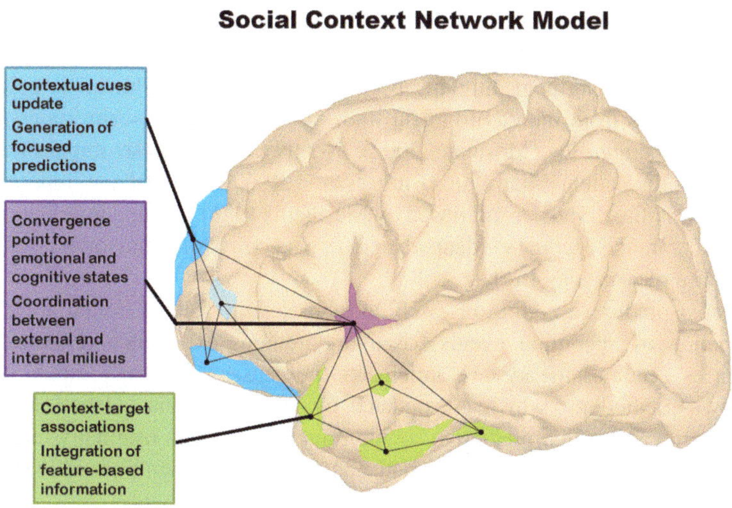

Fig. 1 Social context network model (SCNM). Lateral view of the left hemisphere showing the fronto-temporo-insular network. In this network, prefrontal areas would be involved in the generation of focused predictions by updating associations among representations in a specific context. Target–context associations subserved by temporal regions would be integrated with feature-based information processed in frontal regions. Finally, the insular cortex would support the convergence of emotional and cognitive states related to the coordination between external and internal milieus. Connected nodes represent fronto-temporo-insular interactions. Reproduced with authorization from Baez and Ibanez (2014)

context-dependent fashion, independently of the stimuli's physical attributes (Watanabe and Sakagami 2007). Contextual update of visual targets also implicates the superior orbital sulcus, as shown in both human and animal research. This frontal area would be important to generate focused predictions by updating associations among varied forms of information in specific contexts (Bar 2004).

Recent theories have suggested that functions related to metacognition, including mentalizing and self-knowledge, are specific to rostral prefrontal cortex (Burgess and Wu 2013). More specifically, Fleming and Dolan (2012) proposed a model in which the function of the rostral and dorsal lateral prefrontal cortex is important for the accuracy of retrospective metacognitive judgments of performance. In contrast, prospective judgments may depend upon the medial prefrontal cortex. Moreover, the rostro-lateral prefrontal cortex may receive input from interoceptive cortices (insula and anterior cingulate), generating an accurate metacognitive representation. However, unlike contextual updates and predictions mediated by the prefrontal cortex, metacognitive processes are consciously reportable (Fleming and Dolan 2012). Context-driven predictions in social cognition range from implicit to explicit levels (Ibanez and Manes 2012). Metacognitive processes may modulate explicit predictions but not implicit ones, which suggests that these are not critical aspects of the functioning of the frontal hub proposed by the SCNM.

Accordingly, disturbances of frontal lobe function are associated with deficits to recognize how context alters the meaning of stimuli (Mesulam 2002). In patients featuring such impairment, thinking becomes concrete and behavior is guided by superficial cues from the environment. Indeed, some patients seem impervious to contextual incongruity (Mesulam 2002). Notably, while these impairments are manifest in the patients' daily life (Burgess et al. 2009), they behave impeccably in the doctor's office. Thus, as proposed by Mesulam (1986), the major deficits caused by frontal damage become evident when external control of behavior is minimal (as in the case of everyday life).

Finally, predictive coding research corroborates the role of frontal areas in the anticipation of upcoming events based on previous experience (Friston 2012). Inferences occur as the system anticipates the causes of sensory inputs through top-down predictions, which are updated by frontal regions via bottom-up prediction errors. Note that predictive coding principles, though originally intended to account for basic brain function, also apply to social cognition and emotional phenomena (Seth et al. 2011).

3.2 Target–Context Associations in the Temporal Lobes

Contextual learning is rooted in the establishment of target–context associations (Greene et al. 2006). Basic associative processes, such as extinction (Bouton et al. 2006), involve activity in the hippocampus, the amygdala, and related temporal sites (e.g., perirhinal cortex). Associative processing seems to depend on the medial temporal lobes (Bar 2004), which are critical for the representation of contextual markers and their integration with signals coming from frontal regions. In humans, global contextual associations engage the parahippocampal cortex, which receives polysensory and somatosensory information (Bar 2004) and is also implicated in episodic memory.

Moreover, clinical research underscores the role of temporal lobes in contextual processing for social cognition. For instance, a recent study with patients with temporal lobe epilepsy or lobectomy (Cohn et al. 2015) showed relationships between hippocampal atrophy and social inference abilities, and between anterior temporal neocortex atrophy and sarcasm comprehension. By the same token, studies on neurodegenerative diseases—e.g., behavioral variant frontotemporal dementia (bvFTD), Alzheimer's disease, and semantic dementia—have shown that anterior temporal pole atrophy impairs several social cognition domains, including ToM (Torralva et al. 2007), empathy (Baez et al. 2014b), and moral judgment (Baez et al. 2014a, 2015b). In addition, neuroimaging evidence from bvFTD patients (Baez et al. 2015c) and direct electrophysiological recordings in intractable epileptic patients (Hesse et al. 2015) show that the amygdala is critical to detect the intentionality of other's actions, an ability that depends on the appraisal of contextual cues.

3.3 The Insular Cortex as a Convergence Area

The insula seems to act as an integrative hub for signals from the internal and external milieus (Ibanez et al. 2010; Singer et al. 2009). It also brings about global feeling states by merging contextual information together with modality-specific feeling states, uncertainty, and individual preferences (Singer et al. 2009). These experience-guided processes link intentions and motivations for goal-oriented behavior. The synergistic role of the insula seems to depend on its connections with frontotemporal regions (Couto et al. 2013), especially those involved in the regulation of context-dependent behavior (anterior cingulate and orbitofrontal cortex). Subcortical regions connected to the insula (amygdala and striatum) also play an important role in context-dependent responses (Apicella 2007). Thus, contextual effects are largely driven by reciprocal modulations between the insula and frontotemporal structures. Specifically, in the SCNM, the insula would constitute a convergence point for emotional and cognitive operations related to the coordination between external and internal states, facilitating frontotemporal interactions in social context processing.

The insula is also crucial for sensing visceral signals (Ibanez and Manes 2012). This interoceptive function has been directly linked with emotion recognition (Craig 2009), empathy (Singer et al. 2009), and decision making (Furman et al. 2013). In particular, the right anterior insula, as part of the salience network, biases attention toward emotionally salient stimuli (Fox et al. 2006). Also, convergent activity from the insula and other interoceptive regions, such as the anterior cingulate, is implicated in varied emotional and social cognition domains (Kennedy and Adolphs 2012). In this vein, Critchley et al. (2004) showed that neural activity in the anterior insula and the opercular cortex mediates explicit awareness of internal bodily processes. Moreover, negative emotional experiences correlate with interoceptive awareness (Critchley et al. 2004). In sum, interoceptive mechanisms involving the insula modulate contextual effects on social cognition.

4 The Social Context Network Model in Psychiatric and Neurological Diseases

Most psychiatric and neurological conditions are characterized by social cognition deficits and/or abnormal activation of "social brain" areas (Kennedy and Adolphs 2012). Indeed, upon a breakdown of frontotemporal dynamics, these disorders disrupt implicit social interaction (Schilbach et al. 2013) and context–target associations. The SCNM offers a framework to understand such deficits in psychiatric and neurological diseases. By way of illustration, below, we discuss how these impairments manifest in bvFTD, autism spectrum disorder (ASD), and other conditions.

4.1 Behavioral Variant Frontotemporal Dementia

Patients with bvFTD exhibit insidiously progressive changes in personality and social interaction, even before the onset of other cognitive deficits. Episodic memory, visuospatial abilities, and praxias are typically intact or relatively well preserved. Conversely, deficits in social interaction, lack of empathy, disinhibition, and impulsiveness are evident since early stages of the disease (Piguet et al. 2011).

Affected social cognition domains include emotion recognition (Lough et al. 2006), empathy (Baez et al. 2014b, 2015c), decision making (Manes et al. 2011), ToM (Torralva et al. 2007), and moral judgment (Baez et al. 2014a, 2015b). All of these impairments could reflect a general disturbance of social context processing subsequent to alterations in a broad fronto-temporo-insular network (Ibanez and Manes 2012), which is impaired in bvFTD patients (Seeley et al. 2009). For instance, relative to controls, bvFTD patients take less objection to scenarios where the protagonists deliberately inflicted pain on another, and they are less willing to exonerate protagonists for accidentally causing harm (Baez et al. 2014a). Exculpating agents who unwillingly inflict harm require a robust representation of their intentions, as the preponderant negative response to the outcome must be overridden by reference to situational cues. This ability seems to be affected in bvFTD patients.

Additionally, when performing an empathy-for-pain task, bvFTD patients do not easily discriminate between accidental and neutral or intentional situations (Baez et al 2014b, 2015c). This is expected because empathy is a contextual phenomenon affected by stimulus ambiguity (Melloni et al. 2014): since accidental pain situations are less clear and explicit, they imply more ambiguity and greater demands to ascertain the action's intentionality. In healthy subjects, ambiguity resolution benefits from contextual cues (Bar 2004), especially when the scenario involves someone in pain (Melloni et al. 2014). Accordingly, social cognition impairments in bvFTD may reflect underlying deficits in context processing.

Consistent with the brain regions proposed by the SCNM, neuroimaging studies, Viskontas et al. (2007) suggested that bvFTD patients are characterized by frontal, temporal, anterior insular, and anterior cingulate abnormalities, with pronounced orbitofrontal atrophy. Some studies (Baez et al. 2015c; Viskontas et al. 2007) have reported correlations between behavioral symptoms and orbitofrontal cortex volume, suggesting that behavior regulation depends on this region, in connection with a predominantly right-sided network involving the insula and striatum. In addition, voxel-based morphometry studies on bvFTD have revealed significant gray matter loss in the anterior insula and varied frontal areas. Particularly, a recent study (Baez et al. 2015c) showed that atrophy of limbic structures (amygdala and anterior paracingulate cortex) in bvFTD is related to impairments in intentionality comprehension, while atrophy of the orbitofrontal cortex correlates with deficits in empathic concern (see Fig. 2).

Also, resting-state fMRI of bvFTD patients has revealed abnormalities in functional connectivity among hubs of the SCNM. A recent study employing graph

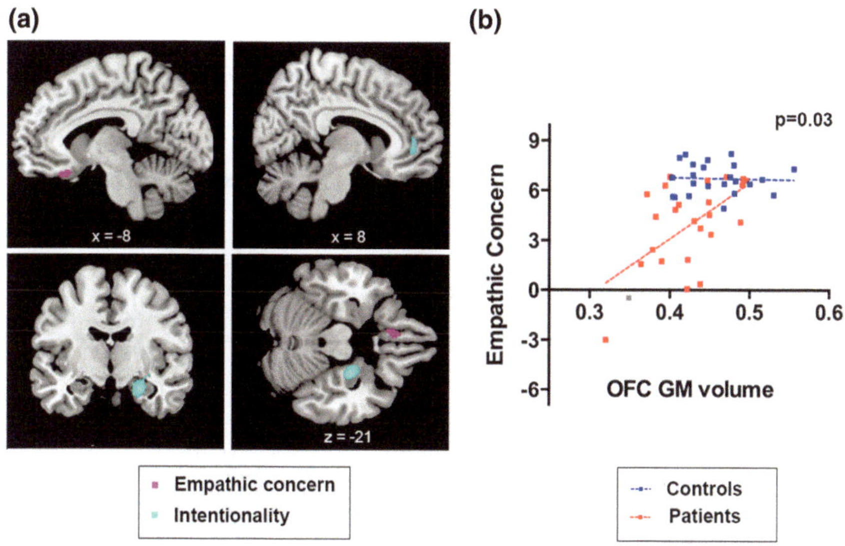

Fig. 2 Atrophied brain regions related to behavioral impairments in bvFTD patients. **a** Regions of reduced GM volume associated with intentionality comprehension of accidental harms and empathic concern for intentional harms. **b** Significant associations between GM volume in the left OFC and ratings of empathic concern for intentional harms. Reproduced with authorization from Baez et al. 2015c

theory analyses (Sedeno et al. 2015) showed that these patients have decreased network centrality in the fronto-temporo-insular network (Fig. 3). Furthermore, in agreement with the SCNM, this aberrant network organization predicted social-executive dysfunction profiles in the patients. The metric used in this study seems useful to distinguish bvFTD patients from patients with fronto-insular strokes and healthy controls. Thus, the SCNM would provide an adequate model to understand social impairments in bvFTD.

4.2 Autism Spectrum Disorder (ASD)

The term ASD describes multiple disorders characterized by impairments in social communication and repetitive behaviors. In particular, individuals with ASD exhibit deficits in emotion recognition (Falkmer et al. 2011), ToM (Cheng et al. 2015), and moral judgment (Moran et al. 2011). Individuals with ASD also showed reduced empathy as measured by self-report questionnaires (Baron-Cohen et al. 2004) and empathy for pain tasks (Baez et al. 2012). However, as suggested by Bird et al. (2010), empathy for pain impairments observed in ASD are explained by the degree of alexithymia, rather than the autism spectrum condition per se. Moreover, a recent study (Krach et al. 2015) showed that although ASD patients may have

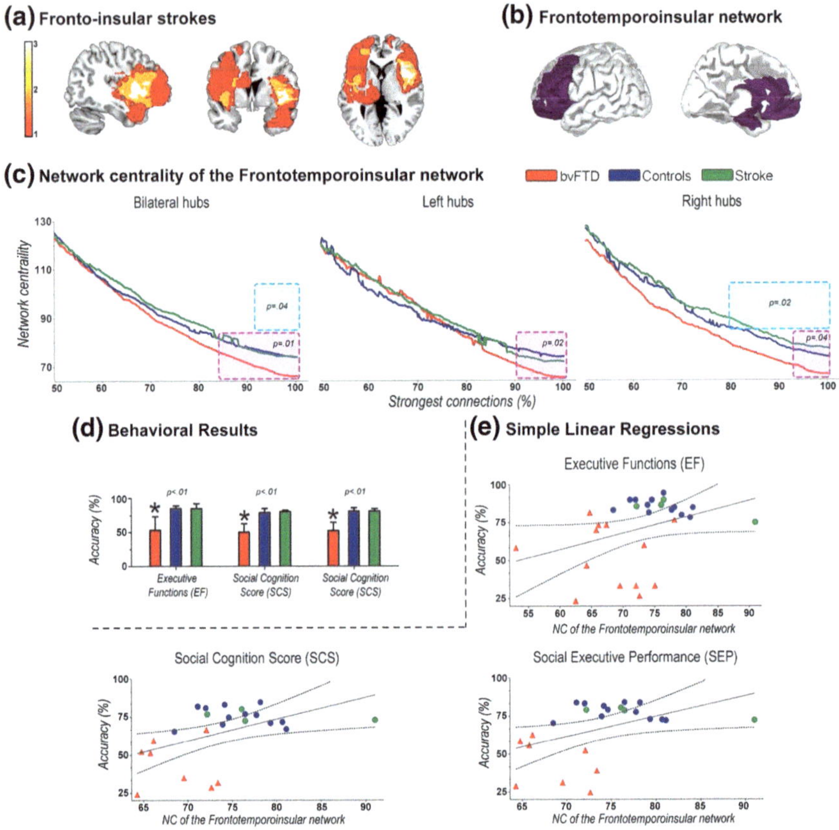

Fig. 3 Decreased centrality in the fronto-temporo-insular network in bvFTD. **a** Frontal and insular structures that were injured in stroke patients. **b** Regions included in the fronto-temporo-insular network. **c** *Pink* boxes indicate the clusters where the bvFTD patients presented decreased network centrality (NC) compared to controls. *Light blue* boxes indicate the clusters where bvFTD patients showed decreased NC compared to the fronto-insular stroke group. **d** Compared with controls and stroke patients, bvFTD patients showed impairments in executive functions and social cognition. **e** The network centrality of the bilateral fronto-temporo-insular network was associated with participants' performance in executive functions and social cognition. Reproduced with authorization from Sedeno et al. (2015)

preserved abilities to share another's physical pain, they have problems in processing more complex scenarios involving social pain. In particular, the patients' behavioral responses for complex social situations, unlike those of controls, lacked correspondence with the anterior cingulate and insular cortex activity. Instead, they presented distinctive hippocampal activity. This finding suggests that in ASD, the evaluation of complex social situations may be based on learned social scripts rather than on an interoceptive assessment of their emotional states (Krach et al. 2015).

Social cognition impairments in ASD are related to contextual sensitivity (Baez and Ibanez 2014; Klin 2000), probably reflecting a single underlying factor: deficits to implicitly encode and integrate contextual information required to construe social meanings (Baez et al. 2012). For instance, Baez et al. (2012) assessed the performance of adults with ASD in social cognition tasks with different levels of contextual dependence and involvement of real-life scenarios. Specifically, the authors assessed the following: (a) emotion recognition through the Awareness of Social Inference Test (TASIT); (b) emotional and cognitive aspects of ToM with the Reading the Mind in the Eyes Test (RMET) and the Faux Pas Test (FPT); (c) the cognitive and affective components of empathy through an Empathy for Pain Task (EPT) and the Interpersonal Reactivity Index; (d) moral judgment with a well-characterized moral task (Baez et al. 2014a); (e) self-monitoring skills by means of the Revised Self-Monitoring Scale; and (f) knowledge of social norms through an explicit (abstract and context independent) instrument, namely the Social Norms Questionnaire (SNQ).

The patients performed poorly on social cognition tasks that involved implicit encoding of socially relevant information and automatic integration of contextual information to interpret social scenarios. In this population, difficulties to implicitly recognize contextual clues may be the trigger behind social cognition impairments (Baez and Ibanez 2014; Baez et al. 2012). This possibility is reinforced by the observation of preserved performance in tasks which feature clearly defined situational elements and can be solved with relatively abstract universal rules (Baez and Ibanez 2014; Schilbach et al. 2012) (Fig. 4). A general executive deficit may be partially related to the patients' difficulties to integrate socially relevant contextual information. However, this possibility is undermined by the evidence of positive

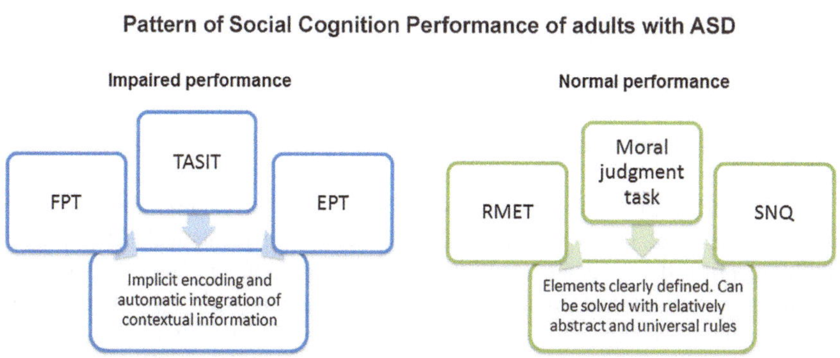

Fig. 4 Pattern of performance of adults with ASD in social cognition tasks. Adults with ASD were impaired in tasks that involved implicit encoding and automatic integration of contextual cues to interpret a given social situation. Conversely, they performed as well as controls in tasks with well defined situational elements that can be solved using relatively abstract, universal rules. *FPT* Faux Pas Test; *TASIT* The Awareness of Social Inference Test; *EPT* Empathy for Pain Task; *RMET* Reading the Mind in the Eyes Test; *SNQ* Social Norms Questionnaire. Reproduced with authorization from Baez and Ibanez (2014)

(e.g., Ozonoff et al. 1991) and null (e.g., Baez et al. 2012; Landa and Goldberg 2005) associations between executive functions and social cognition performance. Further studies are needed to reveal the specific contribution of executive functions to contextual social cognition in ASD.

In neural terms, individuals with ASD present structural and functional abnormalities in several brain structures, including the cingulate gyrus, the temporo-parietal junction, and the precuneus (Via et al. 2011), as well as frontal, temporal, and insular areas (Kosaka et al. 2010). Thus, the contextual social cognition impairments observed in ASD may also be partially explained by the abnormal functioning of the fronto-temporo-insular network proposed by the SCNM.

4.3 Other Neuropsychiatric Disorders

The previous two sections suggest that social cognition impairments may result from a general inability to integrate contextual information, triggered by abnormalities in a fronto-temporo-insular network. Contextual social cognition impairments may be also related to a general executive deficit. However, evidence is not conclusive and further studies should explore the issue in a broad array of neuropsychiatric conditions. Beyond this observation, the interpretation proposed by the SCNM may be extended to other neurological and psychiatric disorders involving deficits in social cognition domains.

Especially relevant are patients with frontal lobe damage (Mesulam 1986), whose deficits become evident in ecological tasks featuring implicit contextual cues (Burgess et al. 2009). In the same vein, short-term contextual processing tasks reveal abnormal behavioral and electrophysiological responses in patients with prefrontal compromise (Fogelson et al. 2009). Evidence from other brain diseases further emphasizes the role of context-processing skills in social cognition. In Huntington's disease, for instance, difficulties to appraise disgust and other negative emotions are reduced when target faces are accompanied by contextual clues (Baez et al. 2015a).

Similar difficulties have been observed in other psychiatric disorders. For instance, Baez et al. (2013) assessed the performance of patients with schizophrenia and bipolar disorder in social cognition tasks including different levels of contextual dependency and real-life involvement. Similar to adults with ASD, both patient groups exhibited deficits when such levels were high. The deficits were particularly severe in patients with schizophrenia, who also exhibit reduced affect recognition when faces are integrated within broader social contexts (Sasson et al. 2015). Moreover, relative to controls, these patients did not notably improve emotion recognition accuracy when faces appeared within congruent contexts, suggesting reduced benefit from complementary contextual information. In addition, although patients with schizophrenia preserve the ability to identify archetypal gestures, they are more likely than controls to perceive other people's gestures as self-referential when contextual information is ambiguous (White et al. 2016). Note that, in this population, inefficient integration of situational information may influence other deficits, supporting the idea that social context processing is impaired. Notably, such deficits, which affect both social and nonsocial domains, become more marked in ecological tasks (Baez et al. 2013), highlighting the need for novel context-sensitive approaches.

In anatomical terms, patients with schizophrenia (Wong et al. 2003) and, to a lesser degree, patients with bipolar disorder (Bearden et al. 2001) exhibit disturbances in the temporal and frontal areas proposed by the SCNM. These differences in the degree of frontotemporal disruption may explain the more severe social context-processing impairments observed in the former group. Finally, increased attentional demands in complex scenarios worsen emotion-related deficits in psychopaths and young offenders (Gonzalez-Gadea et al. 2014).

In summary, social cognition deficits present in psychiatric and neurological disorders may be partially explained by a general social context-processing impairment produced by fronto-temporo-insular network atrophy. The above-mentioned findings provide preliminary evidence for this hypothesis; however, future research should empirically test the assumptions of the SCNM, providing more refined evidence on processes and regions critically involved in contextual modulation of social cognition, and testing the model against other alternative accounts.

5 Toward an Ecological Assessment of Social Context

Altogether, the above-mentioned findings highlight the contextual appraisal disturbances as one further commonality between neurological and psychiatric conditions. As predicted by the SCNM, such deficits are caused by damage to prefrontal, insular, or temporal regions (Ibanez and Manes 2012). However, most tasks available to explore the cognitive profile of relevant populations fail to tap the ability to process implicit contextual information. Most of the traditional tests are not good "models of the world" (Burgess et al. 2009) because of their abstract, decontextualized nature. Instead, the comparatively few tasks involving real-life social scenarios have shown greater sensitivity in the clinical assessment of psychiatric and neurological populations (Baez et al. 2012, 2013, 2014b; Ibanez et al. 2014). These observations emphasize the importance of developing more ecological measures assessing contextual sensitivity.

Most social cognition experiments employ isolation paradigms (Schilbach et al. 2013), in which individual participants view pictures, words, or videos and categorize/judge their contents or infer their protagonists' mental/emotional states. In these conditions, subjects act as detached spectators of artificial situations (Garcia and Ibanez 2014). Despite its valuable contributions, this widespread approach fails to address the core of social relations, namely the deployment of interactive processes between emotionally engaged participants in contextually dynamic environments. A promising avenue of development for the field is immanent in the principles of two-person neuroscience (Schilbach et al. 2013), which suggests that interpersonal understanding is primarily a matter of social interaction and emotional engagement with others. This approach seeks to bridge "the spectatorial gap" by exploring inter-brain communication in multi-participant experimental settings, as shown in a number of studies (Garcia and Ibanez 2014). Supporting this approach, recent methodological advances (e.g., Redcay et al. 2013) have favored increased ecological validity through the study of social cognition processes in real time, looking at how people actively engage and interact with one another in social encounters.

In consequence, future studies in psychiatric and neurological populations should strictly control for context-dependent levels in social cognition tasks, ranging from context-free to context-rich paradigms with varied manipulations of situational cues. Ecological validity could be increased through methods assessing social cognition processes in real-time (e.g., Redcay et al. 2013) and spontaneous interactions between socially engaged participants (Garcia and Ibanez 2014; Schilbach et al. 2013). Ideally, protocols should also explore possible dissociations between social and nonsocial domains. In this sense, two key issues to be addressed in future research are the role of contextual information in social interactions and predictions, and the neural basis of mechanisms integrating information from social context frames. Methodologically speaking, challenges include the identification of extant social cognition tests which prove sensitive to contextual disturbances across neuropsychiatric conditions, and the development of paradigms aimed to test the three components of the SCNM. This would improve the operationalization of the model's hypotheses to be tested in psychiatric and neurological disorders compromising frontal, insular, and temporal regions. Breakthroughs in these directions could afford a robust framework to assess contextual social cognition in a wide range of populations (Ibanez and Manes 2012).

6 Conclusions

Contextual effects are inherent in daily-life social situations. The SCNM proposes a fronto-temporo-insular network responsible for processing social contextual effects. Patients with bvFTD and ASD clearly illustrate how the social cognition deficits observed in psychiatric and neurological disorders may be explained by a general social context-processing impairment triggered by abnormalities in the network postulated by the model. However, the interpretation proposed by the SCNM may be extended to other neurologic and psychiatric disorders such as stroke, Huntington's disease, schizophrenia, and bipolar disorder, among others.

Although context-processing impairments occur similarly across neurological and psychiatric conditions, most current tasks fail to capture the influence of implicit contextual information on social cognitive processes. Evidence suggests that tasks involving real-life social scenarios are more sensitive for the clinical assessment of both types of populations. This observation highlights the importance of assessing social cognition by means of ecological instruments tapping contextual sensitivity.

Acknowledgments This work was partially supported by grants from CONICET, CONICYT/FONDECYT Regular (1,130,920), FONCyT-PICT 2012-0412, 2012-1309, and the INECO Foundation FONDAP 15150012.

References

Adolphs R (2009) The social brain: neural basis of social knowledge. Annu Rev Psychol 60:693–716. doi:10.1146/annurev.psych.60.110707.163514

Apicella P (2007) Leading tonically active neurons of the striatum from reward detection to context recognition. Trends Neurosci 30:299–306. doi:10.1016/j.tins.2007.03.011

Baez S, Ibanez A (2014) The effects of context processing on social cognition impairments in adults with Asperger's syndrome. Front Neurosci 8:270. doi:10.3389/fnins.2014.00270

Baez S, Rattazzi A, Gonzalez-Gadea ML, Torralva T, Vigliecca NS, Decety J, Ibanez A (2012) Integrating intention and context: assessing social cognition in adults with Asperger syndrome. Front Hum Neurosci 6:302. doi:10.3389/fnhum.2012.00302

Baez S, Herrera E, Villarin L, Theil D, Gonzalez-Gadea ML, Gomez P, Ibanez AM (2013) Contextual social cognition impairments in schizophrenia and bipolar disorder. PLoS ONE 8: e57664. doi:10.1371/journal.pone.0057664

Baez S, Couto B, Torralva T, Sposato LA, Huepe D, Montanes P, Ibanez A (2014a) Comparing moral judgments of patients with frontotemporal dementia and frontal stroke. JAMA Neurol 71:1172–1176. doi:10.1001/jamaneurol.2014.347

Baez S, Manes F, Huepe D, Torralva T, Fiorentino N, Richter F, Ibanez A (2014b) Primary empathy deficits in frontotemporal dementia. Front Aging Neurosci 6:262. doi:10.3389/fnagi.2014.00262

Baez S, Herrera E, Gershanik O, Garcia AM, Bocanegra Y, Kargieman L, Ibanez A (2015a) Impairments in negative emotion recognition and empathy for pain in Huntington's disease families. Neuropsychologia 68:158–167. doi:10.1016/j.neuropsychologia.2015.01.012

Baez S, Kanske P, Matallana D, Montanes P, Reyes P, Slachevsky A, Ibanez A (2015b) Integration of intention and outcome for moral judgment in frontotemporal dementia: brain structural signatures. Neurodegenerative Dis. doi:10.1159/000441918

Baez S, Morales JP, Slachevsky A, Torralva T, Matus C, Manes F, Ibanez A (2015c) Orbitofrontal and limbic signatures of empathic concern and intentional harm in the behavioral variant frontotemporal dementia. Cortex J Devoted Study Nerv Syst Behav 75:20–32. doi:10.1016/j.cortex.2015.11.007

Bar M (2004) Visual objects in context. Nat Rev Neurosci 5:617–629. doi:10.1038/nrn1476

Baron-Cohen S, Wheelwright S (2004) The empathy quotient: an investigation of adults with Asperger syndrome or high functioning autism, and normal sex differences. J Autism Dev Disord 34:163–175

Barrett LF, Mesquita B, Gendron M (2011) Context in emotion perception. Curr Dir Psychol Sci 20:286–290

Bearden CE, Hoffman KM, Cannon TD (2001) The neuropsychology and neuroanatomy of bipolar affective disorder: a critical review. Bipolar Disord 3:106–150 (discussion 151-103)

Bird G, Silani G, Brindley R, White S, Frith U, Singer T (2010) Empathic brain responses in insula are modulated by levels of alexithymia but not autism. Brain J Neurol 133:1515–1525. doi:10.1093/brain/awq060

Bouton ME, Westbrook RF, Corcoran KA, Maren S (2006) Contextual and temporal modulation of extinction: behavioral and biological mechanisms. Biol Psychiatry 60:352–360. doi:10.1016/S0006-3223(06)00099-0

Burgess P, Wu H (2013) Rostral prefrontal cortex (brodmann area 10): metacognition in the brain. In: Stuss D, Knight RA (eds) Principles frontal lobe function. Oxford University Press, New York, pp 524–534

Burgess PW, Alderman N, Volle E, Benoit RG, Gilbert SJ (2009) Mesulam's frontal lobe mystery re-examined. Restor Neurol Neurosci 27:493–506. pii: 3645W28847411K64)

Cheng W, Rolls ET, Gu H, Zhang J, Feng J (2015) Autism: reduced connectivity between cortical areas involved in face expression, theory of mind, and the sense of self. Brain J Neurol 138:1382–1393. doi:10.1093/brain/awv051

Cohn M, St-Laurent M, Barnett A, McAndrews MP (2015) Social inference deficits in temporal lobe epilepsy and lobectomy: risk factors and neural substrates. Social Cogn Affect Neurosci 10:636–644. doi:10.1093/scan/nsu101

Couto B, Sedeno L, Sposato LA, Sigman M, Riccio PM, Salles A, Ibanez A (2013) Insular networks for emotional processing and social cognition: comparison of two case reports with either cortical or subcortical involvement. Cortex J Devoted Study Nerv Syst Behav 49:1420–1434. doi:10.1016/j.cortex.2012.08.006

Craig AD (2009) How do you feel–now? The anterior insula and human awareness. Nat Rev Neurosci 10:59–70. doi:10.1038/nrn2555

Critchley HD, Wiens S, Rotshtein P, Ohman A, Dolan RJ (2004) Neural systems supporting interoceptive awareness. Nat Neurosci 7:189–195. doi:10.1038/nn1176

Falkmer M, Bjällmark A, Larsson M, Falkmer T (2011) Recognition of facially expressed emotions and visual search strategies in adults with Asperger syndrome. Res Autism Spectrum Disord 5:210–217. doi:10.1016/j.rasd.2010.03.013

Fleming SM, Dolan RJ (2012) The neural basis of metacognitive ability. Philosophical transactions of the Royal Society of London. Ser B Biol sci 367:1338–1349. doi:10.1098/rstb.2011.0417

Fogelson N, Shah M, Scabini D, Knight RT (2009) Prefrontal cortex is critical for contextual processing: evidence from brain lesions. Brain 132:3002–3010. doi:10.1093/brain/awp230

Fox MD, Corbetta M, Snyder AZ, Vincent JL, Raichle ME (2006) Spontaneous neuronal activity distinguishes human dorsal and ventral attention systems. Proc Natl Acad Sci USA 103:10046–10051. doi:10.1073/pnas.0604187103

Friston KJ (2012) The history of the future of the Bayesian brain. Neuroimage 62:1230–1233. doi:10.1016/j.neuroimage.2011.10.004

Furman DJ, Waugh CE, Bhattacharjee K, Thompson RJ, Gotlib IH (2013) Interoceptive awareness, positive affect, and decision making in major depressive disorder. J Affect Disord 151:780–785. doi:10.1016/j.jad.2013.06.044

Garcia AM, Ibanez A (2014) Two-person neuroscience and naturalistic social communication: the role of language and linguistic variables in brain-coupling research. Front Psychiatry 5:124. doi:10.3389/fpsyt.2014.00124

Gonzalez-Gadea ML, Herrera E, Parra M, Gomez Mendez P, Baez S, Manes F, Ibanez A (2014) Emotion recognition and cognitive empathy deficits in adolescent offenders revealed by context-sensitive tasks. Front Hum Neurosci 8:850. doi:10.3389/fnhum.2014.00850

Greene AJ, Gross WL, Elsinger CL, Rao SM (2006) An FMRI analysis of the human hippocampus: inference, context, and task awareness. J Cogn Neurosci 18:1156–1173. doi:10.1162/jocn.2006.18.7.1156

Hesse E, Mikulan E, Decety J, Sigman M, Garcia MD, Silva W, Ibanez A (2015) Early detection of intentional harm in the human amygdala. Brain J Neurol. doi:10.1093/brain/awv336

Ibanez A, Manes F (2012) Contextual social cognition and the behavioral variant of frontotemporal dementia. Neurology 78:1354–1362

Ibanez A, Gleichgerrcht E, Manes F (2010) Clinical effects of insular damage in humans. Brain Struct Funct 214:397–410. doi:10.1007/s00429-010-0256-y

Ibanez A, Kuljis RO, Matallana D, Manes F (2014) Bridging psychiatry and neurology through social neuroscience. World Psychiatry 13:148–149. doi:10.1002/wps.20125
Johns M, Schmader T, Lickel B (2005) Ashamed to be an American? The role of identification in predicting vicarious shame for anti-arab prejudice after 9-11. Self Identity 4:331–348. doi:10.1080/15298860500145822
Kennedy DP, Adolphs R (2012) The social brain in psychiatric and neurological disorders. Trends Cogn Sci 16:559–572. doi:10.1016/j.tics.2012.09.006
Klin A (2000) Attributing social meaning to ambiguous visual stimuli in higher-functioning autism and Asperger syndrome: the social attribution task. J Child Psychol Psychiatry 41:831–846
Kosaka H, Omori M, Munesue T, Ishitobi M, Matsumura Y, Takahashi T, Wada Y (2010) Smaller insula and inferior frontal volumes in young adults with pervasive developmental disorders. Neuroimage 50:1357–1363. doi:10.1016/j.neuroimage.2010.01.085
Krach S, Kamp-Becker I, Einhauser W, Sommer J, Frassle S, Jansen A, Paulus FM (2015) Evidence from pupillometry and fMRI indicates reduced neural response during vicarious social pain but not physical pain in autism. Hum Brain Mapp 36:4730–4744. doi:10.1002/hbm.22949
Lamm C, Batson CD, Decety J (2007) The neural substrate of hum empathy: effects of perspective-taking and cognitive appraisal. J Cogn Neurosci 19:42–58. doi:10.1162/jocn.2007.19.1.42
Landa RJ, Goldberg MC (2005) Language, social, and executive functions in high functioning autism: a continuum of performance. J Autism Dev Disord 35:557–573. doi:10.1007/s10803-005-0001-1
Lough S, Kipps CM, Treise C, Watson P, Blair JR, Hodges JR (2006) Social reasoning, emotion and empathy in frontotemporal dementia. Neuropsychologia 44:950–958. doi:10.1016/j.neuropsychologia.2005.08.009
Manes F, Torralva T, Ibanez A, Roca M, Bekinschtein T, Gleichgerrcht E (2011) Decision-making in frontotemporal dementia: clinical, theoretical and legal implications. Dement Geriatr Cogn Disord 32:11–17. doi:10.1159/000329912
Masten CL, Morelli SA, Eisenberger NI (2011) An fMRI investigation of empathy for 'social pain' and subsequent prosocial behavior. NeuroImage 55:381–388. doi:10.1016/j.neuroimage.2010.11.060
Melloni M, Lopez V, Ibanez A (2014) Empathy and contextual social cognition. Cogn Affect Behav Neurosci 14:407–425. doi:10.3758/s13415-013-0205-3
Mesulam MM (1986) Frontal cortex and behavior. Ann Neurol 19:320–325. doi:10.1002/ana.410190403
Mesulam MM (2002) The human frontal lobes: transcending the default mode through contingent encoding. Principle Frontal Lobe Funct 1:8–31
Moran JM, Young LL, Saxe R, Lee SM, O'Young D, Mavros PL, Gabrieli JD (2011) Impaired theory of mind for moral judgment in high-functioning autism. Proc Natl Acad Sci USA 108:2688–2692. doi:10.1073/pnas.1011734108
Muller-Pinzler L, Rademacher L, Paulus FM, Krach S (2015) When your friends make you cringe: social closeness modulates vicarious embarrassment-related neural activity. Soc Cogn Affect Neurosci. doi:10.1093/scan/nsv130
Ozonoff S, Pennington BF, Rogers SJ (1991) Executive function deficits in high-functioning autistic individuals: relationship to theory of mind. J Child Psychol Psychiatry 32:1081–1105
Paulus FM, Muller-Pinzler L, Jansen A, Gazzola V, Krach S (2015) Mentalizing and the role of the posterior superior temporal sulcus in sharing others' embarrassment. Cereb Cortex 25:2065–2075. doi:10.1093/cercor/bhu011
Piguet O, Hornberger M, Mioshi E, Hodges JR (2011) Behavioural-variant frontotemporal dementia: diagnosis, clinical staging, and management. Lancet Neurol 10:162–172. doi:10.1016/S1474-4422(10)70299-4
Redcay E, Dodell-Feder D, Mavros PL, Kleiner M, Pearrow MJ, Triantafyllou C, Saxe R (2013) Atypical brain activation patterns during a face-to-face joint attention game in adults with autism spectrum disorder. Hum Brain Mapp 34:2511–2523. doi:10.1002/hbm.22086

Sasson NJ, Pinkham AE, Weittenhiller LP, Faso DJ, Simpson C (2015) Context effects on facial affect recognition in schizophrenia and autism: behavioral and eye-tracking evidence. Schizophr Bull. doi:10.1093/schbul/sbv176

Schilbach L, Eickhoff SB, Cieslik EC, Kuzmanovic B, Vogeley K (2012) Shall we do this together? Social gaze influences action control in a comparison group, but not in individuals with high-functioning autism. Autism Int J Res Pract 16:151–162. doi:10.1177/1362361311409258

Schilbach L, Timmermans B, Reddy V, Costall A, Bente G, Schlicht T, Vogeley K (2013) Toward a second-person neuroscience. Behav Brain Sci 36:393–414. doi:10.1017/S0140525X12000660

Sedeno L, Couto B, Garcia-Cordero I, Melloni M, Baez S, Morales JP, … Ibanez, A (2015) Brain network organization and social executive performance in frontotemporal dementia. J Int Neuropsychol Soc JINS 22:250–262. doi:10.1017/S1355617715000703

Seeley WW, Crawford RK, Zhou J, Miller BL, Greicius MD (2009) Neurodegenerative diseases target large-scale human brain networks. Neuron 62:42–52. doi:10.1016/j.neuron.2009.03.024

Seth AK, Suzuki K, Critchley HD (2011) An interoceptive predictive coding model of conscious presence. Front Psychol 2:395. doi:10.3389/fpsyg.2011.00395

Sigala N, Kusunoki M, Nimmo-Smith I, Gaffan D, Duncan J (2008) Hierarchical coding for sequential task events in the monkey prefrontal cortex. Proc Natl Acad Sci USA 105:11969–11974. doi:10.1073/pnas.0802569105

Singer T, Seymour B, O'Doherty JP, Stephan KE, Dolan RJ, Frith CD (2006) Empathic neural responses are modulated by the perceived fairness of others. Nature 439:466–469. doi:10.1038/nature04271

Singer T, Critchley HD, Preuschoff K (2009) A common role of insula in feelings, empathy and uncertainty. Trends Cogn Sci 13:334–340. doi:10.1016/S1364-6613(09)00135-1

Todorov A, Harris LT, Fiske ST (2006) Toward socially inspired social neuroscience. Brain Res 1079:76–85. doi:10.1016/j.brainres.2005.12.114

Torralva T, Kipps CM, Hodges JR, Clark L, Bekinschtein T, Roca M, Manes F (2007) The relationship between affective decision-making and theory of mind in the frontal variant of fronto-temporal dementia. Neuropsychologia 45:342–349. doi:10.1016/j.neuropsychologia.2006.05.031

Via E, Radua J, Cardoner N, Happe F, Mataix-Cols D (2011) Meta-analysis of gray matter abnormalities in autism spectrum disorder: should Asperger disorder be subsumed under a broader umbrella of autistic spectrum disorder? Arch Gen Psychiatry 68:409–418. doi:10.1001/archgenpsychiatry.2011.27

Viskontas IV, Possin KL, Miller BL (2007) Symptoms of frontotemporal dementia provide insights into orbitofrontal cortex function and social behavior. Ann NY Acad Sci 1121:528–545. doi:10.1196/annals.1401.025

Watanabe M, Sakagami M (2007) Integration of cognitive and motivational context information in the primate prefrontal cortex. Cereb Cortex 17(Suppl 1):101–109. doi:10.1093/cercor/bhm067

White TP, Borgan F, Ralley O, Shergill SS (2016) You looking at me?: Interpreting social cues in schizophrenia. Psychol Med 46:149–160. doi:10.1017/S0033291715001622

Wong AH, Van Tol HH (2003) Schizophrenia: from phenomenology to neurobiology. Neurosci Biobehav Rev 27:269–306

Social-Cognitive Deficits in Schizophrenia

Daniela Mier and Peter Kirsch

Abstract Patients with schizophrenia not only suffer from prototypical psychotic symptoms such as delusions and hallucinations and from cognitive deficits, but also from tremendous deficits in social functioning. However, little is known about the interplay between the cognitive and the social-cognitive deficits in schizophrenia. Our chapter gives an overview on behavioral, as well as functional imaging studies on social cognition in schizophrenia. Main findings on cognitive and motivational deficits in schizophrenia are reviewed and introduced within the context of the dopamine hypothesis of schizophrenia. The reviewed findings suggest that disturbed "social brain" functioning in schizophrenia, depending on the specific context, can either lead to a neglect of the emotions and intentions of others or to the false attribution of these emotions and intentions in an emotionally neutral social content. We integrate these findings with the current knowledge about aberrant dopaminergic firing in schizophrenia by presenting a comprehensive model explaining core symptoms of the disorder. The main implication of the presented model is that neither cognitive-motivational, nor social-cognitive deficits alone cause schizophrenia symptoms, but that symptoms only emerge by the interplay of disturbed social brain functioning with aberrant dopaminergic firing.

Keywords Schizophrenia · Social cognition · Social brain · Amygdala · Superior temporal sulcus · Dopamine hypothesis · Aberrant salience · Ventral striatum · Prefrontal cortex

D. Mier (✉) · P. Kirsch
Department of Clinical Psychology, Central Institute of Mental Health, Medical Faculty Mannheim, University of Heidelberg, Mannheim, Germany
e-mail: Daniela.Mier@zi-mannheim.de

P. Kirsch
e-mail: Peter.Kirsch@zi-mannheim.de

Curr Topics Behav Neurosci (2016) 30: 397–410
DOI 10.1007/7854_2015_427
Published Online: 01 April 2016

Contents

1 Introduction 398
2 Schizophrenia 398
3 Social Cognition: Behavior 399
4 Social Cognition: Brain Functioning 400
5 Dopamine, Cognition, and Social Cognition 402
6 Social Cognition: Model 405
References 406

1 Introduction

In the neurosciences, social cognition became popular with the seminal paper by Brothers from 1990 that established the concept of a "social brain" in the field. She summarized basic processes of social perception as so-called social cognition and located them in distinct brain regions, namely the fusiform gyrus, superior temporal sulcus, and amygdala (Brothers 1990). In this article, Brother also argued that disturbances in the "social brain" lead to mental disorders such as paranoid psychoses and autism. While Brothers' work inspired more research on social processing in psychiatric diseases in general, the "social brain" theory of Burns fired the interest particularly in the area of schizophrenia research. According to Burns, schizophrenia is a costly by-product of the evolution of the social brain (Burns 2004). According to this very clear and provocative hypothesis, the "social brain" is central to the emergence of schizophrenia and with this should also be the central focus of research.

The present chapter provides a brief review of studies investigating the "social brain" in schizophrenia and combines the findings in a model with the dopamine hypothesis, a very prominent heuristic for the occurrence of schizophrenia.

2 Schizophrenia

Schizophrenia is a severe psychiatric disorder that is highly heritable (Tandon et al. 2008), often presents with a chronic course (Tandon et al. 2009), and leads to massive psychosocial impairments for the patients. The core of the most common etiology models of schizophrenia is the vulnerability–stress hypothesis (van Winkel et al. 2008). It is assumed that a combination of pre-, peri-, and postnatal stressors results in a vulnerability to develop a psychosis. A psychosis can be defined as a tremendous change in perception, cognition, behavior, and feeling that can occur in different psychiatric disorders, and is clearly linked to the positive syndrome of schizophrenia. If a person with a high vulnerability encounters high levels of stress in later life, such a psychosis can be triggered. The dominant theory to explain the symptoms that can occur in the course of schizophrenia is the dopamine hypothesis (Abi-Dargham 2004). The dopamine hypothesis states that schizophrenia is associated with a tonic prefrontal dopamine deficiency and a state-dependent subcortical dopamine excess (Grace 1991).

In general, the disorder can be characterized by the syndrome triad of positive and negative symptoms and disorganization. The positive syndrome comprises hallucinations and delusions, the negative syndrome includes symptoms such as anhedonia, apathy, and attention deficits, and disorganization is characterized by bizarre behavior and disturbances of the form of thought (Liddle 1987). While the positive symptoms are predominant in the acute psychotic state and can be related to a hyperactive subcortical dopaminergic system, negative symptoms often persist for years, independent of an acute psychosis. Disorganization is mostly exaggerated during an acute psychosis, and often persists in the chronic state, too. Both negative symptoms and disorganization can be linked to hypoactive prefrontal dopamine transmission.

In the following, the literature on social-cognitive deficits in schizophrenia is presented. In addition to Burns (2004) who described the "social brain" dysfunction in schizophrenia as an innate characteristic, Frith and Corcoran suggested that social-cognitive deficits in schizophrenia can vary and depend on the state of the disease (Frith and Corcoran 1996). In their work, the authors focused on the ability to mentalize, or in other words to build, a Theory of Mind (ToM), i.e., the ability to infer wishes, desires, and intentions of oneself and of others (Premack and Woodruff 1978). In this context, Frith and Corcoran proposed that positive symptoms are linked to hypermentalizing and negative symptoms to hypomentalizing. According to this categorization, positive symptoms should result in a pattern of additional emotion and intention attributions to other people, while negative symptoms should result in a lack of emotion and intention attributions.

3 Social Cognition: Behavior

Social cognition was described by Brothers as all perceptual processes that lead to the accurate recognition of dispositions and intentions of oneself and of others (Brothers 1990). Derived from the work of Brothers, face processing, emotion recognition, and ToM can be defined as the core processes of social cognition (Mier et al. 2010b). Schizophrenia has been associated with deficits in all of these processes. Impaired emotion recognition in schizophrenia has already been described in the 1980s (Novic et al. 1984), and publications on this topic show an increasing frequency since then. Patients with schizophrenia were found to perform worse in the recognition of all kinds of emotions, as well as in the recognition of neutral facial expressions (Edwards et al. 2002; Kohler et al. 2003). Interestingly, recent findings suggest a negative bias, i.e., the tendency to perceive neutral or positive stimuli as negative, as causal to the deficit in the recognition of neutral facial expressions (Mier et al. 2014). Moreover, it was shown that patients with schizophrenia have impairments in complex social-cognitive abilities, such as ToM (Mier et al. 2010c) and empathy (Lehmann et al. 2014). Studies on empathy reveal that patients with schizophrenia are impaired in the cognitive domain, but unaffected with regard to the affective domain (Lehmann et al. 2014), a pattern comparable to recent findings in autism (Krach et al. 2015) (for a definition of the different empathy processes please see Chap. Models, Mechanisms and Moderators Dissociating Empathy and Theory of Mind of this book).

However, until now, little evidence exists for the hypothesis of Frith and Corcoran: hypomentalizing in patients with negative symptoms and hypermentalizing in patients with positive symptoms. While the authors themselves provided evidence for this phenomenon across several studies, most of the behavioral studies from other research groups do not present evidence for this dichotomy of mentalizing deficits. The reason is not necessarily that the classification would not be valid, but that most studies use smaller sample sizes and focus on general social-cognitive deficits rather than investigating the specific social-cognitive deficits in a large sample with subgroups of patients.

A very interesting finding that adds to the idea of hypermentalizing in schizophrenia is the negative bias in response to neutral social stimuli. While it was shown for decades that patients with schizophrenia have a deficit in the recognition of neutral facial expressions, only more recently the pattern of misinterpretations has gained interest. Haralanova and colleagues showed that patients with schizophrenia rate neutral social scenes as more arousal eliciting (Haralanova et al. 2012). Pinkham and colleagues found schizophrenia patients with acute paranoia to attribute anger more often to neutral facial expressions than patients with schizophrenia without acute paranoia (Pinkham et al. 2011). Hooker and colleagues showed in a study on emotional priming that patients with schizophrenia rate neutral facial expressions more often as untrustworthy when primed with negative emotional stimuli (Hooker et al. 2011). Finally, our group found a negative bias to neutral facial expressions in remitted patients with schizophrenia. Patients rated neutral and positive facial expressions more often as anger, fear, or disgust than healthy controls (Mier et al. 2014). These results are in line with the idea of hypermentalizing, because they indicate that patients with schizophrenia tend to perceive (negative) emotions in stimuli that are emotionally neutral.

Complementing the behavioral studies, results from functional brain imaging can clearly add evidence to the idea of negative bias and hypermentalizing in schizophrenia.

4 Social Cognition: Brain Functioning

While in the beginning of functional imaging studies of schizophrenia the "social brain" was rarely addressed, these days many studies are focusing on emotional and social processing, either merely reflecting a general trend in the field, or also reflecting the growing awareness of the relevance of dysfunctions of the "social brain" in schizophrenia for understanding and finally treating the disease.

At this point, it should be noted that there is for sure nothing like a "social brain". Each brain region has multiple functions and is involved in a vast number of processes (attention, sensory processing, motor processing, etc.). However, there are core areas that are consistently found to be involved in social processing, and these are the fusiform gyrus, the amygdala, the superior temporal sulcus, the inferior prefrontal gyrus, and the medial prefrontal gyrus.

First of all, it should be mentioned that disturbances in terms of activity and/or connectivity have been found in each of the areas of the "social brain" in schizophrenia. The fusiform gyrus has been found to be hypoactive during emotion recognition, suggesting deficits in face processing (Quintana et al. 2003). Moreover, the amygdala was demonstrated to be hypoactive during emotion recognition tasks (Anticevic et al. 2012), a pattern that was also found for the superior temporal sulcus (Habel et al. 2010). In the context of ToM studies, inferior prefrontal gyrus (Russell et al. 2000) and medial prefrontal gyrus (Lee et al. 2011) hypoactivation was found.

Interestingly, while evidence from facial emotion processing, as well as from ToM tasks, suggests hypoactivation in the "social brain", this pattern seems to be reversed for social scenes that convey no emotional or intentional meaning, as well as for the processing of neutral facial expressions. Backasch and colleagues demonstrated that patients with schizophrenia have increased superior temporal sulcus activity for noncooperative actions (two actors both handling an object), but not for cooperative behavior (two actors handling an object together (Backasch et al. 2013)). Walter and colleagues demonstrated enhanced superior temporal sulcus activity in patients with schizophrenia in the control condition of a ToM task (Walter et al. 2009). In a recent study from Ciaramidaro, this finding was replicated (Ciaramidaro et al. 2015). In addition, Ciaramidaro and colleagues revealed that patients have not only increased activation of the superior temporal sulcus in the control task, but also increased connectivity between right superior temporal sulcus and ventromedial prefrontal cortex (Ciaramidaro et al. 2015). Interestingly, the authors also presented a comparison with patients with autism and demonstrated that this hyperengagement of the superior temporal sulcus is specific for patients with schizophrenia.

These studies relied on intentional and neutral social scenes. Studies investigating face processing found a comparable picture. With a social-cognitive paradigm, we revealed hyperactivation in the superior temporal sulcus for neutral face processing and emotion recognition, but not for ToM in patients with schizophrenia (Mier et al. 2010c). A similar pattern was found by Pinkham and colleagues who showed that the superior temporal sulcus was hyperactive for age judgments of neutral faces, but not for trustworthiness judgments and that this effect was specific for patients with schizophrenia (in comparison with patients with autism) (Pinkham et al. 2008).

Anticevic and colleagues showed in their meta-analysis on emotion processing in schizophrenia that patients with schizophrenia have a hypoactive amygdala in response to negative emotional stimuli, but that the opposite is true for neutral stimulus material, i.e., that patients rather show amygdala hyperactivation to these stimuli (Anticevic et al. 2012). In line with the findings from Anticevic and colleagues, several groups investigating the processing of neutral facial expressions in an emotion recognition context found amygdala hyperactivation in schizophrenia. For example, Holt et al. 2006 used a passive viewing task with facial expressions and found amygdala hyperactivation, in particular for the neutral expressions in patients with schizophrenia (Holt et al. 2006). During neutral face processing, our patient sample not only showed hyperactivation of the superior temporal sulcus, but also of the amygdala (Mier et al. 2010c). In addition, using another experimental design with an adaptive emotion recognition task, we found a significant correlation between amygdala activation in response to neutral facial expressions and negative bias in patients with schizophrenia (Mier et al. 2014).

In summary, in comparison to healthy controls, patients with schizophrenia tend to have increased activation of amygdala and superior temporal sulcus in response to social stimuli without an emotional (or intentional) meaning and rather decreased activation in these and frontal areas when the task affords active cognitive processing of emotions or intentions (such as differentiating between negative emotions, or ToM). Taking all these findings together, it could be argued that activation of the "social brain" in schizophrenia is context dependent. While hyperactivation in non-emotional or non-social contexts produces hypermentalizing leading to positive symptoms like delusions, hypoactivation in an emotional or social context leads to hypomentalizing and negative symptoms. Interestingly, both hypo- and hyperactivation can be observed in the same patients at the same time (Mier et al. 2010c), suggesting that (a) there is more than a simple up and down of activation that characterizes the "social brain" in schizophrenia and (b) hypoactivation, as well as hyperactivation, is a schizophrenia trait that makes patients prone to the development of positive or negative symptoms (see Sect. 6 and Fig. 1 for a comprehensive overview).

However, in schizophrenia, not only the "social brain" is affected, but patients also suffer from severe motivational and cognitive deficits. To increase our knowledge of social dysfunctions in schizophrenia, it is important to understand how social cognition interacts with motivational and cognitive processing, leading to the massive social-cognitive impairments that occur in schizophrenia.

5 Dopamine, Cognition, and Social Cognition

Before social cognition came into the focus of schizophrenia research, most of the research focused on cognitive deficits. Evidence exists that the cognitive deficits occur early in the development of the disorder and are progredient in the course of the disease (Napal et al. 2012). These cognitive deficits range from impairments in attention and memory to affected executive functions. Moreover, in agreement with the first description of schizophrenia as "dementia praecox," already the earliest

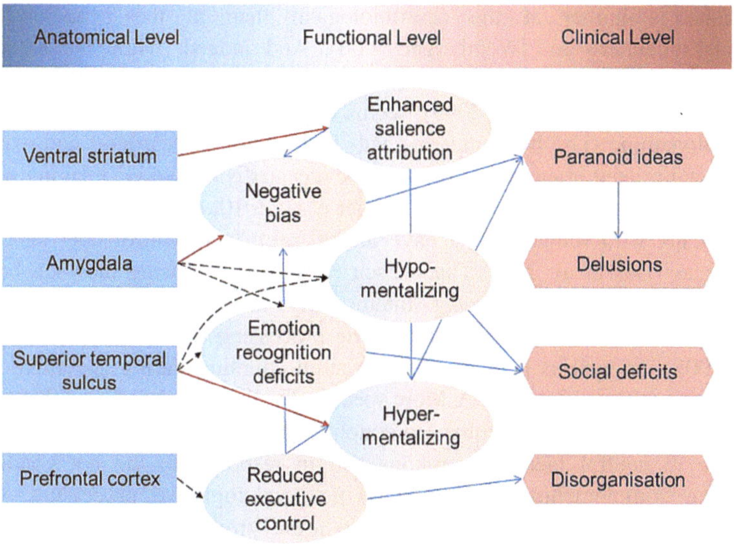

Fig. 1 Model combining the social brain with the dopamine hypothesis to explain the development of schizophrenia pathology. *Note*: *Blue arrows* Leads to, *Red arrows* Hyperactivation, *Dotted arrows* Hypoactivation

findings on schizophrenia demonstrated a cognitive decline in terms of intelligence (Aylward et al. 1984) and general brain volume (Shenton et al. 2001).

The cognitive deficits in schizophrenia, and especially those within the context of decision making and executive functions, have been associated with alterations in dopaminergic processing (Abi-Dargham et al. 2002; Rausch et al. 2014). One of the most prominent theories is based on the assumption that schizophrenia is not only associated with "too much" subcortical dopamine, but also with "too little" prefrontal dopamine (Knable and Weinberger 1997). The hyperdopaminergic subcortical state should cause positive symptoms, and the hypodopaminergic prefrontal state should cause negative symptoms, including deficits in executive functions. In line with this theory, Kapur (2003, 2005) described psychosis as a state of aberrant salience (Kapur 2003; Kapur et al. 2005). According to him, psychosis and, in particular, delusions develop in consequence of an attempt to make sense of the environment and the actual (internal) state. The random firing of dopaminergic neurons would lead to faulty attributions of salience to (neutral) objects or situations. This perceived salience of something that is usually not of importance causes a conflict that has to be cognitively solved. And this solution could be the first delusional thought. In case this process is repeated in a vicious circle, it can eventually result in a full delusion.

One option to investigate the association between dopamine and brain functioning is the imaging genetics approach. Imaging genetics studies are based on the intermediate phenotype concept that assumes that the penetrance of genetic variations is higher at the neurobiological than at the behavioral level (Meyer-Lindenberg and Weinberger 2006). And indeed, while the association between *COMT* variations and executive functions (Barnett et al. 2008), or even schizophrenia (Okochi et al. 2009), seems to be weak, the link between less prefrontal dopamine (i.e., by comparing val- with met-allele carriers) and executive functioning-related brain activation could be repeatedly shown via imaging genetics studies focusing on *COMT* variations (Mier et al. 2010a). Moreover, it was shown that patients with schizophrenia have altered prefrontal control over subcortical brain areas (Das et al. 2007; Diaconescu et al. 2011), closing the loop between prefrontal and limbic functioning. Interestingly, these studies give evidence for altered PFC–amygdala coupling during the processing of facial expressions (Das et al. 2007), as well as between PFC–striatum coupling during appetitive conditioning (Diaconescu et al. 2011). More specifically, these studies found that patients with schizophrenia showed enhanced coupling in the non-salient condition (i.e., in response to neutral facial expressions and in response to the not conditioned stimulus), complementing the findings of social-cognitive processing, with a stronger coupling in schizophrenia patients in the "irrelevant" task condition.

Since findings on cognitive and social-cognitive deficits in schizophrenia are mainly separated, the question remains how cognition and social cognition are related. First of all, on a neural level, there is no substantial overlap between "executive" and "social" networks. While executive functions have been mainly associated with dorsolateral and parietal functioning, the processing of social stimuli has been associated with the posterior superior temporal sulcus, fusiform gyrus, and amygdala. Both processes are associated with inferior prefrontal functioning, but with evidence for separable parts (Liakakis et al. 2011). One area that has been associated with both processes is the medial prefrontal cortex/anterior cingulate cortex (Amodio and Frith 2006; Paus 2001). Activation in the anterior cingulate cortex might reflect the cognitive efforts involved to solve social-cognitive tasks (Mier et al. 2010b), suggesting more social-cognitive deficits in patients with pronounced cognitive deficits. And indeed, patients with strong disorganization were found to be more impaired in social-cognitive tasks (Brune and Schaub 2012). Studies statistically controlling cognitive dysfunction, however, still find a strong association between social-cognitive deficits and schizophrenia (Pickup 2008). In conclusion, there is evidence for aberrant salience processing and for deficient cognitive and social-cognitive processing in schizophrenia, but the link between these processes is completely understood neither on the neural nor on the behavioral level. Hence, the question how these processes influence each other is open for future research. Moreover, it is still unresolved how these processes together cause psychosis. The last paragraph of this chapter will present a model showing how these processes in interaction might lead to the emergence of psychosis.

6 Social Cognition: Model

Summarizing the findings reviewed in this chapter, there are several main processes on the behavioral level where disturbances in schizophrenia can be identified and related to a dysfunctions in an associated neural system:

(a) Emotion recognition ↔ Amygdala;
(b) Intention recognition ↔ Superior temporal sulcus;
(c) Salience processing ↔ Ventral striatum;
(d) Executive functioning ↔ Prefrontal cortex.

Interestingly, the pattern of hypo- or hyperactivation of these structures is depending on the context of the task. As reviewed above, patients with schizophrenia show amygdala and superior temporal sulcus hyperactivation for emotionally neutral social stimuli, and amygdala and prefrontal hypoactivation for social stimuli that are presented in a mentalizing relevant context. Staying in the context of the (admittedly very phrenological) classification of functions to brain regions, we propose the following model that integrates evidence from the cognitive and social-cognitive domain (Fig. 1).

Patients with schizophrenia are prone to attributing negative emotions (amygdala) and intentions (superior temporal sulcus) to neutral social stimuli. Chaotic dopamine transmission in the ventral striatal areas randomly results in aberrant salience when these stimuli are encountered. Reduced prefrontal activation (inferior prefrontal gyrus and medial prefrontal gyrus) and connectivity to the limbic system result in reduced cognitive control over the emotional processing of these stimuli. Together, these processes lead to paranoid misperceptions of neutral social stimuli. Importantly, there is evidence that deficient social cognition is a schizophrenia trait and might present an intermediate phenotype (Derntl and Habel 2011). In the context of our model, the negative bias and the hypermentalizing, however, would not lead to paranoid ideas in the absence of the dysfunctioning of the dopaminergic system. Hence, depending on the state of the disease and the context, there is more or less negative bias and hypermentalizing, and depending on whether reinforced by dopamine or not and the amount of prefrontal control, these social-cognitive misinterpretations cause psychosis or not. In addition, the model not only covers positive psychopathology, but also parts of the negative syndrome and disorganization by relating emotion recognition deficits and hypomentalization to social deficits and reduced executive control to disorganization.

This model has several parts that can be experimentally investigated in patients with schizophrenia, as well as in healthy participants varying in dopaminergic polymorphisms. For example, the relationship between ventral striatum dopamine and negative bias or hypermentalizing can be nicely investigated applying positron emission tomography (PET) or an imaging genetics approach. The application of PET would allow assessing the binding of dopamine in the ventral striatum and correlating the binding strength with behavior in a social-cognitive task, while imaging genetics would allow investigating the association between variations in polymorphisms of the dopamine transporter and activation in the ventral striatum during a social-cognitive task. Further, path models could be applied for investigating the association between amygdala activity, emotion recognition performance, and social deficits.

Importantly, concluded from the model, deficits in the "social brain" are the basis for paranoid thoughts but lead only via dopaminergic dysfunction to psychosis. While this model is in line with the dopamine hypothesis of schizophrenia, it expands the view by including social cognition as an additional target. Hence, targeting social cognition in therapy can be of tremendous importance to reduce the risk for psychosis. Current studies investigating the effect of social-cognitive trainings already reveal promising results with regard to activation of the "social brain" (Hooker et al. 2013), as well as with regard to social functioning and delusional symptoms (Moritz et al. 2011). However, most of these reports are preliminary studies, requiring further research to answer the question which components of social-cognitive trainings are the most efficient (Roberts et al. 2014). Moreover, psychopharmacological interventions explicitly focusing on amygdala and superior temporal sulcus dysfunctions should be developed and might present a valuable addition not only to reduce the symptoms of an acute psychosis but also to prevent psychosis.

References

Abi-Dargham A (2004) Do we still believe in the dopamine hypothesis? New data bring new evidence. Int J Neuropsychopharmacol 7(Suppl 1):S1–S5

Abi-Dargham A, Mawlawi O, Lombardo I, Gil R, Martinez D, Huang Y, Hwang DR, Keilp J, Kochan L, Van Heertum R, Gorman JM, Laruelle M (2002) Prefrontal dopamine D1 receptors and working memory in schizophrenia. J Neurosci Off J Soc Neurosci 22:3708–3719

Amodio DM, Frith CD (2006) Meeting of minds: the medial frontal cortex and social cognition. Nat Rev Neurosci 7:268–277

Anticevic A, Van Snellenberg JX, Cohen RE, Repovs G, Dowd EC, Barch DM (2012) Amygdala recruitment in schizophrenia in response to aversive emotional material: a meta-analysis of neuroimaging studies. Schizophr Bull 38:608–621

Aylward E, Walker E, Bettes B (1984) Intelligence in schizophrenia: meta-analysis of the research. Schizophr Bull 10:430–459

Backasch B, Straube B, Pyka M, Klohn-Saghatolislam F, Muller MJ, Kircher TT, Leube DT (2013) Hyperintentionality during automatic perception of naturalistic cooperative behavior in patients with schizophrenia. Soc Neurosci 8:489–504

Barnett JH, Scoriels L, Munafo MR (2008) Meta-analysis of the cognitive effects of the catechol-O-methyltransferase gene Val158/108Met polymorphism. Biol Psychiatry 64:137–144

Brothers L (1990) The social brain: a project for integrating primate behavior and neurophysiology in a new domain. Concepts Neurosci 1:27–51

Brune M, Schaub D (2012) Mental state attribution in schizophrenia: what distinguishes patients with "poor" from patients with "fair" mentalising skills? Eur Psychiatry J Assoc Eur Psychiatr 27:358–364

Burns JK (2004) An evolutionary theory of schizophrenia: cortical connectivity, metarepresentation, and the social brain. Behav Brain Sci 27:831–855; discussion 855–885

Ciaramidaro A, Bolte S, Schlitt S, Hainz D, Poustka F, Weber B, Bara BG, Freitag C, Walter H (2015) Schizophrenia and autism as contrasting minds: neural evidence for the hypo-hyper-intentionality hypothesis. Schizophr Bull 41:171–179

Das P, Kemp AH, Flynn G, Harris AW, Liddell BJ, Whitford TJ, Peduto A, Gordon E, Williams LM (2007) Functional disconnections in the direct and indirect amygdala pathways for fear processing in schizophrenia. Schizophr Res 90:284–294

Derntl B, Habel U (2011) Deficits in social cognition: a marker for psychiatric disorders? Eur Arch Psychiatry Clin Neurosci 261(Suppl 2):S145–S149

Diaconescu AO, Jensen J, Wang H, Willeit M, Menon M, Kapur S, McIntosh AR (2011) Aberrant effective connectivity in schizophrenia patients during appetitive conditioning. Front Hum Neurosci 4:239

Edwards J, Jackson HJ, Pattison PE (2002) Emotion recognition via facial expression and affective prosody in schizophrenia: a methodological review. Clin Psychol Rev 22:789–832

Frith CD, Corcoran R (1996) Exploring theory of mind in people with schizophrenia. Psychol Med 26:521–530

Grace AA (1991) Phasic versus tonic dopamine release and the modulation of dopamine system responsivity: a hypothesis for the etiology of schizophrenia. Neuroscience 41:1–24

Habel U, Chechko N, Pauly K, Koch K, Backes V, Seiferth N, Shah NJ, Stocker T, Schneider F, Kellermann T (2010) Neural correlates of emotion recognition in schizophrenia. Schizophr Res 122:113–123

Haralanova E, Haralanov S, Beraldi A, Moller HJ, Hennig-Fast K (2012) Subjective emotional over-arousal to neutral social scenes in paranoid schizophrenia. Eur Arch Psychiatry Clin Neurosci 262:59–68

Holt DJ, Kunkel L, Weiss AP, Goff DC, Wright CI, Shin LM, Rauch SL, Hootnick J, Heckers S (2006) Increased medial temporal lobe activation during the passive viewing of emotional and neutral facial expressions in schizophrenia. Schizophr Res 82:153–162

Hooker CI, Tully LM, Verosky SC, Fisher M, Holland C, Vinogradov S (2011) Can I trust you? Negative affective priming influences social judgments in schizophrenia. J Abnorm Psychol 120:98–107

Hooker CI, Bruce L, Fisher M, Verosky SC, Miyakawa A, D'Esposito M, Vinogradov S (2013) The influence of combined cognitive plus social-cognitive training on amygdala response during face emotion recognition in schizophrenia. Psychiatry Res 213:99–107

Kapur S (2003) Psychosis as a state of aberrant salience: a framework linking biology, phenomenology, and pharmacology in schizophrenia. Am J Psychiatry 160:13–23

Kapur S, Mizrahi R, Li M (2005) From dopamine to salience to psychosis–linking biology, pharmacology and phenomenology of psychosis. Schizophr Res 79:59–68

Knable MB, Weinberger DR (1997) Dopamine, the prefrontal cortex and schizophrenia. J Psychopharmacol 11:123–131

Kohler CG, Turner TH, Bilker WB, Brensinger CM, Siegel SJ, Kanes SJ, Gur RE, Gur RC (2003) Facial emotion recognition in schizophrenia: intensity effects and error pattern. Am J Psychiatry 160:1768–1774

Krach S, Kamp-Becker I, Einhauser W, Sommer J, Frassle S, Jansen A, Rademacher L, Muller-Pinzler L, Gazzola V, Paulus FM (2015) Evidence from pupillometry and fMRI indicates reduced neural response during vicarious social pain but not physical pain in autism. Hum Brain Mapp 36(11):4730–4744

Lee J, Quintana J, Nori P, Green MF (2011) Theory of mind in schizophrenia: exploring neural mechanisms of belief attribution. Soc Neurosci 6:569–581

Lehmann A, Bahcesular K, Brockmann EM, Biederbick SE, Dziobek I, Gallinat J, Montag C (2014) Subjective experience of emotions and emotional empathy in paranoid schizophrenia. Psychiatry Res 220:825–833

Liakakis G, Nickel J, Seitz RJ (2011) Diversity of the inferior frontal gyrus–a meta-analysis of neuroimaging studies. Behav Brain Res 225:341–347

Liddle PF (1987) The symptoms of chronic schizophrenia: a re-examination of the positive-negative dichotomy. Br J Psychiatry J Ment Sci 151:145–151

Meyer-Lindenberg A, Weinberger DR (2006) Intermediate phenotypes and genetic mechanisms of psychiatric disorders. Nat Rev Neurosci 7:818–827

Mier D, Kirsch P, Meyer-Lindenberg A (2010a) Neural substrates of pleiotropic action of genetic variation in COMT: a meta-analysis. Mol Psychiatry 15:918–927

Mier D, Lis S, Neuthe K, Sauer C, Esslinger C, Gallhofer B, Kirsch P (2010b) The involvement of emotion recognition in affective theory of mind. Psychophysiology 47:1028–1039

Mier D, Sauer C, Lis S, Esslinger C, Wilhelm J, Gallhofer B, Kirsch P (2010c) Neuronal correlates of affective theory of mind in schizophrenia out-patients: evidence for a baseline deficit. Psychol Med 40:1607–1617

Mier D, Lis S, Zygrodnik K, Sauer C, Ulferts J, Gallhofer B, Kirsch P (2014) Evidence for altered amygdala activation in schizophrenia in an adaptive emotion recognition task. Psychiatry Res 221:195–203

Moritz S, Kerstan A, Veckenstedt R, Randjbar S, Vitzthum F, Schmidt C, Heise M, Woodward TS (2011) Further evidence for the efficacy of a metacognitive group training in schizophrenia. Behav Res Ther 49:151–157

Napal O, Ojeda N, Sanchez P, Elizagarate E, Pena J, Ezcurra J, Gutierrez M (2012) The course of the schizophrenia and its impact on cognition: a review of literature. Actas espanolas de psiquiatria 40:198–220

Novic J, Luchins DJ, Perline R (1984) Facial affect recognition in schizophrenia. Is there a differential deficit? Br J Psychiatry J Ment Sci 144:533–537

Okochi T, Ikeda M, Kishi T, Kawashima K, Kinoshita Y, Kitajima T, Yamanouchi Y, Tomita M, Inada T, Ozaki N, Iwata N (2009) Meta-analysis of association between genetic variants in COMT and schizophrenia: an update. Schizophr Res 110:140–148

Paus T (2001) Primate anterior cingulate cortex: where motor control, drive and cognition interface. Nat Rev Neurosci 2:417–424

Pickup GJ (2008) Relationship between theory of mind and executive function in schizophrenia: a systematic review. Psychopathology 41:206–213

Pinkham AE, Hopfinger JB, Pelphrey KA, Piven J, Penn DL (2008) Neural bases for impaired social cognition in schizophrenia and autism spectrum disorders. Schizophr Res 99:164–175

Pinkham AE, Brensinger C, Kohler C, Gur RE, Gur RC (2011) Actively paranoid patients with schizophrenia over attribute anger to neutral faces. Schizophr Res 125:174–178

Premack D, Woodruff G (1978) Does the chimpanzee have a theory of mind? Behav Brain Sci 4:515–526

Quintana J, Wong T, Ortiz-Portillo E, Marder SR, Mazziotta JC (2003) Right lateral fusiform gyrus dysfunction during facial information processing in schizophrenia. Biol Psychiatry 53:1099–1112

Rausch F, Mier D, Eifler S, Esslinger C, Schilling C, Schirmbeck F, Englisch S, Meyer-Lindenberg A, Kirsch P, Zink M (2014) Reduced activation in ventral striatum and ventral tegmental area during probabilistic decision-making in schizophrenia. Schizophr Res 156:143–149

Roberts DL, Combs DR, Willoughby M, Mintz J, Gibson C, Rupp B, Penn DL (2014) A randomized, controlled trial of social cognition and interaction training (SCIT) for outpatients with schizophrenia spectrum disorders. Br J Clin Psychol Br Psychol Soc 53:281–298

Russell TA, Rubia K, Bullmore ET, Soni W, Suckling J, Brammer MJ, Simmons A, Williams SC, Sharma T (2000) Exploring the social brain in schizophrenia: left prefrontal underactivation during mental state attribution. Am J Psychiatry 157:2040–2042

Shenton ME, Dickey CC, Frumin M, McCarley RW (2001) A review of MRI findings in schizophrenia. Schizophr Res 49:1–52

Tandon R, Keshavan MS, Nasrallah HA (2008) Schizophrenia, "just the facts" what we know in 2008. 2. Epidemiol Etiol Schizophr Res 102:1–18

Tandon R, Nasrallah HA, Keshavan MS (2009) Schizophrenia, "just the facts" 4. Clin Features Concept Schizophr Res 110:1–23 van Winkel R, Stefanis NC, Myin-Germeys I (2008) Psychosocial stress and psychosis: a review of the neurobiological mechanisms and the evidence for gene-stress interaction. Schizophr Bull 34:1095–1105

Walter H, Ciaramidaro A, Adenzato M, Vasic N, Ardito RB, Erk S, Bara BG (2009) Dysfunction of the social brain in schizophrenia is modulated by intention type: an fMRI study. Soc Cogn Affect Neurosci 4:166–176

The Programming of the Social Brain by Stress During Childhood and Adolescence: From Rodents to Humans

Stamatina Tzanoulinou and Carmen Sandi

Abstract The quality and quantity of social experience is fundamental to an individual's health and well-being. Early life stress is known to be an important factor in the programming of the social brain that exerts detrimental effects on social behaviors. The peri-adolescent period, comprising late childhood and adolescence, represents a critical developmental window with regard to the programming effects of stress on the social brain. Here, we discuss social behavior and the physiological and neurobiological consequences of stress during peri-adolescence in the context of rodent paradigms that model human adversity, including social neglect and isolation, social abuse, and exposure to fearful experiences. Furthermore, we discuss peri-adolescent stress as a potent component that influences the social behaviors of individuals in close contact with stressed individuals and that can also influence future generations. We also discuss the temporal dynamics programmed by stress on the social brain and debate whether social behavior alterations are adaptive or maladaptive. By revising the existing literature and defining open questions, we aim to expand the framework in which interactions among peri-adolescent stress, the social brain, and behavior can be better conceptualized.

Keywords Social behavior · Aggression · Early life stress · Puberty · Peri-adolescent stress · Peripubertal stress · Cycle of violence · Brain programming

---

S. Tzanoulinou
Department of Fundamental Neurosciences, University of Lausanne, Rue Du Bugnon 9, CH-1005 Lausanne, Switzerland
e-mail: stamatina.tzanoulinou@unil.ch

C. Sandi (✉)
Brain Mind Institute, School of Life Sciences, Ecole Polytechnique Federale de Lausanne (EPFL), CH-1050 Lausanne, Switzerland
e-mail: carmen.sandi@epfl.ch

Curr Topics Behav Neurosci (2016) 30: 411–430
DOI 10.1007/7854_2015_430
Published Online: 04 January 2016

Contents

1 Introduction 412
2 The Peri-Adolescent Period as a Critical Developmental Window 414
3 Long-Term Programming of the Social Brain in Rodent Models of Peri-Adolescent Stress 415
3.1 Post-weaning Social Isolation 416
3.2 Peripubertal Stress 416
3.3 Social Subjugation 418
3.4 Physiological Alterations Induced by Peri-Adolescent Stress 418
3.5 Neurobiological Mechanisms Affected by Peri-Adolescent Stress 419
4 Programming of Pathological Aggression by Peri-Adolescent Stress: A Vicious Cycle? 421
5 Programming of Pathological Aggression by Peri-Adolescent Stress: Delayed or Immediate Effects? 422
6 Programming of Pathological Aggression by Peri-Adolescent Stress: Adaptive or Maladaptive? 424
7 Open Questions and Future Perspectives 425
References 426

1 Introduction

Humans are a social species that lives in organized societies. The quality of the social world is increasingly recognized as fundamental to an individual's health and well-being. In particular, its impact has been exemplified in many reports that reveal the negative consequences, for individuals and societies, of social deviations such as loneliness and poor social relationships. It is therefore important to understand the factors and mechanisms that regulate each individual's social behaviors.

Social behaviors occur at different levels, starting in infancy with attachment, particularly to the primary caregiver, which is followed by peer interactions that peak during adolescence and the formation of pair-bonds and parental behaviors later in life (Insel and Fernald 2004). Deviant social behaviors can be manifested in many ways, as, for example, a lack of sociability or integration into social groups, aggression toward the self and others, as well as the absence of or insufficient parental care. Aberrant social behavior is frequently present in most psychiatric and neurological disorders (Kennedy and Adolphs 2012), which hints at its vulnerability and fragility.

During development, there are critical time windows of plasticity and maturation, when experiences from the environment can exert programming effects for future outcomes (Andersen and Teicher 2008; Heim and Binder 2012; Casey et al. 2014). While positive or normative experiences are necessary for developing and refining one's future behavioral repertoire, along with the corresponding neurobiological adaptations, negative experiences can be exceptionally detrimental to the developing individual and can lead to psychopathology (Heim and Binder 2012). The fact that stress during childhood and adolescence exerts programming effects that may lead to psychopathology in adult life is accepted and has been studied in particular with regard to depression, cognitive abilities, and social behaviors (Buwalda et al. 2011; Sandi and Haller 2015).

Early life stress is gaining attention as a predisposing factor for the development of aberrant social behaviors later in life (Haller et al. 2014) and as a potential factor that affects the social life of future generations (Franklin et al. 2011). Investigating the programming effects of early life stress and their underlying neurobiological mechanisms will help to clarify the processes that guide the development of social behaviors and to potentially aid in the conception of novel mechanistically founded treatment strategies. Although recent technological advancements are providing unprecedented insights into the neurobiological factors that play a role in human behaviors—e.g., genetic predispositions that interact with early adversity and alter brain structure and function—the search for ultimate causal mechanisms currently requires additional insights that are provided by studies in animal models, such as rodents (Buwalda et al. 2011; Haller et al. 2014; Sandi and Haller 2015). In addition to the high degree of experimental control that can be achieved by using these model systems, rodent studies allow researchers to assess with high specificity the mechanistic links between the cellular and molecular correlates of early adversity and particular behavioral phenotypes, especially with the development of tools such as in vivo optogenetics, viral approaches in combination with specific lines of animals, and chemogenetic approaches to name but a few (Tye and Deisseroth 2012; Urban and Roth 2015). The application of these techniques in rodents has started identifying some of the molecular, cellular, and circuit components critically involved in the control of social behaviors and aggression (e.g., (Lin et al. 2011; Yizhar et al. 2011; Kohl et al. 2013; Lee et al. 2014; Felix-Ortiz and Tye 2014; van der Kooij et al. 2014a, 2014b; Challis and Berton 2015; Kohl et al. 2015).

Among the different critical developmental windows in which an individual is highly sensitive to the effects of stress (Casey et al. 2010; Heim and Binder 2012), a large number of studies have focused on stress that occurs during the early postnatal period [e.g., maternal separation models; for a review see Haller et al. (2014)]. Recently, the late childhood and adolescent period (referred to from here on as peri-adolescence) have attracted considerable attention as emerging evidence identifies this as a period of heightened vulnerability to the effects of stress and a strong modulator of social behaviors in adulthood. Additionally, many psychiatric disorders seem to appear during this period (Paus et al. 2008). Here, we will review rodent models that are used to investigate the programming effects of stress during this particular time window on social behaviors, and we will discuss the underlying neurobiological mechanisms that have been revealed so far. Note that given that most studies have investigated aggression outcomes of early life stress, this topic is particularly represented in this review. We will also draw parallels with evidence from human studies. Finally, we will discuss the open questions in the field and how basic research can be used to expand its translational potential in the future.

2 The Peri-Adolescent Period as a Critical Developmental Window

The peri-adolescent period, which represents the transition from childhood to adulthood, includes puberty and sexual maturation as some of its central physiological changes (Spear 2000). It is marked by hormonal, physiological, and neuronal changes that make individuals of this age group unique compared to younger or older members of their species (Spear 2000). More specifically, a prequel to puberty is the increase of androgens from the adrenals, which is followed in puberty by the increased release of gonadal hormones (Spear 2000). The nomenclature for these periods is slightly different for different species. In humans, adolescence can be broadly considered to range from the 10th or 12th to the 18th year (Spear 2000; Eiland and Romeo 2013). In rodents, adolescence has been proposed to start at postnatal day 28 (P28) and to end at approximately P42, with P34–36 corresponding to the period of puberty in female rats and P40–42 marking puberty in male rats (Spear 2000). Here, we will consider stress models that start post-weaning, i.e., from P21 onward, which comprises the period from childhood up until late adolescence.

During this period, peer interactions and social behaviors are particularly relevant. Individuals exhibit changes in their social behavioral patterns that will later develop into adult behavioral phenotypes (Eiland and Romeo 2013). Adolescent humans increase their social interactions with their peers while decreasing contact with their seniors (Spear 2000; Buwalda et al. 2011), and peak play behavior occurs during adolescence (Spear 2000). Social play in rats starts as early as P18, peaks between P32 and P40, and afterward tends to decrease as the animals are reared into adulthood and to be replaced by an adult aggressive behavioral repertoire (Panksepp 1981). Akin to human peer-to-peer relationships, social play or play-fighting behavior in young rats involves particular patterns of social interaction that do not lead to injury and do not have adult-like aggressive characteristics (Pellis and Pellis 2007). These patterns can be categorized into distinct behaviors, such as pouncing (the attempt to nose against the other animal's nape of the neck), boxing (pushing and grabbing each other with the forepaws in an upright posture) and pinning (with one animal laying on its back and the other one on top of it) (Panksepp 1981). Although mice show some patterns of play behavior, these seem to be lesser in quality and quantity than those observed in rats (Pellis and Pasztor 1999). Play behaviors in rodents have been shown to be rewarding [reviewed in Trezza et al. (2010)], and playing has been hypothesized to be important for learning motor sequences and behaviors that will be essential to the animals as adults (the motor-training hypothesis) (Bell et al. 2010). Moreover, adolescents show augmented risk-taking and novelty-seeking behaviors that, along with peer influences, can often lead to delinquent behaviors (Spear 2000). Altogether, these changes in behavior are believed to facilitate and prepare the organism for the transition from childhood to adulthood (Spear 2000; Casey et al. 2010).

Different brain regions exhibit differential vulnerability in their responses to stressors in an age-dependent manner (Buwalda et al. 2011). Notably, during adolescence, changes in synaptic pruning and myelination rates have been reported, in addition to the extensive shaping of connectivity between brain regions (Spear 2000; Andersen and Teicher 2008). More specifically, in adolescents, there is an increase in white matter volume that can be attributed to a variety of developmental events, such as ongoing myelination and radial axonal growth (Paus et al. 2008; Keshavan et al. 2014). Gray matter volume changes also occur in an inverted U-shaped fashion, and this process is variable in an age- and brain region-dependent manner (Keshavan et al. 2014). It is important to note that some of the brain areas that undergo significant changes during childhood and puberty are involved in the processing of socioemotional cues, and hence, they form the putative social brain network (Keshavan et al. 2014). Along these lines of evidence, a nonlinear model of brain development has been proposed in which higher-order regions, such as the prefrontal cortex, show delayed development and follow a different trajectory than subcortical areas (e.g., the amygdala and the striatum) (Casey et al. 2010). This partly accounts for the altered emotional and affective processing seen in adolescents.

Moreover, the increased reactivity of the hypothalamic–pituitary–adrenocortical (HPA) axis, under both basal and stress conditions, has been noted in humans during adolescence (Lupien et al. 2009; Keshavan et al. 2014). Similarly, in adolescent rodents that are exposed to stress, a greater reactivity—both in terms of duration and magnitude—of the HPA axis has been reported that has been attributed to the immaturity of the regions and systems involved in the negative feedback loop needed to halt the stress response (Lupien et al. 2009). Finally, it should be noted that although there are important differences between the developmental trajectories of rodents and humans, adolescence is marked by extensive physiological and neurobiological changes in both species (Lupien et al. 2009).

3 Long-Term Programming of the Social Brain in Rodent Models of Peri-Adolescent Stress

Different rodent models have attempted to recapitulate a variety of adverse experiences that are commonly experienced by humans during peri-adolescence. The modeled human adversity ranges from situations of social neglect and isolation to exposure to fearful experiences or social abuse, and the corresponding models that we will review here are known as post-weaning social isolation, peripubertal stress, and social subjugation (Fig. 1). We will first introduce each of these models, including their respective effects on the social behaviors displayed in adulthood, and then, we will summarize the main findings implicating alterations in physiological and neurobiological systems as potential mediators of the behavioral effects.

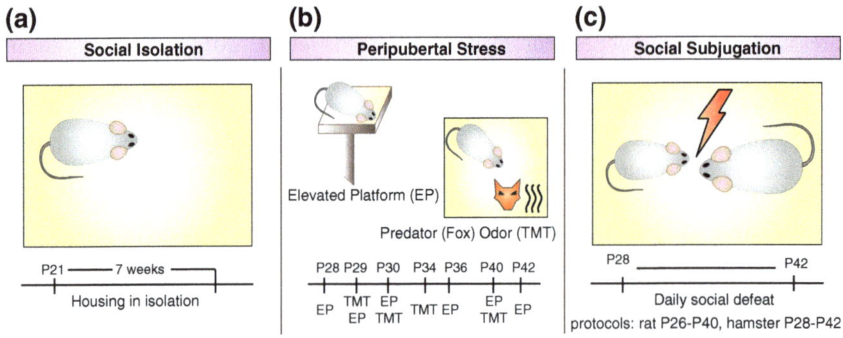

Fig. 1 Schematic of the three peri-adolescent stress protocols reviewed in this article. *TMT* trimethylthiazoline; a chemical component of fox feces, *P* postnatal day

3.1 Post-weaning Social Isolation

Social neglect, particularly when experienced early in life, can lead to persistent behavioral and hormonal alterations, including altered social behavior, augmented emotional aggression, and altered hormonal responses (Chapple et al. 2005; Pesonen et al. 2010). One of the best-characterized experimental protocols of social neglect involves housing isolation in rats starting immediately after weaning (P21) and lasting for 7 weeks (Fig. 1a). As adults, these animals have difficulty integrating into groups (Tulogdi et al. 2014) and they display quantitative and qualitative alterations in aggressive behaviors as measured using a resident-intruder test (Haller et al. 2014). More precisely, they perform what is considered to be pathological aggressive behavior because it is opposite to the natural pattern observed in the species; e.g., they direct their attacks to the head, throat, and belly, which are considered the most vulnerable body parts, and moreover, they do not prime or warn their opponent of upcoming attacks with prior signaling by displaying offensive threats, as non-isolated animals normally do. Isolated animals' behavior is significantly more fragmented and is interchanged more often and more rapidly than is observed in control animals between different behaviors (Toth et al. 2008). Notably, these abnormal aggressive patterns were not alleviated by 3 weeks of resocialization, which highlights the perseverative nature of social isolation early in life (Tulogdi et al. 2014).

3.2 Peripubertal Stress

The peripubertal stress paradigm, which has so far been mainly applied to rats, aims to model exposure to fear-induction experiences of a nonsocial nature during childhood and puberty. Specifically, on 7 scattered days during the period from P28 to P42, animals are exposed to two different stressors: an elevated platform and a predator odor, both of which are innately stressful to rats (Fig. 1b) (Marquez et al. 2013).

Adult animals that have been stressed during peripuberty show decreased exploration of younger conspecifics in the 3-chambered test of social interaction, indicating a reduction in social motivation (Marquez et al. 2013; Tzanoulinou et al. 2014). They also show abnormal aggressive behaviors. Similarly, in adult male rats that are submitted to post-weaning social isolation (see above), peripubertally stressed rats display increased inter-male aggression in a resident-intruder assay, with different variations of the protocol revealing a profile of abnormal aggressive behavior, including more attacks and bites on vulnerable parts of the body (Marquez et al. 2013). Additionally, when exposed to nonthreatening intruders, such as females (Cordero et al. 2012), juveniles, or even anesthetized male rats that do not oppose a real threat, peripubertally stressed animals continue to show increased levels of aggression. Moreover, they are more aggressive toward more powerful opponents, i.e., bigger intruders, against which the supposedly optimal strategy would be to show signs of submission (Marquez et al. 2013). When stress during peripuberty is applied to female rats, their adult behaviors also become more aggressive, both against another female and against males (Cordero et al. 2013). Interestingly, peripubertally stressed females displayed increased aggression toward their male partner during the first cohabitation day and peripubertally stressed mothers showed shorter latencies to attack a male intruder during a maternal aggression test (Cordero et al. 2013) (see Table 1 for a summary of peripubertal

Table 1 Main consequences of peripubertal stress in social behaviors

Behavioral consequences of peripubertal stress in rats			
Test	Observation	Age at observation	References
Social interaction (three-chamber test)	↓ Social exploration	Adulthood (P90+)	Marquez et al. (2013) and Tzanoulinou et al. (2014)
Resident-intruder	↑ Aggression	Adulthood (P90+)	Marquez et al. (2013) and Tzanoulinou et al. (2014)
Resident-intruder	Abnormal aggression (↑ attacks to vulnerable body parts, ↑ aggression toward nonthreatening opponents)	Adulthood (P90+)	Marquez et al. (2013)
Resident-intruder (females)	↑ Aggression	Adulthood (P90+)	Cordero et al. (2013)
Male–female cohabitation	↑ Aggression of males toward females	Adulthood (P90+)	Cordero et al. (2012)

All studies concern males unless otherwise indicated

stress long-term effects in the social domain). In the context of differences in stress reactivity according to sex, it would be important to characterize in future studies whether the aggressive behavior demonstrated by peripubertally stressed females resembles that of the males, not only in quantity, but also in the above-mentioned qualitative characteristics (i.e. against bigger, smaller or anesthetized intruders).

3.3 Social Subjugation

Children and adolescents are sometimes subjected to physical abuse and aggression by individuals who are either within or outside the family. In rodents, social subjugation during the peri-adolescent period has been modeled in golden hamsters and rats. Specifically, juvenile hamsters from P28 to P42 are placed in the cage of an adult male for 30 min (Fig. 1c) (Delville et al. 1998). During this time, the experimental animals become the recipients of bites and attacks by the older hamsters. Unlike the aggressive pattern exhibited by peripubertally stressed rats, which attack regardless of the opponent's size (Marquez et al. 2013), hamsters that were exposed to subjugation during peripuberty show increased aggression toward smaller/weaker intruders but decreased aggression toward similar/equal opponents, highlighting the context-dependent nature of aggressive behavior (Delville et al. 1998). However, when social subjugation was studied in rats (protocol applied from P26 to P40), increased aggression that was observed following stress provocation of either the experimental or the opponent rat (i.e., tail pinch applied every 60 s during a 10-minute test) was also displayed toward stronger/heavier intruders, and this effect continued to be evident at 17 weeks after the last social subjugation session (Cunningham and McGinnis 2008).

An important consideration when evaluating the specific impact of a stress-induction protocol is the difficulty in exclusively confining the observed effects to social subjugation, as animals are kept in isolation from weaning until aggressive behavior assessment. As discussed above, isolation per se can also program individuals to show altered aggressive behaviors.

3.4 Physiological Alterations Induced by Peri-Adolescent Stress

The three animal models of peri-adolescent stress reviewed here lead to different patterns of alterations in the HPA axis, which is in agreement with existing evidence related to abnormal social behaviors that shows they can be associated to both over- and hypo-activation of the HPA axis (Sandi and Haller 2015). The two models that involve a social component, social deprivation (post-weaning social isolation model), and social subjugation induce heightened glucocorticoid responses to social challenges. This was reported as part of the HPA response to the social subjugation protocol when it was applied during adolescence in hamsters (Wommack and Delville 2003). In the aftermath of the stress manipulation, on the first stress-induction day (P28), both control animals (exposed to an empty new cage instead of an aggressive encounter) and socially subjugated hamsters showed increased cortisol levels compared to baseline. However, the cortisol response was reduced in control animals by P42, suggesting habituation, but was not reduced in subjugated hamsters (Wommack and Delville 2003), indicating the robustness of this manipulation for inducing a sustained stress response throughout different experimental days. Heightened glucocorticoid responses were also observed following an aggressive encounter in adulthood in animals that had been exposed to the post-weaning social isolation protocol. This increase was accompanied by an increased heart rate upon the encounter (Toth et al. 2011).

On the contrary, the peripubertal stress model was found to lead to a reduction in plasma corticosterone when animals were exposed to additional stressors at adulthood (Marquez et al. 2013). A reduction in the corticosterone response to novel stressors in adulthood was also observed in a pharmacological-mimicking protocol that was based on the systemic injection of a corticosterone dose known to elicit stress-equivalent plasma steroid doses on the same peripubertal days that stress was applied in the peripubertal stress protocol (Veenit et al. 2013). This pharmacological protocol led to the same alterations that were observed in the peripubertal stress animals in the social domain but not in other behaviors (e.g., anxiety-like behaviors) (Veenit et al. 2013). These data suggest that elevated glucocorticoid levels during peri-adolescent exposure to stress might be critically involved in the neurobiological programming of changes in the social brain and in behavior (see Table 2 for a summary of physiological consequences following peripubertal stress). Given that there are significant sex-dependent differences in stress responses (Kudielka and Kirschbaum 2005), it would be interesting to address HPA axis reactivity following peripubertal stress in females to be able to compare in males and females how neuroendocrine reactivity to stress relates to behavioral and neurobiological outcomes.

3.5 Neurobiological Mechanisms Affected by Peri-Adolescent Stress

The study of neurobiological mechanisms that translate the effects of different peri-adolescent stress experiences is still rather incipient, with existing data from the different models scattered. This fragmentation does not allow the drawing of an emerging coherent picture. However, from these studies, a few common features have emerged. For example, the amygdala, in particular its medial nucleus, appears to be a target of stress in all of the models. Analyses of brain activity using

Table 2 Main physiological and neurobiological consequences observed in adult rats submitted to peripubertal stress

Physiological and neurobiological consequences of peripubertal stress in rats			
Assay	Observation	Age at observation/condition	References
Plasma corticosterone	↓ HPA axis response to stress	Adolescence (P42)/ after stress exposure	Marquez et al. (2013)
Plasma hormonal levels	↑ Testosterone/corticosterone ratio	Adulthood (P90+)/ after resident-intruder	Marquez et al. (2013)
Quantitative real-time PCR	↑ Corticotropin-releasing hormone receptor 1 in CeA and hippocampal CA1	Adulthood (P90+)/ baseline	Veenit et al. (2014)
2-deoxy-glucose autoradiography	↑ Brain energy metabolism in CeA, MeA, BLA, LA	Adulthood (P90+)/ baseline	Marquez et al. (2013)
C-fos immunohistochemistry	↑ Activation of CeA, MeA ↓ Activation of MO	Adulthood (P90+)/ after Resident-intruder	Marquez et al. (2013)
Quantitative real-time PCR	↑ GluN1 mRNA in CeA ↓ GAD67 mRNA in CeA	Adulthood (P90+)/ baseline	Tzanoulinou et al. (2014)
Quantitative real-time PCR and ChIP	↑ MAOA, 5HTT mRNA and histone 3 acetylation in the PFC	Adulthood (P90+)/ baseline	Marquez et al. (2013)

HPA Hypothalamic–pituitary–adrenal axis, *PCR* polymerase chain reaction, *ChIP* chromatin immunoprecipitation, *CeA* central amygdala, *MeA* medial amygdala, *BLA* basolateral amygdala, *LA* lateral amygdala, *MO* medial orbital cortex, *GAD67* glutamic acid decarboxylase 67, *MAOA* monoamine oxidase A, *5HTT* serotonin transporter, *PFC* prefrontal cortex, and *GluN1* glutamate (NMDA) receptor subunit 1

immunohistochemistry in order to observe immediate early gene expression have shown enhanced activation in the medial amygdala in rat models of both post-weaning social isolation (Toth et al. 2012) and peripubertal stress (Marquez et al. 2013). The former model also led to the activation of several other regions that are implicated in the control of aggressive behavior (i.e., the prefrontal cortex, especially the lateral and medial orbitofrontal cortex, and the cingulate cortex 1, the bed nucleus of *stria terminalis*, the basolateral amygdala nucleus, the hypothalamic attack area, the paraventricular nucleus, and the locus coeruleus) (Toth et al. 2012). Contrary to these findings, the prefrontal cortex, specifically the medial orbitofrontal cortex, was inhibited in the peripubertal stress model in parallel with the increased expression of the monoamine oxidase A (MAOA) and serotonin transporter (5HTT) genes and the activation of histone 3 acetylation of the MAOA gene (Marquez et al. 2013). The causal implication of these changes in the stress-induced phenotype was suggested by the successful inhibition of peripubertal stress-induced aggression that was observed following chronic treatment with the MAOA inhibitor clorgyline in adulthood (Marquez et al. 2013). The social subjugation model was shown to lead to an increased number of immunoreactive tyrosine hydroxylase (TH) cells in the medial amygdala and the bed nucleus of stria terminalis during stress exposure; these changes are transient, in that they were not observed at P70 (Wommack et al. 2004). Moreover, the animals in this model also showed decreased levels of arginine vasopressin (AVP) in the anterior hypothalamus in parallel with an increased density of serotonin (5-HT)-immunoreactive varicosities in both the anterior hypothalamus and the lateral septum (Delville et al. 1998). Given the general view that AVP is a key player in promoting aggression and that 5-HT inhibits aggression, these correlative results hint at the differential regulation of aggression by different neuromodulators and brain areas, which may be responsible for the ambiguous context-dependent aggressive behavior of the subjugated hamsters.

Other relevant mechanisms include alterations in markers of excitation/inhibition in the amygdala, particularly in the central nucleus (Tzanoulinou et al. 2014), and in the prefrontal cortex (Tzanoulinou et al. 2015) that have been reported in the peripubertal stress model and that are suggestive of an increase in excitation over inhibition. Similarly, inhibitory-related changes, particularly in the amygdala, have been shown following another model of juvenile stress in rats (Jacobson-Pick et al. 2008; Jacobson-Pick and Richter-Levin 2012). Evidence was also reported for the involvement of the corticotropin-releasing hormone (CRH) system in the behavioral programming induced by peripubertal stress exposure (Veenit et al. 2014). (See Table 2 for a summary of neurobiological long-term effects after peripubertal stress)

4 Programming of Pathological Aggression by Peri-Adolescent Stress: A Vicious Cycle?

One of the potential consequences of the findings discussed here is that experiences involving strong adversity (e.g., neglect, abuse, or intense fear) during the transition from childhood to adolescence can create a vicious cycle of dysfunction in the social domain. In humans, although not all individuals are equally affected, there is an increased risk that the 'victims' of early life adversity will become the perpetrators of violence when they become adults (Jonson-Reid et al. 2010). Until recently, the prevalent views put forward to explain this phenomenon included cultural (e.g., male dominance in society) and social learning (children learn from their early role models, such as their parents) explanations for the reproduction of patterns of aggressive behaviors that developed in adulthood by those that were targets of early life adversity. The current explosion of neuroscientific evidence shows that neurodevelopmental trajectories can be drastically affected by early life stress (Paus et al. 2008; Lupien et al. 2009; Keshavan et al. 2014), which brings an important novel perspective to this topic that suggests that to a large extent, it is the 'hardware' of the individual that is affected, not just the 'software.'

In this context, recent studies in the animal literature support the view that the biological programming of pathological aggression by peripubertal stress may consequently have a dramatic impact on how individuals interact with someone who was originally the victim of adversity. Female rats that cohabitated with adult peripubertally stressed male rats received sustained attacks from their partners despite their enhanced display of defensive/submissive behaviors (Cordero et al. 2012). As a consequence of cohabitation with aggressive peripubertally stressed males, these females showed clear signs of anxiety- and depression-like behaviors, including hormonal (i.e., blunted corticosterone responses) and neurobiological (increased activity of serotonergic neurons in the dorsal raphe upon exposure to an unfamiliar male) alterations that have been typically reported in human depression (Cordero et al. 2012). Vulnerability to developing these symptoms was greatest in highly anxious female rats (Poirier et al. 2014). Most importantly, their male offspring at adulthood also showed increased aggression against their female partners, even though the offspring had not been exposed during their life to their father (Cordero et al. 2012). Different mechanisms can account for the transmission of the aggressive phenotype from peripubertally stressed males to their subsequent generation/s (e.g., epigenetic changes in the germ line and prenatal or postnatal maternal effects) (Monk et al. 2012; Saab and Mansuy 2014), and further work is needed to determine the key factors involved.

5 Programming of Pathological Aggression by Peri-Adolescent Stress: Delayed or Immediate Effects?

An important question in the context of this review is whether the effects of early adversity on social behaviors emerge with time, as individuals become adults, in which case we would be dealing with the specific programming of adult behavior, or alternatively, whether the effects are immediate and are already observed during the peri-adolescent period, in which case the adult phenotype is the result of the sustained shaping of the individual's behavior throughout life. Although this question, as formulated here, has not been systematically addressed, and many of the relevant studies were designed to directly measure adult behavioral changes under the assumption that any effects would be due to delayed regulation, evidence from a few studies of rodent models of peri-adolescent stress on play behavior strongly favor the possibility that immediate social effects might also play a role.

For example, in the model of social subjugation in hamsters (see Sect. 3.3), stress altered the dynamics of the transition between play-fighting behavior and adult-like aggressive patterns (Wommack et al. 2003). In non-stressed hamsters, play-fighting evolved from a first phase (P26–33), when attacks targeted the face and the cheeks, to a transition phase (P40–47), when attacks shifted toward the sides, belly and rear, to the emergence of adult aggressive behavior around P54, and when most attacks targeted the belly or rear. However, socially subjugated animals were more likely to perform more belly and rear attacks than their non-subjugated counterparts even during the first phase, indicating an accelerated transition from play-fighting to adult-like aggressive patterns (Wommack et al. 2003).

Interestingly, this was particularly the case in animals that showed less submissive behavior (Wommack and Delville 2003). In the peripubertal corticosterone administration model (see Sects. 3.2 and 3.4), more aggressive play-fighting was observed shortly after puberty (P44) (Veenit et al. 2013). Enhanced play-fighting was also observed following 5 days of social isolation in the peri-adolescent period (Varlinskaya and Spear 2008). In fact, an important consideration when discussing models of social isolation (note that the social subjugation model also involves social isolation) is the fact that the timing of social deprivation coincides with the peak of play behavior in rodents. In rats, it has been shown that week 4, but not week 5, is particularly important for adult programming of social interactions (Hol et al. 1999; van den Berg et al. 1999). Deprivation from play behavior during this critical period leads to social incompetence later in life (Pellis and Pellis 2007). Similarly, in humans, social isolation or even perceived isolation has been linked with a broad spectrum of outcomes, ranging from general health and physiological problems, such as elevated cortisol and blood pressure, to impaired cognitive abilities and altered perception of social cues and threats (Cacioppo and Hawkley 2009).

Likewise, in humans, most studies of the impact of early life adversity have focused on the impact of adult aggression, which has influenced thinking about these effects potentially having a delayed onset. However, a few reports that have examined the more immediate impact of early life stress on child and/or adolescent behaviors indicate the emergence of antisocial behaviors at this early age when levels of experienced trauma have been very high (Weder et al. 2009). Although aggression triggered by early trauma has repeatedly been shown to depend on an interaction with the MAOA genotype, children with the most severe traumatic experiences were found to have high aggression scores regardless of MAOA genotype (Weder et al. 2009).

In this context, it is important to note that in children, toxic stress is also frequently part of the nature of peer interactions. For example, bullying is a very common phenomenon during peri-adolescence. Bullying is defined as the *unilateral* expression of aggressive behavior toward an individual (with the assaulted party not fighting back); it thus entails an imbalance of power that can be real or perceived (Holt et al. 2007; Wolke and Lereya 2015). The aggression need not always be physical and direct; it can also be indirect or relational, e.g., when a person is excluded from group activities, or verbal (Shetgiri 2013; Wolke and Lereya 2015). Cyberbullying is increasingly recognized as a type of bullying that involves aggressive behavior via instant messages and the social media, e.g., by spreading false rumors about someone or publically exposing aspects of his or her life (Shetgiri 2013). Longitudinal studies have shown that bullies, victims, and bullies/victims show some forms of social behavior problems later in life (Wolke and Lereya 2015) and, strikingly, they tend to have high levels of victimization outside of school, such as sexual abuse and child maltreatment, with bullies/victims scoring higher in all the victimization categories (Holt et al. 2007). As mentioned above, rodent social subjugation is considered to model adverse circumstances comparable to bullying. Similar to the human condition, social subjugation in rodents leads to aberrant social behaviors later in life (Delville et al. 1998; Wommack and Delville 2003).

6 Programming of Pathological Aggression by Peri-Adolescent Stress: Adaptive or Maladaptive?

So far, and in agreement with a broad neuroscientific literature, we have implied that the long-term programming of enhanced aggression by early life stress is a maladaptive process. It is important to note that such an interpretation is based on the assumption that enhanced displays of aggression are disadvantageous. From an evolutionary perspective, heightened aggression can be considered to be advantageous in that it can facilitate access to resources and be used to defend against attacks in both animals and humans (Buss and Shackelford 1997).

However, although aggressive displays can be adaptive for immediate self-protection and for the establishment of social hierarchies, with its protracted effects, in the long-term, heightened aggressiveness might become maladaptive. This is particularly the case in humans, as aggression could be, to a certain point, advantageous in children, but it becomes a drawback as they start aging and encountering retaliation from slightly older or stronger adolescents and adults. Moreover, behaviors that are adaptive in one social context are not necessarily adaptive in another, and our society has established clear rules to penalize indiscriminately aggressive individuals. Therefore, the problem in our society might be that whereas biology programs individuals to develop behavioral responses and styles in accordance with the context within which the individual was reared (which in many cases may lead to what would be biologically considered 'adaptive'), societal contexts for human adults in a large proportion of our societies require an exquisite control of explicit aggressive displays. Thus, aggressive behavior could be characterized as maladaptive when it diverges from the normative levels (i.e., in the context of psychopathology), while it is important to keep in mind that these behavioral manifestations could be adaptive in other contexts or under certain conditions (Nederhof and Schmidt 2012).

Within this framework, there are two seemingly contradictory hypotheses that are relevant to the programming of behavior by early life adversity: the 'cumulative stress' and the 'mismatch' hypotheses (Schmidt 2011; Nederhof and Schmidt 2012). While the former theory proposes that the accumulation of stress or adversity early in life renders an individual more sensitive to future challenges, the second supports the idea that early life events prepare the individual for later challenges, and therefore, an individual that has already been exposed to stress may adapt more successfully to a stressful than to a stress-free environment and vice versa (Nederhof and Schmidt 2012; Santarelli et al. 2014). In fact, it has been proposed that the two hypotheses can be reconciled, depending on the individual's sensitivity to programming effects (Nederhof and Schmidt 2012; Daskalakis et al. 2012). In this context, individual differences (due to genetic background, early postnatal environment, etc.), as well as the timing of the stressors, can be critical factors that determine an individual's sensitivity to the programming effects of stress. Thus, given that peri-adolescence is a critical developmental window that is characterized by high sensitivity to the programming effects of stress, possible interactions with other factors, such as genetic background and prior life history (e.g., housing conditions), should be taken into account to understand the resulting behavioral adaptations or maladaptations.

7 Open Questions and Future Perspectives

Problematic social behaviors and interactions have been found in a plethora of psychiatric and neurologic disorders (Kennedy and Adolphs 2012). Moreover, lack of social contact has been shown to lead to a variety of physiological, behavioral, and neurobiological health problems (Cacioppo and Hawkley 2009), suggesting that social interactions can play a positive role in buffering adversity. Early life stress and impoverished social interactions can interact, distancing individuals from their well-being and functionality in society. Even though reasonable progress has been made regarding the neurobiological mechanisms that guide the trajectory from peri-adolescent stress to aberrant social behaviors, much remains to be understood before efficient treatment strategies to the negative consequences of stress (see above) can be offered.

Animal models with face, construct, and predictive validity have proven useful in determining the programming effects of stress on the social brain. Nevertheless, several questions remain open. What are the temporal dynamics that contribute to the development of these aberrant social behaviors? Do these behavioral changes appear gradually, soon after the first stress exposure, or are they more pronounced, with time indicating incubation effects? Moreover, how persistent or transient are these behavioral consequences? The answers to these questions are needed to obtain a better comprehension of the time window when intervention strategies can be effectively applied. Moreover, are social deficits expressed independently or in the context of deficiencies in other behavioral domains? Are they due to general motivation issues or to cognitive deficits that precede them, and is the learning of social rules and norms perturbed early on? Are peri-adolescence-induced decreased motivation for social interactions and increased aggression due to an altered perception of the qualities of each experience? For example, employing ultrasonic vocalization (USV) measurements would be helpful for investigating whether stressed rodents perceive social interactions as pleasurable or aversive as illustrated in a recent report indicating that post-weaning—but not post-adolescence—social isolation leads to a reduction in approach behaviors toward high-frequency, appetitive, 50-kHz USV (Seffer et al. 2015). The behavioral dissection of the consequences of peri-adolescent stress will presumably increase the translational potential of available and novel animal models.

As discussed above, the programming of the social brain by early adversity can be readily affected by individual differences due to genetic background, among other factors. Therefore, rodent studies that take into account individual variation and vulnerability would be valuable for expanding our understanding of stress-induced programming effects. In this regard, it is noticeable that whereas the vast majority of rat studies use outbred strains, most mouse studies are executed with inbred lines which from a genetic perspective gives a stronger generalization power to findings obtained with outbred rather than inbred approaches. Along the same line, human studies that present causal links between stress and later outcomes, e.g., twin studies, would be very informative in these regards. Moreover, when examining the transgenerational effects of stress, it is important that the investigated behavioral domains are expanded to include not only aggression and violence but also other aberrant social interactions in general, such as social withdrawal and loneliness. Thus, we would like to propose that the cycle of violence discussed above could be extended to include other behaviors and to thereby constitute a broader vicious cycle.

Finally, rodent models can address the above questions in a circuit- and synapse-specific manner. In recent decades, important technological advancements (such as in vivo optogenetic and viral approaches that can be combined with genetically modified lines of rats and mice) have made it possible to interrogate and tease out the essential cell types and projections that control behavior in real-time (Tye and Deisseroth 2012), including the neural control of social behaviors (Lin et al. 2011; Yizhar et al. 2011; Lee et al. 2014; Felix-Ortiz and Tye 2014; Challis and Berton 2015). In combination with animal models as the ones described in this review, such methodologies can be valuable for increasing our understanding of critical periods and for tackling the mechanisms via which stress programs the social brain.

Acknowledgments This work was supported by grants from the Swiss National Science Foundation (31003A-152614 and the NCCR Synapsy), the EU FP7 project MATRICS (no 603016), Oak Foundation, and intramural funding from the EPFL.

References

Andersen SL, Teicher MH (2008) Stress, sensitive periods and maturational events in adolescent depression. Trends Neurosci 31:183–191

Bell HC, Pellis SM, Kolb B (2010) Juvenile peer play experience and the development of the orbitofrontal and medial prefrontal cortices. Behav Brain Res 207:7–13

Buss DM, Shackelford TK (1997) Human aggression in evolutionary psychological perspective. Clin Psychol Rev 17:605–619

Buwalda B, Geerdink M, Vidal J, Koolhaas JM (2011) Social behavior and social stress in adolescence: a focus on animal models. Neurosci Biobehav Rev 35:1713–1721

Cacioppo JT, Hawkley LC (2009) Perceived social isolation and cognition. Trends Cogn Sci 13:447–454
Casey BJ, Jones RM, Levita L, Libby V, Pattwell SS, Ruberry EJ, Soliman F, Somerville LH (2010) The storm and stress of adolescence: insights from human imaging and mouse genetics. Dev Psychobiol 52:225–235
Casey BJ, Oliveri ME, Insel T (2014) A neurodevelopmental perspective on the research domain criteria (RDoC) framework. Biol Psychiatry 76:350–353
Challis C, Berton O (2015) Top-down control of serotonin systems by the prefrontal cortex: a path toward restored socioemotional function in depression. ACS Chem Neurosci 6:1040–1054
Chapple CL, Tyler KA, Bersani BE (2005) Child neglect and adolescent violence: examining the effects of self-control and peer rejection. Violence Vict 20:39–53
Cordero MI, Ansermet F, Sandi C (2013) Long-term programming of enhanced aggression by peripuberty stress in female rats. Psychoneuroendocrinology 38:2758–2769
Cordero MI, Poirier GL, Marquez C, Veenit V, Fontana X, Salehi B, Ansermet F, Sandi C (2012) Evidence for biological roots in the transgenerational transmission of intimate partner violence. Transl Psychiatry 2:e106
Cunningham RL, McGinnis MY (2008) Prepubertal social subjugation and anabolic androgenic steroid-induced aggression in male rats. J Neuroendocrinol 20:997–1005
Daskalakis NP, Oitzl MS, Schachinger H, Champagne DL, de Kloet ER (2012) Testing the cumulative stress and mismatch hypotheses of psychopathology in a rat model of early-life adversity. Physiol Behav 106:707–721
Delville Y, Melloni RH, Ferris CF Jr (1998) Behavioral and neurobiological consequences of social subjugation during puberty in golden hamsters. J Neurosci 18:2667–2672
Eiland L, Romeo RD (2013) Stress and the developing adolescent brain. Neuroscience 249:162–171
Felix-Ortiz AC, Tye KM (2014) Amygdala inputs to the ventral hippocampus bidirectionally modulate social behavior. J Neurosci 34:586–595
Franklin TB, Linder N, Russig H, Thony B, Mansuy IM (2011) Influence of early stress on social abilities and serotonergic functions across generations in mice. PLoS ONE 6:e21842
Haller J, Harold G, Sandi C, Neumann ID (2014) Effects of adverse early-life events on aggression and anti-social behaviours in animals and humans. J Neuroendocrinol 26:724–738
Heim C, Binder EB (2012) Current research trends in early life stress and depression: review of human studies on sensitive periods, gene-environment interactions, and epigenetics. Exp Neurol 233:102–111
Hol T, van den Berg CL, van Ree JM, Spruijt BM (1999) Isolation during the play period in infancy decreases adult social interactions in rats. Behav Brain Res 100:91–97
Holt MK, Finkelhor D, Kantor GK (2007) Hidden forms of victimization in elementary students involved in bullying. Sch Psychol Rev 36:345–360
Insel TR, Fernald RD (2004) How the brain processes social information: searching for the social brain. Annu Rev Neurosci 27:697–722
Jacobson-Pick S, Elkobi A, Vander S, Rosenblum K, Richter-Levin G (2008) Juvenile stress-induced alteration of maturation of the GABAA receptor alpha subunit in the rat. Int J Neuropsychopharmacol 11:891–903
Jacobson-Pick S, Richter-Levin G (2012) Short- and long-term effects of juvenile stressor exposure on the expression of GABAA receptor subunits in rats. Stress 15:416–424
Jonson-Reid M, Presnall N, Drake B, Fox L, Bierut L, Reich W, Kane P, Todd RD, Constantino JN (2010) Effects of child maltreatment and inherited liability on antisocial development: an official records study. J Am Acad Child Adolesc Psychiatry 49:321–332 quiz 431
Kennedy DP, Adolphs R (2012) The social brain in psychiatric and neurological disorders. Trends Cogn Sci 16:559–572
Keshavan MS, Giedd J, Lau JYF, Lewis DA, Paus T (2014) Changes in the adolescent brain and the pathophysiology of psychotic disorders. Lancet Psychiatry 1:549–558
Kohl C, Riccio O, Grosse J, Zanoletti O, Fournier C, Schmidt MV, Sandi C (2013) Hippocampal neuroligin-2 overexpression leads to reduced aggression and inhibited novelty reactivity in rats. PLoS ONE 8:e56871

Kohl C, Wang XD, Grosse J, Fournier C, Harbich D, Westerholz S, Li JT, Bacq A, Sippel C, Hausch F, Sandi C, Schmidt MV (2015) Hippocampal neuroligin-2 links early-life stress with impaired social recognition and increased aggression in adult mice. Psychoneuroendocrinology 55:128–143
Kudielka BM, Kirschbaum C (2005) Sex differences in HPA axis responses to stress: a review. Biol Psychol 69:113–132
Lee H, Kim DW, Remedios R, Anthony TE, Chang A, Madisen L, Zeng H, Anderson DJ (2014) Scalable control of mounting and attack by Esr1+ neurons in the ventromedial hypothalamus. Nature 509:627–632
Lin D, Boyle MP, Dollar P, Lee H, Lein ES, Perona P, Anderson DJ (2011) Functional identification of an aggression locus in the mouse hypothalamus. Nature 470:221–226
Lupien SJ, McEwen BS, Gunnar MR, Heim C (2009) Effects of stress throughout the lifespan on the brain, behaviour and cognition. Nat Rev Neurosci 10:434–445
Marquez C, Poirier GL, Cordero MI, Larsen MH, Groner A, Marquis J, Magistretti PJ, Trono D, Sandi C (2013) Peripuberty stress leads to abnormal aggression, altered amygdala and orbitofrontal reactivity and increased prefrontal MAOA gene expression. Transl Psychiatry 3: e216
Monk C, Spicer J, Champagne FA (2012) Linking prenatal maternal adversity to developmental outcomes in infants: the role of epigenetic pathways. Dev Psychopathol 24:1361–1376
Nederhof E, Schmidt MV (2012) Mismatch or cumulative stress: toward an integrated hypothesis of programming effects. Physiol Behav 106:691–700
Panksepp J (1981) The ontogeny of play in rats. Dev Psychobiol 14:327–332
Paus T, Keshavan M, Giedd JN (2008) Why do many psychiatric disorders emerge during adolescence? Nat Rev Neurosci 9:947–957
Pellis SM, Pasztor TJ (1999) The developmental onset of a rudimentary form of play fighting in C57 mice. Dev Psychobiol 34:175–182
Pellis SM, Pellis VC (2007) Rough-and-tumble play and the development of the social brain. Curr Dir Psychol Sci 16:95–98
Pesonen AK, Raikkonen K, Feldt K, Heinonen K, Osmond C, Phillips DI, Barker DJ, Eriksson JG, Kajantie E (2010) Childhood separation experience predicts HPA axis hormonal responses in late adulthood: a natural experiment of World War II. Psychoneuroendocrinology 35:758–767
Poirier GL, Imamura N, Zanoletti O, Sandi C (2014) Social deficits induced by peripubertal stress in rats are reversed by resveratrol. J Psychiatr Res 57:157–164
Saab BJ, Mansuy IM (2014) Neurobiological disease etiology and inheritance: an epigenetic perspective. J Exp Biol 217:94–101
Sandi C, Haller J (2015) Stress and the social brain: behavioural effects and neurobiological mechanisms. Nat Rev Neurosci 16:290–304
Santarelli S, Lesuis SL, Wang XD, Wagner KV, Hartmann J, Labermaier C, Scharf SH, Muller MB, Holsboer F, Schmidt MV (2014) Evidence supporting the match/mismatch hypothesis of psychiatric disorders. Eur Neuropsychopharmacol 24:907–918
Schmidt MV (2011) Animal models for depression and the mismatch hypothesis of disease. Psychoneuroendocrinology 36:330–338
Seffer D, Rippberger H, Schwarting RK, Wohr M (2015) Pro-social 50-kHz ultrasonic communication in rats: post-weaning but not post-adolescent social isolation leads to social impairments-phenotypic rescue by re-socialization. Front Behav Neurosci 9:102
Shetgiri R (2013) Bullying and victimization among children. Adv Pediatr 60:33–51
Spear LP (2000) The adolescent brain and age-related behavioral manifestations. Neurosci Biobehav Rev 24:417–463
Toth M, Halasz J, Mikics E, Barsy B, Haller J (2008) Early social deprivation induces disturbed social communication and violent aggression in adulthood. Behav Neurosci 122:849–854
Toth M, Mikics E, Tulogdi A, Aliczki M, Haller J (2011) Post-weaning social isolation induces abnormal forms of aggression in conjunction with increased glucocorticoid and autonomic stress responses. Horm Behav 60:28–36

Toth M, Tulogdi A, Biro L, Soros P, Mikics E, Haller J (2012) The neural background of hyper-emotional aggression induced by post-weaning social isolation. Behav Brain Res 233:120–129

Trezza V, Baarendse PJ, Vanderschuren LJ (2010) The pleasures of play: pharmacological insights into social reward mechanisms. Trends Pharmacol Sci 31:463–469

Tulogdi A, Toth M, Barsvari B, Biro L, Mikics E, Haller J (2014) Effects of resocialization on post-weaning social isolation-induced abnormal aggression and social deficits in rats. Dev Psychobiol 56:49–57

Tye KM, Deisseroth K (2012) Optogenetic investigation of neural circuits underlying brain disease in animal models. Nat Rev Neurosci 13:251–266

Tzanoulinou S, Garcia-Mompo C, Castillo-Gomez E, Veenit V, Nacher J, Sandi C (2014) Long-term behavioral programming induced by peripuberty stress in rats is accompanied by GABAergic-related alterations in the Amygdala. PLoS ONE 9:e94666

Tzanoulinou S, Garcia-Mompo C, Riccio O, Grosse J, Zanoletti O, Dedousis P, Nacher J, Sandi C (2015) Neuroligin-2 expression in the prefrontal cortex is involved in attention deficits induced by peripubertal stress. *Neuropsychopharmacology*

Urban DJ, Roth BL (2015) DREADDs (designer receptors exclusively activated by designer drugs): chemogenetic tools with therapeutic utility. Annu Rev Pharmacol Toxicol 55:399–417

van den Berg CL, Pijlman FT, Koning HA, Diergaarde L, van Ree JM, Spruijt BM (1999) Isolation changes the incentive value of sucrose and social behaviour in juvenile and adult rats. Behav Brain Res 106:133–142

van der Kooij MA, Fantin M, Kraev I, Korshunova I, Grosse J, Zanoletti O, Guirado R, Garcia-Mompo C, Nacher J, Stewart MG, Berezin V, Sandi C (2014a) Impaired hippocampal neuroligin-2 function by chronic stress or synthetic peptide treatment is linked to social deficits and increased aggression. Neuropsychopharmacology 39:1148–1158

van der Kooij MA, Fantin M, Rejmak E, Grosse J, Zanoletti O, Fournier C, Ganguly K, Kalita K, Kaczmarek L, Sandi C (2014b) Role for MMP-9 in stress-induced downregulation of nectin-3 in hippocampal CA1 and associated behavioural alterations. Nat Commun 5:4995

Varlinskaya EI, Spear LP (2008) Social interactions in adolescent and adult sprague-dawley rats: impact of social deprivation and test context familiarity. Behav Brain Res 188:398–405

Veenit V, Cordero MI, Tzanoulinou S, Sandi C (2013) Increased corticosterone in peripubertal rats leads to long-lasting alterations in social exploration and aggression. Front Behav Neurosci 7:26

Veenit V, Riccio O, Sandi C (2014) CRHR1 links peripuberty stress with deficits in social and stress-coping behaviors. J Psychiatr Res 53:1–7

Weder N, Yang BZ, Douglas-Palumberi H, Massey J, Krystal JH, Gelernter J, Kaufman J (2009) MAOA genotype, maltreatment, and aggressive behavior: the changing impact of genotype at varying levels of trauma. Biol Psychiatry 65:417–424

Wolke D, Lereya ST (2015) Long-term effects of bullying. *Arch Dis Child*

Wommack JC, Delville Y (2003) Repeated social stress and the development of agonistic behavior: individual differences in coping responses in male golden hamsters. Physiol Behav 80:303–308

Wommack JC, Salinas A, Melloni RH Jr, Delville Y (2004) Behavioural and neuroendocrine adaptations to repeated stress during puberty in male golden hamsters. J Neuroendocrinol 16:767–775

Wommack JC, Taravosh-Lahn K, David JT, Delville Y (2003) Repeated exposure to social stress alters the development of agonistic behavior in male golden hamsters. Horm Behav 43:229–236

Yizhar O, Fenno LE, Prigge M, Schneider F, Davidson TJ, O'Shea DJ, Sohal VS, Goshen I, Finkelstein J, Paz JT, Stehfest K, Fudim R, Ramakrishnan C, Huguenard JR, Hegemann P, Deisseroth K (2011) Neocortical excitation/inhibition balance in information processing and social dysfunction. Nature 477:171–178

MIX
Papier aus verantwortungsvollen Quellen
Paper from responsible sources
FSC® C105338

If you have any concerns about our products,
you can contact us on
ProductSafety@springernature.com

In case Publisher is established outside the EU,
the EU authorized representative is:
Springer Nature Customer Service Center GmbH
Europaplatz 3, 69115 Heidelberg, Germany

Printed by Libri Plureos GmbH
in Hamburg, Germany